FIELDBUS SYSTEMS AND THEIR APPLICATIONS 2005

A Proceedings Volume from the 6th IFAC International Conference,
Puebla, Mexico
14–25 November 2005

Edited by

MIGUEL LEÓN CHÁVEZ
Facultad de Ciencias de la Computacion
BUAP
14 sur y Av. San Claudio, CU
Puebla,CP 72570
MEXICO

Published for the
International Federation of Automatic Control
By
ELSEVIER LTD

Elsevier
The Boulevard, Langford Lane, Kidlington, Oxford OX5 1GB, UK
30 Corporate Drive, Suite 400, Burlington, MA 01803, USA

First edition 2006

British Library Cataloguing in Publication Data
A catalogue record for this book is available from the British Library

Library of Congress Cataloging-in-Publication Data
A catalog record for this book is available from the Library of Congress

ISBN–13: 978-0-08-045364-4
ISBN–10: 0-08-045364-3

Printed and bound in Great Britain
06 07 08 09 10 10 9 8 7 6 5 4 3 2 1

MESSAGE FROM THE PROGRAM COMMITTEE CHAIRMAN

The series of FeT – Fieldbus Systems and their Applications Conferences started in 1995 in Vienna, Austria. The two first editions, both held in Vienna, used German as the working language. In 1999 FeT was organized in Magdeburg, Germany, but became effectively an international conference using already the English as working language. Since FeT'2001 in Nancy, France, the conference became an IFAC – International Federation of Automatic Control sponsored event. This enabled the fifth edition of FeT, FeT'2003, to be held in Aveiro, Portugal, in conjunction with the fifth edition of SICICA, the IFAC Symposium on Intelligent Components and Instruments for Control Applications. In Aveiro, it was then decided to accept the proposal from Miguel León-Chávez to organize FeT'2005 in Puebla, Mexico. This has been really a challenge to test the strength of the research community that supports the FeT conferences. It seems at the moment of writing this text that, like the previous editions, the first American edition of FeT will also be a success.

FeT'2005 was supported by a large International Program Committee with about 40 members from 15 countries, 6 of them out of Europe. This enabled a revision process where each of the more than 50 papers submitted was evaluated by 4 IPC members selected from different countries, whenever possible. From those, 36 high quality papers from 13 different countries were selected for presentation at the conference.

The scope of the 2005 issue of FeT addressed a set of topics such as fieldbus based systems, services, protocols and profiles, system integration with heterogeneous networks, management, real-time, safety, dependability and security, distributed embedded systems, wireless networking for field applications, education and emerging trends. After the paper selection it was possible to organize 13 sessions with the following topics: safety, dependability and security, real-time and distributed systems, building automation, automation networks and real-time Ethernet, applications (all with 2 sessions), networked control systems, wireless networks and automotive.

In order to strength the international nature of the conference, two keynote speeches from experts outside Europe were also organized. The first one entitled "Bandwidth Allocation Scheme in Fieldbuses" and presented by Prof. Seung Ho Hong, from the School of Electrical Engineering and Computer Science of Hanyang University, Korea, reflects the usual focus of the conference on fieldbuses. On the second day, Prof. Ian F. Akyildiz, from the Broadband and Wireless Networking Lab of the School of Electrical and Computer Engineering of the Georgia Institute of Technology, USA, will speak about "Key Technologies for Wireless Networking in the Next Decade". The choice of this latter subject reflects the current interest of wireless communications for industrial applications.

In order to reinforce the research community that supports the FeT series of conferences, the steering committee has taken the decision of recognizing the contribution of authors that were able to keep a constant and large participation. A diploma will be awarded during the conference to the authors with more papers published in the five first editions of the FeT conferences.

As usual, the organization of a conference like this one is a challenging job that depends strongly on the effort of several people. Firstly, I would like to thank Miguel León-Chávez and his Mexican team by the work developed in the dissemination of the conference announcement and in the logistics of the paper submission, selection and proceedings pre-prints organization. I'm sure he will also succeed in the event organization. I would also like to thank the steering committee for their support in several difficult decisions that were taken during this process and the IPC members for their prompt answer during the reviewing process, providing high quality reviews that highly facilitated the paper selection.

Finally, I would leave a word of appreciation for the conference participants. I'm sure they will enjoy the conference both at the technical and at the personal level. I think that this community will be reinforced by the opportunities for personal contact and technical discussion during the conference. I also believe that this will be a marvelous opportunity to discover Mexico in general and the charming city of Puebla in particular, thus raising the interest in possible co-operation with Latin America.

Welcome to Puebla and enjoy FeT'2005 and Mexico.

José A. Fonseca

Technical Program

6th IFAC International Conference on Fieldbus Systems and their Applications

Benemérita Universidad Autónoma de Puebla, México

FeT'2005

Proceedings Preprints

Sponsored by: IFAC International Federation of Automatic Control

Hosted by:

Co-sponsored by: CINVESTAV UDLA PUEBLA

Organized by:

Facultad de Ciencias de la Computación
Facultad de Ciencias de la Electrónica
BUAP

November 14-15, 2005.

General Chair
Miguel León Chávez

Program Chair
José Alberto Fonseca

National Organizing Committee

Fabiola López y López
Luis Gerardo de la Fraga
Jaime Cid Monjaraz
Vicente Alarcón Aquino
Arturo Díaz Pérez
Saúl Pomares Hernández
Francisco Rodríguez Henríquez

International Program Committee

Luis Almeida (Portugal)
Jean-Dominique Decotignie (Switz.)
Christian Diedrich (Germany)
Peter Fischer (Germany)
Lucia Regina Franco (Brasil)
Gerhard P. Hancke (South Africa)
Juergen Jasperneite (Germany)
Sergio Junco (Argentina)
Wolfgang Kastner (Austria)
Wook Hyun Kwon (Coreia)
Lucia Lo Bello (Italy)
Stefan Mahlknecht (Austria)
Nicolas Navet (France)
Christer Norström (Sweden)
Juan Pimentel (USA)
Thilo Sauter (Austria)
Françoise Simonot-Lion (France)
Jean-Pierre Thomesse (France)
Yvon Trinquet (France)
Francisco Vasques (Portugal)
Takahiro Yakoh (Japon)
Carlos Cardeira (Portugal)
Claudio Demartini (Italy)
Dietmar Dietrich (Austria)
José Alberto Fonseca (Portugal)
Wolfgang Halang (Germany)
Hans Hansson (Sweden)
Guy Juanole (France)
Klaus Kabitzsch (Germany)
Philip Koopman (USA)
Francis Lepage (France)
Dietmar Loy (Austria)
Marga Marcos (Spain)
Peter Neumann (Germany)
Carlos Pereira (Brasil)
Gerhard Russ (Austria)
Herbert Schweinzer (Austria)
YeQiong Song (France)
Eduardo Tovar (Portugal)
Hanns-Karl Tronnier (Belgium)
Martin Wollschlaeger (Germany)
Holger Zeltwanger (Germany)

List of reviewers

Alarcón Aquino, V.
Almeida, L.
Decotignie, J.-D.
Díaz Pérez, A.
Dietrich, D.
Fischer, P.
Fonseca, J-A.
Franco, L.
Halang, W.
Hansson, H.
Juanole, G.
Junco, S.
Kabitzsch, K.
Kastner, W.
Kwon, W.H.
León Chávez, M.
Lepage, F.
Lo Bello, L.
López y López, F.
Loy, D.
Marcos, M.
Navet, N.
Neumann, P.
Pereira, C.
Pimentel, J.
Rodríguez Henríquez, F.
Russ, G.
Sauter, T.
Schweinzer, H.
Simonot-Lion, F.
Song, Y.
Thomesse, J.-P.
Tovar, E.
Trinquet, Y.
Vasques, F.
Wollschlaeger, M.

ELSEVIER
IFAC
PUBLICATIONS

Integration of Scalable Safety and Security Actions in IEC 61499 Control Applications

Christian Schwab, Marcus Tangermann, Arndt Lüder
Univ. of Magdeburg, Center Distributed Systems – CVS@IAF
Universitaetsplatz 2, 39106 Magdeburg, Germany
{christian.schwab|marcus.tangermann|arndt.lueder}@mb.uni-magdeburg.de

Abstract

In the last decade, manufacturing automation systems have experienced a strong trend towards an increasing complexity, variability and flexibility caused by growing product spectra and declining lot sizes. To cope with the growing requirements of today's automation system, the application of modular mechatronical units have been explored resulting in distributed control systems. An upcoming standard, the IEC 61499 Function Blocks, provides the according programming technologies for the engineer to handle such systems.

With the application of distributed control systems and the integration of new communication technologies as for example Industrial Ethernet communication systems the communication within control systems grows up and reaches a new level were system safety and security get a higher impact on the applicability of manufacturing systems. These two aspects are not yet directly covered by the IEC 61499 standard. Within this paper a possible way of integration of necessary activities to ensure system safety and security (so called Safety and Security Actions) in a standard conform way in a distributed control system is described.

1. Introduction

In the last decade, process and control systems have experienced a strong trend towards an increasing complexity, variability and flexibility which can be seen in the following facts:

- product spectra have got a bigger size,
- lot sizes became smaller,
- production systems are now more flexible, and
- used manpower has decreased [1].

To cope with the increasing complexity, variability and flexibility of process and control systems, three main aspects have to be considered.

At first, the complex hardware structure will be based on small, intelligent, network-connected, and plug-and-play enabled control devices. These devices will form a community of sensors and actuators enabling the application of system design principles forming and applying mechatronical units without the need of central supervisor entities like Programmable Logic Controllers (PLCs). This improves the variability, flexibility, scalability of control systems, supports the management of the complexity by allowing concepts such as resource sharing, and enables capabilities such as self-configuration, failure device replacement scenarios etc. [2, 3, 4].

At second, the network-connected devices are connected to an Ethernet based communication system using the internet as a WAN.

Finally, new software approaches to control and manage a network of intelligent devices are and will be developed. With respect to this, an important milestone to cope with distributed resources is the international standard IEC 61499 [5, 6] whose event-driven architecture defines a generic software architecture to deal with and to manage distributed Industrial Process Measurement and Control Systems (IPMCSs).

Following these trends the international research project TORERO [7, 8] (Total life cycle web-integrated control) – funded by the European Commission – has aimed at specifying a Distributed Control System (DCS) applied in factory automation which is called TORERO system and which is based on intelligent, networked devices and IEC 61499 based distributed control applications.

Within the TORERO system different devices will interact by crossing sometimes unknown or even critical communication paths. For example, application data may be exchanged between two manufacturing cells using the Internet. This communication may cause problems with respect to security and safety of the manufacturing system. Data may be read along or - more critical - changed during the transmission. The malicious hacker may access devices to "play a little bit with the robot".

Up to now there are no means to handle necessary activities to ensure safety and security of Distributed Control Systems in an adequate way using the IEC 61499.

Within this paper this problem will be tackled. An IEC 61499 Function Block based architecture enabling a efficient and sufficient integration of necessary safety and security mechanisms by providing Safety Actions and Security Actions is described. Therefore the paper will describe at first the TORERO approach and the Distributed Control Systems Design Methodology developed for the TORERO systems. Then the possible influences and harms as well as the appropriate safety and security mechanisms for manufacturing systems will be presented. Based on both considerations, Safety and Security Actions ensuring safety and security of manufacturing systems and its automatic integration in IEC 61499 Function Block Systems will be described.

2. TORERO Approach

The TORERO approach aims at specifying both an Integrated Development Environment (TORERO IDE) and an architectural platform on specific devices (TORERO Devices) which together allow for the distribution of control applications based on event triggered Function Blocks (FBs) related to IEC 61499 standard and Ethernet TCP/IP based industrial communication protocols.

The smallest part of a Distributed Control System in TORERO is an intelligent control device called TORERO Device (TD). It is a mechatronical component that can interact in a TORERO environment.

In collaboration, all TDs of a TORERO system can realize the required control functionality.

Basically, a TD is a sensing or actuating component equipped with suitable hardware and software. The hardware (sensing, actuating component) will be accessed by the control application, using so called device functions, an encapsulation of hardware-related functionality within an IEC 61499 Function Block which is called TORERO Proxy Block.

As a basic feature, the TD provides plug-and-play mechanisms which are based on the open Universal Plug-an-Play standard UPnP [9, 10] providing information about its device functions as well as further device related information by a device description file formatted in XML.

The TORERO IDE (TIDE) provides a tool for implementing control applications for Distributed Control Systems (DCSs) using IEC 61499 Function Blocks. Beside the FB Editor itself, the TIDE contains a central data model storing information about the available TORERO Devices and their Device Functions which have been provided by the device descriptions as well as information about the control application by describing the Function Block System and its relations to the TDs.

Using the two integral parts described above (TDs and TIDE), the user can follow the engineering process which is defined in the TORERO project leading to an implementation of a DCS' control application applying the following steps:

1. *Device/Network model creation:* Based on the information gathered via UPnP a model of the available devices in the network is build. Within this model the available TORERO Device Functions are described by TORERO Proxy Blocks.
2. *Application Modeling and Programming:* Using IEC 61499 Function Block models, the application for the DCS is developed independently from the distribution of the Function Blocks to devices but under integration of the TORERO Proxy Blocks.
3. *Allocation of Application:* Based on the capabilities and real-time requirements of the event/data connections between the FBs, the FBs are assigned to the available TORERO Devices in the system.
4. *Weaving and Compilation:* Based on the results of step 3, the necessary communication related code to ensure the communication between blocks on different devices is generated automatically.
5. *Deployment and startup:* The compiled code is downloaded to the different devices using FTP and the system is started using UPnP mechanisms.

The specification can be found in [7, 8, 11].

3. Safety and Security Aspects for Distributed Control Systems

Derived from the requirements for industrial automation and based on the TORERO approach, the main safety and security related aspects applied for DCSs will be considered within this chapter which result in a Safety and Security Matrix facing errors and protection risks versus appropriate countermeasures.

3.1. Safety

The aim of safety systems is to protect humans and animals life, machines and environment. In industrial automation, technical systems are required which allow for the safeguarding of these systems in an automatic way. As soon as an error is detected in a safety system, the fail safe system remains in a safe position or changes to a safe position depending on the specific application.

The international standard IEC 61508 [12] describes the functional safety of a Programmable Electronic System (PES). Currently under development are the IEC 61511 (process industry) and the IEC 62061 (factory automation). Within IEC 61508 standard the so called Safety Integrity Levels (SILs) are described. These levels are ranging from SIL1 for the lowest level to SIL4 for the highest level and describing the degree of trust in an automation system to fulfill its work in a proper way based on the remaining average failure rate. For the main number of applications, SIL3 is appropriate which is calculated as continuous/high demand mode, which is between 10^{-8} and 10^{-7} for the overall system.

Since communication is an integral part of DCS in safety related systems depend on the safety of the communication system. Assuming that communication among control devices of a DCS consumes one per cent of the probability of failures we have a probability of failures between 10^{-9} and 10^{-10} for the communication part for such systems [13].

Errors resulting in a loss of safety related to the application part are all possible errors of messages which could be the following.

3.1.1. Iteration of data

Messages iteration describes the situation where the same message arrives multiple times at the receiver. The receiver will react each time on the message. This may cause a delay in the intended activities or even a system break down.

3.1.2. Loss of data

During a communication activity some control relevant data can be lost. Thereby, maybe some safety relevant data as the shutdown event of a system to a safe state can be lost resulting in a state where the system is not safe.

3.1.3. Insertion of data

Special data can be included in the message during the transmission by a failure of a device within the communication path. Thereby, the control relevant semantic error may occur making the message wrong with respect to the controlled behavior.

3.1.4. Wrong order of data

Different messages between the two devices may use different communication paths. Thereby, the second message may reach the receiver before the first one. Thereby, sensor data within the messages may be interpreted in a wrong way by the receiver.

3.1.5. Message corruption

Data within a message can be changed by technical reasons. This may result in the same problems as in the case of data insertion.

3.1.6. Message delay

A message may require more time for its transmission in the communication system than expected. Thereby, the data in the message may get outdated with respect to the control application.

All this failures can result in an unexpected behavior of the DCS. Thereby, mainly the integrity of the overall system, i.e. the consistency of data used to come to a control decision, is lost.

3.2. Security

Security with respect to communication systems is one of the currently upcoming topics. It is discussed in a very controversial way.

Security is usually understood as ensuring the integrity and privacy of devices and data with respect unrequested access.

Security within automation networks can be defined based on existing criteria and protection goals from office networks [15, 16]. In the following the main criteria used in today's networks will be defined.

3.2.1. Integrity

Transmitted data will not be modified on the transmission path, are complete, and reach the target in the same order as transmitted by the sender. For example, the data of an ftp file transfer are not exchanged by a third person during the transmission.

3.2.2. Non-reputability

It can be verified at any time who has initiated a connection and who has transmitted which data at which point in time. In practice, this means e.g. that the data of log files are explicit and fraud resistant. This is especially useful for remote maintenance scenarios where manufacturers access their components in an existing facility, e.g. for updating the software. In case of the failure of the facility caused by this maintenance activities the manufacturer can be hold responsible, based on the fraud resistant log files.

3.2.3. Confidentiality

Sent data cannot be accessed by a third person on the transmission path. For example, this goal can be reached by using appropriate cryptographic algorithms to that extent to which they may be applicable. The application of such algorithms can be problematic due to the high amount of processing capabilities needed, especially regarding real-time communication and embedded devices with their restricted CPUs.

3.2.4. Availability

The network and connected devices can send and process data at any time within a given timeframe. Availability forms a very intractable point regarding network security of automation systems. As a result of the restricted resources of embedded devices the access to these devices can be prevented by overloading the network (denial of service).

3.2.5. Authentication

During the authentication process, the identity of a communication partner is determined and additionally it is checked, whether this partner has the required access rights for a given network service. In practice, the user/password combination (e.g. for an FTP transfer) or the digital signature (e.g. for e-mail communication) fall in this category.

Based on these criteria protection needs can be categorized and protection goals can be defined [20]. The protection needs will be vary for all fife areas between none and very high by 4 levels. Based on this classification a set of necessary activities can be derived enabling the security goals. The main goals are:

- Protection against unauthorized information gain (loss of confidentiality),
- Protection against unauthorized modification of information (loss of integrity), and
- Protection against unauthorized interference of functionality (loss of availability).

3.3. Methods for Safety Communication

Different mechanisms are possible in order to eliminate or reduce safety relevant errors. Table 1 presents the Safety Matrix containing possible errors and appropriate methods to eliminate or reduce them related to the following methods and mechanisms [13, 14].

Table 1. Safety Matrix

	Sequential Number	Timestamp/ Timeout	Identification of Sender/Receiver	Safety Code	Data Backup
Iteration of data	X	X			
Loss of data	X				X
Insertion of data	X		X	X	
Wrong order of data	X	X			
Message corruption				X	X
Message delay		X			

3.3.1. Sequential Number

Each message has a consecutive number. In most simple case the message includes a toggle bit. This allows the receiver to check the sequence of messages provided by the sender. The method is reducing risk caused by message iteration, loss, insertion and if the messages are sent in incorrect sequence.

3.3.2. Timestamp / Timeout

Information sent in wrong time can be useless, harmless and in some cases even dangerous for the user. Each message has a time code, which describes the sending time. This kind of information can be used instead of or combined with sequence numbers. The method is reducing risk caused by message iterations and if the messages are sent in incorrect sequence, too late, too early or if the transfer times vary too much. Receiver accepts messages only when they arrive in time or during a predefined time window (Timeout, e.g. Watchdog). Usually exception handling is used to react upon delayed message.

3.3.3. Identification of sender and receiver,

Each message has a source and/or destination address or other code. Inclusion of a source identifier in messages enable users of the messages to verify that messages are from intended source. Inclusion of a destination identifier in messages enable users of the messages to verify that messages are intended for them.

3.3.4. Safety Code

The method adds into the message a checking code; also other type of data consistency checks are available. The simplest code is the parity checking. The characters are encoded so that an additional bit is added to each character. The method will only detect one bit error bursts in each codeword. Other checks are checksum and CRC (cyclic redundancy check), which is the most effective.

3.3.5. Data backup

Data backup in order to recover all information in case of loss of data or message corruption.

3.4. Methods for Secure Communication

The security criteria described in the section above can be mapped to concrete security measures for a more secure communication as shown in the Security Matrix in Table 2.

Table 2. Security Matrix

	Encrypting	Hashing	Authentication	Filtering
Integrity	X	X		
Non-repudiability	X		X	
Confidentiality	X			
Availability				X
Authentication	X		X	

3.4.1. Encrypting

Encryption is a mechanism to hide the meaning of data within a message. The original digits of a message will be encrypted using for strangers unknown algorithms.

As an example of encryption, which is mostly used to ensure the integrity of messages, are mechanism based on public key infrastructures (PKI). PKI systems use an asynchronous encryption mechanism. Therefore, each communication partner has to provide a pair of keys:

- A well known public and
- A hidden private key

3.4.2. Hashing

Hashing is a method to calculate and integrate a checksum in a message. Very often hashing is used together with encryption.

For each messages, a hash is computed based on the message to send from device 1 to device 2. This hash can then be encrypted using the private key of device 1 and added to the message. On device 2 the same hash algorithm as on device 1 is used to calculate the hash sum of the received message. Using the public key of device 1, the sent hash is decrypted and compared with the newly calculated hash. If both hashes are equal, the message was not changed on transmission. It is obviously that this mechanisms can also be used to ensure Non-reputability.

It is important to note that the applied algorithms are not reversible. This means that an attacker cannot take advantage of knowing the public key.

3.4.3. Authentication

For authentication of communication partners, various methods could be used. Beside the mechanisms of the PKI described above for authenticating per message, other mechanisms such as the RADIUS authentication can be used.

To ensure confidentiality of data, encryption mechanisms of various types can be used. Beside the PKI systems described above, also synchronous encryption is possible and due to the high amount of computation resources necessary for computation of PKI algorithms more suitable for embedded system.

3.4.4. Filtering

Filtering of messages means the extraction of messages out of the communication path by a certain device or device part based on given message content, message sender and message receiver criteria.

Due to the limited capabilities such as lesser computational power and memory, control devices are for example more vulnerable to denial of service attacks (DOS). During a DOS, the attacker tries to prevent the device from serving normal requests by generating a high network load on the network attacking either a specific service (e.g. generating lots of connection requests to a specific port) or the stack itself by sending e.g. a high amount of wrong packets. A solution for a device can be to reduce the amount of connection request per second, to filter specific parts of packets, or to filter all messages not send from special sender set.

It is obvious that thereby the availability of the network and the connected devices can be reached.

4. Integration of Safety and Security Aspects in IEC 61499 Applications

Based on the safety and security related activities defined above, in this chapter their application for a DCS defined in TORERO will be considered.

4.1. Definition of Safety and Security Actions for DCSs

After the design phase of the IEC 61499 based control application (Step 2 of the Engineering Process) and the allocation of Function Blocks to devices (Step 3 of the Engineering Process) the user can decide which safety and security aspects he wants to take into account for the application et all, for devices, and for individual FB connections. Therefore the user can define Safety and Security Actions.

After the definition of Safety and Security Actions the allocation of Function Blocks to devices has to be verified. It may happen that the involved devices are not able to execute the necessary Safety and Security Actions or the capacity of a device is overloaded. This can be detected based on the information of the device description belonging to the device. In this case the allocation of the Function Block system has to be changed and the procedure has to be repeated.

Depending on the selected Safety and Security Actions, specific Function Blocks containing the appropriate safety and security functionalities will be generated automatically during the Weaving and Compilation phase (Step 4 of the Engineering Process) for the selected or for all communication connections between the allocated Function Blocks located on different devices within the network. For a TORERO Distributed Control System, the followings actions are defined:

Safety Action:

1. Sequential Number,
2. Timestamp / Timeout,
3. Identification of Sender and Receiver,
4. Safety Code.

Security Action:

1. Encrypting of User Data,
2. Encrypting of complete Message,
3. Hashing of User Data,
4. Hashing of complete Message,
5. Filtering,
6. Authentication.

Based on the definition of Safety and Security Actions, in the next section the integration of the appropriate functionalities in IEC 61499 based applications will be introduced.

4.2. Automatic Integration of Safety and Security Actions in IEC 61499 Applications

As an example, a simple network of FBs containing Drive1 and Drive2 located on devices 1 and 2 as shown in Figure 1 will be considered. Both drives need to be synchronized in a certain way by the application (The intended real-time conditions of the synchronization are not relevant within this paper). For a better understanding the main relevant parts of the DCS design process will be repeated.

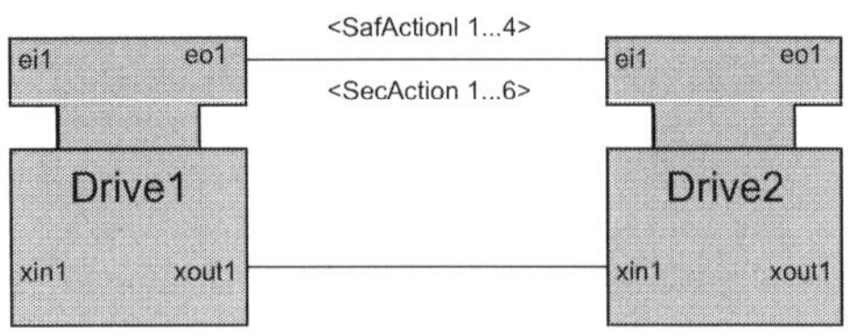

Figure 1. Simple FB Network Example

The TORERO approach aims at designing IEC 61499 Function Block based control applications for Distributed Control Systems (DCS) within three steps (Step 2 to Step 4 of the Engineering Process) containing the necessary Safety and Security Function Blocks.

Within the **first step** of the design process the control application is designed in a hardware independent way by using interconnected IEC 61499 Function Blocks containing the necessary hardware independent control applications (Figure 1). The Function Blocks Drive1 and Drive2 are Composite Function Blocks containing beside of the actual control application Function Blocks the TORERO Proxy Blocks derived from the specific device functions of the device for accessing of the control hardware.

In the **second step** of the design process the FBs will be allocated to the TORERO Devices which have been detected applying UPnP mechanisms during the start-up phase. The allocation in this case is trivial: Function Block Drive1 will be allocated to Device 1, the same is true for FB Drive2 and Device 2. Both devices are equipped with an Ethernet interface supporting Modbus/TCP.

In the **third step**, the necessary Communication Service Blocks will be integrated. This will be done automatically during the weaving process with respect to the Hard Real-Time classification [17], proposed by TORERO, and the defined Safety and Security Actions will be integrated in the Communication Service Blocks (Figure 2). For the automatic generation of communication related for IEC 61499 based applications please refer to [18, 19]. All communication related mechanisms (protocol implementation, e.g. Modbus/TCP, safety and security functions) are hidden for the user within the Communication Service Blocks (CSBs).

In general, the CSB provides the interface to the underlying automation network (Ethernet) containing the automation protocol and appropriate safety and security functionalities. The mapping of the event triggered IEC 61499 Function Blocks to the underlying automation protocols covering both event triggered and time triggered architectures has been specified in [18].

For the correct addressing and handling of event/data combinations an additional data connection from Function Block Drive1 to the Communication Service Block is introduced (for clearness not depicted in Figure 2) which contains the address of the event destination. This connection will be generated depending on the allocation of the FBs within the network. For the reverse direction the same mechanisms will be applied with the according event and data inputs/outputs.

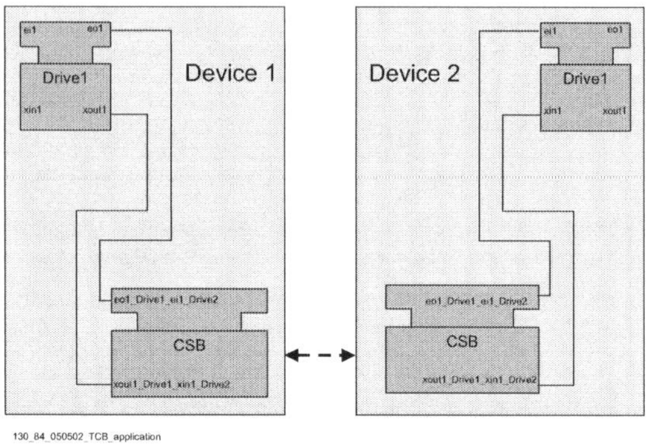

Figure 2. Generation of Communication Service Blocks

The Communication Service Block (CSB) contains beside of the actual implementation of the automation protocol, several service function blocks as follows, see also Figure 3:

- Security Function Block for SecAction 1/3 responsible for Encrypting/Hashing the User Data,
- Security Function Block for SecAction 2/4 responsible for Encrypting/Hashing the whole message,
- Security Function Blocks for SecAction 5 and 6 responsible for Filtering and Authentication, and
- Safety FB for all Safety relevant actions SafAction 1…4.

After the allocation phase of the control application to the device set as described above, the user can specify SafActions and SecActions of selected connections by guarding these connection as depicted in Figure 1 (<SafAction 1…4> and <SecAction 1…6>) or he can define an overall SafAction/SecAction combination for the complete application or all communication relations of devices. This will be supported by a TIDE prompts.

The additional data connection which will already used for the correct addressing of the event connection, will be used to enable the trigger of the appropriate algorithms in the safety and security Function Blocks, and thus applying the selected safety and security mechanisms.

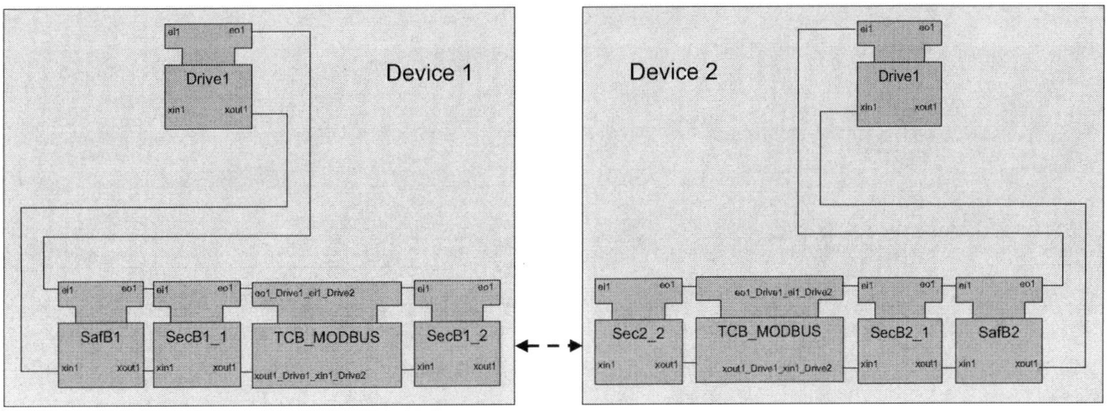

130_84_050502_TCB_application

Figure 3. Generation of CSB including Automation Protocol and Safety/Security FBs

Considering Figure 3, the event connection with the appropriate data will be replaced by:

- an event connection chain connecting all relevant function blocks (automation protocol and safety/security functions) within the composite Communication Service Block, and
- two appropriate data connections, one containing the original data, and the other (not depicted) containing beside of the event destination the selected Safety and Security Action including authentication information if necessary.

In Function Block SafB1 all safety relevant functions and mechanisms – adding of a sequential number (SafAction 1), watchdog functionality (SafAction 2), identification mechanisms of sender and receiver (SafAction 3), and application of a safety code such as CRC, (SafAction 4) – are implemented.

Function Block SecB1_1 supports SecAction 1 and 3 corresponding to Encrypting and Hashing of User Data.

Function Block TCP_MODBUS contains the actual automation protocol, here for example the Modbus/TCP protocol. Here, the message will be transformed into the specific Modbus/TCP telegram structure – the appropriate Modbus/TCP Function Code (e.g. FC 02 corresponding to Read Input Values) with specific User Data (in case of FC 02 the input values which have to be read) – in form of adding Modbus/TCP header and tail data to the message.

Finally, and applying Function Block SecB1_2, security related functions corresponding to SecAction 2 and 4, and 5 and 6 will be implemented.

Selecting SecAction 2 and 4, the complete message will be encrypted and hashed. By applying SecAction 5 specific filtering mechanisms can be realized. For that, the Modbus/TCP Function Code will be analyzed to allow only reading functionalities (e.g. Modbus/TCP FC 02) and disable write functions (e.g. Modbus/TCP FC 05) for instance. The authentication function (SecAction 6) allows for a selected access of functionalities located on different devices.

All functions, except the communication protocol Function Block are optional and can be selected for each connection within the network. SecAction 1/3 and 2/4 are mutually exclusive, either you can chose encrypting/hashing only the User Data or the whole message. In the first case the actual Modbus/TCP telegram structure will be retained while the User Data will be coded, in the latter case it is not possible anymore to identify the transmitted message as a Modbus/TCP telegram.

The resulting telegram structure can be seen in Figure 4 whereas the actual User Data (Event + Data connection) are embedded in SecAction 1 header and tail, which is embedded in the Modbus/TCP Frame (MBus), which itself is encapsulated in a SecAction 2 and SafAction 1 frame. The resulting message in turn will be transported in case of Modbus/TCP via TCP/IP or UDP/IP and the underlying Ethernet hardware.

Figure 4. Resulting Telegram Structure

On the receiver side all functionalities and service function blocks – ranging from the automation protocol to safety/security functions are implemented in reverse order and with reverse functionalities in order to extract the pure event/data information.

So far, only functions which add headers and tails are considered. Special messages based on special protocols are implementation dependent.

5. Synopsis and Outlook

This paper has shown an approach for the automatic integration of safety and security related functionalities in IEC 61499 based control applications. Based on a concept for an automatic generation of communication related code depending on the previous allocation of the FBs to the underlying hardware, a scalable concept for the integration of Safety and Security functionalities have been presented. A test implementation for the automatic integration process for the use of event-triggered protocols (e.g. Modbus/TCP) in combination with different Safety (SafAction 2) and Security (SecAction 2 and 5) Actions is under development. First prototypical tests have shown the principal applicability of the described architecture. Future work will include the implementation of all functional mechanisms and covering other Ethernet based communication protocols.

Acknowledgement

The work presented within this paper was/is partially supported by the European project TORERO (IST) and the Integrated Project VAN (IST NMP).

References

[1] Lüder, A., A. Klostermeyer, J. Peschke, A. Bratoukhine, and T. Sauter: Distributed Automation: PABADIS versus HMS. In: Journal IEEE Transactions on Industrial Informatics, Vol. 1, No. 1, February 2005.

[2] Mercer: Strukturwandel fordert neue Strategien – Studie Automatisierung 2010. Mercer Management Consulting München, www.mercer.de, December 2004.

[3] ARC: Market Analysis and Forecast Through 2007. ARC Advisory Group, Dedham, USA, www.arcweb.com,. June 2004.

[4] ARC: Industrial Ethernet Devices Worldwide Market Forecasted to Grow 51,4% Annually, ARC Advisory Group, Dedham, USA, www.arcweb.com, April 2005.

[5] IEC: International Standard IEC 61499-1 Function Blocks – Part 1: Architecture, Geneva, Switzerland, www.iec.ch, 2005.

[6] Lewis, R.W.: Modelling Control Systems Using IEC 61499: Applying Function Blocks to Distributed Systems (IEE Control Series, 59), 2001.

[7] The TORERO Consortium: www.uni-magdeburg.de/iaf/cvs/torero, 2005.

[8] Schwab, C., M. Tangermann and L. ferrarini: Web Based Methodology for Engineering and Maintenance of Distributed Control Systems: The TORERO Approach. In: Proceedings IEEE International Conference on Industrial Informatics – INDIN, Perth, Australia, August 2005.

[9] The TORERO Consortium: Deliverable 1.2 – Detailed Specification (Control Device, etc.), Magdeburg, www.uni-magdeburg.de/iaf/cvs/torero, August 2003.

[10] Universal Plug-and-Play, www.upnp.org, 2005.

[11] Ferrarini, L., C. Veber, C. Schwab, M. Tangermann and A. Prayati: Control Functions Development for Distributed Automation Systems Using the TORERO Approach. In: Proceedings International Federation of Automatic Control – IFAC World Congress, Prague, Czech Republic, July 2005.

[12] IEC: Functional safety of electrical/electronic/programmable electronic safety-related systems – ALL PARTS, Geneva, Switzerland, www.iec.ch, 2005.

[13] Ditting, S.: Safe Communication on Ethernet. In: IAONA Handbook – Industrial Ethernet, Magdeburg, Germany, 2. Edition, April 2005.

[14] Hérard J. et.al: NORDTEST Report – Validation of Communication in Safety Critical Control Systems, Espoo, Finland, 2003.

[15] N.N.: Information Technology Security Evaluation Criteria (ITSEC): Provisional Harmonized Criteria, Bundesanzeiger Verlagsges. mbH.

[16] M. Merkow, J. Breithaupt: Security Assurance Through the Common Criteria, New Riders Publishing, 2003.

[17] IAONA e.V.: IAONA Real-Time Categories & Classes, Draft, Magdeburg, Germany, www.iaona.org, 2003.

[18] Schwab, C., M. Tangermann, A. Lüder, A. Kalogeras, and L. Ferrarini: Mapping of IEC 61499 Function Blocks to Automation Protocols within the Torero Approach. In: Proceedings IEEE International Conference on Industrial Informatics – INDIN, Berlin, June 2004.

[19] Tangermann, M., Schwab, C., Kalogeras, A., Lorentz, K., Prayati, A.: Aspect-Orientation of Control Application Code for Distributed Automation Systems: The TORERO Approach". First Workshop on Java Technologies for Real-Time and Embedded Systems – JTRES 2003, Catania, Italy, November 2003, in: Meersman, R., Z. Tari, A. Corsaro et al. (Editors): On the Move to Meaningful Internet Systems 2003: OTM 2003 Workshops, Lecture Notes in Computer Sciences 2889, Springer, Berlin, 2003.

[20] M. Tangermann: Thinking in classes – IAONA security data sheet and security classes, Praxis Profiline – Industrial Ethernet, April 2005, Vogel Verlag, pp. 14-15.

ELSEVIER
IFAC
PUBLICATIONS

A FIELDBUS FOR SAFETY-RELATED REAL-TIME OPERATION SUPPORTING FORWARD RECOVERY FROM REDUNDANT NODES

Martin Skambraks * **Peter Neumann** **

* *FernUniversität, 58084 Hagen, Germany, martin.skambraks@fernuni-hagen.de*
** *ifak Institut f. Automation und Kommunikation Magdeburg, Germany, peter.neumann@ifak-md.de*

Abstract: A concept for a ring-based fieldbus serving both data exchange with process peripherals and between multiple redundant processing nodes is presented. The concept fulfills real-time demands and supports state restoration of error-affected or replaced processing nodes at runtime. Its tight integration into a holistic architecture of a safety-related programmable electronic system features this form of forward recovery without the drawback of inadequately long system response times. A multi-ring connection scheme combines high, scalable reliability and low wiring expenses with an especially simple structure. Communication is performed in discrete intervals and in synchrony with the cyclically organised application processing, leading to a simple and easy-to-model temporal behaviour of the entire system. This eases verification and reduces the cost for safety licensing. *Copyright© 2005 IFAC.*

Keywords: Real-time communication, ring topology, safety licensing, fault-tolerance, forward recovery, state restoration, hardware redundancy.

1. INTRODUCTION

The most important design principle for safety-related systems is simplicity. First, design simplicity prevents engineering errors and, later, it eases safety licensing (Liddell 1994). In terms of Programmable Electronic Systems (PESs) to be employed in safety-related real-time applications, design simplicity is desirable in various aspects, e. g., programming style, operating principle, temporal behaviour, and architecture of hard- and software. Since minimising the complexity in some aspects might affect others, following '*Design for Simplicity*' as major design guideline requires a holistic system view. In other words: *Simple solutions require complete intellectual penetration of a subject matter* (Biedenkopf 1994).

Fault-tolerance is usually realised by configuring multiple PES-instances redundantly and, additionally, employing redundant processing structures inside each PES. The latter approach is generally more appropriate to handle transient hardware failures caused by electromagnetic disturbances, since it is capable of detecting and correcting processing errors and, thus, preserves the correct internal state. The ever-increasing manufacturing densities of integrated circuits increases the susceptibility to electromagnetic disturbances and causes growing importance of redundant information processing inside a PES. Some techniques recommended by the safety standard IEC 61508 are, e. g., multiple processors combined with majority voting or RAM monitoring with a modified Hamming code. Typically, a combination of several techniques is necessary to cope with all failure possibilities. As a result, system complexity increases significantly. What is more significant, these techniques can only cover a limited number of simultaneously occurring failures. Once this limit has been exceeded, the associated PES might enter a faulty state and must be excluded from the redundantly operating PES configuration. This redundance attrition must be taken into account by design, either by increasing the number of applied PES-instances so that a complete outage of the redundant configuration is improbable or by enabling to restore the internal state of a PES-instance at runtime. In most contemporary systems, the former approach is used since realising the latter is problematic: either the computing performance is decreased to an impracticable level or the high complexity causes an unacceptable verification effort. This is why a new PES concept was developed, which follows a rather unusual but more integrative approach to realise state restoration at runtime with an acceptable computing performance and a minimum increase of system complexity. In a redundant configuration, each PES outputs a Serial Data Stream (SDS) that provides full information about the internal processing states. This SDS technique unifies three safety functions: *non-intrusive monitoring*, *recording process activities*, *detecting processing errors*, and *state restoration at runtime*.

Following the intention to combine high reliability and low wiring expenses without infringing the strategy of designing for simplicity, a special communication concept was devised, which serves both data exchange with the process peripherals (e. g., sensors, actuators) and the exchange of SDSs between the instances of a redundant configuration of PESs. Thus, the transfer of I/O data and SDSs is combined in one interface.

The main part of this paper is structured as follows. Section 2 discusses the main problems related to state restoration at runtime and lists some benefits and drawbacks of previously published approaches. Based on this, the succeeding section explains the operating principle of the PES to which the proposed fieldbus concept is dedicated. Thereby, the recovery process and the transfer schedule of the SDSs are described in detail. Section 4 starts by listing some requirements that were considered fundamental during the development of the proposed fieldbus concept. Then, the advantages and drawbacks of the cyclic ring-based communication principle of INTERBUS are discussed and its similarities to the proposed concept are outlined. This is followed by a description of the multi-channel technique and the multi-ring technique, which were implemented to minimise design complexity and to achieve a high level of fault-tolerance. A short summary at the end recapitulates the most essential aspects, mentions the current state of our work, and states open issues.

2. THE CONCEPT OF SERIAL DATA STREAMS

The proposed fieldbus is part of a holistic safety concept for a novel PES architecture, which has already been partly introduced in (Skambraks and Hamuda 2004). Instead of processing redundant information inside, the PES itself is designed to be configured redundantly. Each PES instance outputs a Serial Data Stream (SDS) that provides full information about internal processing. In a redundant configuration, these SDSs are exchanged between PES instances to serve the following four safety functions.

Non-intrusive monitoring: Since the SDSs inform about the internal processing, they enable non-intrusive monitoring by external devices.

Recording process activities: The SDSs can be utilised to externally record the system behaviour for later program flow analysis.

Detecting processing errors: Each PES can detect processing errors by comparing its SDS with the SDSs of other PESs.

State restoration at runtime: In case a PES is affected by a transient hardware fault, the SDSs of redundant PESs enable to copy the internal state and to resume processing at runtime.

The recovery function does not only require a special transmission schedule for the exchange of SDSs, it also necessitates certain characteristics regarding the operating principle of the dedicated PES. That is why some common characteristics of previously published restoration techniques need to be discussed first. Then, the task execution policy of the PES needs to be introduced, before the SDS transmission schedule can be explained comprehensively.

2.1 State restoration at runtime

The problem of state restoration at runtime is that, while a PES copies the internal state of other running PES-instances, these PES instances change their internal state continuously. Various techniques for state restoration of processing nodes have been published in the past and nearly all of them have a common characteristic: they base on software execution in discrete intervals (cmp. e. g. (Adams and Sims 1990), (Bondavalli *et al.* 1998) and (D. Basu 2000)). While the application specific software is executed, some information about the internal state are transferred to redundant processing nodes. The execution of the application specific software does not take place till the end of an interval; a small fraction at the end of each interval is solely utilised to transfer information about the most recent internal state changes, typically the data memory changes of the current interval, to redundant processing nodes.

The main problem is that the state restoration techniques decrease the computational performance considerably. This problem can be minimised by using special hardware instead conventional hardware. For example, the amount of data that needs to be transferred to the redundant nodes is usually reduced by determining the data words that have been changed since their last transference. As shown in (Adams and Sims 1990), the time spend for identifying these modified data words can be considerably decreased by using a special hardware structures. Of course, there are also some software-based strategies to minimise the decrease of computational performance (e. g. by defining recovery blocks as in (D. Basu 2000)), but these tend to increase the system complexity – especially the temporal behaviour – considerably and, thus, conflict with the policy 'Design for simpliciy'.

To sum up, realising a PES that supports state restoration at runtime and achieves acceptable performance with a minimum of system complexity requires a holistic approach that builds up on a special operating principle, a special hard- and software architecture and a specially dedicated communication scheme.

2.2 Operating Principle of the Dedicated PESs

The PES architecture to which the proposed fieldbus is dedicated bases on physical separation of *task execution* and *task administration*. Time is quantised into discrete *Execution Intervals*, and tasks are partitioned into a number of *Execution Blocks* each. The Execution Intervals have a fixed duration, and are defined for the cyclically and synchronously operating *Task Administration Unit (TAU)* and *Task Processing Unit* (TPU). The Execution Blocks have the following characteristics.

- Each Execution Block is executable within a single Execution Interval.
- The execution of a block is not pre-emptable.
- Data exchange between blocks is only possible via the TPU memory; the content of the processor registers is lost at the end of each interval.
- A task's Execution Blocks are indexed for identification.
- A task's Execution Blocks do not need to be executed in consecutive order. For each task, the TAU stores a parameter called *NextBlock* in the Task List, which identifies the subsequent Execution Block.

The operating principle can be roughly described as follows. At the beginning of each Execution Interval, the TAU outputs the *ID of Block to Execute*, which identifies the next Execution Block of the task that must be executed according to the scheduling algorithm. The ID corresponds to the task's *NextBlock* parameter stored in the Task List. After the TPU has read this ID, it processes the associated Execution Block. When the TPU completes the block at the end of the Execution Interval, it outputs the *ID of Next Block* identifying the task's Execution Block that needs to be executed next. The TAU reads the ID and stores it in the Task List as new NextBlock parameter. The flow chart in Fig. 1 illustrates this mode of operation in more detail.

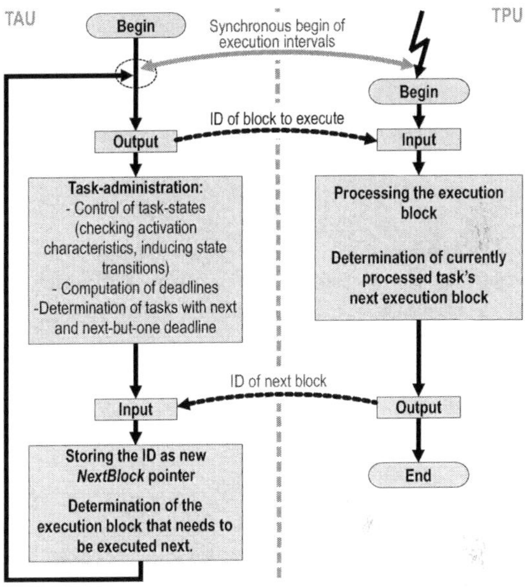

Fig. 1. Operating principle of the PES

If the executed block was a task's last one, i. e., if the task has been executed completely, the TPU outputs the block ID 'Nil'. In this case, the TAU induces an appropriate state transfer that causes the task to be no longer in the state *Activated.* Additionally, the TAU takes this completion into account to compute the *ID of Block to Execute* for the next Execution Interval. That is why the TAU does not only determine the task with the earliest, but also the task with the earliest-but-one deadline. This enables the TAU to immediately output the NextBlock identifier of the task with the next-but-one deadline, in case the task with the next deadline corresponds to the task just been processed *and* just been completed by the TPU.

Although – like other systems that support state restoration – operating in discrete cycles, this operating principle allows the realisation of arbitrary program flows. Its main advantage is the low extent of the task-administration functions which allows to implement the TAU in form of a digital logic circuit (Skambraks and Hamuda 2004).

2.3 SDS Transfer Schedule

The SDS concept bases on the fact that the maximum amount of data changes inside a PES is indirectly limited by the task execution characteristics *Worst Case Execution Time*, *Maximum Response Time*, and *Minimum Activation Period* (Buttazzo 2002). The transfer cycles of the proposed fieldbus match the Execution Intervals of the PES. Within each cycle, the SDS transfers information about a fixed number of internal data changes. This number sets the limit of data changes permitted.

Data changes can be categorised into ones inside the TAU (TAUDCs) and changes inside the TPU (TPUDCs). The frequency of TPUDCs is limited by restricting the amount of an Execution Block's write accesses to a number that is transferable via SDS within each transfer cycle. The number of TAUDCs depends on the frequency of task state changes and modifications of task parameters (e. g., activation characteristics). This number is directly limited by the execution characteristics of all tasks. In contrast to the TPUDCs, the TAUDCs of one Execution Interval are not necessarily transferable within one cycle. This is because, in theory, it is possible that all tasks perform a state transition simultaneously. This would cause a huge amount of data changes inside the TAU (e. g., storing the activation time of each task), and the capability to transfer them within each Execution Interval would require an undesirably high bandwidth. That is why an integer number is assigned to each TAU data word that represents the *Age* of the stored value. By default, each Age parameter is set to '0'. Any time a data word is modified, the associated Age integer is set to the maximum representable value. If an Age value does not equal '0' or '1', it is decremented by one at the beginning of every Execution Interval. Thus, the lowest integer values (except '0') identify the 'oldest' modified data words. By restricting to a distinct number of oldest data words, the amount of data that must be transferred within each interval can be bounded.

The previous paragraph only describes how the transfer of *data changes* inside TAU and TPU is realised via SDSs. In order to enable complete copying of the internal PES state by observing its SDS over a pre-defined period of time, it is also necessary to transfer the data that were not modified recently (e. g., RAM data words written a few minutes ago). A small extension of the concepts described above realises this *complete* state transfer. Fig. 2 illustrates the data transfer via SDS and the subsequent descriptions.

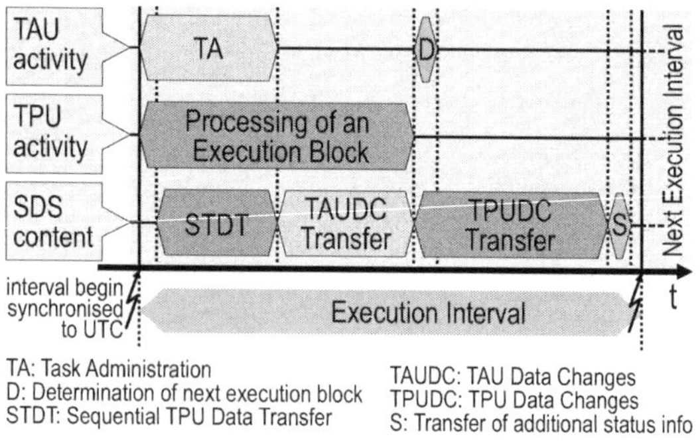

Fig. 2. Transfer schedule of the SDSs

While the TAU administrates the tasks, the SDS transfers a subset of the TPU's RAM content. The entire RAM content is transferred within a number of consecutive Execution Intervals, and in synchrony with UTC. When the task administration finishes, the TAUDCs are transferred. Therefore, the TAU inserts a subset of its oldest data words in the SDS and sets the associated Age integers to '0'. Complete transfer of the TAU data is achieved by changing the 'Ages' of all data words at UTC-synchronous instants. After the TPU completes the current Execution Block, the SDS transfers all its write accesses. Finally, some additional status information like, e. g., the *ID of Block to Execute* is transferred via SDS.

The advantage of this SDS transfer schedule is its efficient usage of the SDS transfer channel. This was only realisable by implementing the SDS generation algorithms – together with the TAU – as digital logic circuit. Implementing the algorithms in hardware allowed to minimise the number of clock cycles needed for completion of the Task-Administration (TA) and to perform the age-variable handling in parallel. As a result, the oldest TAU data words are directly accessible one clock cycle after the TA completes so that the TAUDC transfer can start immediately. Thus, in contrast to other state restoration approaches, this technique can already transfer the latest TAUDCs while the application software is still processed in the current interval.

Obviously, the performance achievable with the proposed PES concept is strongly limited by the data transfer bandwidth of the communication interface. However, computational performance is not the major concern for safety-related systems, and the high bandwidths of modern transfer technologies allow for a performance more than sufficient for most applications of highest safety-criticality.

3. FIELDBUS CONCEPT

Predictability and *availability* belong to the most important characteristics of fieldbus systems for safety-critical real-time applications. Beside these two characteristics, the effort to verify a given fieldbus interface is significant, since this effort corresponds to the cost for safety licensing. Certainly, the total transfer bandwidth is also of interest, but in terms of highly safety-critical real-time systems of less importance than the other aspects mentioned. That is why the following three characteristics have been considered most essential during the development of the proposed fieldbus concept.

Predictable temporal behaviour: The fieldbus concept must provide for predictable temporal behaviour in order to fulfill real-time constraints. This demand for predictability actually forbids repeated transfer of data-frames in case of stochastic transfer errors. Only non-repeating transfer protocols are appropriate; reception failures can only be taken into account in a predictable way on a higher level, e. g. by appropriate exception routines in the application software. That is why minimising the transfer error probability on the physical layer and the data-link layer of the ISO/OSI Reference Model (e. g. by error-detection-correction codes) is especially important for fieldbuses in hard real-time applications.

Fault-tolerance: Since the fieldbus concept should be appropriate even for applications of the highest safety integrity levels (SIL 3 and SIL 4), a particularly low probability of outages is demanded by IEC 61508. Thus, the fieldbus must provide such a high level of fault-tolerance that complete communication outages due to physical failures are nearly impossible. Reliable communication must be guaranteed even if several faults occur simultaneously.

Simple architecture and temporal behaviour: For systems belonging to the two highest safety classes (SIL 3 and SIL 4), IEC 61508 *highly recommends* the use of semi-formal methods to avoid faults during requirement specification, design and development (Part 2, Table B.1 and B.2), and formal methods to verify software specications (Part 3, Table A.1). However, the use of formal methods is practically feasible for systems of limited complexity (Halang and Frigeri 1998), only. Thus, a simple hard- and software architecture as well as a simple temporal behaviour are desirable to ease formal verification. Moreover, this reduces the effort — and the costs — for safety licensing.

The proposed fieldbus concept has been derived from the fieldbus 'Interbus', because of its low hardware requirements, inherent simplicity and its beneficial characteristics regarding real-time requirements. Interbus operates like a spatially distributed closed-loop shift-register. All communication nodes use a shift-register as bus interface, and all nodes' shift-register inputs and outputs are connected to a ring. A master controls the transfer cycles. It has two shift-registers, one for data output and one for data input. Within each cycle, the content of the output register is transferred to the slaves, and the content of the slave registers is transferred to the input register. This is done by shifting the data bits through all nodes of the ring.

The nodes do not need a microcontroller, and no software is necessary to realise transfer protocol functions. This results in an inherently simple architecture of the bus interfaces (Baginski and Müller 1998, Langmann 1999). In (Erdner 2003) it is shown that the cyclical shifting of data through a ring-topology of communication nodes is most appropriate to realise fieldbuses with predictable temporal behaviour.

Like Interbus, the proposed fieldbus connects all system nodes, i. e., the redundant PES instances as well as sensors and actuators, to a ring, and data are transferred from node to node as in a shift-register. However, some extensions to the Interbus concept are necessary in order to enable exchange of SDSs between redundant PES instances and to achieve the level of fault-tolerance required.

3.1 Clock Synchronisation

Interbus requires a master to control shifting data around the distributed shift-register. A failure of the master causes the entire communication to fail. Obviously, the demand for a high level of fault-tolerance forbids this strong dependence on a single master. One way to avoid this strong dependence would be a redundant configuration of bus masters. However, the proposed fieldbus uses a different approach, which does not need a master.

The transfer cycles are carried out in synchrony to a common time base: the international reference Universal Time Co-ordinated (UTC), which is available worldwide via, for instance, GPS, GLONASS and — in the future — GNSS. Therefore, each communication node has an internal clock which is synchronised to UTC. The time signals for synchronisation are provided by special clocks (e. g., GPS time receivers), and transferred via dedicated time channels around the communication ring. Each node acts as a repeater for the incoming time channels.

The clocks providing the time signals for synchronisation of the nodes' internal clocks need to be configured redundantly, as otherwise the outage of one clock can cause the entire system to fail. The proposed fieldbus concept uses multiple channels to distribute the time signals. On the physical layer, the time channels are preferably bundled to one physical line by an FDM technique, since only a low number of data bits needs to be transferred but their time of arrival is of importance.

The time signal channels are ordered by priority. A node always uses the available time signal with the highest priority to synchronise its internal clock. If the clock with the highest priority is affected by a spontaneous failure, all nodes immediately use the time signals of the clock with the next lower priority. If a clock cannot receive its own time signal after passing all nodes of the ring, the clock sets itself out of order. In combination with the multi-ring technique described subsequently, this ensures that all nodes use the same time channel for synchronisation. At runtime and as long as the deviations are small, synchronisation is carried out in a deterministic and smooth way by slight acceleration or deceleration of the nodes' internal clocks. This avoids malfunctions due to abrupt time changes caused by switching to an other time channel.

3.2 Multi-Channel Technique

All nodes are physically connected to one ring, but – unlike Interbus – the proposed fieldbus employs multiple transfer channels. There are one channel for general I/O communication, one channel for each PES and – as already mentioned – redundant time channels. These channels behave logically separate; to keep the wiring effort low they should be bundled to few physical transfer lines by TDM or FDM techniques. Fig. 3 illustrates the multi-channel technique for a PES configuration with three redundant PES instances and two time channels.

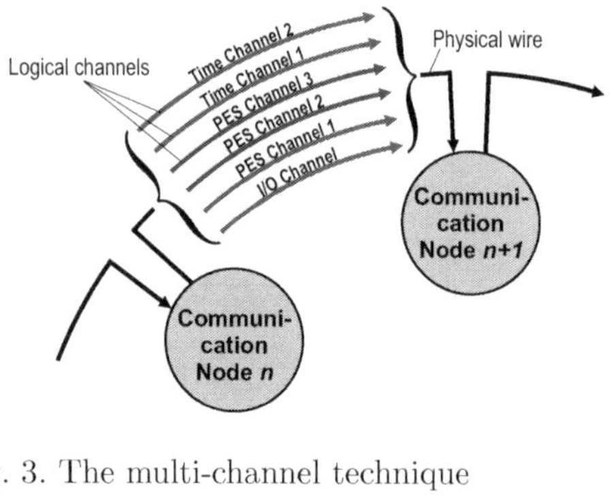

Fig. 3. The multi-channel technique

The multi-channel technique causes only small delays when the SDSs pass through all nodes of the ring. If, e.g., all PES would share only one logic channel for transferring their SDS, the nodes would receive the SDSs of different PESs successively. Using several channels enables that all nodes receive the SDSs of all PESs almost simultaneously, only the repetition of each node causes slight delays of the SDS transfer. This nearly simultaneously reception is essential for the SDS transfer to detect processing errors and to enable the *State restoration at runtime*.

Each PES channel is assigned to one PES, and only this PES sends out data via this channel. The other communication nodes that the PES channel passes through the ring must only repeat the transferred data. Thus, their bus interfaces simply consists of repeaters. If a PES fails, there is no SDS to be transferred via the associated PES channel.

The operating principle of the I/O channel differs slightly from the one of the Interbus, only. Each node determines begin and end of a transfer cycle based on its internal clock; the data bits are serially output in relation to this time base. Since all internal clocks are synchronised to a common time base, all nodes output their data in synchrony. Within a cycle, a node always outputs as many data bits as it receives, just like in the closed-loop shift-register of the Interbus. However, small synchronisation deviations must be taken into account, which may cause some nodes to output their data bits a little bit earlier than others. This problem is solved by using FIFO-structures, cp. Fig. 4, instead of shift-registers. Unlike the shift-registers, the FIFO-structures allow to perform data input and output asynchronously. This enables a node to output its own I/O channel data bits before it received the first data bits from its predecessor. Thus, as long as a node receives the first I/O channel data bit before it sends out its last own I/O data bit, small synchronisation deviations can be compensated. Each bus interface contains two FIFO-structures which are alternately used; while the node writes its new I/O channel data bits in one structure, the other node shifts all I/O data bits around the ring (First, it clocks out the node's I/O channel data bits, then, it clocks out the data bits that it received from the predecessor).

The absence of a master necessitates another deviation from the Interbus concept. Interbus features only data transfer from master to slave and from slave to master; slave nodes cannot communicate with each other. In the proposed fieldbus concept, each node can directly communicate with all other nodes via the I/O channel. Therefore, each node collects all data bits that are transferred via the I/O channel within a cycle as process image in one of two alternately used memory blocks. The process image of a cycle is readable for internal processing during the subsequent cycle. This is

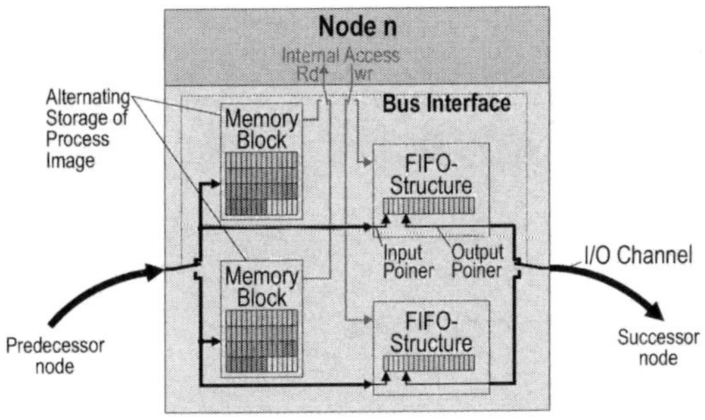

Fig. 4. The nodes' I/O Channel interfaces consist of two memory blocks and two FIFO-structures that alternate their role at begin of each transfer cycle

also shown in Fig. 4. Certainly, this operating principle consumes more hardware resources than Interbus, but its hardware realisation is still very simple.

3.3 Multi-Ring Topology

Conventional ring-based communication techniques increase safety by using both possible data transfer directions. This approach cannot guarantee system availability in case of more than one ring interruption or device outage. That is why the proposed fieldbus concept makes use of a different approach, which bases on multiple communication rings. The output of a node is not only connected to the next node of the ring, but also to the next-but-one node. The additional wires are arranged in a physically separated communication ring which is subsequently called Reserve Communication Ring (RCR). Similarly, the first communication ring is named Primary Communication Ring (PCR). The wiring to the next-but-one ring node via the RCR is also linked back to the predecessor node via the PCR. Thus, one node is connected via two physically separated wires to its successor. With this connection scheme, communication does not fail even if more than one wire cut or complete node outage occurs simultaneously. Fig. 5 illustrates the connection scheme and exemplifies the simultaneous occurrence of a wire cut and a node outage. The simple bi-directional approach would fail in this example.

As long as no wire is broken and all nodes operate properly, all communication can be performed via the PCR. If a wire of the RCR brakes, the entire system maintains normal operation. In case one section of the PCR brakes, the destination node of this section receives the data of its predecessor node via the *RCR/reverse-PCR* detour. If a node fails completely, communication can be maintained via the RCR. As long as a node does not add own data to the I/O channel of the

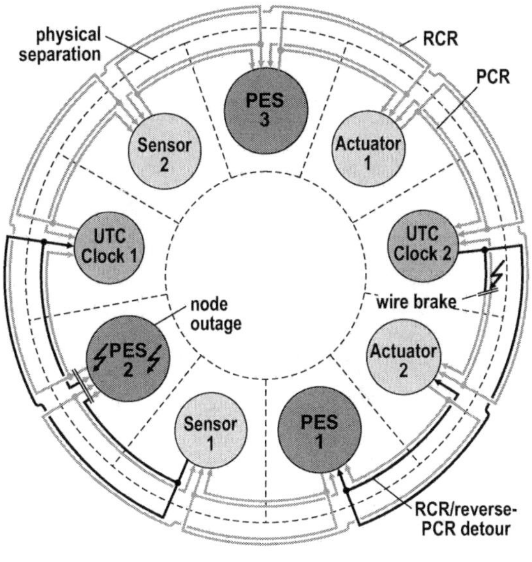

Fig. 5. Connection scheme with three redundant PESs

communication ring (e. g., simple actuator nodes), such a node outage does not affect the entire communication. If a node that adds data to the I/O channel fails completely (e. g., PESs, sensors, clocks), the next ring node must not only pass the incoming I/O channel data of the RCR, but also needs to fill in appropriate data bits which substitute the missing bits of the previous node and indicate its outage.

The 'multi-ring' technique allows for scalable safety by adding further rings and, moreover, can even be combined with bi-directional operation. Furthermore, as long as no outage occurs, the redundant data transfer via RCR can be evaluated to decrease the transfer error probability.

4. CONCLUSION

A novel fieldbus concept which is part of a holistic safety scheme for a new PES architecture was introduced. The concept was designed following the policy 'Progress is the road from the primitive via the complicated to the simple' (Biedenkopf 1994). Both the fieldbus' hardware architecture as well as its communication principle are of remarkable simplicity, which eases verification and – what is even more important – lowers the cost for safety-licensing.

The proposed fieldbus was devised for a new safety-related real-time PES, which features task-oriented real-time execution even though it operates in discrete cycles. The fieldbus serves both data exchange with process peripherals and between multiple redundant processing nodes. Similar to the fieldbus 'Interbus', all system nodes, i. e., the redundant PES instances as well as sensors and actuators, are connected to a ring, and data are transferred from node to node as in a shift-register. The transfer is organised in cycles that match the execution cycles of the dedicated PES architecture. Thus, all system components operate cyclically and in synchrony, leading to a simple and easy-to-model temporal behaviour of the entire system.

The new fieldbus concept bases on the communication strategy of Interbus, not only because of its beneficial characteristics regarding real-time demands, but also because of its remarkably low hardware requirements and inherent simplicity. Unlike Interbus, the proposed concept does not distinguish between master and slave nodes. All nodes have the same hierarchical order; the communication is not under control of a single master which can fail. Instead, all nodes output their data in relation to a common time base. Therefore, each node contains a clock, and the transfer cycles are synchronised to the international reference UTC (Universal Time Co-ordinated). Special clock nodes (e. g. GPS time receivers) provide time signals for synchronisation of the nodes' internal clocks.

Two special techniques were implemented in order to support State restoration of PES instances and to increase the level of fault-tolerance. The *multi-channel technique* enables independent transfer of several data streams round the communication ring and the *multi-ring technique* guarantees uninterrupted operation in case of physical failures like wire breaks or node outages. The latter technique allows to scale the degree of fault-tolerance by adding further communication rings and avoids an impracticable expansion of the wiring effort.

The PES for which the fieldbus was devised supports non-intrusive monitoring, recording of process activities, detecting processing errors and State restoration at runtime in a unified way. This functional unification, which bases on exchanging Serial Data Streams (SDSs) between redundant PES instances, results in a hardware design of minimum complexity. Moreover, the capability of State restoration at runtime leads to a remarkable level of fault-tolerance, since error-affected PES instances can be replaced or – in case of processing failures due to temporary electro-magnetic disturbances – restarted without the need to stop operation. However, this form of recovery imposes special requirements on the exchange of SDSs. The proposed fieldbus suffices these demands and combines I/O communication and SDS exchange in one concept.

So far, a VHDL description has been prepared that realises the bus interface of the proposed fieldbus concept. The VHDL design has extensively been tested by simulation and has been combined with a previously developed VHDL description of the dedicated PES architecture. This System-on-Chip PES has already been successfully implemented in an FPGA. Until now, only electrical wiring is used for data transfer. Our current research focusses on optimising the data transfer by using one optical wire for all channels.

REFERENCES

Adams, D.J. and T. Sims (1990). A tagged memory technique for recovery from transient errors in fault-tolerant systems. In: *Proc. Teal-Time Systems Symp.*. pp. 312–321.

Baginski, A. and M. Müller (1998). *INTERBUS.* Hüthig Verlag. Heidelberg.

Biedenkopf, K. (1994). Komplexität und kompliziertheit (complexity and complicateness). *Informatik Spektrum* **17**, 82–86.

Bondavalli, A., F. Di Giandomenico, F. Grandoni, D. Powell and C. Rabéjac (1998). State restoration in a COTS-based N-modular architecture. In: *1st IEEE Int. Symposium on Object-oriented Real-time distributed Computing (ISORC '98).* Kyoto, Japan. pp. 174–183.

Buttazzo, Giorgio C. (2002). *Hard Real Time Systems, Predictable Scheduling Algorithms and Applications.* Kluwer Academic Publishers. Boston.

D. Basu, R. Paramasivam (2000). An approach to software assisted recovery from hardware transient faults for real time systems. In: *Computer Safety, Reliability and Security – Safecomp 2000 Conference Proceedings.* Springer-Verlag. Berlin Heidelberg. pp. 264 – 274.

Erdner, T. (2003). *Entwurf eines realzeitfhigen fehlertoleranten Feldbussystems (Concept of a fault-tolerant fieldbus with real-time behaviour).* VDI Verlag GmbH. Düsseldorf.

Halang, W. A. and A. H. Frigeri (1998). Methods and languages for safety related real time programming. In: *Computer Safety, Reliability and Security, SAFECOMP Conference Proceedings.* Springer Verlag. Berlin, Germany. pp. 196 – 208.

Langmann, R. (1999). *INTERBUS: Technologie zur Automation.* Carl Hanser Verlag. Munich.

Liddell, David (1994). Simple design makes reliable computers. In: *Revised Papers from a Workshop on Hardware and Software Architectures for Fault Tolerance.* Springer-Verlag. London, UK. pp. 91–94.

Skambraks, M. and G. Hamuda (2004). A task-based real-time pes for safety-critical applications. In: *Proceedings of IFAC Workshop on Programmable Devices and Systems.* IFAC. Cracow, Poland.

Copyright © Fieldbus Systems and Their Applications
Puebla, Mexico, 2005

ELSEVIER
IFAC
PUBLICATIONS

DEPENDABILITY EVALUATION OF REAL-TIME APPLICATIONS DISTRIBUTED ON TDMA-BASED NETWORKS

François Simonot[1], Françoise Simonot-Lion[2], YeQiong Song[2]

(1) IECN – ESSTIN, (2) LORIA – INPL – UHP Nancy 1
Campus Scientifique, BP 239 – 54506 Vandœuvre-lès-Nancy (France)
simonot@esstin.uhp-nancy.fr ; {simonot,song}@loria.fr

Abstract: TDMA is largely adopted by many networks for providing to real-time applications with determinism. However, this determinism could be jeopardized under the transient environmental perturbations, causing application failures. In this paper, we investigate the impact of the transient perturbations (especially due to Electromagnetic Interferences) on the application dependability. We give a contribution on the method for evaluating the application failure probability in function of the TDMA cycle error probability within an EMI zone. This method extends the existing one on the "consecutive-k-out-of-n:F" systems to including variable probability. Numerical applications have shown its efficiency for computing the application failure probability within a given EMI area. *Copyright © 2005 IFAC*

Keywords: dependability, reliability, TDMA, TTP/C, FlexRay, EMI perturbation, transient fault.

1 INTRODUCTION

For achieving deterministic communication properties, many real-time networks use TDMA-based protocols. Such a protocol eases the design of time-triggered applications. TDMA slots are assigned to each data producers in a periodic (or cyclic) way with a fixed TDMA cycle duration. A receiver node periodically receives therefore data it consumes at fixed TDMA time slots so that associated actions can be executed at the right time. Moreover an absence of data production or transfer can be easily detected by the system making the consumer to take a right decision if necessary. In practice, for providing more reliability, the designer of time-triggered applications generally makes data producers to send data to the consumer with period much smaller than necessary (i.e., Nyquist frequency), tolerating thus occasional production or transmission errors to some extend. This is particularly interesting for systems operating at harsh conditions (i.e., subjected to environmental perturbations) provoking transient errors.

For instance, when TDMA-based networks are subjected to EMI (Electromagnetic Interference) perturbations, which are, for example, the typical case of automotive networks, message transmissions can be erroneous. In most of networks such as LIN, TTP/C and FlexRay, transmission errors are detected but erroneous messages are not systematically retransmitted.

In this paper we focus on the analysis of the dependability of a TDMA-based network faced to EMI perturbations with respect to the dependability properties required by a safety critical application distributed on this network. In particular, examples of such an application are in-vehicle embedded systems. With the increasing deployment of electronic embedded components in vehicles, one hears in the media more and more vehicle failures due to the malfunctioning of electronic components.

If today's identified problems find rapidly their solutions, some non-identified problems, making random behaviours of the vehicle, remains unsolved (e.g. vehicle speed blocked by the cruise control system). It seems that EMI perturbations are the main cause of those random behaviours of the vehicle. The Wall Street Journal, edited 8 September 1997, quoted in Risk Digest, (1997) talked about accidents involving vehicles, which to all appearances were caused by EMI. More recently in France, it happens that the cruise control system of certain cars behaves randomly (e.g. impossible to reduce the speed!).

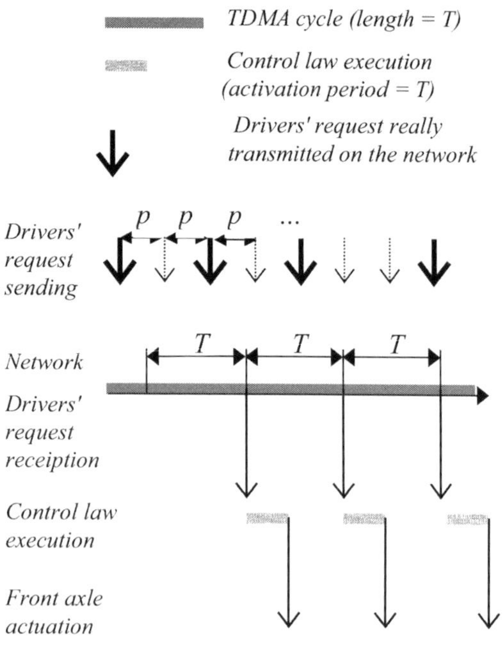

Fig. 1 - Drivers' request transmission

This problem can become even more critical with future integration of X-by-wire systems in a vehicle. So the behaviour of such a system face to EMI has to be evaluated and in particular the impact of the network on the vehicle dependability has to be analysed.

Let us take as an example a Steer-by-Wire system. Critical functions such as front axle control are assumed to be executed on a micro controller (ECU – Electronic Control Unit) periodically taking into account the output of the physical system to control, environmental information and the driver request given at the hand wheel level. This last information is produced periodically by another ECU and transmitted through a TDMA-based network to control law (Fig. 1). Obviously, for this application, the length of the TDMA cycle is equal to the activation period of the control law (T in Fig. 1)). The lack of this drivers' request, due for example to transmission errors, during a short-term, that is for a limited number of control law consecutive executions, can be tolerated. However, it is obvious that long-term absence of input data at the consumer side can lead to dangerous situations. So it is important to be able to evaluate the risk that *consecutive erroneous TDMA cycles* exceed the application-tolerating threshold (in terms of a given TDMA cycle number or equivalently the time length). This risk is termed in this paper the *application failure probability,* which gives a metric to measure the application robustness. Obviously this probability will depend on the perturbation model.
In this paper, we propose a method for the evaluation of the *application failure probability* for critical systems distributed on a TDMA-based network and subject to several profiles of EMI perturbations. This evaluation is based on the assumption that one single non-erroneous cycle is enough for bringing the system to its normal state. This hypothesis is verified for the given Steer-by-wire architecture that we use as example in the following.

However, only little work has been done to deal with the *application failure probability* under EMI perturbations and more generally under transient faults. In (Navet *et al.*, 2000) we first addressed the problem for applications distributed over CAN and introduced the worst-case deadline failure probability as one of the possible application dependability metrics. In (Broster *at al.*, 2004), the impact of the EMI on the real-time delivery capability of CAN and TTCAN has been evaluated. These results cannot be readily applied to the TDMA-based networks. In fact, CAN retransmits whenever a transmission error occurs whereas in TDMA-based networks, there is no systematic retransmission upon transmission errors.

In our previous work (see Wilwert *et al.*, 2003 and Wilwert *et al.*, 2005), we focused either on the evaluation of the X-by-wire application-tolerating threshold or the method to evaluate the application failure probability under restrictive hypothesis on the perturbation model. For instance, the probability for a TDMA cycle to be erroneous, termed in the following "TDMA cycle error probability", has been assumed constant. In fact it is not the general case. A given source can cause EMI that varies in time and / or the distance between a perturbation source and the vehicle can vary in time. It is shown in (Chao *et al.*, 1995) that the evaluation of the application failure probability with constant TDMA cycle error probability can be performed based on the classic results on the "*consecutive-k-out-of-n:F*" systems. In this paper, we extend these results by dealing with the application failure probability evaluation with variable TDMA cycle error probability. So the previous theoretic results on the "*consecutive-k-out-of-n:F*" systems have to be extended.

In what follows, we will give in section 2 a description of the EMI perturbations that an in-vehicle application can meet; such perturbations may provoke TDMA cycle errors (because of either transmission errors or producing errors). Section 3 is devoted to our main contribution that is a method for evaluating the application failure probability with variable TDMA cycle error probability. Section 4 shows numerical applications for some typical error models (TDMA cycle error probability profiles) and that are obtained on a Steer-by-wire system. Section 5 gives the concluding remarks.

2 EMI PERTURBATIONS AND ERROR MODELS

Electromagnetic interferences are mainly caused by radio-communication transmitters, radars, and high voltage lines. Their influence on electronic components depends on the frequency and strength of the electromagnetic fields. In automotive industry, each carmaker specifies an internal regulatory policy that imposes the robustness level of electronic device

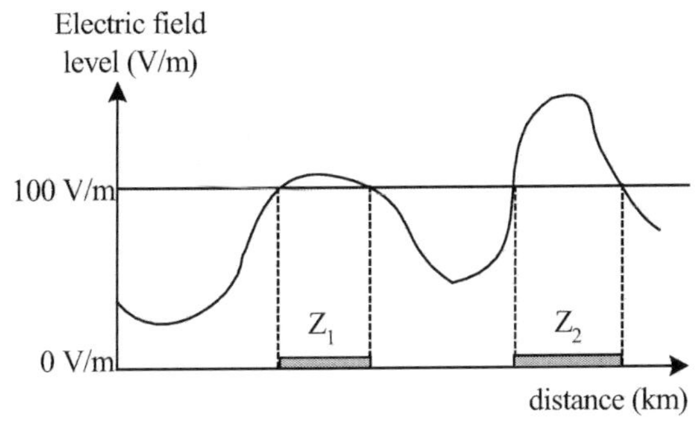

Fig. 2 Example of electric field level of a reference road

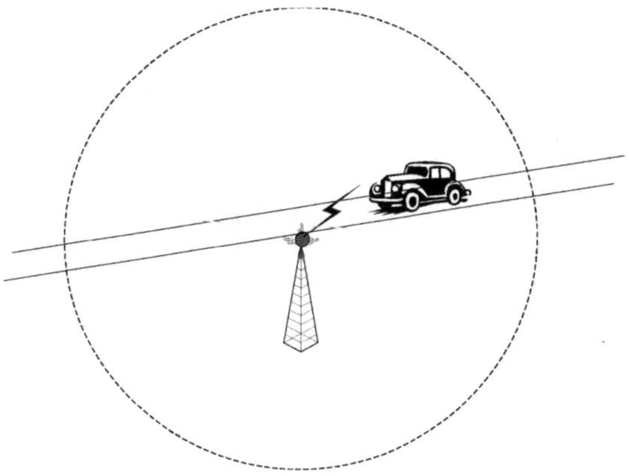

Fig. 3 Example of an EMI zone with variable perturbation levels

with electromagnetic interference sources under a given voltage level and for a given interval of frequencies. So a test process is applied on each electronic component in order to verify its conformity to the specific carmaker standard. Nevertheless, this conformity is just proved for given frequencies and voltage level. In fact, it is established that the testing condition are not met everywhere; it exists some traffic areas, for example near airports, where a vehicle can go through an area subject to a higher level of voltage and / or other frequencies than the specified ones and therefore, the probability that an in-vehicle embedded system can be corrupted by electromagnetic interferences is not zero. For example, carmakers often consider that the upper limit for the robustness assessment of electronic components is 100 V/m. This is to say that when a car goes through an EMI perturbed zone with a force higher than 100 volts per meter, its embedded electronic systems may exhibit errors.

Some sources of electromagnetic interferences are statically disposed along the road (for example, radars or high voltage lines). CEERF, a French project, funded by Ministry of Transport, proposed a characterization of the electromagnetic pollution for the French road system (Predit-CEERF, 2003). This project targeted mainly the automotive industry by proposing a cartography of the EMI sources and electromagnetic field levels in France and a method for its updating. These results are obtained thanks to a monitor embedded in a car and whose role is to record the frequency and the level of the ambient electromagnetic field during a journey along several representative roads. From this recording, we are able to select the length (in km) of each area under EMI perturbation of higher than 100 V/m (see Fig. 2); on the represented trajectory, two parts of this trajectory, areas Z_1 and Z_2, are subject to perturbations of more than 100V/m).

Without losing generality, in what follows, we will only focus on the analysis within *one* EMI zone. In fact, as assumed in section 1, for the targeted application, one single correct reception of the input data at the consumer side before the application-tolerating threshold is required for bringing the system to its normal state. So multiple zones (as they are independent each other) can be treated separately. In (Wilwert *et al.*, 2005), we proposed how to evaluate the failure probability for a specific vehicle trajectory such as what is shown in Fig. 2.

For passing through a given EMI area, a vehicle (assumed with a constant speed) will take a certain time called hereafter the *passing through time* and denoted by T_z. If one knows the TDMA cycle duration T_{cyc}, the passing through time can also be represented in terms of the number of TDMA cycles n. In (Wilwert *et al.*, 2003) and (Wilwert *et al.*, 2005), for taking into account the worst-case protocol recovery overhead, we evaluated this value:

$$n = \left\lceil \frac{T_z}{T_{cyc}} \right\rceil + 2 \qquad (1)$$

We assume furthermore that each TDMA cycle has a non-null error probability within an EMI perturbation area. In general this probability is variable and depends on the position of the vehicle within the zone. Intuitively the stronger the electric field (measured in V/m) is, the higher the probability to have an erroneous TDMA cycle is. Fig.3 shows a typical situation where a vehicle passes through an EMI area generated by a radio transmitter. If a rough approximation by the free space propagation model (Rappaport, 1995) can be used, the electric field strength at a point will be inversely proportional to the square of its distance to the source.

Unfortunately, the exact characterization of the TDMA cycle error probabilities within a zone has not been realized because of large measurements data needed for being statistically confident.

In this study, to get a general idea about the impact of EMI on the application robustness, we will evaluate the application failure probability by analysing some typical profiles of $P = (p_1, p_2, \ldots, p_n)$ called hereafter "error models".

2.1 *Constant-P model*

This first error model describes a constant perturbation. For the total passing through time of *n* TDMA cycles of a given EMI zone, we assume that each TDMA cycle has a same error probability, i.e. $p_i = p$ for all $i=1, 2, ..., n$.

Of course, this may not correspond to an actual situation. However, it is often the case of the laboratory tests. We keep it as a reference model for further comparisons.

2.2 *Radio-P model*

The second error model is a *Radio-P* model. It is designed to represent the error model of a vehicle passing through an EMI area of *n* TDMA cycles generated by a radio transmitter (e.g. Fig. 3). Assuming that the free space propagation model (Rappaport, 1997) is adopted and the error probability of a TDMA cycle is somehow proportional[1] to the received electric field strength, for a given TMDA cycle *i* ($i = 1, 2, ..., n$) we give its error probability by:

$$p_i = \frac{a}{(\frac{n+1}{2} - i)^2 + b} \qquad (2)$$

Where *a* and *b* (with $a \leq b$) are free parameters which can be adjusted for fitting to a concrete situation. This model is illustrated in Fig. 4.

2.3 *Radar-P model*

The third error model is a *Radar-P* model. It is proposed to represent the error model of a vehicle passing through an EMI zone of *n* TDMA cycles generated by a radar source (e.g. near to an airport). The electric field varies periodically with the radar-scanning period of *T* (*with* $T < T_z$). Again we assume that the error probability of a TDMA cycle is proportional to the received electric field strength.

For a given TMDA cycle *i* ($i = 1, 2, ..., n$) we give its error probability by:

$$p_i = a + b \sin \frac{2\pi}{T} i \qquad (3)$$

Where *a* and *b* (with $a - b > 0$ and $a + b \leq 1$) are free parameters which can be adjusted for fitting to a concrete situation. Fig.4 depicts the trends of a *Radar-P* model.

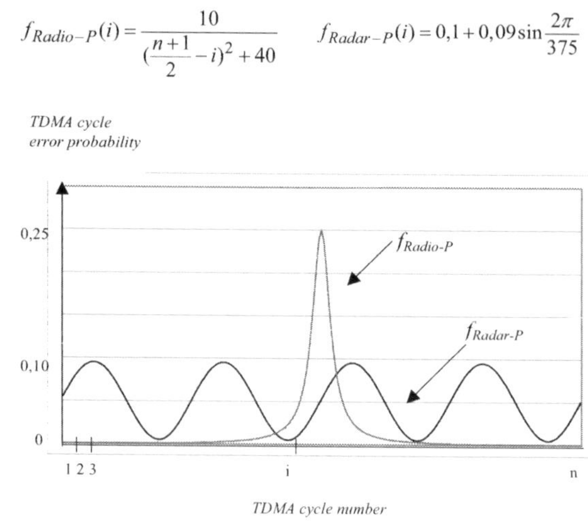

Fig. 4 Trends of *Radio-P* and *Radar-P* models

3 APPLICATION FAILURE PROBABILITY EVALUATION

3.1 *Existing results for Constant-P model*

Knowing a zone of *n* TDMA cycles and the application-tolerating threshold of *k* TDMA cycles, for constant *p*, the problem can be treated using existing results on the reliability of a system composed of an ordered sequence of *n* components and such that the system fails if and only if at least *k* consecutive components fail.

This kind of systems are termed "*consecutive-k-out-of-n:F*" systems and denoted by $C(k,n:F)$. For such a system, we note *n* the number of components, *p* the probability that a component fails, L_n a number of consecutive failed components and *k-1* the largest tolerable number of consecutive failed components; the reliability of the system is evaluated by the probability that $L_n < k$, denoted by $P(L_n < k) = R(k,n;p)$, or equivalently the failure probability $P_{fail} = P(L_n \geq k) = 1 - R(k,n;p)$. The following formula was proposed first in (Burr and Cane, 1961) and then simplified in (Lambris and Papastavridis, 1985) and (Hwang, 1986).

$$R(k,n;p) = \sum_{m=0}^{\lfloor (n+1)/(k+1) \rfloor} (-1)^m p^{mk} q^{m-1} \left(\binom{n-mk}{m-1} + q \binom{n-mk}{m} \right)$$

where $q = 1 - p$.

The numerical evaluation of $R(k,n;p)$ via this formula is quite complex; so, we have developed in (Wilwert *et al.*, 2005) a recurrent relation permitting to compute the failure probability for any *n* and *k*.

For $P = (p_1, p_2, ..., p_n)$, as there does not exist closed form solutions, we give in the sequel a solution for $R(k,n;P)$.

---

[1] This is only a general assumption. In practice, node and communication channel redundancy is often used to reduce the impact of EMI perturbations. The error probability estimation for a given perturbation could be more complex and field tests are necessary. In (Wilwert *et al.*, 2005), we discussed in more detail the ways to estimate it.

3.2 *New results for P = (p1, p2, ..., pn) with non constant pi*

Now this *P*, without loss of generality, let us consider an infinite sequence of independent Bernoulli trials $X_1, X_2, ... X_n, ...$ defined on the probability space (Ω, A, P) with $p_i = P(X_i = 1)$ for $i \geq 1$.

We call "word" a sequence of consecutive successes of Bernoulli trials (when $X_i = 1$).

The goal of our work is to investigate the probability law of the random variable L_n, the length of the longest word known during the *n* first trials. To our great surprise, it seems that this problem, in case of the non-identically distributed random variables, has never been addressed.

We define:

- T_k = the first instant where a sequence of k consecutive successes appear.
- $u_n(k) = P(L_n < k)$ for $n \geq 0$ and $k \geq 1$. For fixed $k \geq 1$, the sequence $u_n(k) = P(L_n < k)$ is decreasing and lower bounded by 0.
- $\lambda_n(k) = q_{n-k} p_{n-k+1} p_{n-k+2} \cdots p_n$ for $n \geq k$ with $q_0 = 1$ and $q_n = 1 - p_n$ if $n \geq 1$.

Property1

The sequence $u_n(k) = P(L_n < k)$ verifies the following relation:

For $k \geq 1$ and $n \geq k+1$,

$$u_n(k) = u_{n-1}(k) - \lambda_n(k) u_{n-k-1}(k) \qquad (4)$$

with initial conditions :

$u_n(k) = 1$ for $0 \leq n \leq k-1$ and $u_k(k) = 1 - \lambda_k(k)$.

Proof :

We have :

$$P(T_k = n) = p_n p_{n-1} \cdots p_{n-k+1} q_{n-k} P(L_{n-k-1} < k)$$

and

$$P(L_n \geq k) = P(T_k \leq n),$$

which give the following relation :

$u_n(k) = u_{n-1}(k) - \lambda_n(k) u_{n-k-1}(k)$ for $k \geq 1$ and $n \geq k+1$.

End of proof

Property1 contains complete information on the behavior of $u_n(k) = P(L_n < k)$ and allows an exact calculation of the probability law of L_n and provides an efficient algorithm for computing $u_n(k)$. Moreover, the following useful monotonic property of $u_n(k)$ can be established.

Table 1 Application failure probability under *Constant-P* error model

TDMA Cycle Length T_{cyc}	Application failure P_{fail}	Number of TDMA Cycles N	Maximum tolerable number of consecutive erroneous cycles k
4	3.30E-09	377	10
4.25	3.12E-08	355	9
4.5	2.95E-07	336	8
4.75	2.79E-07	318	8
5	2.65E-07	302	8
5.25	2.53E-06	288	7
5.5	2.41E-06	275	7
5.75	2.31E-05	263	6
6	2.21E-05	252	6
6.25	2.12E-05	242	6
6.5	2.04E-05	233	6
6.75	1.98E-04	225	5
7	1.91E-04	217	5
7.25	1.84E-04	209	5
7.5	1.77E-04	202	5
7.75	1.72E-04	196	5
8	1.67E-04	190	5
8.25	0.00161977	184	4
8.5	0.00157484	179	4
8.75	0.0015299	174	4
9	0.00148497	169	4
9.25	0.00144902	165	4
9.5	0.00140408	160	4
9.75	0.00136813	156	4
10	0.00133218	152	4

Property2

If $p'_j \geq p_j$ for all $j \geq 1$, then $u_n(k) \geq u'_n(k)$ for all n and all $k \geq 1$.

That is to say : $p'_j \geq p_j$ for all $j \geq 1$ implies that L_n is stochastically less or equal to L'_n for all n.

It turns out that $p'_j \geq p_j \geq p''_j$ for all $j \geq 1$ implies $u''_n(k) \geq u_n(k) \geq u'_n(k)$ for all n and all $k \geq 1$.

4 NUMERICAL RESULTS

In this section we will apply the previously established algorithms (the complexity of the program varies with n) to the three typical error models called *Constant-P*, *Radio-P* and *Radar-P* described above.

We focus on an EMI perturbation with passing through time $T_z=1500$ms and application-tolerating threshold $T_{max}=40$ms. Note that, for the Steer-by-wire system that we presented formerly, these values are extracted on the one hand, from the cartography of EMI sources and electromagnetic field levels in France (Predit-CERF, 2003) and, on the other hand, by executing a Matlab / Simulink model that integrates the control law, the physical system and the vehicle characteristics (Wilwert *et al.*, 2005).

For a given TDMA cycle duration T_{cyc}, n is given by equation 1, whereas the application-tolerating threshold in terms of the number of TMDA cycles, k, is given by:

$$k = \left\lfloor \frac{T_{\max}}{T_{cyc}} \right\rfloor \qquad (5)$$

Our objective is to evaluate the application failure probability P_{fail} for a given TDMA cycle duration T_{cyc}.

In order to also analyze the influence of the TDMA cycle duration on P_{fail}, we make vary T_{cyc} and, consequently, the activation period of the control law from 4ms to 10ms with step of 0.25ms (these values have to be specified both by automatic control specialists and by system architect designer). The obtained results provide guidelines for system designer to correctly dimensioning T_{cyc} for meeting a specific requirement on P_{fail}.

4.1 *Constant-P model*

Let $p = 0.1$, by using the algorithm given by property1, we get the following failure probability P_{fail} (Table 1).

In view of the equations 1 and 5, when T_{cyc} increases, n and k both decrease. As was seen in section 3.2, for a fixed k value, the failure probability is an increasing function of n.

4.2 *Radio-P model*

Let a = 10 and 11, b = 20 and 19 respectively, p_i are given according to equation 2. The failure probability P_{fail} for different T_{cyc} is given in Table 2.

In addition to the general comments we have already made for the *Constant-P* case, we can also observe the effect of property 2.

Table 2 Application failure probability under *Radio-P* error model

TDMA Cycle Length	Application failure (a=10, b=20)	Application failure (a=11, b=19)	Number of TDMA Cycles	
T_{cyc}	P_{fail}	P'_{fail}	n	k
4	2.22E-08	8.19E-08	377	10
4.25	2.94E-07	9.73E-07	355	9
4.5	3.30E-06	9.82E-06	336	8
4.75	3.30E-06	9.82E-06	318	8
5	3.30E-06	9.82E-06	302	8
5.25	3.12E-05	8.32E-05	288	7
5.5	3.12E-05	8.32E-05	275	7
5.75	2.46E-04	5.86E-04	263	6
6	2.46E-04	5.86E-04	252	6
6.25	2.46E-04	5.86E-04	242	6
6.5	2.46E-04	5.86E-04	233	6
6.75	0.001609891	0.00340238	225	5
7	0.001609891	0.00340238	217	5
7.25	0.001609891	0.00340238	209	5
7.5	0.001609891	0.00340238	202	5
7.75	0.001609891	0.00340238	196	5
8	0.001609891	0.00340238	190	5
8.25	0.008690406	0.01621666	184	4
8.5	0.008690406	0.01621666	179	4
8.75	0.008690406	0.01621666	174	4
9	0.008690406	0.01621666	169	4
9.25	0.008690406	0.01621666	165	4
9.5	0.008690406	0.01621666	160	4
9.75	0.008690406	0.01621666	156	4
10	0.008690406	0.01621666	152	4

In fact, if we note:

$P = (p_1, p_2, ..., p_n)$ for $a = 10$ and $b = 20$,

$P' = (p'_1, p'_2, ..., p'_n)$ for $a = 11$ and $b = 19$,

according to equation 2, it turns out that $p'_i > p_i$ for all $i = 1, 2, ..., n$. So it is not surprising that $P'_{fail} > P_{fail}$ in Table 2.

4.3 *Radar-P model*

p_i are given according to equation 3 with $a = 0.1$, $b = 0.09$ and radar scanning period $T = 375$ms (i.e. $4T = T_z$),.

The failure probability P_{fail} for different T_{cyc} is given in Table 3.

Table 3 Application failure probability under *Radar-P* error model

TDMA Cycle Length T_{cyc}	Application failure P_{fail}	Number of TDMA Cycles N	Maximum tolerable number of consecutive erroneous cycles k
4	5.55E-07	377	10
4.25	2.93E-06	355	9
4.5	1.57E-05	336	8
4.75	1.47E-05	318	8
5	1.38E-05	302	8
5.25	7.53E-05	288	7
5.5	7.14E-05	275	7
5.75	3.92E-04	263	6
6	3.74E-04	252	6
6.25	3.57E-04	242	6
6.5	3.42E-04	233	6
6.75	0.00192067	225	5
7	0.00184522	217	5
7.25	0.0017695	209	5
7.5	0.00170302	202	5
7.75	0.00164584	196	5
8	0.00158847	190	5
8.25	0.00907411	184	4
8.5	0.008806	179	4
8.75	0.00853722	174	4
9	0.00826772	169	4
9.25	0.00805156	165	4
9.5	0.0077806	160	4
9.75	0.0075632	156	4
10	0.00734519	152	4

This kind of numerical results can be used to verify whether a given application, distributed on TDMA-based networks and under a known EMI zone, can still meet the dependability constraint in terms of failure probability. It can also provide to a designer with guidelines for correctly dimensioning T_{cyc} for meeting a specific requirement on application failure probability.

5 CONCLUSION

In this paper we have investigated the impact of the EMI perturbations on the dependability of applications distributed around TDMA-base networks where we assumed that application failure occurs when consecutive erroneous TDMA cycles exceed a certain threshold. This problem is of prime importance, especially for automotive industry as on the one hand, the most adopted embedded networks such as LIN, CAN and FlexRay are based on TDMA, and on the other hand many embedded applications (e.g. X-by-Wire systems) have to meet stringent dependability constraints, this even under EMI perturbations. We contributed to the method for evaluating the application failure probability. For this, we have proposed an important theoretic result which extends the existing one on the "*consecutive-k-out-of-n:F*" systems to including variable probability. Although we have only analyzed several typical error models, our method is still available whatever the profile of p_i may be. These probabilities can be obtained in practice by measurements. This method can also be used to study the system dependability of any transient perturbations.

REFERENCES

Broster, I., A. Burns, and G. Rodríguez-Navas (2004). Comparing real-time communication under electromagnetic interference, In *Proceedings of the 16th IEEE Euromicro Conference on Real-Time Systems*, pages 45–52, Catania, Italy, July 2004.

Burr, E. J. and G. Cane (1961). End-toEnd Arguments in system design. In *Biometrika*, Vol. 48, pp.461-465.

Chao, M.T., J-C. Fu and M-V. Koutras (1995). Survey of reliability studies of consecutive-k-out-of-n:F and related systems. In *IEEE Transactions on reliability*, 44(1): 120-127, march 1995.

Hwang, F.-K. (1986). Simplified reliabilities for consecutive-k-out-of-n:F systems. In *Algebraic Discrete Methods*, Vol. 7, pp. 258-264, 1986.

Lambris, M. and S. G. Papastavridis (1985). Exact reliability formulas for linear and circular consecutive-k-out-of-n:F systems. In *IEEE Transactions on Reliability*, Vol. 34, pp. 124-126, 1985.

Navet, N., Y-Q. Song, and F. Simonot (2000). Worst-case deadline failure probability in real-time applications distributed over CAN. In *Journal of Systems Architecture*, 46(7).

PREDIT-CEERF (2005). Caractérisation de l'environnement électromagnétique routier en France. *Technical report* (in French), 2003.

Rappaport, T.S. (1995). *Wireless Communications: Principles and Practice*, Prentice Hall, 1995.

The Risks Digest (1997). GM car acceleration due to EMI. http://catless.ncl.ac.uk/Risks/19.38.html, 19 (38).

Wilwert, C., F. Simonot-Lion, Y.Q. Song and F. Simonot (2005). Quantitative Evaluation of the Safety of X-by-Wire Architecture subject to EMI Perturbations. In proceedings 10th *IEEE Conference of Emerging Technology for Factory Automation, ETFA'2005*, September 2005, Catania, Sicily, Italy.

Wilwert, C., Y.Q. Song, F. Simonot-Lion and T. Clément (2003). Evaluating Quality of Service and Behavioral Reliability of Steer-by-Wire Systems. In Proceedings *9th IEEE Conference of Emerging Technology for Factory Automation, ETFA'2003*, Lisbonne, Portugal.

ELSEVIER
IFAC
PUBLICATIONS

REAL-TIME PROCEDURES IN DISTRIBUTED SYSTEMS

Mário J.B. Calha [1], Valter F. Silva [2], José A.G. Fonseca [3]

[1] *ESTCB - Instituto Politécnico de Castelo Branco*
Av. do Empresário
6000-767 Castelo Branco, Portugal

[2] *ESTGA – Universidade de Aveiro*
Campus Universitário de Santiago
3810-193 Aveiro, Portugal

[3] *DET / IEETA - Universidade de Aveiro*
Campus Universitário de Santiago
3810-193 Aveiro, Portugal

Abstract: The design of real-time systems can be accomplished using various modelling techniques in which the timing specification has to be included. At such level of abstraction, the functionality might be described using real-time procedures. These procedures specify the actions to be accomplished without any knowledge about the system architecture, in terms of nodes and network. When a procedure maps to various tasks, the determination of the tasks' and messages' parameters is not trivial. The technique presented in this paper builds upon previous work by the authors and helps to reduce the time from design to implementation in real-time systems. *Copyright © 2005 IFAC*

Keywords: Real-time systems, Real-time tasks, Distributed models, Parameter estimation, Speed control.

1. INTRODUCTION

The design of real-time software must incorporate all of the fundamental concepts associated with high-quality software like abstraction and modularity. In addition, real-time software poses a set of unique problems for the designer like:

- Representation of interrupts and context switching,
- Concurrency as manifested by multitasking and multiprocessing,
- Intertask communication and synchronization,
- Wide variations in data and communication rates,
- Representation of timing constraints,
- Asynchronous processing,
- Necessary and unavoidable coupling with operating systems, hardware, and other external system elements.

Several real-time software design methodologies have been proposed to grapple with some or all of the problems noted above. Some design methods extend one of the three classes of design, namely: data flow (Ward, 1986), data structure (Jackson, 1983), or object-oriented (Booch, 1987) methodologies. Others introduce an entirely separate approach, using finite state machine models or message passing systems (Witt, 1985), Petri nets (Vidondo, 1983), or a specialized language (Steusloff, 1984) as a basis.

More recently the use of time Petri nets in real-time systems is proposed by Okawa and Yoneda (1996), Toussaint, *et al.* (1997), Roux, *et al.* (2002) and Barreto, *et al.* (2004).

Data flow-oriented design methods are the most widely used in the industry. Data-flow models are used to show how data flows through a sequence of processing steps. The data is transformed at each step before moving on to the next stage. These processing steps or transformations are program functions when data-flow diagrams are used to document a software design. Extensions to data flow representations that provide the mechanics for real-time software design have been proposed. Gomaa (1984) proposed one extension called Design Method for Real-Time Systems (DARTS). This extension allows real-time system designers to adapt data flow techniques to the special needs of real-time applications.

Software development based on common architectural idioms has its focus shifted from lines-of-code to coarser-grained architectural elements (software components and connectors) and their overall interconnection structure (Medidovic, 2000). Generally, software architectures are composed of components, connectors and configurations. Components are the set of computation units. Connectors are architectural building blocks used to model interactions among components and rules that govern those interactions. Finally, configurations are connected graphs of components and connectors that describe architectural structure.

To support architecture-based development, formal modelling notations and analysis and development tools that operate on architectural specifications are needed. Architecture Description Languages (ADLs) and their accompanying toolsets have been proposed as the answer. Loosely defined, "an ADL for software applications focuses on the high-level structure of the overall application rather than the implementation details of any specific source module" (Vestal, 1993).

A number of ADLs have been proposed for modelling software architectures both within a particular domain and as general-purpose architecture modelling languages. These ADLs can be divided into first-generation ADLs and XML-based ADLs (Dashofy, *et al.*, 2001). First-generation ADLs are characterized by proprietary language syntaxes, while XML-based ADLs benefit from the extensible nature of XML standard. Representatives of the first-generation software ADLs are: ACME, Rapide, Unicon and Wright. The first-generation software ADLs are thoroughly classified and compared by Medidovic and Taylor (2000). Some examples of XML-based ADLs are: ADML and xADL. The more recent XML-based ADLs are presented by Spencer (2000) and (Dashofy, *et al.*, 2001).

The Unified Modelling Language (UML) (Rumbaugh, *et al.*, 1998) can also be used to model systems. But UML is a rather heavyweight design notation, modelling the full structure and semantics of a software system in seven separate views.

Another approach to the design of real-time systems is the use of real-time procedures that include timing constraints. These procedures may be organized with a data flow-oriented method based on simple data streams. A data stream shows the information flow between tasks. The tasks follow the producer-consumer model. Data flows from one producer task to one or more consumer tasks. Consumer tasks may also produce data to other tasks thus becoming consumer/producer tasks. This way, a data stream represents only the tasks and data that make use of critical resources like the processing units and the network. This method differs from other approaches, like the Petri nets and other behavioural diagrams, mainly in the way the transitions between tasks are specified. These transitions don't specify the triggering conditions but follow a simpler approach more suited to distributed control systems. In this approach only the possible sets of messages to be transmitted has to be specified. This way, the internal computational aspects of tasks remain encapsulated.

The result of the proposed approach is a set of data streams, where tasks and messages have some parameters derived. These parameters will make possible a holistic system scheduling of all entities. Also, the data stream analysis is shown to be beneficial in different phases of a distributed system planning.

Various reasons were behind the choice of this approach, instead of one of the presented ADLs.

First, the modification, or extension, of software ADLs can be hard due to one of two reasons, either it is custom-tailored to support only one great feature, but not covering the needs of this study, or it is so general that a lot of changes are required in order to have only the required features.

Second, a component in an ADL refers to a unit of computation, while in this study both the tasks and the messages are the main entities, therefore bringing messages to the same level of importance.

Third, this approach is only focused in aspects relevant to the parameter determination of entities towards an automated scheduling.

Fourth, when comparing to the UML, the proposed approach is geared toward lightweight experimentation and easy extension.

This paper is organized in four further sections. In section 2 the interactions between tasks are studied together with the data streams. The real-time procedures are presented in section 3. In section 3 a case study, the Robchair, is used to demonstrate the benefits of using the combination of real-time procedures and data streams. Finally, some conclusions and further work are discussed in the last section.

2. INTERACTION BETWEEN TASKS

A job is usually defined as a sequence of instructions to be executed by a processor. It is also known as a thread of execution or a (processor) scheduling unit. In real-time systems, a job has, at least, two parameters that are the release instant *ri* and the worst-case computation time C_i. A task is a potentially infinite sequence of jobs where each job is a task instance. A task is periodic if it is time-triggered, with a regular release. The length of time between releases of successive instances of task τ_i is a constant, T_i, which is called the period of the task. Therefore, a periodic task τ_i can be completely characterized by the following three parameters: its worst-case computation time C_i, its period T_i and its relative phase Ph_i, which determines the first release instant. Real-time jobs also have a temporal limit for finishing the execution that is called deadline D_i that is relative to the job's release time. All jobs from a task have a common deadline. Most of the tasks executed in a distributed system either need data from other tasks or generate results to be used somewhere in the system, or both. A task that generates some data is called a producer task and a task that uses that data for any purpose is called a consumer task.

According to the communication paradigm selected for this study, every interactive task uses messages to exchange data with other tasks. Particularly, interactive tasks communicate through periodic messages. A periodic message is a potentially infinite sequence of message instances. In real-time systems, a message has, at least, two parameters that are the release instant r_j and the worst-case transmission time C_j. A periodic message σ_j can be completely characterized by the following three parameters: its worst-case transmission time C_j, its period T_j and its relative phase Ph_j, which determines the first release instant. Also a message has a deadline measured relatively to the release instant D_j.

The transmission window of a message and the execution window of a task both refer to the interval between the release instant and the absolute deadline.

In a system where tasks and messages have precedence constraints, the definition of parameters like the initial phase and the relative deadlines of these entities depend upon the choice of whether, or not, the windows of consecutive entities, with a precedence relation, overlap. In this study, tasks alternate with messages; therefore, an execution window may only be preceded by, or precedes, a transmission window. The parameters are determined so that scheduling is possible with the current set of tasks, messages and their predefined parameters. If there is any change in the system, then a scheduling has to be attempted in order to accept the proposed changes. If the scheduling does not succeed, then a new parameter determination, that guarantees the scheduling, must be accomplished or, otherwise, the changes must be rejected.

The most basic form of intertask interaction in distributed systems is message exchange. This enables a sending task, producer task, to transmit a single message to a receiving task, consumer task. This is the communication paradigm used for intertask interaction. Message passing between a pair of tasks can be supported by two message communication operations: send and receive, defined in terms of destinations and messages. In order for one task to communicate with another, one task sends a message (a sequence of bytes) to a destination and another task at the destination receives the message. This activity involves the communication of data from the sending task to the receiving task and may involve the synchronization of the two tasks.

Due to the interaction between tasks, several scenarios may occur in an ordinary system. A common restriction to all scenarios is that any message consumed has to be available in the beginning of the task execution and any produced message is only available after the end of the task execution. Several scenarios were identified. These were arranged in two groups: the basic scenarios and the expansion scenarios. The basic scenarios represent the simplest form of interactions where the messages to be transmitted are defined in the diagrams. These scenarios comprise both unicast and multicast communication. The expansion scenarios are built upon the basic scenarios, where the set of possible messages to be transmitted is defined. The actual messages to be transmitted are only defined during system functioning. A set of four basic scenarios and a set of eight expansion scenarios are now presented.

2.1 Basic scenarios

The first two scenarios refer to unicast and are depicted in fig. 1.

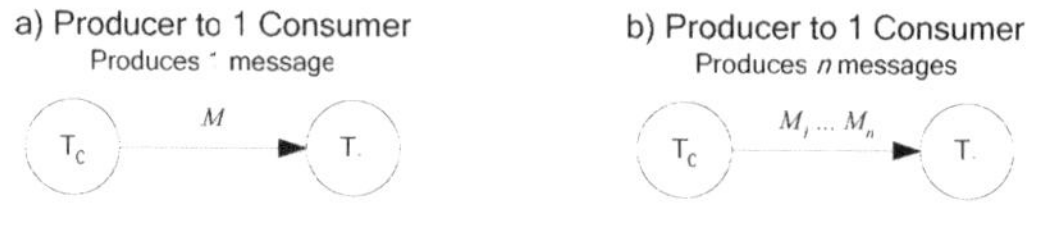

Fig. 1. Unicast of a single message or multiple messages.

The basic interaction between tasks is shown in fig. 1-a). In this scenario, task T_0 produces message M to task T_1.

A producer task can unicast several messages, M_1 to M_n, to another task as shown in fig. 1-b). If all messages have the same properties, apart from the message size, than this scenario is just an extension of the previous one. The total transmission time is the cumulative transmission time of each message.

Fig. 2 depicts two scenarios that refer to multicast.

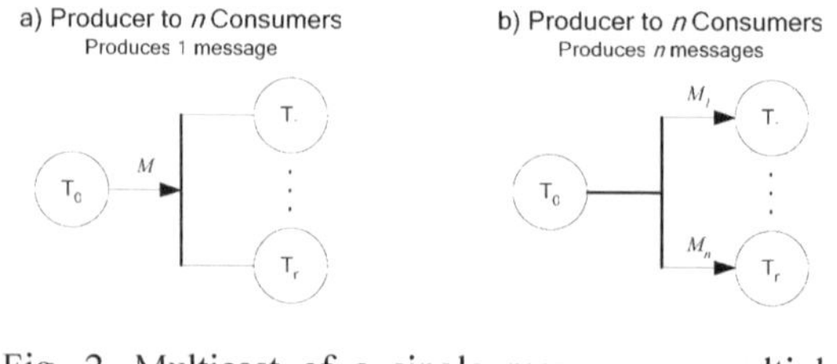

Fig. 2. Multicast of a single message or multiple messages.

Multicasting of a single message to several tasks is shown in fig. 2-a). The combination of the multicast property with the possibility of sending several messages results in the situation shown in fig. 2-b). This scenario makes possible sending different messages to different tasks. Again, the transmission time is the cumulative transmission time of each message.

2.2 Expansion scenarios

The expansion scenarios comprise four unicast scenarios and four multicast scenarios. The first two unicast scenarios that show the transmission of a message from a set of messages are depicted in fig. 3.

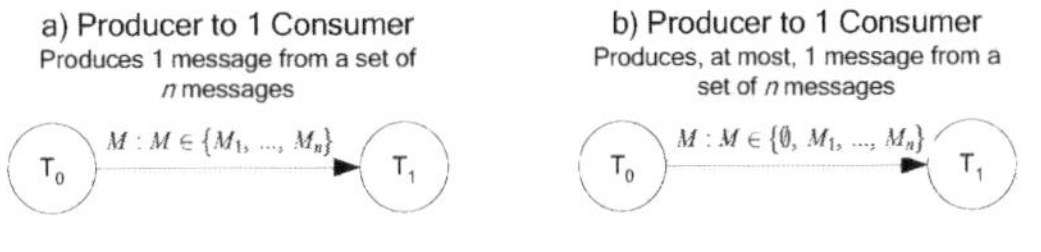

Fig. 3. Unicast of a single message from a set of messages.

Fig. 3-a) shows task T_0 producing one message to task T_1. This message is one of a set of possible messages to be transmitted. In this scenario it is reasonable to demand that all messages have the same properties apart from the message size. This way the transmission time is calculated using the message with the largest size.

The possibility of not transmitting any message is considered in fig. 3-b). In this situation, the producer task can produce one message from the set $\{0, M_1,\ldots,M_n\}$. The null element in the beginning of the set means that a message might not be produced. Because this scenario has a non-periodic produced message it should be considered as asynchronous communication.

For the definition of these scenarios, it is only required the definition of the possible sets of messages and their properties. The computational aspects of the involved tasks, which lead to each message set, do not need to be considered. Therefore, at this level of abstraction, the transitions don't have associated conditions that need to be met, because the question is not "What conditions lead to the production of each set of messages?", but simply "What sets of messages can be produced?". Therefore this approach leads to the encapsulation of tasks reducing the complexity of the analysis.

The other two unicast scenarios are depicted in fig. 4.

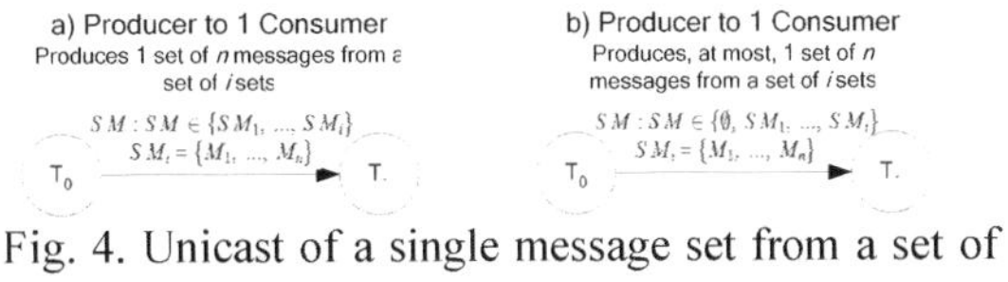

Fig. 4. Unicast of a single message set from a set of message sets.

Fig. 4-a) shows the unicast transmission of a set of messages, *SM*, where task T_0 produces one set of messages to task T_1. This set of messages is selected during system functioning from the possible sets of messages. The possibility of not transmitting any set of messages is considered in fig. 4-b).

The first two multicast scenarios are depicted in fig. 5.

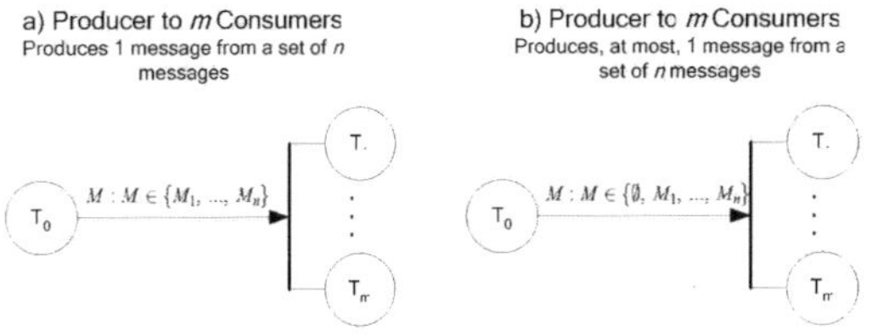

Fig. 5. Multicast of a single message from a set of messages.

Fig. 5-a) shows the situation of a multicast of a single message *M* from a set of messages, $M_1 \ldots M_m$. Only one of these messages is transmitted after each execution of task T_0. This scenario is basically the union of the scenarios depicted in fig. 2-a) and fig. 3-a).

On the other hand, fig. 5-b) considers the possibility of the task either transmit a message from the set of messages, $M_1 \ldots M_m$ or not transmit a message at all. Because this scenario has a non-periodic produced message it should be considered as asynchronous communication.

The other two multicast scenarios are depicted in fig. 6.

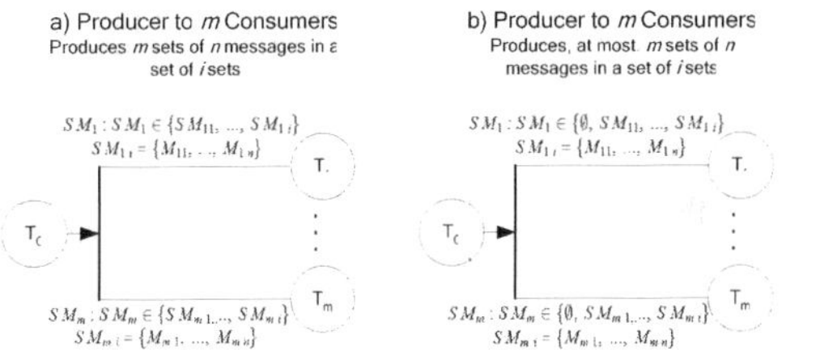

Fig. 6. Multicast of multiple message sets from a set of message sets.

Fig. 6-a) shows the multicast of multiple sets of messages from a set of message sets. On the other hand, fig. 6-b) considers the possibility of the task not transmit a set of messages.

With these scenarios many other can be constructed.

2.3 Data streams

A data stream may begin with a producer task and end with a consumer task. According to the scenario, various data streams might be identified. For example the scenario in fig. 7 depicts a task that produces four messages. Message M_1 is consumed by two different consumer tasks, T_1 and T_2, message M_2 is consumed by task T_3, message M_3 is consumed by task T_4 and message M_4 is consumed by task T_5. From this scenario, five data streams are identified as shown in fig. 7.

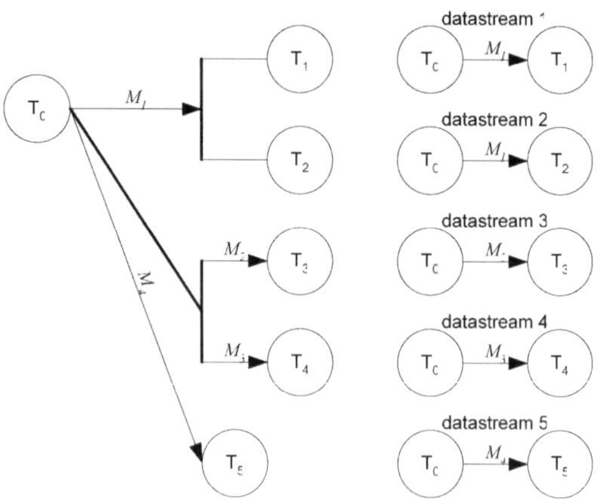

Fig. 7. Multicast of multiple message sets from a set of message sets.

In a broader view of the data flows, a data stream can also begin with a message, or even sets of messages, and end with a message, or even sets of messages, see fig. 8.

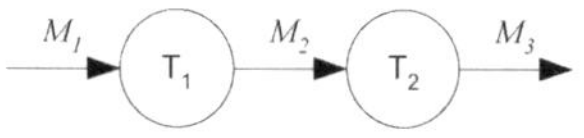

Fig. 8. Example of a data stream that begins and ends with messages.

3. AN APPROACH TO REAL-TIME PROCEDURES IN DISTRIBUTED SYSTEMS

A distributed system is built upon a set of computational nodes interconnected through a network. A procedure is a conceptual set of operations that might be implemented through a set of tasks, where each operation does not necessarily map to each task. This partitioning should take into consideration aspects like: separation of functionality, location of needed resources and load balancing between the nodes. Due to the partitioning, some tasks might be allocated to specific nodes in the case where specific resources are required, and these resources are only available at a certain node. Other tasks might be less stringent and their allocation is possible within a set of nodes. Due to this distribution, the message-passing paradigm is used to exchange data between tasks.

A real-time procedure is expected to be executed within a time frame, know as deadline. If a real-time procedure is mapped to a single task then the deadline of the procedure becomes the deadline of the task. If, on the contrary, a real-time procedure is mapped to various communicating tasks, then the issue is how to define the deadlines of the involved tasks and messages. A possible way to specify the interactions between multiple tasks is through the use of data streams, as previously presented.

In a system with multiple tasks, two approaches to determine the parameters, according to the precedence relations, have been identified. In the first one the execution window and the transmission windows do not overlap, while on the second these windows may overlap. In the first approach not only the tasks and messages have precedence restrictions but also their execution, or transmission, windows. This means that the execution window of a producer task has to precede the transmission window of the produced message. In the second approach these windows do not have this restriction.

Considering the mapping of a procedure to tasks and the overlapping of the previous approaches, four combinations are identified:

- A procedure maps to a single task and there is no overlapping,
- A procedure maps to a single task and there is overlapping,
- A procedure maps to various tasks and there is no overlapping,
- A procedure maps to various tasks and there is overlapping.

The first combination was already explored in previous works and two approaches were derived: the message/task deadline approach and the message/task maximum finishing approach. The second combination will be covered in future works. The other two approaches are covered by Calha and Fonseca (2005).

3.1 Parameter derivation for procedures that map to multiple tasks

A fundamental parameter of a real-time procedure is its deadline. Real-time synchronous procedures also have an initial phase and a period. Considering synchronous real-time procedures that map to multiple tasks, several approaches can be considered. An approach that maximizes the deadlines of the tasks and messages is to define the various absolute deadlines, relative to the release instant of the procedure, so that the deadline of the procedure is always met. An example of this approach is shown in fig. 9.

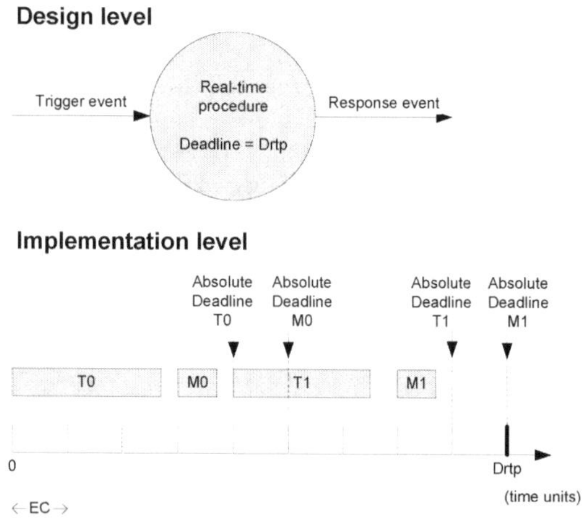

Fig. 9. Example of the definition of deadlines for each entity of a real-time procedure.

In this example, at the design level, a real-time procedure is started due to a trigger event and has to generate a response event within a time window defined by D_{rtp}. This procedure is implemented by two tasks, T_0 and T_1, and two messages, M_0 and M_1, so that T_0 produces M_0 and T_1 consumes M_0 and produces M_1, the response event. In the figure, the absolute deadlines of the entities are also represented. These absolute deadlines are the temporal limits for each entity that guarantee the execution of the procedure within its own deadline, D_{rtp}.

Considering the previous example with a single branch data stream, the equations that give the absolute deadlines of the various entities are:

- $D_{M1} = D_{rtp}$
- $D_{T1} = D_{M1} - WCTT_{M1}$
- $D_{M0} = D_{T1} - WCET_{T1}$
- $D_{T0} = D_{M0} - WCTT_{M0}$

Where the Worst Case Execution Time is represented by WCET and the Worst Case Transmission Time is represented by WCTT.

For more complex data streams this approach has to be considered for each branch of the data stream and for each entity, which participates in more than one branch, it should be considered the smallest value for its absolute deadline.

The initial phases and the relative deadlines can only be determined after choosing if there is overlapping between the windows, or not.

3.2 Using real-time procedures in closed-loop control applications

Closed-loop control applications have three basic stages, namely: sensing, control and actuation. This loop might have a dead time that is equal or less than the period of the loop. In fact, this dead time might be imposed by the controlled process or simply given by the interval between the sensing and the actuation. This type of system requires another parameter that defines the release instant of the actuation procedure so that it is periodic.

4. CASE STUDY: CONTROL OF A WHEELCHAIR

One example of a system where the concepts derived above are beneficial is the robotic wheelchair RobChair (Pires and Nunes, 2002) depicted in fig. 10.

The authors of this paper are involved in the design of the distributed embedded system that is used for the RobChair operation. A brief description of its sensory and actuation system is presented in order to permit an analysis of tasks performed by the control loops.

Fig. 10. The RobChair.

RobChair is based on a mechanical structure built by Vector Mobility, equipped with two driving wheels powered by two 24V Permanent Magnet DC Motors with a nominal torque of 29.3 Nm, and three caster wheels to assure stability. The motors are driven by two power amplifiers 80A8T from Advanced Motion Controls, which allow voltage and current control modes with resolutions above 12bits. Each motor has an optical encoder with a resolution of 20000 pulses per wheel revolution that is used for odometry calculations. A 2-axis inductive analog joystick allows a HMI between the user and the RobChair. To obtain surrounding obstacle information 12 IR proximity sensors (On/Off), 12 Sharp GP2D12 analog IR range sensors and an ultrasonic based range finder system ME-EERUF (Moita and Nunes, 2001) are used. Additionally a SICK LMS 200 laser measurement system and low resolution firewire cameras are being integrated. To provide absolute positioning a magnet sensor ruler developed at ISR (Bento, *et al.*, 2005) is also being installed.

In Maia, *et al.* (2003), a global navigation architecture for this autonomous mobile robot is presented. It consists in a layered approach that can be mapped on a distributed architecture where one or more functional units provide one or more functionalities. In the RobChair project, the first functionalities implemented were the interfaces with the sensory and actuation systems, the low-level motion control and a reactive shared-control behaviour (Pires and Nunes, 2002).

In order to illustrate the previous concept of real-time procedures, we focus on two examples: the control loops for the speed control of each wheel and the collision alarm.

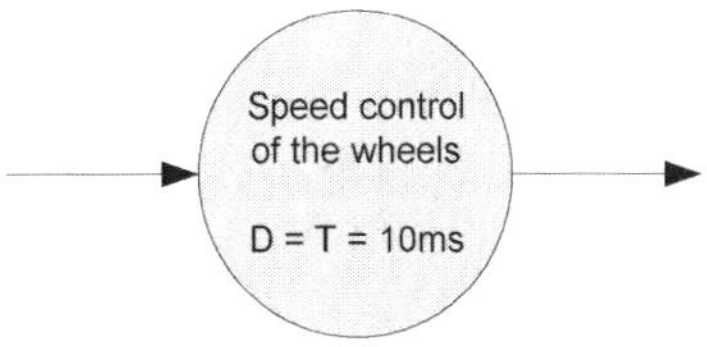

Fig. 11. Real-time procedure of the speed control of the wheels.

The speed control of the wheels (fig. 11) consists in a set of tasks, namely, encoder count acquisition, displacement calculation, position determination, new set point determination and actuation, see fig. 12. Other tasks can also be identified such as the system information and the parameter change. The sampling period, i.e., the reading of the encoders is of 10ms. A further requirement is the delay between sampling and actuation which must be either a small fraction of the sampling period or as close as possible of a full sampling period (Cervin, 1999).

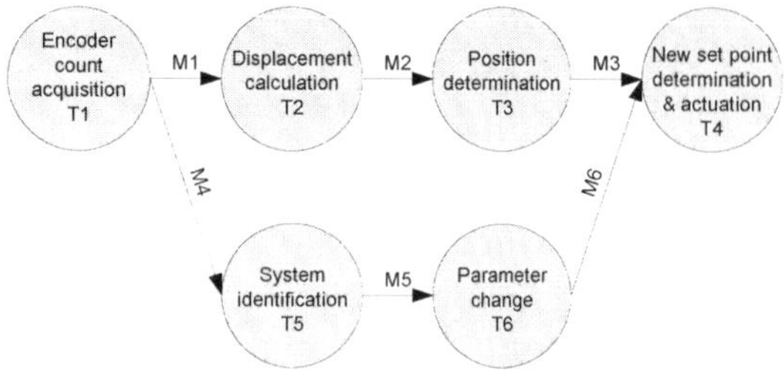

Fig. 12. Set of tasks from the real-time procedure of the speed control of the wheels.

The other example can be the detection of a collision by, e.g., the infrared sensors and the consequent reaction.

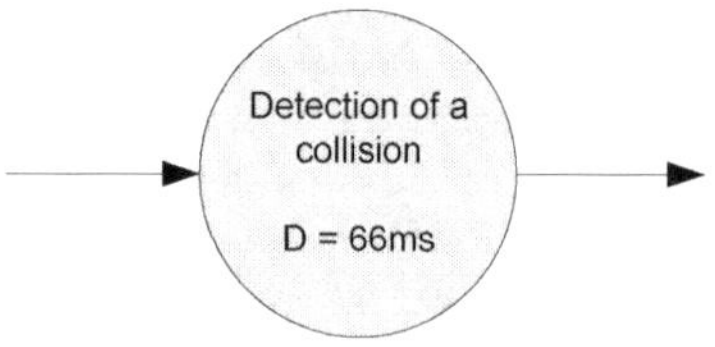

Fig. 13. Real-time procedure of the detection of a collision.

This procedure can also be divided into different tasks such as the event triggered detection at the sensors node, the data processing to decide if the event is significant and, if a decision to react is taken, the issuing of the commands to stop the motion. Here the real-time requirements are the maximum end-to-end delay. In this case it can be 66 ms (considering a maximum speed of 1,5 m/s, a detection range of 10 cm, sensors mounted in the periphery of the RobChair, and ignoring inertia).

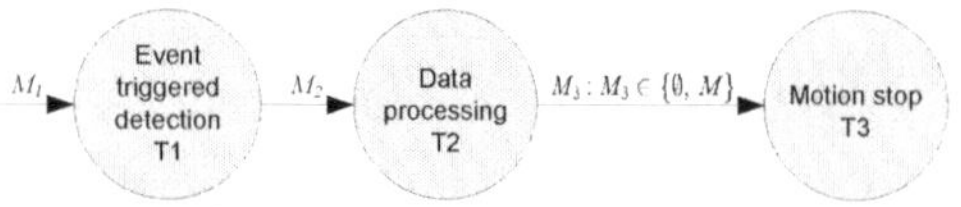

Fig. 14. Set of tasks from the real-time procedure of the detection of a collision.

These two examples show clearly that the timeliness requirements found in a real-time system such as the one described are defined at a higher level than the task level. Real-time procedures described in the previous section are well suited to this definition level.

The first procedure is synchronous and imposes two important timing constraints upon the actuation, namely a dead time and a deadline. Considering an actuation close to the period, both should be defined so that the release instant of task T_4 defines the smallest execution window that is possible. This execution window is given by the interval between the release instant and the deadline.

The second procedure is asynchronous and imposes a deadline. This deadline becomes the deadline of task T_3.

It should be noticed that this task is only executed if there is a message transmitted by task T_2.

The resulting data streams are still independent of the system partitioning but help in a first definition of constraints for tasks and messages. After the definition of the system architecture, and having the tasks allocated to the nodes, these constraints might be further refined. The definition of these constraints can be automated in order to produce a schedulable system. An approach to the scheduling of tasks and messages was presented by Calha and Fonseca (2004) while the dispatching was covered in Calha and Fonseca (2002).

5. CONCLUSIONS AND FUTURE WORK

Real-time procedures allow closing the gap between the system design and the implementation. At the system design level, procedures and their parameters are specified. In dependence of the information gathered during the analysis phase, the procedures might be more, or less, complex in terms of the tasks involved. But when these procedures map to various tasks, the determination of the tasks' and messages' parameters is not trivial. This technique helps to reduce the time from design to implementation in real-time systems.

In this paper, a possible approach to the holistic scheduling and dispatching in real-time systems was presented. In this approach, data streams that give support to different types of analysis have been used to specify real-time procedures. The combination between real-time procedures and data streams is valuable in helping to achieve a schedulable holistic system where timeliness is a must.

The concept of real-time procedures is now being integrated into the simulator SIMHOL. This simulator supports the joint scheduling and dispatching of messages and tasks and was presented by Calha and Fonseca (2003).

REFERENCES

Barreto, R., S. Cavalcante and P. Maciel (2004). A Time Petri Net Approach for Finding Pre-Runtime Schedules in Embedded Hard Real-Time Systems. *24th International Conference on Distributed Computing Systems Workshop (ICDCSW'04)*, Tokyo, Japan.

Bento, L., U. Nunes, F. Moita and A. Surrecio (2005). Sensor Fusion for Precise Autonomous

Vehicle Navigation in Outdoor Semi-structured Environments. Submitted to the *IEEE International Conference on Intelligent Transportation Systems (ITS 05)*.

Booch, G. (1987). Software Engineering with Ada. *Benjamin-Cummings*, 2nd ed.

Cervin, A. (1999). Improving scheduling of control tasks. *Proceedings of the 11th Euromicro conference on Real-Time Systems*. pp. 4-10.

Calha, M.J. and J.A. Fonseca (2002). Adapting FTT-CAN for the joint dispatching of tasks and messages. *Proceedings of the 4th IEEE International Workshop on Factory Communication Systems (WFCS'02)*, Vesteras, Sweden.

Calha, M.J. and J.A. Fonseca (2003). SIMHOL – A graphical simulator for the joint scheduling of messages and tasks in distributed embedded systems. *Proceedings of the 5th IFAC International Conference on Fieldbus Systems and their Applications (FeT 2003)*, Aveiro, Portugal.

Calha, M.J. and J.A. Fonseca (2004). Approaches to the FTT-based scheduling of tasks and messages. *Proceedings of the 5th IEEE International Workshop on Factory Communication Systems (WFCS'04)*, Vienna, Austria.

Calha, M.J. and J.A. Fonseca (2005). Data streams – an analysis of the interactions between real-time tasks. *Proceedings of the 10th IEEE International Conference on Emerging Technologies and Factory Automation (ETFA'05)*. Catania, Italy.

Dashofy, E.M. A. Hoek and R.N. Taylor (2001). A Highly-Extensible, XML-Based Architecture Description Language. *Proceedings of the Working IEEE/IFIP Conference on Software Architectures (WICSA 2001)*. Amsterdam, Netherlands.

Gomaa, H. (1984). A Software Design Method for Real Time Systems. *CACM*, vol. 27, no. 9, pp. 938-949.

Jackson, M. (1983). System Development. *Van Nostrand Reinhold.*

Maia, R., R. Cortesão, U. Nunes, V. Silva and J. Fonseca (2003). Robust low-level motion-control of WMR with active observers. In *Proceedings 2001 IEEE International Conference on Advanced Robotics (ICAR03)* vol. 2, 876-882.

Medidovic, N. and R.N. Taylor (2000). A Classification and Comparison Framework for Software Architecture Description Languages. *IEEE Transactions on Software Engineering*. 26(1):70-93.

Moita, F. and U. Nunes (2001). Multi-echo technique for feature detection and identification using simple sonar configurations. *In Proceedings 2001 IEEE/ASME International Conference on Advanced Intelligent Mechatronics (AIM01)* vol. 1, 389-394.

Okawa Y. and Yoneda T. (1996). Schedulability verification of real-time systems with extended time Petri nets. *International Journal of Mini and Microcomputers*, vol. 18, n° 3, p. 148-156.

Pires, G. and U. Nunes (2002). A Wheelchair steered through voice commands and assisted by a reactive fuzzy-logic controller. *International Journal of Intelligent and Robotic Systems*, vol. 34, n. 3, 301-314.

Rumbaugh, J., I. Jacobson and G. Booch (1998). The Unified Moddeling Language Reference Manual. *Addison-Wesley*.

Roux, O.H. and A.M. Déplanche (2002). A t-time petri net extension for real time-task scheduling modeling. *European Journal of Automation (JESA)*.

Spencer, J. (2000). Architecture Description Markup Language (ADML): Creating an Open Market for IT Architecture Tools. *Open Group White Paper*.

Sprunt, B., L. Sha and J. Lehoczky (1989). Aperiodic task scheduling for hard real-time systems. *Journal of Real-Time Systems*, 1(1):27-60.

Steusloff, H.U. (1984). Advanced Real-Time Languages for Distributed Industrial Process Control. *IEEE Computer*, vol. 17, no. 2, pp. 37-46.

Toussaint, J., F. Simonot-Lion and Jean-Pierre Thomesse (1997). Time constraint verifications methods based time petri nets. In: *6th Workshop on Future Trends in Distributed Computing Systems (FT-DCS'97)*. Tunis, Tunisia. pp. 262–267.

Vestal, S. (1993). A Cursory Overview and Comparison of Four Architecture Description Languages. *Technical Report*, Honeywell Technology Center.

Vidondo, F. (1983). GALILEO: Design Language for Real-Time Systems. *Proc. ITT Conf. on Programming Productivity and Quality*, ITT Corporation, pp. 198-210.

Ward, P.T. and S.J. Mellor (1986). Structured Development for Real-Time Systems. 3 volumes, *Yourdon Press*.

Witt, B.I. (1985). Communication Modules: A Software Design Model for Concurrent Distributed Systems. *IEEE Computer*, vol. 18, no. 1, pp. 67-77.

ELSEVIER IFAC PUBLICATIONS

REAL TIME COMPUTING IN A HIGH PERFORMANCE CLUSTER.

Palomera-Pérez, M.
Maestría en Ciencias e Ingeniería de la Computación.
Universidad Nacional Autónoma de México.
Email: mapp@uxmcc2.iimas.unam.mx

Almeida, L.
Departamento de Electronica e Telecomunicações.
Universidade de Aveiro.
Email: lda@det.ua.pt

Benítez-Pérez, H.
Departamento de Ingeniería de Sistemas Computacionales y Automatización.
IIMAS, UNAM.
Email: hector@uxdea4.iimas.unam.mx

Abstract: Nowadays, cluster computing is a common strategy to achieve high performance computing and resolve computationally intensive tasks such as flight simulation, medical imaging, etc. Two powerful tools have been used to facilitate the development of cluster oriented applications, namely the message passing interface (MPI) and the parallel virtual machine (PVM). However, there are situations in which the application execution is time constrained, e.g., complex real-time image processing, but neither MPI or PVM take time constraints in consideration. Therefore, this paper proposes a cluster integration methodology that is amenable to real-time analysis and implementation, based on Linux with improved temporal resolution and a high-speed TDMA interconnection bus built on top of SCI.

The main contribution of the paper is a worst-case response time analysis, its characterization using randomly generated task sets and its verification using a system simulator. The proposed analysis can be applied to a class of distributed real-time systems in which a set of nodes, running tasks under fixed priorities, communicate over a TDMA bus, but it is overly pessimistic in the presence of a significant percentage of communicating tasks.

Keywords: Middleware, system design and architecture,support for distributed applications; Real time, scheduling, operating system.

1 INTRODUCTION

In spite of the recent accelerated growth in computing power there are still many applications, e.g., online optimization, flight simulation or medical imaging, where the computing needs are often not covered.

In those cases there are two choices: using a super computer having two or more processors or using a computer cluster interconnected with a high speed communication network. From the user point-of-view, the main difference between these two alternatives is the way tasks are allocated to the processors. In the former case this is carried out by the operating system of the super computer while in the latter case it is the user that does it manually.

In the field of cluster technology, the distribution of tasks has received considerable attention. For instance, within the Condor project (Condor) a full-featured batch system was developed that provides a global task queueing mechanism and chooses when and where to run the tasks based upon a given policy. When there are parallel tasks and communication is needed, libraries as MPI and PVM play an important role to perform this goal.

However, when tasks are time constrained, e.g., for real-time complex image processing applications (EVENTS, 2001), neither MPI or PVM allow determining an upper bound to the task finishing time. Therefore, adequate execution methodologies are needed that allow determining such upper bound. This problem has been addressed, for example, in (Heimfarth, et al, 2003) where a cluster is developed to perform real-time image processing in the scope of the Events project (EVENTS, 2001). They considered nodes executing tasks with fixed priorities, using the Linux-RTAI real-time operating system, and a TDMA bus communication built on top of SCI (IEEE-P1596, 1992). Previous attempts to achieve real-time cluster performance included the definition of a real-time profile for SCI, known as SCI/RT (IEEE-P1596.6, 1992), which however, was dropped a few years after due to lack of commercial support.

In (Palomera, 2005) we propose using a similar architecture to the one followed in (Heimfarth, et al, 2003) and a prototype has already been implemented, including part of the MPI standard. The target application domain is mainly real-time complex image processing as required, for example, for certain medical and industrial imaging applications.

In this paper we present a worst-case response time analysis adequate for that architecture, which considers the mutual interference caused by tasks and messages. We also characterize the efficiency of the analysis using randomly generated task sets and we verify the analysis using a system simulator.

In our model, a set of nodes execute tasks under fixed priorities and communicate with each other by means of a high-speed TDMA bus. We believe this architecture is rather general and is often found in practical real-time distributed embedded systems thus enlarging its applicability and interest.

The rest of this paper is organized as follows. The related work is discussed in section 2. Section 3 presents the system model. Section 4 describes the response-time analysis of tasks and section 5 shows the response-time analysis for messages Section 6 presents a method to derive the task/message release jitter. The results from several experiments to characterize the analysis are shown in section 7. Finally, section 8 presents the conclusion.

2 RELATED WORK.

During the last decade several results concerning uniprocessor schedulability analysis for fixed priority tasks were developed (Cheng, 2002). For example,assigning priorities inversely to the tasks deadlines is optimal when the deadlines are less than or equal to tasks periods(Leung and Whitehead, 1982). This is called Deadline Monotonic (DM) scheduling. (Audsley, et al, 1991) developed a schedulability test for this kind of tasks, and later extended this analysis to consider the case in which the tasks access shared resources and suffer from release jitter(Audsley, et al, 1993).

The problem of analyzing the schedulability of a distributed system considering both the tasks executing in the nodes and the messages transmitted on the bus is commonly referred to as holistic schedulability analysis(Tindell and Clark, 1994). This analysis is extended in (Palencia and Harbour, 1999) increasing significantly the maximum schedulable utilization. (Chevochot and Puaut,2000) include in the analysis the run-time support in charge of executing applications . In (Calha and Fonseca, 2002) a network-centric approach is presented in which both tasks and messages are triggered by a central scheduler. A simulator for this execution environment is described in (Calha and Fonseca, 2003).

In previous works, an adequate protocol eliminates contention at the bus access, e.g., using TDMA (Tindell and Clark, 1994, Chevochot and Puaut,2000, Palencia and Harbour, 1999) , but the run-time support overhead, such as the TDMA driver that dispatches the messages is not taken into account. On the other hand, the use of CAN as in (Calha and Fonseca, 2002) is not adequate for high performance computing applications. In this work, a TDMA bus access method is also used, which can be easily implemented over SCI, as in (Heimfarth, et al, 2003) . However, a specific task is considered in each node just for dispatching messages to the bus, i.e., the message manager. This task carries out the transmission of queued messages within the node's TDMA slot, only. The time used by this task is also taken into account in the schedulability analysis.

This paper is mainly based on (Tindell and Clark, 1994) and(Audsley, et al, 1993) in what concerns the tasks schedulability analysis. As for messages, their schedulability analysis is based on (Almeida and Pedreiras, 2004) with the necessary adaptations for this framework. Moreover, no offsets among tasks are considered but there is a synchronization mechanism that is necessary to manage the TDMA bus, enforcing mutual exclusion access to the bus slots.

The main features of the analysis herein proposed are:

- A new method to find the message transfer time.
- More accurate computing overhead analysis.
- System model is adequate to a high level interface such as MPI.

3 SYSTEM MODEL.

In this work, we consider a computer cluster where N nodes communicate with each other by means of a distributed shared memory bus, namely SCI. However, for the purpose of our analysis, the specificities of SCI will be abstracted away and we will refer simply to a generic TDMA bus. Each node η has a set Π_η of N_η fully preemptive periodic tasks, $\Pi_\eta(J_i(C_i, T_i, D_i, P_i), i = 1..N_\eta)$, with each task J_i being characterized by a period T_i, worst-case execution time C_i, relative deadline D_i less than or equal to the respective period, and a fixed priority P_i that is derived from task period, deadline, or any suboptimal criteria.

Tasks can communicate with other tasks in the same node or different nodes. Communication is carried out by message passing, only. Each message uses an integer number of fixed length packets, where C_p is the time to transmit a single packet. In this sense, message transmission is preemptive since it can be interrupted between packets, to transfer a ready higher priority message. The set of messages transmitted by node η is Θ_η containing M_η messages, $\Theta_\eta(\mu_i(\varsigma_i, \tau_i, \delta_i, \phi_i, S_i, E_i), i = 1..M_\eta)$. Message μ_i has a length ς_i in number of packets, and it is sent from the sender task S_i to the receiver task E_i. Its period τ_i as well as its priority ϕ_i are inherited from the respective parameters of the sender task S_i. On the other hand, the deadline δ_i must be less than the deadline of the receiver task E_i.

The whole system uses a tick base scheduling, i.e., a timer interrupts the processor in each node, periodically, to run the task manager, tm. This manager verifies whether a task with priority higher than the currently running one becomes ready. In that case, the running task is preempted while the higher priority one is dispatched. Internally, the task manager holds 2 priority queues, the pending queue for tasks that are waiting for an event, and the ready queue for ready tasks. There are two kinds of events, periodic activation and message arrival. The periodic activation event occurs when the actual time is a multiple of the task period, indicating that the task is ready again. The message arrival event is triggered when a new message arrives so that if a task in the pending queue is waiting for it, the task is moved to the ready queue. T_{clk} is the clock tick, which corresponds to the period of tm. This is set at pre-run-time to the greatest common divisor of all task periods in all nodes. Consequently, the tasks periods can then be expressed as an integer number of ticks. Moreover, all ticks are synchronized in all nodes so that every tm is activated synchronously. In the prototype implementation, this is achieved with a barrier, without the need for clock synchronization.

Message transmission is handled by a task called message manager, mm, that is unique in each node. It implements a priority queue for messages generated within the node. Moreover, each mm task is activated after a node-specific offset with respect to the barrier that synchronizes the ticks, so that its activation coincides with the respective slot in the TDMA round.The period of mm, i.e., T_m, is set to T_{clk}, the minimum period in the system, in order to provide the highest promptness for message transmission. The execution time of mm, i.e., C_m, is obtained from Equation 1 considering N nodes. Notice that an extra slot is used for the execution of the tm task. Also, with very high-speed buses such as SCI, the time to transmit is negligible when compared with the time to prepare the data at the network driver and card interface. This is why the transmission time is considered equal to the execution time.Therefore, C_m is the TDMA slot size while T_m is the TDMA round. Finally, in each instance the message manager may execute for less than C_m depending on the current status of the messages ready queue.

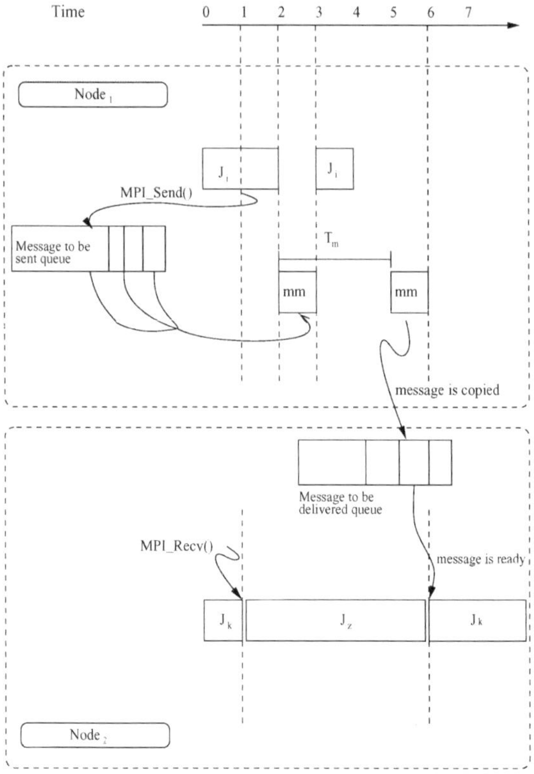

Figure 1: Example of communication with MPI primitives.

$$C_m = \frac{T_m}{N+1} \quad (1)$$

Figure 1 shows a communication example between two tasks in two nodes using the MPI primitives $MPI\_Send()$ and $MPI\_Recv()$. In $node_1$ at time 1 J_i sends a message to J_k by invoking the primitive $MPI\_Send()$. This primitive fragments the message in packets and places them in the message ready queue according to the their priority. At approximately the same time, in $node_2$, J_k requests the message by invoking $MPI\_Recv()$. However, since the message is not in the receiver message queue,J_k suspends itself indicating to tm_2 that it is waiting for a message. At time 2, in $node_1$, mm_1 starts running but spends all its capacity handling higher priority messages, thus J_i's message is not sent. Meanwhile, in $node_2$ another task J_z is running. At time 5 mm_1 is activated again ($T_m = 3$ time units) and the message is actually transferred to $node_2$. Thus, at time 6, the message is ready and J_k is waiting for it, so tm_2 preempts the current task J_z and resumes executing J_k.

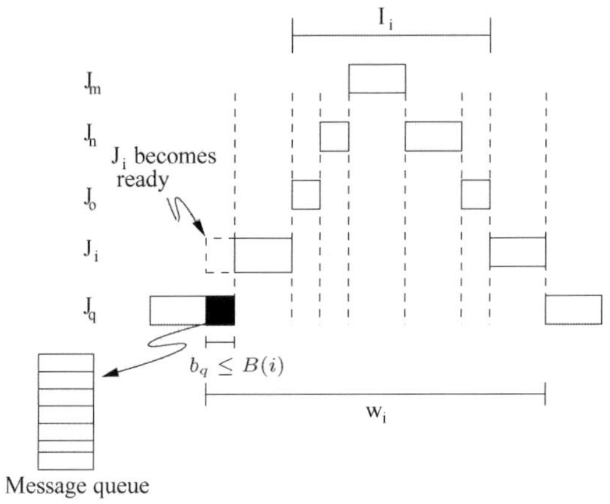

Figure 2: Task interference and task blocking.

4 TASKS RESPONSE-TIME ANALYSIS.

The worst-case response time of task J_i with respect to its periodic, i.e., R_i, is given by its release jitter, ji_i, plus the worst-case response-time after release, w_i, Equation 2.

$$R_i = ji_i + w_i \qquad (2)$$

To find out w_i we must consider all possible contributions to delay the task. Figure 2 shows the contributions that can occur in our system, I_i is the interference caused by higher priority tasks and B_i is the blocking that the task can suffer when starting caused by a lower priority task currently copying a message to or from the message queues. The interference I_i depends on w_i and it can be obtained from Equation 3, which considers that $hp(i)$ is the set of tasks with priority higher than that of task J_i. On the other hand, B_i is obtained from Equation 4, which considers that $lp(i)$ is the set of tasks with lower priority than task J_i.

$$I_i(w_i) = \sum_{\forall j \in hp(i)} \left\lceil \frac{w_i + ji_j}{T_j} \right\rceil C_j \qquad (3)$$

$$B_i = \max_{\forall j \in lp(i)} (b_j) \qquad (4)$$

Another factor to be considered is the time used by the task manager. However, considering the maximum execution time in every tm invocation is too pessimistic, thus we break down the respective execution time in different factors that we account for independently. Firstly, tm must move the ready tasks from the pending queue to the ready queue. This takes a time bounded by C_{pr}. Then, tm checks the ready queue for a task with priority higher than that of the current task. This is accomplished in at most C_{clk}. If one is found, tm causes a context switch, which takes further C_r time units. Notice that while C_{clk} is used every invocation, either C_r as well as C_{pr} are only used depending on the periods and priorities of the other tasks.

Equations 5 define the number of tm occurrences, L_i, the number of tasks that are moved from one queue to another, KT_i, and the number of activations of tasks with higher priority than task J_i, KH_i, all of these within the w_i.

$$\begin{aligned} L_i(w_i) &= \left\lfloor \frac{w_i}{T_{clk}} \right\rfloor \\ KT_i(w_i) &= \sum_{\forall j \in \Pi_\eta} \left\lfloor \frac{w_i + ji_j}{T_j} \right\rfloor \\ KH_i(w_i) &= \sum_{\forall j \in hp(i)} \left\lfloor \frac{w_i + ji_j}{T_j} \right\rfloor \end{aligned} \qquad (5)$$

Moreover, the effective number of higher priority tasks dispatched within w_i, KH_i, is upper bounded to L_i since only one task is dispatched per tm invocation. Hence, the tick-scheduling overhead for task J_i is expressed by Equation 6.

$$\begin{aligned} S_i(w_i) &= L_i(w_i)C_{clk} + KT_i(w_i)C_{pr} + \\ &\quad \min(L_i(w_i), KH_i(w_i))C_r \end{aligned} \qquad (6)$$

A similar analysis to determine the tick-scheduling overhead can be found in (Tindell and Clark, 1994).

Finally, it is necessary to account for the time used to transmit the messages. One simple way to do it is to include a highest priority task with period T_m and cost C_m in Π_η. Alternatively, a less pessimistic way consists of determining the number of messages that have been generated over w_i and accounting for the respective transmission time, only.

Equation 7 allows determining the maximum number of packets generated over w_i, considering that $m_{hp}(i)$ is the set of messages having higher priority than that of message i.

$$G_i(w_i) = \sum_{\forall j \in m_{hp}(i)} \left(\left\lceil \frac{w_i + \iota_j}{T_j} \right\rceil \varsigma_j \right) + \varsigma_i \qquad (7)$$

If the transmission slot has capacity to ψ packets, then $\left\lceil \frac{G_i(w_i)}{\psi} \right\rceil$ slots are required to transmit all. However, during w_i as much as $\left\lceil \frac{w_i}{T_m} \right\rceil$ slots can be used. Thus, the time used to transmit the message is given by

$$M_i(w_i) = \min\left(\left\lceil \frac{G_i(w_i)}{\psi} \right\rceil, \left\lceil \frac{w_i}{T_m} \right\rceil \right) C_m \qquad (8)$$

Equations 3,4,6 and 8 lead to the final result in 9, providing the worst-case response time for a task J_i. Equation 9 can be solved iteratively with $w_i^0 = C_i$,

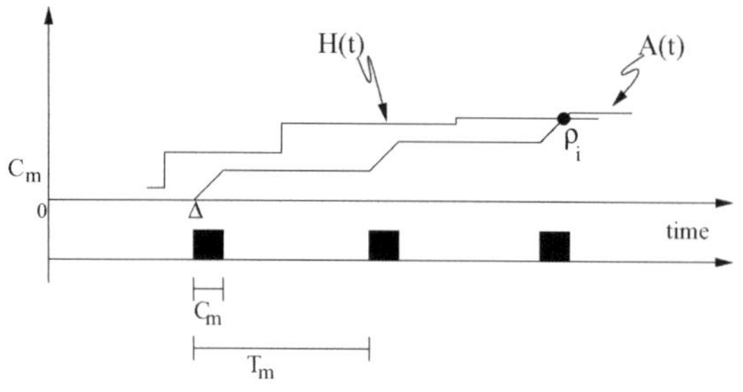

Figure 3: The TDMA slot seen as a server (availability $A(t)$ and submitted load $H_i(t)$ functions).

and it either converges to $w^n = w^{n+1}$ or grows beyond the deadline D_i within a finite number of iterations.

$$w_i^{n+1} = C_i + B_i + I_i(w_i^n) + S_i(w_i^n) + M_i(w_i^n) \quad (9)$$

Finally, the task set Π_η in node η is scheduled if Condition 10 is met for all the tasks in the set.

$$\forall i \in \Pi_\eta \quad : \quad R_i \leq D_i \quad (10)$$

5 MESSAGES RESPONSE-TIME ANALYSIS.

To derive the response-time of messages we consider the TDMA slot of node η as a periodic server with capacity $C_m$and replenishment period T_m, which schedules several periodic message streams. Therefore, we define the server availability function as $A(t)$ (Equation 11), Figure 3. This function returns in each instant t the cumulative server capacity available to process messages up to t.

$$A(t) = \begin{cases} kC_m, & kT_m \leq t < kT_m + \Delta \\ t-(k+1)\Delta, & kT_m + \Delta \leq t < (k+1)T_m \end{cases} \quad (11)$$
$$\Delta = T_m - C_m$$
$$k = \left\lfloor \frac{t}{T_m} \right\rfloor$$

Once the server availability is determined, we need to compute the maximum cumulative higher priority messages load submitted to the server up to instant t with respect to message i, $H_i(t)$, including the packets of message i itself (Equation 12). Notice that messages are periodic and fragmented in fixed size packets.

$$H_i(t) = \left(\sum_{\forall j \in m_{hp}(i)} \left\lceil \frac{t+\iota_j}{T_j} \right\rceil \varsigma_j + \varsigma_i \right) C_p \quad (12)$$

As explained in (Almeida and Pedreiras, 2004), the response-time of message i, tt_i, is given by the earliest instant in which the availability is just enough to cover the submitted load, as depicted in Figure 3.

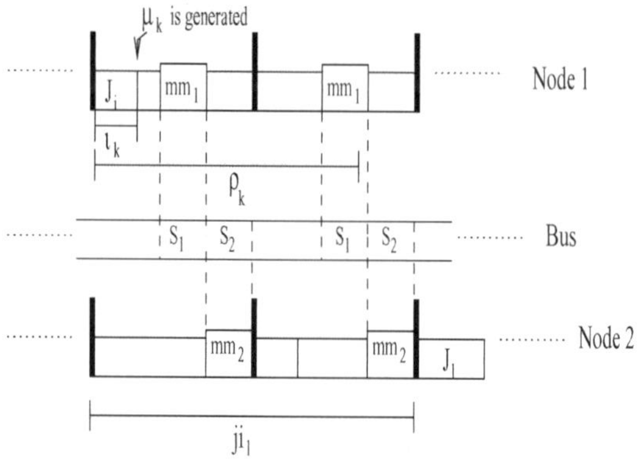

Figure 4: Example of task/message jitter.

The value of tt_i can be computed using Equation 13, which uses of the inverse of the availability function $A(t)$, $A^{inv}(t)$.

$$tt_i^{n+1} = A^{inv}(H_i(tt_i^n)) \quad (13)$$

Equation 13 can be solved iteratively with $tt_i^0 = H_i(0)$ and it either converges to $tt_i^n = tt_i^{n+1}$ or grows beyond the deadline δ_i within a finite number of iterations.

The worst-case response-time of message i, ρ_i is given by the Equation 14, where ι_i is message i release jitter.

$$\rho_i = \iota_i + tt_i \quad (14)$$

Finally the set Θ_η of messages produced by node η is schedulable if Condition 15 is met.

$$\forall i \in \Theta_\eta : \rho_i \leq \delta_i \quad (15)$$

6 TASK/MESSAGE RELEASE JITTER.

In this work we consider that tasks are activated in a strict periodic fashion. However, when they are triggered by the reception of a message, i.e., tasks that receive messages, variations in the message delivery instant will induce release jitter in the task. Similarly, the transmission of messages is triggered by tasks in execution and, because of possible variations in the interference caused by higher priority tasks, the message release instant my vary thus exhibiting release jitter. Figure 4 illustrates the appearance of release jitter. Task J_i on the node 1 sends the message μ_k to task J_l on node 2. The triggering of μ_k can be as early as C_i or as late as R_i. Thus its release jitter is given by Equation 16.

$$\iota_k = R_i - C_i \quad (16)$$

On the other hand, task J_l despite being activated at the same time as task J_i cannot initiate execution before the arrival of message μ_k after a worst-case interval equal to ρ_k. Moreover, since the activation of J_l is carried out by the message manager mm_2, which executes once every T_{clk}, the release jitter that can affect J_l is upper bounded by Equation 17.

$$ji_l = \left\lceil \frac{\rho_k}{T_{clk}} \right\rceil T_{clk} \qquad (17)$$

However, according to Equations 14 and 2, both ρ_k and R_l depend on the respective release jitters values. Therefore, these equations are solved using an iterative approach that starts considering all release jitters equal to zero. Then, the response-time of the tasks and messages are evaluated. With these values, the release jitters are determined using Equations 16 and 17. The analysis is repeated using the new jitter values and so forth until either the jitters in two consecutive iterations are equal or whichever of conditions 10 or 15 becomes false.

This iterative process either converges or grows beyond a task/message deadline within a finite number of iterations because the response-time functions grow monotonically and, from iteration to iteration, the increment in 17 is lower bounded by $\left\lceil \frac{C_{min}}{T_{clk}} \right\rceil T_{clk}$ where C_{min} is the minimum cost of the tasks that send messages.

7 EXPERIMENTAL EVALUATION.

In this section, we describe three experiments using randomly-generated task/message sets to evaluate the efficiency of the analysis. For each task/message set we analysed the respective schedulability by two means, using the schedulability conditions shown in the previous sections and using a system simulator adapted from SimHol (Calha and Fonseca, 2003). In this latter case, the whole system was simulated during a period of time corresponding to two macro-cycles, i.e., two times the least common multiple of all the periods, looking for deadline violations. Such period of time ensured that the worst-case response-time situation would be encountered during the simulation.

In all the experiments, the total utilization per node, U_η, was varied from 0.1 to 1.0 with 0.01 steps. For each value of U_η 100 task/message sets were generated. Table 1 contains several parameters that were kept across all the generated task/message sets.

Further constraints were considered in the generation of the random sets. Namely, the task period was expressed as an integer multiple of T_{clk} in the range [10 100] milliseconds. However, it was guaranteed that a sufficient number of tasks with equal periods existed

B_i	T_{clk}	C_{clk}	C_{pr}	C_r	C_p
$10\mu s$	$10ms$	$10\mu s$	$10\mu s$	$10\mu s$	$1\mu s$

Table 1: Fixed parameters used across the experiments.

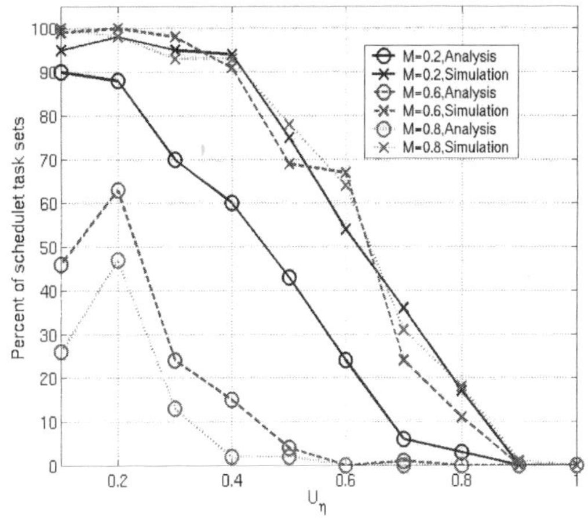

Figure 5: Percentage of schedulable sets under constant bus utilization (30%), two nodes and variable number of messages M (20%, 60%, 80% of the total task number).

in each set in order to allow communication between them. The task deadline D_i was considered equal to T_i and the message deadline δ_m was set equal to the deadline of the receiver task minus a random percentage of the receiver task execution time C_j. The value of the task execution time C_i was generated randomly within bounds that allowed obtaining the desired number of tasks, except for the last task whose execution time was tuned in order to obtain a total processor utilization equal to U_η.

In all the three experiments we observe the variation in the level of schedulability expressed by the percentage of schedulable sets as a function of the node utilization U_η. The results of each experiment were parameterised according to a third parameter, allowing to plot several curves. These curves were plotted with respect to the results of the analysis and of the simulation.

In first experiment we analyse the impact of varying numbers of messages in the sets schedulablity. The number of messages was expressed as a fraction of the number of tasks, M, and it was varied from 0 to 1 with 0.1 steps. However, the bus utilization was kept constant and equal to 0.3. The generation of the message packet lengths was carried out in a way similar to the generation of the tasks execution times. The number of tasks per node was also kept constant and equal to 20. Notice that each value of M includes near 10^4 task sets, covering all values of U_η.

Figure 5 shows the results for two nodes and for M equal to 0.2, 0.6 and 0.8, respectively. We can see that the number of messages has a big impact on the

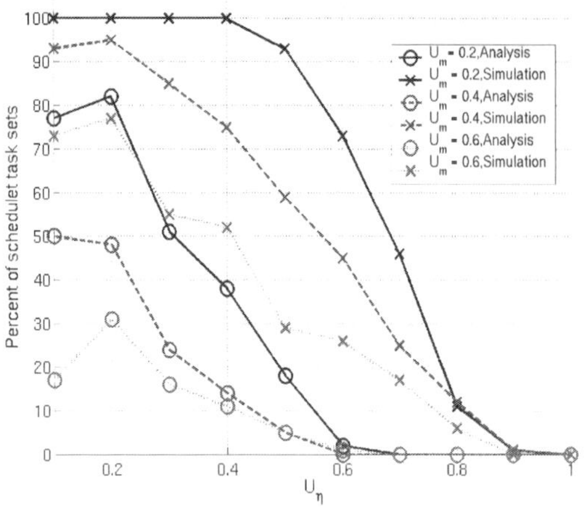

Figure 6: Percentage of schedulable sets under constant message number (50% of the total task number), two nodes and variable bus utilizationU_m(20%, 40%, 60%)

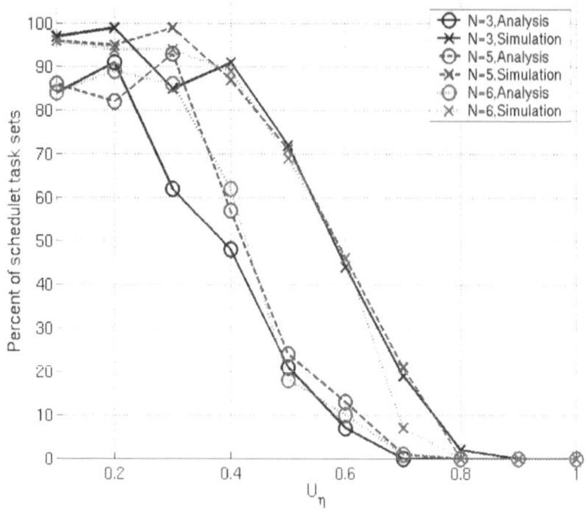

Figure 7: Percentage of schedulable sets under constant message number (30% of the total task number) and bus utilization (30%), with variable number of nodes (3,5,6).

global system schedulability analysis, this can be explained if we consider that inside one node if a task that sends a message has less priority than a receiver task, the jitter of this task affects the response-time of the first one, therefore also the jitter of receiver task on other node is affected and so forth. This situation generates a pessimistic jitter value which is more pessimistic when there are more communicating tasks, i.e., more messages. When the number of communicating tasks reaches 50%, the analysis is already overly pessimistic. In fact, the simulation shows that the number of messages is not so important as long as the total bus bandwidth is kept constant.

In the second experiment we evaluate the impact of varying bus utilization. In this case, the number of tasks and messages per node was kept constant, 20 and 10 respectively. The bus utilization factor U_m, i.e., the bandwidth used by the messages, was varied from 0.1 to 1.0 in steps of 0.1. Again, about 10^4 task sets were generated for each value of U_m. The results for two nodes are displayed in Figure 6. It shows clearly that the communication bus is a bottleneck, with a substantial degradation of the system schedulability as the bus bandwidth utilization increases.

In the last experiment we evaluate the impact of the number of nodes in the system schedulability. The number of tasks was kept constant and equal to 20 per node. The number of messages was equal to 50 percent of the tasks. The bus utilization factor was fixed to 0.3. Figure 7 shows the results for N equal 3, 5, and 6.

The impact of the number of nodes is not significant as long as the bus utilization factor is kept constant.

Finally, we also carried out a brief comparison of the worst-case end to end delays across communicating tasks determined by the analysis and those measured by the simulator. We found that, as expected, the end-to-end delay determined by the analysis is always longer than the one found during simulation. Depending on the particular cases, the latter one varied between 30% and 90% of the former.

8 CONCLUSIONS.

High performance clusters are commonly used, nowadays, in situations that require more computational power than available in a single computer. In some cases, the applications impose real-time constraints on the computations, such as in flight simulators or in real-time medical imaging instrumentation. This paper explores the possibility of real-time computing in such high performance clusters, namely in those where a set of nodes execute tasks according to fixed priorities and a high speed bus is used to interconnect the nodes. The software architecture is developed around MPI and based on Linux with improved temporal resolution and the specific communication protocol used is SCI, over which a TDMA access control is deployed. A prototype implementation has been carried out that shows the practical effectiveness of the proposed architecture to carry out real-time computing over a cluster.

In this paper we focused on the temporal analysis of the computations and communications in order to derive a response-time based schedulability analysis of the whole task and message system. The analysis developed in this paper is based on several previous analysis but it is tuned for architectures of the type referred above and it efficiently considers the overheads related to communication. A simulator of the referred architecture was developed in order to verify the safety of the analysis. Then, several experiments with random task/message sets were carried out to compare the results of the analysis with those of the simulations. We found that the proposed analysis is safe but becomes overly pessimistic when the number of communicating tasks grows up to 50% or more.

Finally, the analysis developed and assessed in this paper is applicable to a class of distributed systems based on a TDMA bus and a fixed priorities task and message models. This class is relatively vast in distributed real-time embedded systems and thus we consider that the analysis has a value beyond the scope in which it was conceived.

In future work we will evaluate the likeliness of worst-case scenarios in order to improve the efficiency of the analysis.

Acknowledgments: The authors would like to thank the UNAM Postgraduate Program on Computer Science for the sponsorship given in relation to this work and Mr. Francisco Cardenas for his technical support.

References

Almeida Luis, Pedreiras Paulo (2004). *Scheduling within temporal partitions: Response-time analysis and server design.* In Proc. of the EMSOFT 2004.

Audsley N.C., Burns A., Richardson M. F., and Wellings A. J. (1993). *Applying New Scheduling Theory to Static Priority Pre-emptive Scheduling.* Software Engineering Journal, Vol. 8, No. 5, pp. 284-292, September 1993.

Audsley N.C., Burns A., Richardson M. F., and Wellings A. J. (1991). *Hard real-time scheduling: the deadlinemonotonic approach.* In Proc. of the Eighth RTOSS , May 1991.

Calha Mário J., Fonseca José Alberto (2003). *SimHol- A Graphical Simulator for the Joint Scheduling of Messages and Tasks in Distributed Embedded Systems.* 5th International IFAC Conference on Fieldbus Systems and their Applications (Fet 2003), Aveiro, Portugal, July 7-9, 2003

Calha Mário J., Fonseca José Alberto (2002). *Addapting FFT-CAN for the hoint dispatching of tasks and messages.* 4th IEEE International Workshop on Factory Communication Systems, 2002.

Cheng Albert M. K. (2002). *Real-Time Systems. Scheduling, Analysis, and Verification.* 2002. John Wiley & Sons.

Chevochot P., Puaut I. (2000). *Holistic Schedulability Analysis of a Fault-tolerant Real-Time Distibuted Run-time Support*, 7th IEEE International Conference on RealTime Computing Systems and Applications (RTCSA'00),Cheju Island, South Korea, December 2000.

Condor team. *Condor: High Throughput Computing.* University of Wisconsin-Madison. http://www.cs.wisc.edu/condor/

European Project EVENTS, http://www.eptron.es/projects/events/

Heimfarth Tales, Götz Marcelo, Ramming Franz J., Wagner Flávio R. (2003). *RTC: A Real-time Communication Middleware on Top of RTAI-Linux* 6th IEEE International Symposium on Object-Oriented Real-Time Distributed Computing (ISORC'03)

IEEE, P1596 (1992). *IEEE Standard for Scalable Coherent Interface (SCI)*, Poscataway, NJ: IEEEService Center, 1992.

IEEE. P1596.6 (1992). *SCI/RT- Scalable Coherent Interface for Real-Time Applications*, 1992

Leung J. and Whitehead J. (1982). *On Complexity of Fixed-Priority Scheduling of Periodic Real-Time Tasks.* Performance Evaluation, 2, pp. 237-250, 1982.

Palencia J.C and Harbour M.G (1999). *Exploiting Precedence Relations in the Schedulability Abalñysis of Distributed Real-Time Systems*, 20th IEEE Real-Time Systems Symposium (RTSS'99),Phoenix,USA, November 1999.

Palomera Miguel A. (2005). *Desarrollo de un Sistema en Tiempo Real para Cómputo de Alto Desempeño.* Master's Thesis, Universidad Nacional Autónoma de México Posgraduated Program on Computer Science, 2005.

Tindell K. and Clark J. (1994). *Holistic schedulability for distributed hard real-time systems.* Microprocessing and Microprogramming, 40:117–134, 1994.

ELSEVIER
IFAC
PUBLICATIONS

CURRENT CHALLENGES IN ABSTRACTING DATA POINTS

Burgstaller W.[1], Soucek S.[2], Palensky P.[1]

[1]Institute of Computer Technology
Vienna University of Technology
Gusshausstr. 27-29/384, A-1040 Vienna, Austria
Tel.: +43-1-58801-38411
Fax: +43-1-58801-38499
E-Mail: burgstaller, palensky@ict.tuwien.ac.at

[2]LOYTEC electronics GmbH
Stolzenthalergasse 24/3, A-1080 Vienna, Austria
Tel.: +43 -1-40208050
Fax: +43-1-402080599
E-Mail: soucek@ieee.org

Abstract: Services in building automation traditionally have been implemented using a variety of fieldbus systems. Recently growing interest has been put into convergence of the various systems. A classical approach is the use of gateways. The engineering effort, however, to build gateways for an arbitrary number of systems is of quadratic order. This paper focuses on a data-centered view, which defines a set of abstract services needed for building services, and discusses the resulting challenges. With this approach an application can be migrated from one system to another, leading to a linear engineering effort for controllers, data servers and multi-protocol nodes. *Copyright © 2005 IFAC*

Keywords: building automation, convergence, fieldbus systems, gateways.

1. INTRODUCTION

Interoperability was and still is an important factor when it comes to engineering. The 1990ies where the time, when standardization committees within ISO, ETSI, IEC and their numerous national counter parts, converted the proprietary landscape of automation networks into a set of standardized technologies (Fischer and Palensky, 2004). Publishing standards and opening technology was an important step towards lower engineering costs, higher quality and more flexibility. Interoperability rules within these standards ensured, that multi-vendor installations worked more or less seamlessly (Loy et al., 2001). The vendors complied with standard data types, standard services, standard semantics and profiles.

This benefit was unfortunately only valid within one technology, within one standard. Transforming and translating data, services and functionality from one world into another is an arbitrarily difficult task.

Related work can be found in the NOAH project (Di Stefano et. al., 2000; Mella and Russo, 1999), a European Project that aimed at the harmonization of device profiles of the EN 50170 standard series (a set of automation networks in the industrial automation domain). (Kastner and Neugschwandtner, 2004) describes the application of the OSGi framework to the popular home automation network EIB (European Installation Bus), in order to get a high level data abstraction of EIB datapoints. OSGi, however, is a rather heavy architecture, relying on multi-threaded environments and usually based on JAVA. (Velasco et. al., 2003) describes an object model specification for Internet-based remote access on control and automation data. The processes are abstracted and accessible via so-called virtual indistrial devices and a virtual industrial protocol, partly based on SOAP, an XLM-based not-too-slim protocol for remote object management. (Fischer, 1999) explains the results of working groups 3 and 4 of CEN TC247 "Controls for Mechanical Building Services". The working groups elaborated "field level objects", where already difficulites in abstracting "higher" services like COV (change of value) or alarming were discovered. Meanwhile, this working group works on the publication of a technical specification on mapping BACnet onto LonWorks and vice-versa.

The need for interconnecting networks of different types stems from the insight that processes of different kind can be combined, leading to new functionality. An example can be found in building automation. Having HVAC (heating, ventilation, air conditioning and cooling) automated with network type A and the lighting system realized with network type B was previously not seen a problem. This changed, when people realized that lights, shading and climate control are physically linked: Sunlight, that shines through the window heats and illuminates the respective room. Sunblind whose lamellae have a reflective coating on one side can increase or prevent this effect. This is a simple example where processes of previously isolated nature contribute to one single function.

This need for interconnectivity and interoperability is not only limited to automation networks. The increased usage of the Internet as a universal source of information and a media for all kinds of business processes lead to numerous attempts to connect "the I/O with the CEO" (Palensky, 2005). This slogan expresses the fact that business software and networks are getting more and more in touch with the processes and networks "in the field". People realize that making process data of a production plant available to the IT-department enables for instance SAP to create a report on the efficiency of the usage of raw materials. So there is a need for linking the controlling department with the production process and the customer relationship management (CRM) software.

The convergence of the existing networks and technologies is a difficult task (Palensky, 2004). Beside numerous serious technological issues, it is also the investment that was done in existing network infrastructure, that currently prevents the usage and development of a universal one-for-all network. Networks of different type that are supposed to fulfill a common goal reveal an unpredicted set of differences.

In general building services are characteristic control tasks of applications in building environments (Kastner, *et al.*, 2005). Typical building services include climate control, visual comfort, safety, security, transportation, or supply and disposal. (Daniels, 2003). One of the main differences to classical methods in process control are the time constants of the controlled, physical processes. This relaxes the requirements of quality of control (QoC) for control algorithms built for network-based control systems (Soucek and Sauter, 2004). This in principle enables the system integrator more levels of freedom to achieve convergence by introducing meta-layers, translators and bridging services to close the gap between different systems. This has to be done carefully, however, as the exchange of services or services that span technologies may face even more problems.

Such considerations are not new, and there are still no satisfying solutions out there. A new player in the domain of control networks is the usage of Internet technology. Transport protocols like the TCP/IP suite, media like Ethernet or GPRS (general packet radio switching) and higher protocols like SOAP (simple object access protocol) or OPC (OLE for process control) are unstoppable technologies that make their way into the world of control networks. This paper describes the problems and the lessons learned from IT-near technologies and how a data-centered design can help lifting data and functionality from one network to another one.

2. PROBLEM DESCRIPTION

The general functionality of BACS can be broken down into three levels, representing the incarnation of the automation pyramid for BACS (BACS-2 2004): field, automation and management level.

At the field level actual interaction with the physical world takes place. Data are collected and submitted to processing entities as well as the environment physically controlled by receiving data. On the automation level controllers operate on data prepared by the field level, establishing logical connections and control loops. This type of process data exchange is referred to as horizontal communication (or process communication relating to the nature of the data). In addition, the automation level prepares (possibly aggregate) values for vertical access by the management level (e.g., trending).

Vertical communication can be divided into services related to accessing and modifying data from within the application, for example adjusting a set point or retrieving trend logs (frequently referred to as management communication), and others concerned with modifying the application itself, for example changing binding information or program transfer (engineering communication).

Historically, convergence between the different network technologies has been implemented by gateways between the systems on any of the above levels. This helps to integrate domains previously being isolated islands. However, application developers still need a steep learning curve to switch between technologies. Another way to achieve convergence is to hide the intricacies of the underlying network technology and provide the application developer with a set of standard services instead. Clearly, a node implemented on a certain system can only be operated within its native environment, but the same control application can be moved to another system.

The key requirement for an abstraction of the underlying control network services is to hide the communication-specific characteristics, using a data representation and a set of services, which allows mapping of the basic functionalities of BACS to different technologies. This is an application-centric approach, where the control application can be easily migrated from one network technology to another. Clearly, the available services in the different levels of communication require different amount of efforts to achieve this.

In the process communication level of the distributed system constituted by a building automation and control application the system represents itself to the application developer as a set of elementary data elements, called data points (or simply points). These data points are the logical representation of the underlying physical process, which control network nodes drive or measure. Each node can be associated with one or more data points. In the logical view, each data point represents a single datum of the application, which can be transferred and processed. This level is using a representation close to the physical process, therefore it is suitable to an abstraction.

For management communication more complex entities than simple data points are widely employed. An abstraction of these services can be possible in BAS for tasks such as aggregated data, trend logs and alarming. Other objects in this category include measures to maintain interoperability within a certain network technology (such as elements required by a device profile), which may even be contradictive between different technologies. In contrast to the basic process communication, services of management communication can vary widely from system to system and constitute a challenge.

Engineering communication is probably the most difficult set of services to abstract. Establishing the communication links for horizontal communication to build the application graph (known as binding) is a typical candidate. However, configuration tasks such as creating data points in the first place is usually a task for network management also. In most instances, this not only regards the applications on the network nodes but also the tools needed to install a complete system. As engineering services they are specific to their technology and cannot be abstracted with reasonable effort.

Clearly, the level of complexity required for abstracting services of the respective communication levels requires a limitation to certain services, leaving others "technology-specific". The objective of this article is to define and discuss the set of services for certain classes of applications in BAS, where the development effort emphasizes the application/control logic and shall be transferable between network technologies as well as implementable on embedded controller designs:

- Fixed controller hardware: Much like a PLC a set of fixed communication objects are pre-configured on the device (no additional communication objects possible), e.g., a fixed number of I/Os.
- Flexible controller hardware: A configuration file describes the relation between communication objects of a field bus and the abstracted services; the application can determine how to behave on the specific controller host.
- Data server application: The application is not controlling a process, but rather monitoring or modifying set points (e.g. touchpad panel, XML data server, visualization.
- Multi-protocol controller: As a generalization of a controller to process data points, the data points may be abstracted elements of different network technologies on the same controller host (e.g., LonWorks and BACnet points).

3. ANALYSIS OF CONTROL NETWORKS

In the area of building automation there are different fieldbus and communication systems using their own characteristic representation of data and services to transfer the data. We compared the specifications of LonWorks (EIA 709 2002), BACnet (BACS-5 2004) and the SOAP/XML-based protocols oBIX (oBIX, 2005), BACnet/WS (BACS-5 2004) and OPC XML-DA (OPC, 2005) in order to develop a data abstraction model.

3.1 Data Representation in BACS

In a data model for a distributed application the data points are connected through a directed graph, distinguishing output points and input points. The application is defined by this graph and a set of processing rules describing the interactions caused by the change of a point value.

Typically, points in the data point graph can be logically grouped to describe specific functions of the system. Such groups forming a coherent subset of the entire application (both data points and the processing rules that belong to them) are referred to as functional blocks. They can also be grouped to form larger functional blocks. This concept is illustrated in Fig. 1. The vertices in the graph represent the data points, the thin-line edges network connections and the bold edges processing data flow connections in a field unit. By grouping certain points by their functional relation, the functional blocks FB1 and FB2 are formed. These may or may not coincide with actual physical nodes. At higher levels of abstraction the application engineer may work with aggregates of functional blocks. The aggregate behaves like its own functional entity with the bold vertices being the interfacing data points.

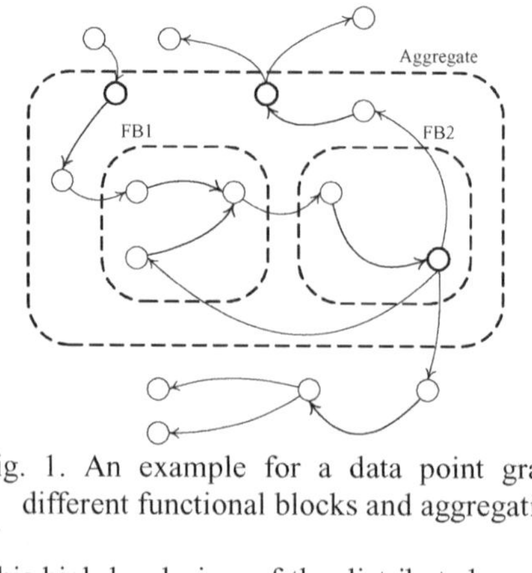

Fig. 1. An example for a data point graph using different functional blocks and aggregation.

This high-level view of the distributed application is an accepted standard for BAS Data points then serve as the external interface for accessing device functionality. While each control network technology for BAS is required to implement process communication according to this model, the specific interpretation of data points varies among them.

Apart from the point itself, the universe of points also needs to be defined. A flat or hierarchical organization may be suitable for different protocols. The abstracted view, however, is only concerned with traversing the organization.

The main characteristic of a data point is its present value, which is the digital representation of a physical unit. In addition, each point is augmented by certain meta-data, which is interpreted in the context of the application (see Table 1). The classes of the value properties are:

- Value characteristics are meta-data, which describe the meaning of the value and enhance interoperability,
- Application characteristics include information, which may influence the behavior of the application (e.g. gain of the controller by adapting to a poll interval),
- User characteristics include information valuable to a human user (e.g. descriptive texts).

Table 1 Classes of value properties

Class	Properties
Value	Data type, Range, Unit, Resolution, Precision, Timestamp
Application	Direction, Status, Poll cycle, Change of value interval
User	Name, Description, Unit Text

The *data type* of the value may be of different kind. Common *data types* are Boolean, integer (8, 16 and 32 bits) and floating point (e.g. single or double precision of the IEEE floating-point standard (IEEE 754 1985)). The *range* describes the upper and lower limit of the value of e.g. the sensor.

The unit of the value of the data point is being described with the parameter *unit*. One possibility is to use units according to the International System of Units SI (ISO 1000, 1992), but almost all investigated systems allow also non-SI units. The parameters *resolution* and *precision* specify the smallest interval measurable by a sensor and the meaningful digits of the value. The *timestamp* indicates the time when the value was generated.

The *direction* specifies whether the data point is an input (readable), an output (readable/writeable) or a calculated value (readable/writeable), which has no physical representation in the field. The *status* provides information on the "health" of a data point. The given information varies from system to system, however, most systems provide at least the status information "normal" and "alarm".

The *poll cycle* is used in case a native communication system does not provide a synchronous service and defines the maximum time between two poll requests to the native data point. Some systems also provide a change of value (COV) service, where the parameter *COV interval* is necessary for analog data points to specify a hysteresis.

To organize data points BACS usually provide a *description,* which is intended to describe a data point in a human readable manner. Unique *handles* and *names* are used to identify a data points.

As already mentioned, not all of these parameters are present in the investigated communication systems. In a few systems, some parameters are only optional and therefore depend on a certain implementation. The above listed parameters may be transmitted in every message (e.g. timestamp) or be the answer to a certain request (e.g. description) to the data point. Some parameters have to be figured out by another source than the data point (e.g. a database).

3.2 Communication Services in BACS

For exchanging data, communication systems use different services. In a first approach we concentrate only on the data access services to reduce complexity. Alarms, schedules, and trend logs, which may also be subject to abstraction, are left for future investigation. Other management or engineering services are not subject to abstraction and considered technology-specific.

Elementary forms of exchanging data are *read data point* and *write data point*. These services are available in every communication system. Additional services are *read multiple* and *write multiple data points*. The *read* and *write data point services* may be synchronous or asynchronous. Although there are more characteristics of the services in fieldbus systems, we did not go more into detail, because our aim was to find an abstraction of data points.

3.3 Gateway Approaches

Nowadays gateways are used to "translate" information on the level of the application layer. They act as proxies between two entirely different worlds. A two-port gateway application therefore consists of a software program that "talks" both protocol stacks and "understands" the services and data types of both sides Fig 2.

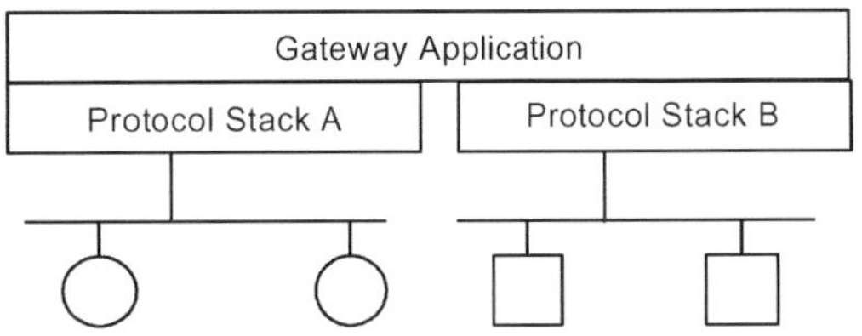

Fig. 2. Typical gateway architecture

The main problem of gateways is network management: Both technologies usually come with a proprietary and highly specialized way of how to assign network addresses, how to find other nodes, how to configure the network and distributed applications, etc. Translating network management services from one side to another one is probably one of the most difficult tasks for a gateway and therefore typically not realized. Instead, a gateway needs management from both sides, if both sides are sophisticated enough.

The ideal case is that the gateway needs no management at all. This implies that it

- get its addresses automatically,
- finds all nodes automatically,
- understands all data structures of all nodes automatically, and that it
- offers all data structures of one side to the other side in a protocol conform and interoperable way.

In reality this needs massive engagement and tuning from operator side. Gateways usually need to be told what to do, and even worse, their configuration is often application-specific. This means, that if the technology-spanning application changes, the gateway must be reconfigured: a very inflexible, annoying and error-prone quality of gateways. An example for and a detailed discussion on a modular gateway architecture can be found in (Sauter and Palensky, 2000).

The classical strategy to cope with such insuperable technological differences is to focus on the common basics of the two networks. In the case of control networks this is single sources and sinks of control or measurement data that have to be found, addressed and translated. While being far from loss-less, this translation at least enables vital information to be shared between to previously isolated systems.

Existing gateway solutions are, however, tied to the technology of the specific networks they connect. This applies to their architecture, their capabilities and their management. This limit becomes visible and turns out to a problem, when more than two networks must be connected. A cascade of gateways – nothing unusual in real-world projects – leads to many problems, if traditional gateways are simply nested. The level of network transparency, needed by an application, can often not be achieved. This very problem is tackled in this paper, the way how important applications can have easy and abstracted access to multiple networks.

4. DATA POINT ABSTRACTION LAYER

To fit the requirements we developed a data point abstraction layer (DPAL) model as shown in Fig. 3.

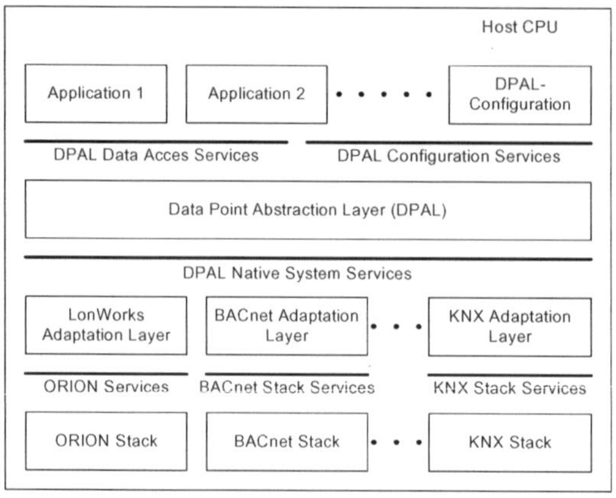

Fig. 3. Data Point Abstraction Layer (DPAL) model

The DPAL plays the central role. It provides three interfaces for data access services, configuration services and the native system interface. The configuration interface is used to manage the DPAL, especially to create and delete a DPAL data point as well as to insert and remove a DPAL data point from the DPAL data point repository. It cannot be used to manage the communication objects in a native system.

The data access interface allows an application to access the value and the properties of a DPAL data point and to browse the repository. The native system interface is designed for the communication between the DPAL and an adaptation layer, which is a kind of gateway between a native communication system and the DPAL. Via the LonWorks interface the ORION protocol stack provides services and data of the EIA-709.1 protocol. The same architecture is also suitable to link a BACnet stack or a KNX stack to the DPAL.

We decided to use the BACnet model as a template for the DPAL model and extended it to the specific requirements of the DPAL rather than develop a new model from scratch. The main reasons for this decision were:

- that the model is already proven in many BACnet implementations,
- that the Annex H of the BACnet standard (BACS-5 2004) contains mappings of BACnet to LonWorks and KNX, which can be used in a similar way as mappings between DPAL, LonWorks and KNX,
- that Addendum C includes a mapping of BACnet/WS attributes and BACnet object properties,
- that the BACnet model can be extended to the defined requirements.

The properties of a DPAL object are divided in general static properties, general dynamic properties and extended (static) properties for analogue, binary and multi-state objects. This concept allows us to implement additional object types as e.g. string objects, data-time objects or node objects, which do not contain a value (used for structuring the data points, see later in this section).

The general static properties include object type, direction, description, data type and the name of the object. Additionally we introduced a property defining the minimum update rate of the DPAL data point. That allows us to implement a quasi-synchronous service for all data points, even if the service is not supported by the native communication system.

The general dynamic properties are the timestamp of the value and the status of the object. The extended (static) properties of e.g. the analogue object holds the properties - value, unit, resolution, precision, low limit, high limit and COV increment. The property unit is a structure, which holds the exponent of the seven SI base units. Using dimensional analysis, every physical unit can be expressed in that way. An additional enumeration is used to distinguish between the dimensionless quantities as e.g. radian or decibel. This representation of units allows converting between units in a standardized and efficient way on hardware, but is not human-readable. Therefore, we introduced a field, which holds the unit as a character string configurable by the application designer.

The general static properties and the extended (static) properties are usually configured once during the commission phase and do not change during normal operation. As the general dynamic properties hold the timestamp and the status of a data point the change during operation.

To use a DPAL object in an application, we use the *create function* of the appropriate adaptation layers, which creates both a DPAL object and the corresponding communication object of the native system (e.g. a BACnet object or a LonWorks network variable). After the new object is configured properly, it can be inserted into the repository as child of any other object in the repository. That allows reflecting the structure of the data points in the field as shown in Fig. 4.

To browse in the repository, the data access interface provides various services. The most important services are to get the child object of a DPAL object, to get objects in the same hierarchy as a specified object or to get the parent object of a DPAL object. Another possibility to figure out the position of a DPAL object in the repository is to encode the position in the string stored in the property DPAL name e.g. Root/P2/P2,2 would be the highlighted object in Fig. 4.

An object can also be removed of the repository. That does not mean, that the object is deleted, but that the object cannot be found by browsing the repository. This is a very useful feature for dynamic communication objects, which appear and disappear during operation and therefore do not have to be deleted and created repeatedly.

To read or write the value of a DAPL object blocking the data access interface provides appropriate services. Additional multiple read and write services are available, which allow an application to read or write the value of more than one object at once. If the native communication system supports atomic read and write functions (e.g. a network variable in LonWorks containing a structure), the multiple read and write functions prevent from getting inconsistent data between an application, the DPAL and the native communication system.

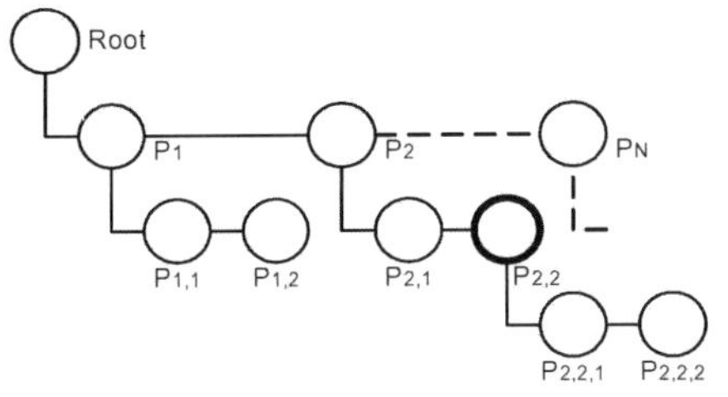

Fig. 4. Organization of data points in the DPAL's repository

Last but not least the data access interface provides a subscription for COV notification and dynamic event notification. Subscribing for a COV notification, the application is informed automatically, if the value of the DPAL object has changed by a defined value. The subscription to the dynamic event notification service informs an application whenever a DPAL object is created, inserted or removed from the repository or deleted.

5. DISCUSSION

The data-centric approach using abstracted services leads to the question, which services have been abstracted. Functions not part of the data point abstraction are left to be "technology-specific".

The main technology-specific functions, which remain especially for BACS are delegated to a "configuration instance". This is a piece of software, which prepares the platform for a DPAL application. The tasks of the configuration instance are:

- Creating static data points: The points have to be created and linked to the native system (e.g. by a configuration file or an automatic point discovery). The DPAL services assume there is a repository to be scanned and the application can figure out, which properties are available.
- Adding/deleting dynamic data points: The configuration instance can create/delete DPAL point on the creation/removal of dynamic native points. The application can register for the changes.

The remainder of this section discusses, how the abstracted services are mapped to LonWorks and BACnet systems and which challenges have to be tackled.

5.1 DPAL Services in LonWorks

The main communication objects in an interoperable LonWorks system are network variables (NVs) predefined in the Standard Network Variable Type (SNVT) List (SNVT, 2003). The mapping between LonWorks objects and DPAL objects is shown in Table 2. In case the value of a property can be derived from the SNVT definition, the corresponding SNVT property is used, configured means that the system integrator has to adopt the value to the certain application. The dynamic properties timestamp and status as well as the DPAL handle are generated by the DPAL and need no configuration or mapping.

Table 2 Mapping LonWokrs objects properties and DPAL properties

DPAL property	LonWorks equivalent
Data type	SNVT Type category, type size
Low limit	SNVT Valid type range
High limit	SNVT Valid type range
Unit	SNVT Units
Resolution	SNVT Type Resolution
Precision	configured
Timestamp	generated by the DPAL
Direction	configured
Status	generated by the DPAL
Poll cycle	not needed if NV has a binding, otherwise configured
COV increment	configured
Description	configured
DPAL handle	generated by the DPAL
DPAL name	configured
Native name	configured

The hierarchical data representation in of structured data may be done in two ways, as it is shown for the widely used network variable type SNVT switch of LonWorks (SNVT, 2003) in Fig. 5. It consists of a value defining a level between zero and hundred percent and of a state, which can be true or false.

On the left side of Fig. 5 the DPAL node object Switch is used as a root element and does not contain a value. The elements of the structure of the SNVT switch are children of the DPAL Switch object. Another possibility mapping the SNVT switch to DPAL objects is shown on the right side of Fig. 5. This variant omits the DPAL object Switch, one of the elements of the structure supersedes it (e.g. Switch.value). The remaining elements (in our example only Switch.state) of the structure are children of this element. The second method of representing structures as DAPL objects is more resource-efficient (one element less), but does not reflect the correlation of the elements of the structure, because they are not in the same hierarchy level.

Using the multiple-read and multiple-write-functions, it is possible to get and set the value of structured network variable in a safe way by addressing all elements of a structure at once. For nodes connected to the LonWorks fieldbus, the data points acts like a LonWorks node. If a DPAL object is an output or a value object, it is represented as an output NV, in case of an input DPAL object, as an input NV. All services including the management services are processed as within a LonWorks node.

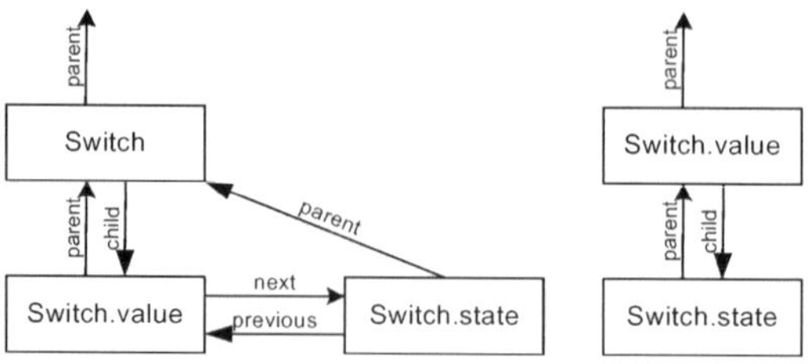

Fig. 5. Two possibilities mapping the SNTV switch to DPAL objects

The LonMark and BACnet objects have common properties but also conceptual differences. Therefore the DPAL object model is capable to abstract on the level of NV, but fails to abstract on the next higher level in LonWorks, the Functional Profiles. They encapsulate a set of network variables and configuration properties for specific applications of certain industries. However, abstracting at this level would mean to abstract the objects themselves (e.g. abstract temperature sensor).

5.2 DPAL Services in BACnet

Since the DPAL model is very similar to the BACnet model, the mapping of both is quite simple. The BACnet protocol defines various types of objects, which differ in number and kind of properties. We limited the mapping to the analogue, multi-state and binary objects. These objects are either input, output or value objects. A BACnet object consist properties, which describe the value, application and user characteristics. The mapping of the BACnet and DPAL properties is shown in Table 3.

Since standard DPAL object only have one value property, communication objects with more than one value are not possible and therefore can be mapped one-to-one to a DAPL object. To read or write the values of different objects at once, BACnet offers multiple-read and multiple-write-services. They have the same functionality as the multiple-read and multiple-write service of the DPAL data access API.

Table 3 Mapping BACnet properties and DPAL properties

DPAL property	BACnet property identifier
Data type	Object_Type
Low limit	Min_Pres_Value
High limit	Max_Pres_Value
Unit	Units
Resolution	Resolution
Precision	configured
Timestamp	generated by the DPAL
Direction	configured
Status	Status_Flags

Poll cycle	Update_Interval
COV increment	COV_Increment
Description	Description
DPAL handle	generated by the DPAL
DPAL name	configured
Native name	Object_Name

6. CONCLUSION

Connecting different types of networks is still challenging. The DPAL effort, a cooperation between industry and academia, tries to ease the topic of cross-technology interoperability or interworking, being inspired by recent developments in the domain of automation and IT. The massive abstraction, done in the DPAL APIs helps to bring together entirely different networks in a way that reduces engineering time and costs.

Since buildings are typically investments for many decades, it is natural that building automation companies regularly have to integrate their systems with existing installations. Nobody removes existing and working technology just because there is a new one that offers a particular new feature. Combining technology might not be the clean and proper way, but is reality. DPAL helps to do this with a minimum amount of pain and time. The level of interoperability will even be increased in future, when the processing power of nodes rise, and more data can lead to more self-documentation, more supporting services and more functionality.

REFERENCES

Dietrich, D., D. Loy and H. Schweinzer (1999). *LON-Technologie*. Hüthig, 2nd ed., Heidelberg.

Palensky, P. (2004). Requirements for the Next Generation of Building Networks. In: *Proceedings of the International Conference on Cybernetics and Information Technologies, Systems and Applications* (ISAS CITSA 2004), p 225-230

Palensky, P. (2005). The JEVis Service Platform - Distributed Energy Data Acquisition and Management. In: *The Industrial Information Technology Handbook*, Zurawski, R. (Ed.), CRC Press, Boca Taton, Florida, 2005, p 111-118

Sauter, T. and Palensky, P. (2000). A closer look into Internet-fieldbus connectivity. In: *Elektrotechnik und Informationstechnik (e&i)*, **5 (2000)**, p314 - 318.

Fischer, P., Palensky, P. (2004), The importance of Being Certified - The role of conformance testing and certification of communication systems in building automation and control devices, In: *Proceedings of the 7th IEEE Africon Conference*, (2004), p 1223-1227

Loy, D., Dietrich, D. and Schweinzer H (2001): Open Control Networks; Kluwer Academic Publishers, Boston/Dordrecht/London, 2001

Daniels, K. (2003) *Advanced Building Systems: A Technical Guide for Architects and Engineers.* Birkhäuser, Basel.

Soucek, S., Sauter, T., (2004). Quality of service concerns in IP-based control systems. In: *IEEE Trans. Ind. Electron.,* **vol. 51, no. 6**, pp. 1249–1258, Dec. 2004

Kastner, W., Neugschwandtner G., Soucek S., and Newman H.M. (2005), Communication Systems for Building Automation and Control. In: *Trans. IEEE,* **vol. 93, no. 6**, Jun. 2005.

BACS-2 (2004), *Building Automation and Control Systems (BACS) – Part 2: Hardware,* ISO Std. 16 484-2, 2004.

OPC (2005), *OPC Foundation.* [Online]. Available: http://www.opcfoundation.org

oBIX (2005), *Open Building Information Exchange.* [Online]. Available: http://www.obix.org

ISO 1000 (1992). *SI units and recommendations for the use of their multiples and of certain other units.* ISO 1000:1992

IEEE 754 (1985). *IEEE Standard for Binary Floating-Point Arithmetic.* ANSI/IEEE Std 754-1985

EIA 709 (2002). *Control Network Protocol Specification.* Electronic Industries Alliance. EIA-709.1-B, January 2002.

BACS-5 (2004) *Building Automation and Control Systems (BACS) -- Part 5: Data communication,* ISO Std. 16484-5, 2004.

SNVT (2003). *The SNVT Master List.* Version 12 Echelon Corporation.

Di Stefano, A.; Bello, L.L.; Bangemann, T. (2000). Harmonized and consistent data management in distributed automation systems: the NOAH approach. *Proceedings of the 2000 IEEE International Symposium on Industrial Electronics*, (ISIE 2000), p 766 - 771

Mella A. and Russo F. (1999). NOAH Project, an Example of Application, in D. Dietrich, P. Neumann and H. Schweinzer (eds.): *System Integration, Networking and Engineering*, Springer, pp54-52

Kastner, W. and Neugschwandtner, G. (2004). Service interfaces for field-level home and building automation, *Proceedings. 2004 IEEE International Workshop on Factory Communication Systems* (WFCS2004), p.103-112

Velasco, M., Fuertes, J.M. and Marti, P. (2003) Process data abstraction/accessibility via internet, in *Proceedings of the IEEE Conference on Emerging Technologies and Factory Automation* (ETFA 2003), p. 357 - 363

Peter Fischer (1999). Mapping of Fieldbus Protocols to Standardised Field Level Objects, in D. Dietrich, P. Neumann and H. Schweinzer (eds.): *System Integration, Networking and Engineering*, Springer, p. 54-52

THE ARTIFICIAL RECOGNITION SYSTEM (ARS): NEW CONCEPTS FOR BUILDING AUTOMATION

Gerhard Pratl, Brigitte Lorenz, Dietmar Dietrich

Institute of Computer Technology
Vienna University of Technology
{pratl, lorenz, dietrich}@ict.tuwien.ac.at

Abstract: Building automation faces a development that leads to more and more sensory information available for processing. Existing approaches are challenged by this abundant amount of data, therefore the authors see a need to introduce new concepts for handling the challenges of the upcoming future. Looking at bionic approaches taken from the field of neurobiology, but also psychoanalysis, a system is created that applies neurological principles to get human-like perception, and psychoanalytical principles to evaluate the perceived scenarios. *Copyright © 2005 IFAC*

Keywords: Fieldbus, Sensors, Autonomous Control, Intelligent Control, Cognitive Systems, Bionics, Knowledge Representation, Image Recognition, Evaluation, Behaviour.

1. INTRODUCTION

By reading journals about Building Automation, like *LonMark* (April 2005: Building a Brighter Future), the problems in the future of fieldbus systems become visible. State of the art is to integrate many thousands of networked sensors, actuators and controllers in one system to control various processes in a building – temperature, light, motion, etc. – based on classic control loops with threshold values or similar principles. Higher functionalities, like energy optimizations or security aspects, are solved by similar control systems. It is hardly imaginable that it will be possible to handle – in particular to integrate and to maintain – a drastically increasing number of integrated, multi-functional components in an efficient and low-cost way. But we will be confronted with such high amounts of diverse sensor data in the not so distant future, especially if we consider sensor networks of thousands of intelligent nodes in products like carpets, walls or similar smart parts. We have to think about new possibilities. Of course the performance of embedded systems is increasing as well, but the question remains: Are the classic methods of automatically processing sensory input data enough? The authors are convinced that in order to answer this question we should analyze nature and try to follow its way by copying its solutions.

This paper reports about the bionic approach and the new results of the *Artificial Recognition System (ARS)* project, which was presented in Brainin, *et al.* (2004), or, more recently, in Pratl, *et al.* (2005). There, the authors argued that within the last decade the psychologists and psychoanalysts could verify that our brain does not compute current situations (images) and scenarios by analyzing *all* the incoming data from the various sensors of our body each time again and again. Instead, the human brain memorizes images and scenarios over a whole lifespan. However, this is not the only aspect we pointed out as decisive difference to the traditional thinking of engineers about human cognitive abilities and how to model them technically. The human brain is not able to fully process all incoming data in real time. The brain is only able to perceive and recognize characteristic forms of the outside world. The rest is associated with the aid of the already memorized images (Dietrich, *et al.*, 2004). The association (comparison) of the new images – which creates a mix of computed data based on inputs from the environment and internal, previously memorized data – releases emotions, which, in turn, trigger and guide the actions and reactions of human beings.

The difficult question is how we can technically implement the observed principles. Classic methods, like neural networks, fuzzy logic, knowledge-based systems, etc. are placed on the market. However, although some solutions are quite brilliant, there is no striking application for the area of fieldbus systems. Therefore the authors follow the ideas and principles of the neurologists and psychoanalysts. Of course, it is impossible to artificially construct a system, which implements every functional aspect of the architecture of the human brain, especially if we want to take over principles of the theories of Sigmund Freud. We have to take the "usual" method of engineers and start with very simple parts. A very good example for this approach is presented in a dissertation dealing with an automatic component part sorter for an assembly line (Brenner, 2001), whose realization has been a great success on the market. The first step was to differentiate between only three different components; today the system is able to differentiate between more than 100 learned parts.

This recalls the story of Kaspar Hauser (Hauser, 1995) who was shielded from the outside world throughout his entire childhood and early youth and was kept alone in a dungeon. There, he could only see and, thus, get to know, a very limited set of objects. When he finally saw the outside world with all its complexity, he had the greatest difficulties to accept this "real" world because he did not know it – in our language: He did not have a memorized image of it, and it took him a lot of time and effort to learn how to interpret all the new impressions.

If we define constraints such that our system only needs to "understand" images similar to already memorized ones, and if we only accept a limited set of possible scenarios (applications), then it will be possible to come to a "simple" solution.

As already explained in Tamarit, *et al.* (2001), the base is a hierarchical system where the inputs are symbolized at different levels. The higher symbols are the input for another processing functional unit, where complex functions are defined and actions/reactions initiated. The information flow is evaluated by emotions, generated by comparison between inputs and memorized images. So, we have to separate the ARS project in two different parts: the hierarchical lower level functions, mainly from the point of view of neurologists, and the higher complex functions, where Freud's model is taken over (Brainin, *et al.,* 2004).

2. APPLICATIONS

We focus on the development of a system, termed the Artificial Recognition System (ARS), which makes use of diverse, redundant sensory inputs in a bionic way to create a perception and consistent representation of the surrounding world. This enables the system to perform the desired functions and to attend to the required applications. In order to demonstrate the approach of our system we have chosen four applications from different domains.

Application 1: Human Surveillance System
By making use of light barriers and detectors, tactile sensors in the floor, door contacts, and stereo cameras the system is enabled to know the position of persons in a building. People are considered to be anonymous, which means that the system has no additional knowledge about their identity. That is to say, unless a person is furthermore provided with an identification mechanism (e.g. due to authentication at a security door). The system is able to provide information about a person's current and past location, so that the path of a person through a building can be tracked and monitored.

Application 2: Child Safety System
The second application we consider is a *child safety* system. The system recognizes whether a particular person is actually a child, and can monitor and guard the actions of the child. When it appears that the safety of the child may be compromised due to a hazardous situation, the system alerts a (human) supervisor.

The decision that a particular person is actually a child is based on diverse sensor information, similar to other mechanisms in the system. This includes the use of camera images to derive height and shape of a person, as well as information from light barriers that are mounted at different heights, and weight information obtained from pressure sensors to support the decision. Example situations that are classified as hazardous are: an open fire, a hot stove plate, a cupboard with open doors, or a child climbing on table. There are additional conditions and criteria that have to be taken into consideration. For example, a situation is only classified as hazardous if a child is alone and unattended (meaning that no adult is in the same vicinity). Other conditions include the facts that the fire has to be burning, or that the stove is indeed hot.

Application 3: Geriatric Care System
The third application is concerned with a *geriatric system* to care for elderly people. In this case the system is able to recognize when an elderly person collapses or faints. Furthermore, the system is able to identify and track the location of predefined objects, such as keys, glasses, and books.

Application 4: Theft Protection
The fourth application shall supervise the whereabouts of classified things. In a given room the system monitors a set of objects (e.g. books in a library) that are not allowed to be removed. In case a person takes away such an object, the system shall inform a (human) supervisor.

These applications share a common layout, in the sense that we define a number of rooms on a floor that has a layout, which is identical for all four applications. The sensors that are used are also identical, and mounted in the same position. In this way the symbolization mechanism shares a common set of symbols (although not all symbols have to be present in all applications).

Obviously the system has only a limited understanding of the world it perceives. The fact that cameras are installed does not automatically imply that the system is able to process all the information in a way that is similar to a human operator observing and evaluating a camera image. For example, suppose that a dog enters the room. This could possible be perceived as a "person" (or, at best, as "child"), since the system has no initial concept of a dog. Hence, the system is bound to make incorrect decisions if it is confronted with facts or images that are outside the scope of its capabilities. This system attribute is intentional, since it does not form part of the task that needs to be fulfilled. If we introduce a new application, which makes it necessary to distinguish animals from persons, the knowledge of the system will have to be extended.

3. SYMBOLIZATION

We have to find a way to cope with the vast amount of diverse sensory input that the system has at its disposal. We want to extract relevant information, ignore everything that is not important and use the existing redundancy in sensory information to get a stable and robust perception of the environment. Therefore we break the available information into small chunks and label them *symbols*. A symbol in this context is a concept that the system has means to operate on. It contains information that can originate either from sensory input or from knowledge of different kinds. If a symbol is created only by sensory input, we refer to it as a *microsymbol* (µSymbol). µSymbols represent small pieces of information that are the base for higher level symbols. Different symbols have different "weight" in the way they represent information. We assume a hierarchy in the symbols, meaning that there are µsymbols with only little weight (i.e. they represent a piece of perception, e.g. a footprint) and there are symbols of higher weight (e.g. there is a symbol that represents a chair leg and there is a symbol that represents a chair – which requires four chair legs and some other symbols to be present in a certain relation to each other). The first step in recognition is to create µSymbols out of sensor input; from then on the system only operates on symbols and not on actual sensor data.

3.1 Microsymbols

For the scope of this paper we limit the variety of different µSymbols and symbols and focus on a set of only three symbols: table, chair and person. The system has the goal to determine the position of persons, and to detect tables and chairs. The sensors we look at are shown in Figure 1.

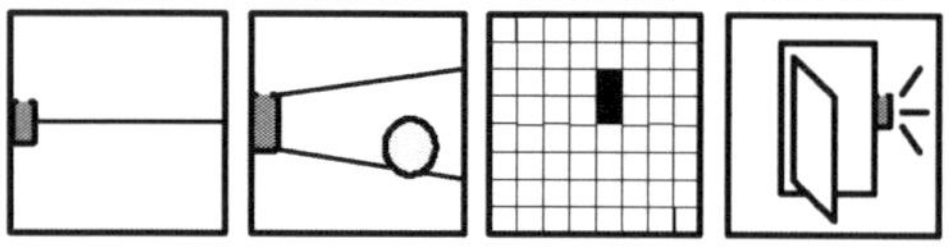

Figure 1. Sensors that provide data on which the perception of the system is based. Cameras are used for multiple purposes.

Cameras play a special role, since they are able to provide lots of information within one sensor (Figure 2). We do not aim at implementing a new set of image processing algorithms, since this is already state of the art. Instead we want to employ these techniques to process camera images and extract different features out of it. This processing happens on different levels of complexity: for once, we want to detect movements in the visible area of the camera: this is done by background subtraction (the cameras are mounted statically).

Figure 2. µSymbols that are created by cameras.

Then we want to extract basic features like lines, edges, corners and areas. Using stereo cameras and feature matching we gain information about the spatial position of these features and can create flat surface areas. In a next step we will use sophisticated camera systems to extract persons that are divided into their body parts, namely head body, legs and arms; this is however not covered in this paper. To match the requirement we introduced before (recognize persons, tables and chairs) we use cameras to detect motion, flat areas and bars (i.e. objects that are long and narrow).

The µSymbols and the respective sensors that we need are listed below. The name of the µSymbol is chosen to provide a human understandable meaning of the quality of this information (of course this principle is not extensible when dealing with larger amounts of diverse sensors; we will have to automate this process when it gets too time consuming). The names in brackets are properties of this µSymbol; µSymbol and properties are an entity, properties cannot exist without a µSymbol.

Table 1 Microsymbols and their associated Sensors

Sensor	Microsymbol
light barrier	personPassesBy(sensorPos, direction)
motion detector	personMoves(sensorPos)
floor sensors	footprint(sensorPos)
door contact	doorOpens(sensorPos)
camera	camMovingPerson(sensorPos, size, direction); camFlatArea(sensorPos, size, position, orientation); camBar(sensorPos, size, position, orientation)

All µSymbols contain information about the location of the sensor and its sensitive area, this is indicated by `sensorPos`. Like in embodied intelligence we need to have metainformation about the sensor to determine how to interpret the actual sensor value. This information needs to be configured at setup time and is then known to the system. Without this metainformation the actual sensory input is useless, therefore we indicate the knowledge of the whereabouts of a sensor by `sensorPos`. What is not shown here is for once a timestamp, indicating when the µSymbol was created. µSymbols have event character, meaning that they exist in only one point in time. Also left out is the sensor type. It is intrinsically contained in the name of the µSymbol, but it is of course possible that more than one sensor can create the same type of µSymbol.

`personPassesBy` indicates that a set of two light barriers has been triggered within a certain time period. `direction` is derived by looking at which light barrier was triggered first. `personMoves` is the information that a motion detector can contribute to the position of a person. The important metainformation is the trigger area of the sensor, i.e. which area the sensor "sees". `footprint` is created by floor sensor, which detects changes in their state. Depending on the size of the floor sensors a footprint µSymbol can depend on only one sensor or on a couple of sensors. `doorOpens` requires a sensor to be installed in a door that is triggered when the door is opened.

`camMovingPerson` is the differential image mentioned earlier (using background subtraction). It indicates that a moving shape, which matches the size of a person, has been detected. It has two properties that describe the shape and the direction in which the shape has moved. `camFlatArea` identifies any kind of flat area; it requires the use of stereo cameras, because the real size (not the size on the camera image), position and orientation are required. `camBar` identifies any objects that have stick-shape, i.e. long and narrow shapes.

3.2 Symbols

The next step is now to take the preprocessed perception information (available as µSymbols) and create symbols. Note that starting from this level there is no more sensor information to be processed, everything that the system knows about the outside world is merely available as µSymbols. We have defined a fixed set of symbol templates, the system only knows symbols that are created from these templates and cannot learn new symbols. Just like with µSymbols this approach will have to be extended in future, when the number of different symbols is vastly extended.

The scheme for symbols is similar to µSymbols. We have a name for each symbol and we have properties. Additionally we define *associations* (see next section); these indicate which symbols can be associated with each other (referring to the type of symbol and µSymbol, not the actual symbol). Symbols can be associated with an arbitrary number of µSymbols, symbols or a mixture of both. And we use a hierarchy of symbols, meaning that symbols can inherit properties and associations of other symbols. The list of symbols we define is:

```
Symbol object

Symbol thing inherits from object
Properties: size, position, orientation,
direction

Symbol person inherits from object
```

The following symbols inherit from symbol `thing` and do not have any additional properties: `table, tableLeg, tabletop, chair, chairLeg, chairBack, chairSeat.`

`Object` is the root symbol in this hierarchy; it has no properties, but is the base for all other symbols, including `thing` and `person` (a person is an object in this context). This eliminates redundancies, because things in general have some common properties and allows to introduce more abstract information processing (knowing that a chair is a thing can be useful knowledge). Note that a lot of symbols do not have additional properties (except for the ones they inherit), still the definition per se is important, because the system has means to process symbols depending on what they represent (indicated by the symbol name). This approach raises a lot of questions, for example how to fill a formal symbol with semantic interpretation that is intrinsic to the system. This *symbol grounding problem* is discussed in (Harnad, 1990).

If we take a look back to the system perception and look only at µSymbols that are created by cameras, we see that there is still a semantic gap that needs to be filled: a camera can find flat areas and bars in the image. These are intended to be used as the building blocks for chairs and tables. However, not every bar makes a chair leg and not every flat area is a tabletop, chair back or chair seat. This gap is filled in the next section.

3.3 Associations

An association defines means to create symbols by combining other symbols or µSymbols. For the goals defined in this paper we need the following associations:

```
person associates with: footprint, doorOpens
personMoves, personPassesBy, camMovingPerson,

table associates with: tableLeg, tabletop

chair associates with: chairleg, chairback,
chairseat
```

The system has to use the µSymbols that have been perceived and combine them in a way that the symbols person, table or chair is recognized. The associations listed here contribute to this process by providing knowledge that is needed. In the case of tables and chairs this knowledge is of structural and spatial kind: a chair has certain geometry, i.e. geometric relations between its parts. All legs shall be parallel to each other and of the same size and the size needs to be in certain boundaries. The seat has to be parallel to the floor and on top of the legs, while the back has to connect orthogonally to the seat. Therefore the perception information that entered the system by means of µSymbols is now analyzed towards matching pieces of information: if there is a flat area of adequate size and orientation, if there are bars that are as well of adequate size and orientation and additionally are connected to the flat area and if there is another flat area that suffices the requirements for a chair back, then the system creates a new symbol `chair`. The µSymbols that are used for this chair (which are of type `camBar` or `camFlatArea`) are converted into the according parts of a chair, `chairLeg`, `chairBack` and `chairSeat`. Tables have similar associations, only that the dimensions are different. Symbols of type `person` are created whenever the associated µSymbols are created. Note that all these µSymbols contribute merely to the position of a person and do not allow any deduction of activities as needed for the applications described in section 2. These are left out in this paper. But they still provide a valuable contribution that is a key design issue of the system: the µSymbols represent redundant information about the position and movement of a person: a `footprint` and a `personPassesBy` µSymbol both contain this information. The system has to be able to use all the available information; therefore we need to classify the contribution of each μSymbol. This requires having a look at the origin where a μSymbol came from. Light barriers (we assume the combination of two light barriers to gain the direction of a movement) that are installed horizontally in a hall have rather precise knowledge of one axis, but rather poor knowledge of the other axis: the position of the light barrier itself gives the first axis, but where on the hall it was interrupted is unknown. A tactile floor sensor exactly knows its position; however, it requires some computational effort to find footsteps that belong to the same person. The direction information of light barriers is useful for this matter.

Each contribution is therefore processed and contributes to the goals of the system. Embodied intelligence teaches us that a lot of "intelligence" is grounded in the way sensors are installed and in the capabilities of the sensors. Therefore the concept and layout of the system and the knowledge of capabilities of each sensor are important when classifying different incoming information.

4. WORLD REPRESENTATION

In the previous sections the flow of information, which originated in sensory input had been processed in a way that allowed the system to catch a "snapshot" of the environment it supervises (we will call this environment the *world*, meaning merely the environment that the system can perceive with its sensors; this "world" will usually be limited to a building). These snapshots exist in one point in time; as soon as one snapshot has been created, the system starts to look at new incoming sensory input and creates the next snapshot.

The sequence of snapshots, however, does not make a continuous representation of the world. If we look at the world representation of a human being, we see that there is a lot of knowledge about the surrounding world, which is not strictly bound to the current sensory input. The world representation consists mainly of memories of older perceptions (which are, of course, regularly updated by more current perceptions). And even when perception does not provide us with additional input, we do not immediately forget what we knew: assume that Alice and Bob are in the same room and Bob leaves the room by walking through a door. Although Alice does not know the exact position of Bob any more (maybe in the next room, maybe somewhere else in the building), she still knows the door he has left through. The same applies, if Alice sees Bob walking behind an obstacle: the gaps in perception while Bob is not visible are filled by knowledge about the continuity of events.

Therefore we introduce the *world representation*. It is based on the "Inner Representation of the Outer World" as presented in (Solms, 2002). It is fed by the information that enters through perception, but applies additional rules to get a continuous representation of the world as the system sees it. Looking at the person and the determination of a persons position we have so far seen a set of snapshots in each of which there might have been a person with a slightly different position. The task of the world representation is now to concatenate all these "person-perceptions" and make a single person out of it – a person that moves. This implies a change in the timing information we have to keep now. In perception everything was bound to a single point in time – reflected by the timestamp. Now we have to extend time information to a time period, because symbols are created and updated during their lifetime.

To ensure the continuity of the world representation, the system has to use the context of the previous world representations. This context helps to create predictions about the future state, which can be verified (or falsified, respectively) later. In the case of persons that are hidden by obstacles, the system knows both position and direction of the person before it loses track. So it can assume that the person is still somewhere around the last position. If briefly afterwards a person is perceived in this vicinity and the assumed direction matches, the gap in the flow of perception can be filled by sheer knowledge (namely that persons can only move continuously) and the path of a person can be reconstructed.

This process is, of course, improvable by increasing the amount of available information. The cameras can be used to not only detect movement, but to also create a color histogram of a person (this histogram would be another μSymbol). When processing disappearing and reappearing person, the histogram is then used to ensure that the system did not lock onto a different person who happened to pass by as well. Again this is based on knowledge, not perception, namely the knowledge that persons cannot change their color distribution (at least not in a short amount of time).

At the end of this process we get a current representation of the world as the system sees it together with a history of this representation. Therefore we now are able to take *time* into consideration – not just the time between adjacent perceptions, but time in the order of several minutes or hours. The system can now look for *scenarios*, that is, sequences of events that are distributed over a certain time period. If it detects a known scenario it can either trigger the creation of a new symbol or indicate this to a human supervisor. Looking at the applications introduced in section 2 we can identify scenarios in geriatric care defining that the system shall detect, if a person lies on the floor and has not moved for some time. Or, in case of the theft protection application, we can trace objects of interest and identify the person that posses such an object and attempts to leave a predefined area (e.g. a library).

5. EVALUATION OF SCENARIOS

The system has to evaluate the perceived situations and scenarios. It has to "judge" whether there is an out-of-normal situation which requires action or not. Examples for actions the system could have to perform are: to save a person from harm in case of a dangerous situation, to assist a person when he or she needs help, to protect objects from being stolen or to keep the room temperature adjusted within a comfortable range. Some actions are very time critical – meaning that they need to be performed immediately – others not. This is the first thing the evaluation mechanism has to detect: whether there is an emergency situation or not.

To identify hazardous and urgent situations a kind of alarm mechanism will be implemented. An alarm is a rather "naked" signal with no or only very little internal structure (Johnson-Laird, 1988). It triggers an immediate reaction in a mostly hard-wired way – a reflex circuit. This purely reactive control loop is the simplest one in a hierarchy of control loops. To use such a hierarchy is biologically motivated.

5.1 *Bionic Evaluation Principles*

In nature, living beings have evolved a cascade of more and more complex control levels for behavior selection, all of them aimed at ensuring survival in highly demanding and constantly changing environments. Each newly emerged control level is built up of parts of the lower levels (Damasio, 2003). However, the different levels are not clearly separated, but rather very much intertwined, with a lot of back-and-forth communication going on over the linkages connecting the various modules. This makes it so hard to analyze their functionality. Moreover, as each level is crucially based on a *feedback loop*, it cannot be decomposed in its parts without losing some of its emergent features. This is because feedback leads to non-linearity and thus the whole process cannot be described as a simple linear superposition of its parts.

Feedback is essential for evaluation. In biology, it can always be described as a loop between experience and expectation whereby both terms are used in a very general way. Experience refers to the input the system receives from the outside world, its environment, and expectation refers to its inner representations and constraints. Outer world and inner representation have to be related to each other. The better this is achieved, the more intelligent the autonomous system will be able to behave. Ideally, the process of relating outer and inner world will be done in a dynamic, adaptive way. This is especially important for constantly changing environments. However, the technical design and implementation of such a flexible association mechanism is a quite difficult task.

Nature starts with simple mechanisms like bodily reflexes, unconditional reactions to appetizing stimuli, initiatives for active behavior like drives and motivations, mechanisms for the intensification or inhibition of specific behaviors mediated via lust or pain, etc. Conditional associations represent an important step forward towards a more adaptive and context-sensitive control of behavior. Conscious reasoning and decision making, guided by ideas, theories, and even a general world view, stand on top of the behavioral control pyramid (Riedl, 1984). This bottom-up, evolutionary derived view of our cognitive abilities is compatible with the picture Sigmund Freud has worked out about the functioning of our brain.

5.2 *Technical analogy*

We focus at investigated in which way the above mentioned bionic principles can inspire the technical design of the evaluation and behavior selection unit of intelligent autonomous systems. The field of building automation is insofar special, as in this field the autonomous system cannot "move around" and actively explore its environment like for instance a mobile robot or a software agent can do. A building automation system is comprised of sensors with a fixed position, together with a communication structure and one or more reasoning units. However, there are quite a lot of sensors and they are very much distributed. Thus, the system will not only have access to central knowledge. Instead, it will rather rely on decentralized sources of information. Moreover, the fact that the spread out sensors are getting more and more intelligent raises the amount of locally available knowledge. Another particular aspect of autonomous systems in the field of building automation is that they should be "invisible" most of the time. This means, that, ideally, they will be exerting a more passive than active behavior. Actions will only be taken in case something goes wrong.

What we will take up from biology is a hierarchical, modular structure of control loops. Although, at the beginning, we will try to keep things as simple as possible. Only some parts of the whole complex biological control architecture will be worked out in more detail, and for this, only a selected set of design principles inspired by nature will be picked up.

Ortony, *et al,* (2002) distinguish three levels of information processing for behavior selection for cognitive architectures:

- *Reactive processing:* This is the simplest form of producing behavior. Especially in emergency situations, hard-wired action patterns shall be released with as little delay as possible. Those actions can be for example alarm signals.
- *Routine processing:* On this level, primitive, automated behaviors are executed. They are initialized by stimuli which trigger them. In the simplest form, this happens in an unconditional way.
- *Reflective processing:* Here, explicit hypothetical representations of possible alternative behaviors are compared and the most suitable one for fulfilling the current goal is selected.

At all these levels, the concept of *emotion* – in all its variants – can be of great use for the evaluation and action selection process. According to Arbib and Fellous (2004) a key function of emotion is to communicate simplified but high impact information.

Simple signals with no or only very little semantic content, like the *alarm signals* of the reactive layer, can be viewed as the starting point for emotions. They are the most rudimentary form of an affect, communicating among the parts of a system the occurrence of a special situation and preparing them for the release of an immediate, coordinated action. For an "alarm system" to work in a reliable way, there must be some characteristic sensor inputs for each alarm condition. Otherwise, those conditions cannot be detected and set off correctly in time.

In biology, creatures can reduce the probability of being surprised by an unpleasant situation by drawing their attention to a possible source of danger as soon as the fist hints of its occurrence show up. Whether in nature or in engineering, resources are always limited and not just processing time. Therefore, a good mechanism for *attention allocation* can be very useful, if not essential. To focus the system's attention on the currently most important aspect of the environment helps to reduce the large amounts of sensory data continuously streaming into the system. In this case, "emotions" can be functionally seen as state-evaluative processes that adjust the relative weighting of different behavioral modes, raising the urgency level for one mode while lowering it for others, depending on the context. Again, this shift of mode is triggered by some stimuli, but here things are more complex than in the purely reactive case. First, a complete routine sequence of actions can be activated with one single stimulus. Second, emotions on this level are no longer simple signals any more. Instead, there is a more elaborated control loop, based on various *drives* or *motivations*, each of them possessing a range of possible values, indicating the respective current activation level. Depending on this, some modes of behavior are favored compared to others. Successful behavioral commitments are rewarded by an intensifying mechanism – referred to as "lust" – inappropriate ones are punished with "pain", which leads to their avoidance. This mechanism introduces internal "sensations" into the system. They can be compared to sensory inputs stemming from the outside, only that they are providing the system with information about internal constraints, not external.

At the level of reflective processing, the behavior selection control loop is even more sophisticated. External and internal states are monitored, evaluated and associated to each other. Some aspects of these processes have already been addressed in section 4. This is because the stages of perception, representation and evaluation cannot always be clearly separated and run through in a subsequent way. They are continuously updated in a cyclic way and influence each other. So, the symbolic representation of the currently perceived scenario is compared with former scenarios stored in an *episodic memory*. Additionally, knowledge from a *semantic memory* is taken into account. Thereby, not only information which is currently active influences the scenario evaluation process, but also general knowledge about the relations of the world. The aim of the comparison process is to find similar previous situations. Again, there has to be a mechanism narrowing the focus of the ongoing processes. Important features have to be highlighted, unimportant ones ignored. Not every detail of the potentially available data is considered. The mechanism looks for characteristic forms, and after having detected some of them, the rest is associated by using previously memorized scenarios.

Moreover, memories are not stored neutrally. Instead, every memory is afflicted with an "emotional" value. These ratings help to guide the evaluation process. They enable a simple, short-cut form of cognitive reasoning. To equip the system with emotional ratings allows it to distinguish between "desirable" and "harmful" situations and behaviors without the necessity of having all of them explicitly experienced or carried out before. It is a mechanism for a rough, predefined rating and selection of necessary actions, which works with some probability within a variety of complex and unforeseen situations.

When talking about "desirable" and "harmful", these terms do not necessarily have to refer to the system *itself*. They can, and will in most cases, refer to the tasks the system has to fulfill. For example, if it has to take care of a person, as in some of our applications, every situation putting the protected person into danger will be classified as "not so good" and get a negative emotional rating.

6. IMPLEMENTATION

The system, as we have pictured it in the previous sections, contains several different modules, which have to cooperate. The central component is a database, which acts as an information sink for all kinds of information. For once there is a database that collects all data that is generated by the sensors. This is necessary because we assume that we are not able to process all data in real-time, at least not in the prototype implementation. To free development of the requirement of timeliness, all sensor data is stored in one database scheme. There is a software module that takes this sensor data and creates μSymbols and symbols. These μSymbols and symbols are also stored in a database scheme (logically this is another database that has no knowledge of the sensor database; just for reasons of administration, both are running on the same physical machine). Because permanent polling of the database for new sensor data is impractical, we have introduced a module that receives updates, writes it to the database and at the same time informs all registered modules about the new data. This way we avoid polling, reduce data traffic, but still maintain consistency, since the database is still the central data storage. This module is transparent to applications, since they only receive data updates, no matter if they were polled or pushed.

Aside of the update pushing the modules can obtain the current and historical state of the world representation by using database queries. This way we get a defined and decoupled interface between the different modules. Perception is done separately for each of the four applications mentioned in section 2. This will have to be changed and the perception module will generate a common set of symbols, which can be used by all applications.

Since the system shall be able to deal with a big amount of sensor information, we have to provide the system with this amount of information. Unfortunately a real world installation with such a great number of sensors is currently not feasible. Therefore, we have decided to introduce a *simulator* (Rösener, *et al.*, 2004). This module serves two purposes: For once, it is able to generate a virtual environment, where sensors can be installed in big numbers. This environment is then filled with objects and persons that interact with the environment. The simulator calculates these interactions and creates virtual sensor data out of it. This data is stored in the database, so that other modules notice no difference between simulated and real world data. The second purpose of the simulator is visualization: It shall not only display the virtual environment with all its ongoing actions, but also visualize the system's world representation of this virtual environment. This way we can judge the quality of the world representation and detect gaps between what the system "perceives" and how the world "really" is.

7. CONCLUSION

We have presented a system that is envisioned to bring automation to the next level by employing bionic models for the control of a technical automation system. The concepts for this system have been defined, and a first implementation is currently in prototype phase. Using existing algorithms for classical image processing and other basic functionalities of the system, we are able to rely on state of the art technology to build a system that implements perception, scenario recognition and evaluation by the means of "emotions" to judge and qualify scenarios.

The steps for the near future include an increase in predefined symbols to extend the perception of the system. The current set is very limited and will be enhanced. Moreover, the existing symbols need to be extended by adding more properties and more associations. Following this, we need to find means to automate symbol generation and association. Currently this knowledge is "hard coded" into the system, but another group of the ARS project is currently doing research on this topic.

Additionally, it will be necessary to introduce context-dependent perception. So far, there is no influence between perception and higher-level mechanisms of the system – the system always perceives the same, no matter what its current goals are. We know however that perception in humans is strongly influenced by the current goals and we will have to model this.

What will remain is that the system will make errors when confronted with unforeseen events. As long as we do not introduce, for example, dogs into the system, it will always mistake a dog for something else (a person) or simply ignore it. The same applies, if a person quickly changes its outer appearance. The system will mistake this person for a new person, assuming that it strongly relies on color histograms. This behavior is intended and can also be found in humans – only on a much more sophisticated level.

8. REFERENCES

Arbib, A. and J.-M. Fellous (2004). Emotions: from brain to robot. *TRENDS in Cognitive Sciences,* **8**, pp. 554-561.

Brainin, E.; D. Dietrich, W. Kastner, P. Palensky, C. Roesener (2004). Neuro-bionic Architecture of Automation – Obstacles and Challenges. In: *IEEE Africon 2004, Gaborone, Botswana*. pp. 1219-1222.

Brenner, A. (2001). *Performanceverbesserung bei der automatisierten Einzelsortierung mit Zeilenkameras in Echtzeit.* Dissertation at the Vienna University of Technology, Institute for Automatisierungstechnik.

Dietrich, D., W. Kastner, H. Schweinzer (2004). *Wahrnehmungsbewusstsein in der Automation - ein neuer bionischer Dekansatz.* In: Automatisierungstechnik, **52.**, 3, pp. 107-116.

Damasio, A. (2003). *Looking for Spinoza: Joy, Sorrow, and the Feeling Brain.* Harvest Books.

Hauser, K., A. Bittleston, A.D. Gibson, W.B. Forward (1995). *Kaspar Hauser Speaks for Himself: Kaspar's Own Writings.* Camp Hill Press.

Johnson-Laird, P.N. (1988). *The Computer and the Mind: An Introduction to Cognitive Science.* Harvard University Press.

LonMark (2005). *LonMark Magazine, European Edition.* TEMA Technologie Marketing AG, Germany, **2**, 4.

Pratl, G., W. T. Penzhorn, D. Dietrich, W. Burgstaller (2005). Perceptive Awareness in Building Automation. In: *Proceedings of the 2005 IEEE International Conference on Computational Cybernetics (ICCC 2005), Mauritius.*

Tamarit, C., D. Dietrich, K. Dimond, G. Russ (2001). A Definition and a Model of a Perceptive Awareness System (PAS). In: *Proceedings of the IFAC FeT 2001.* pp. 177-182.

Harnad, S.: The Symbol Grounding Problem. Physica D 42. pp. 335 – 346, 1990.

Ortony, A., D.A. Norman and W. Revelle (2005). Affect and proto-affect in effective functioning. In: *Who Needs Emotions? The Brain Meets the Robot* (J.M. Fellous and M.A. Arbib (Ed)). New York, Oxford University Press.

Riedl, R. (1984). *Biology of Knowledge: The Evolutionary Basis of Reason.* John Wiley & Sons.

Solms, M. and O. Turnbull (2002). The Brain and the Inner World. An introduction to the neuroscience of subjective experience. Karnac Press.

ELSEVIER
IFAC
PUBLICATIONS

SIMULATION AND VISUALIZATION SYSTEM FOR SENSOR AND ACTUATOR DATA GENERATION

Harald Hareter[1], Gerhard Pratl[2], Dietmar Bruckner[3]

[1] *harald.hareter@arcs.ac.at, Information Technologies, ARC Seibersdorf Research GmbH*
[2] *pratl@ict.tuwien.ac.at, Institute of Computer Technology, Vienna University of Technology*
[3] *bruckner@ict.tuwien.ac.at, Institute of Computer Technology, Vienna University of Technology*

Abstract: The evolution of automation systems has reached a point where the increasing complexity requires new mechanisms to keep control of design and operation. In future all control systems will have to cope with massive data quantities that are collected from the environment. Such systems have to react based on the recognition of whole scenarios. In this paper we present a simulation tool, which is able to generate big amounts of sensor data in an animated, virtual environment. This data is used by two different systems to verify their accuracy. *Copyright © 2005 IFAC*

Keywords: Sensors, Simulators, Modelling, Redundancy, Data processing

1. INTRODUCTION

The number of sensors within control systems has increased dramatically in the last years. Nowadays office and industrial buildings are equipped with up to 70.000 nodes (Kastner, *et al.*, 2004). We expect this development to continue, assuming that prices for networked sensors will decrease and the request for more data will increase. In almost the same way the different types of sensors will also increase. In future control systems need to cope with massive data quantities collected from their environment.

Unfortunately, present automation models do not seem to be ready to handle such masses of information. In order to process the expected quantities of data we see the need for new data processing models that have to be integrated into a control and automation system.

Such a system shall not operate on single values, but has to react based on the recognition of whole scenarios. At ARC Seibersdorf research and the Vienna University of Technology two teams are working on two different approaches that should be able to cope with this task. The first one, called Artificial Recognition System (ARS), is based on a biological archetype, which abstracts sensor information to symbols (Pratl, 2005). The second approach, named Building Assistance system for Safety and Energy efficiency (BASE), processes sensor data with statistical methods.

As shown in (Russ, 2003) such systems require massive amounts of data; however, it is not feasible in today's situation to build a real world installation with some thousands of sensors that provides the required quantities of data for proving the concept. Sensor systems containing these amounts of sensors are very difficult and costly to realize; thus simulation is a reasonable alternative to get suitable inputs for studies of the system behaviour. Additionally, simulation may not only predict the behaviour of a model, but might also provide additional information about how to improve an installation later on (Bratley, *et al.*, 1987).

In this paper we focus on the description of a tool, which is able to simulate sensor outputs from millions of sensors and actuators triggered by different kinds of scenarios and to find the optimal amount and placement of different kinds of sensors and actuators.

The simulation used here is basically divided in models of local interaction for e.g. motion detection and models of global interaction for e.g. temperature monitoring.
The rest of the paper is organized as follows: in section 2 we describe the software setup that builds the basis for the simulator, section 3 describes the procedures and algorithms of the simulation. Section 4 describes two applications for which the simulator is currently used, in section 5 we demonstrate first results and section 6 rounds up the paper by giving the conclusions and providing a short outlook of further developments.

2. SIMULATION ENVIRONMENT

2.1 Requirements

The goal of this simulation tool is to generate the output of different types of sensors that are triggered by predefined scenarios.
Therefore the tool requires knowledge about the environment, which in the first step should be an office building or parts of it. Furthermore type, position, direction and other physical properties of the installed sensors have to be given. Finally, the scenarios have to be defined. This means that the position and physical properties of objects within the building, positions, speed and attributes of persons moving around in the building and the change of physical properties like temperature, humidity or lightning have to be specified.
To arrange these settings we developed a scenario generator which is used for sensor placement and activity settings as well as manipulating physical properties in the building. The simulation tool then calculates the sensor outputs and stores these data in a database. This sensor data is used by different applications (see section 4) which, for example, condense data to symbols, which are then also stored in the database. The last part in this system is the visualization tool which presents sensor data and symbols to the user.
Furthermore the operator should be able to interact with the simulator. This requires real time calculation of sensor data and therefore algorithms with reduced calculation effort, an efficiently designed database-schema and a user interface, which is able to provide only data that are currently relevant.

2.2 Learning about the environment

The basis for the simulation environment is a plant layout of a building or a factory hall, which is in most cases provided by the architect of the building. Because of the diversity of architectural Computer Aided Design (CAD) programs and the multitude of file formats in this special field it was impossible to use the plan layout directly. It has to be converted into a suitable file format. We have decided to use the .X-file format defined by Microsoft (Microsoft, 2004) because it is widespread, easy to parse and there are lots of tools to convert other CAD-files into the .X-file format.
The .X-file consists of templates, which contain 3D mesh data, position data, material data and so on for each object (e.g. walls, windows, doors and other content). Furthermore it is possible to define templates for additional data.
Unfortunately the generated .X-files contain only information of objects like walls, windows, doors and so forth which are the building's constituents. It does not contain any information about the building's structure like floors or rooms. We assume that the .X-file contains only information about objects belonging to the building itself and does not include any furnishing objects.
The first step is to extract structural information about the building from the given object information. This is done by dividing the parts to different floors using compare and sort algorithms. Thereafter the parts are combined to rooms by finding neighboring edges. Finally objects, which belong to walls like windows and doors are integrated into the structure.
The information about rooms will be stored in different data structures. The first one consists of two vectors, which contain the location of two opposing corners of a cuboid which is parallel to the axis of the building and encloses a room. The second structure contains information about the centre, the dimension and the direction of a cuboid enclosing a room. The latter is a linked list, which includes all objects belonging to that room. This diversity of data storage is used to save calculation time during different sections in the simulation.
Unfortunately the algorithm is not able to find all necessary structural information about the building itself. Therefore, in a first approach, the scenario generator (described in section 3.1) has to be used to verify, update and correct the information extracted from the plant layout. This information is stored in another .X-file which contains all additional data that is used for the simulation of the building. Additionally, the whole building data is stored in a database to provide the environment information for the applications using the simulation output.

2.3 An object oriented approach

Every item in the building has different kinds of properties like position, size, material and so forth. Therefore an object oriented approach was chosen for internal representation.
We distinguish between different kinds of objects. The global object in the simulation is the building itself which features global properties like time, environment temperature, outdoor lightning and so on. This main object also contains objects like floors, rooms, walls, etc. Additionally, furnishing and people are also represented by objects but do not belong to the building object.
Each object has different kinds of properties to represent its physical conditions. Universal properties like position belong to every object within the simulation. The availability of other properties like temperature

or weight depends on the physical model and the required simulator output. If, for example, the physical model in the scenario does not include the calculation of temperature there is no need for managing these properties. That is to say that depending on the physical conditions of the scenario the properties of objects were extended on runtime.

2.4 Physical model and sensor types

In the first step the simulation is able to handle the following types of sensors: motion detectors, light barriers, door contacts, door opener buttons, pressure sensors, temperature sensors and luminance sensors.
This sensor types are divided into sensors which are mainly affected by movement of objects, like motion detection or light barriers, and sensors which are affected by the change of physical properties like temperature or luminance sensors.
This is done because of the high amount of sensors. It is not feasible to calculate the output of each sensor in each simulation step. Another matter for this separation is that calculation of sensor output based on movement can be done by comparing the position of the moving object and the sphere the sensor covers. It does not need any physical model for calculation.
Let's first focus on sensors affected by movement. To reduce computation effort the environment is divided into areas of interest. This means that only regions of the building which contain movable objects are involved in the calculation process. Each area belongs to a single room, the base is rectangular and parallel to the x-y-Axes and covers the whole height of the room. The base's size is given by the operator (this means that the size is constant for each area in a room) or could be dynamically calculated depending on the size and the shape of the room and the sensors integrated into the room.
Another method of reducing the amount of sensors involved in the calculation is to give the moving objects an area of influence. This means, each object is surrounded by an active area and sensors with active spheres within this area are included into the calculation process. This method turned out to be unsuitable for simulation a big amount of moving objects because of the high effort for updating of areas.
The second kind of sensors like temperature and brightness sensors require a physical model for calculation. These models have to be as simple as possible for fast calculation, but on the other side must be accurate enough to meet the demands of the application. Therefore the models are separated from the main program and were included into discrete modules. These modules were linked on demand and an interface handles the communication between the main process and the modules. This technique allows the use of different models for different requirements to handle the trade off between computation speed and precision. Furthermore this is an easy method to extend the number of the physical models without recompiling the simulator.
The formalism to calculate thermal conduction in a wall using fouriers law, for example, is stored in a module. For more precisee calculation another module based on finite element calculation can be used by linking the temperature model with this module.
A second separation is done by using the time basis of the sensors. Some types of sensors have to be updated more often than others. The refresh times for sensors, which are triggered by movement have to be fast enough to recognize all movement. The change of the room temperature caused by a heating element in the room is slow-going and therefore the recalculation rate can be low. In the simulator the time base for each sensor type could be defined by the operator or calculated by the program. The lower bound for recalculation of fast sensors depends on the amount of sensors and objects in the building. This time is approximated by a calculation step on start-up by repeating a calculation of an active area including an above-average number of sensors for each movable object. Additionally, the time for recalculating the active areas has to be added.
For slow changing properties like temperature the time basis for each room is derived from the given gradient. It is also possible to define local areas within rooms with other recalculation times. These areas were associated to objects like a stove or other heating elements.
During each calculation cycle the output of affected sensors is stored together with a timestamp in the database. In the same way the environment properties as well as the properties of other objects are stored in the database.

2.5 The user interface

One of the main questions in this project was how to present such an amount of sensors and sensor data to the user. We have decided to use different kinds of views of the building and the sensor network.
The basic view is a 3D-view of the whole building in which the operator is able to rotate the building and zoom into different floors and rooms down to single sensors or sensor groups. Starting from this view the operator is able to select parts of the building and handle them as one complete structure (e.g. by selecting a single floor the rest of the building will be hidden). It is also possible to change to 2D-view (for example a top view) or to interactively move through the building by using a camera view. The presentation of the building can be done in solid mode or in wire frame mode. The first one gives a nearly photorealistic picture of the environment including textures and shade effects. This kind of visualization is mainly designed for demo purpose. The second kind gives the object as wire frame only. For an easier differentiation of the miscellaneous objects each object group is represented by a different color.
In addition a tree view of all objects is included. The root is the environment object, which is divided into room and floor objects. The bottom layer consists of sensor groups and sensors. The object's properties can be displayed and if necessary modified.

In 2D or 3D view the operator is able to check the positions of sensors or sensor groups and can hide sensor types that are currently out of interest. If objects in the simulation interact with sensors and the sensor output changes there will be a short highlight at the sensor position to show the change. Other physical properties like temperature pattern are shown by false color representation. Additionally, all sensor data can be listed in table format to get an overview of all changes.
Another feature of the visualization is to display symbols which are derived from sensor data by different applications. The graphical representations of each symbol are integrated into the visualization by overlaying them.
Simulation and visualization do not necessarily have to run on the same computer. On start-up the operator can select the operation mode: simulation only, visualization only, or both. This separation increases the calculation power and decreases calculation time.

3. MODELLING AND SIMULATION TOOLS

Before the simulation process can be started the building has to be equipped with sensors and actuators and the scenarios, which shall be simulated, have to be defined. The operator is able to change properties of the environment at runtime and shall also be able to easily monitor and check the output.
Therefore the whole simulation system consists of a scenario generation interface for sensor placing and scenario definition, a simulation interface and a visualization interface. In the next section a closer look on each of these interfaces is provided.

3.1. The scenario generator interface

As described in section 2.2 the main object in the simulation process is an office building. The structure of the building is provided by a plant layout. Since the algorithms used so far are not able to recognize the building structure completely the scenario generator is used to remodel the structure.
After loading the .X-file the scenario generator offers a 3D-view of the building and a list of all identified objects like rooms and floors. The operator is able to select objects from the list, verify properties and change it. If, for example, a room object contains a wall object, which does not belong to the room the operator can delete the wall object from the room's object list.
The next step is to place equipment into the building. Objects are loaded from an equipment library, which at this stage includes objects like desks, chairs, cupboards and so on. The library objects can be built with CAD or render applications, which are able to export or convert the object information into .X-Files. This library will increase with every scenario which needed new equipment.
After the building is furnished the scenarios can be defined. This includes the definition of placement and moving paths of objects, the initial value and change of physical properties of the environment and objects as well as the interaction between objects. If there are few objects and properties these settings can be done with the scenario generator. A person, for example, could be placed into a room using drag and drop and the movement path could be established by mouse movement.
To provide a slightly easier way for placing a big amount of objects a set of instructions (similar to a programming language) was defined. With this instruction set a placement file can be created with a text editor including all necessary information.
Movement of objects can thus be defined by assigning the starting point, the starting time, the direction and the speed of the object. Another possibility is to define the starting point, start and end time of the movement and the direction. A further variety is to define start and endpoint, the starting time and the speed of the object. In this stage of implementation this works well if the moving object does not leave the starting rooms. To solve this problem, other path finding algorithms have to be implemented.
The change of environment properties can be done in the same way.

3.2. Sensor placing

The applications described in chapter 4 try to perceive information about what is going on in a building by processing data supplied by sensors. Therefore the sensors have to be placed in such a way that they are able to gather a maximum of information out of the environment. In present installations sensors are placed by experts with reasonable know-how and sometimes it is a cost-intensive trial and error process to get an optimal placement. With this simulation tool the optimal placement of sensors can be easily determined. Furthermore, the minimal number of sensors for detecting environment behaviors can be acquired.
As mentioned above this simulation tool is able to handle hundreds of thousands of sensors and therefore a placement by hand is impossible. The vision is that in near future a great number of sensors is integrated into the furnishing, carpets or wallpapers (Vorwerk, 2005). Therefore one possibility to easily place many sensors is to define sensors placed on a grid and assign these sensors to an object. This could be done by selecting an object (or parts of it) in the scenario generator interface, add a sensor property and set the intervals in x, y and z direction. The sensors are than attached to the surface of the object. In this way the floor of a room for example can easily be equipped with pressure sensors.
The placement of motion detectors, door contacts or light barriers needs other methods because they have to be directed to an area of interest. In most cases these kinds of sensors are not so numerous and therefore a placement by hand is feasible. If, for example, every door within the building should be equipped with door contacts an easier workflow has to be designed. There are few thoughts how this could be done, but in the current stage of implementation this features are not built-in. The placement algorithms for

sensors on surfaces have also to be overworked because the arrangement of the sensors is not optimal. Primarily on uneven surfaces and on edges the placement could be improved. Therefore the use of other placing techniques like grid based algorithms (Lin, Chiu 2005; Dhillon, Chakrabarty 2003) has to be researched.
Another feature, which will be implemented in near future is to change the granularity of the sensor fields in areas of special interest. This could be done by a simple tessellate algorithm. This algorithm is useful to find the optimal amount of sensors to recognize different effects.
Furthermore the implementation of a virtual sensor is planed. This tool will allow the operator to try out the effect of sensors on different positions without changing the configuration stored in the database

3.3. The simulation and visualisation interface

After the scenarios were created they are loaded into the simulation environment. Here all objects are listed in a tree view and the sequence of actions take place during simulation time are displayed on a time line. Here the operator is able to make changes on start up properties of objects or reschedule the order of actions.
If initialization is finished the operator can chose to start the visualisation and simulation on one computer or source out the visualisation. The IP address of the visualisation computer and the computer hosting the database can also be modified.
During simulation the operator is able to switch between the different types of visualisation.

3.4. Process communication

As mentioned above the simulation and the visualization process do not have to run on the same computer. The data processed by the simulation is stored in a database and the visualization or another application (such as ARS or BASE) gets the data from the database.
Every time the simulation writes new data into the database the other applications have to be informed to fetch these new data. The database is not able to inform the applications and polling the database is inefficient. Therefore we use a synchronization module, which can also run on a separate computer; it handles all the communication between the applications. This means that the simulator sends the data to the synchronization module, which again hands the data to the database and informs the other applications. The applications send a request to the synchronization module and receive the data in return.

4. APPLICATIONS

Two examples shall be given to demonstrate the necessity of sensor input generation in large quantities. The first one is the Artificial Recognition System (ARS), a cooperation project between ARCS Seibersdorf and Vienna University of Technology. ARS makes use of diverse, redundant sensory inputs in a bionic way to create a perception of the surrounding world (Roesener, *et al.*, 2004).
In order to create a system that is able to perceive its environment in a way that is similar to the processes that occur in the human mind, a broad base of sensory input is required. Perception at first depends on vast redundancy. Therefore we need to take all available sensor data and extract the information that is important for the functionality of the system. This data input is condensed by means of *symbolization*. A symbol is a piece of information, for which the system has a concept and is therefore able to process.

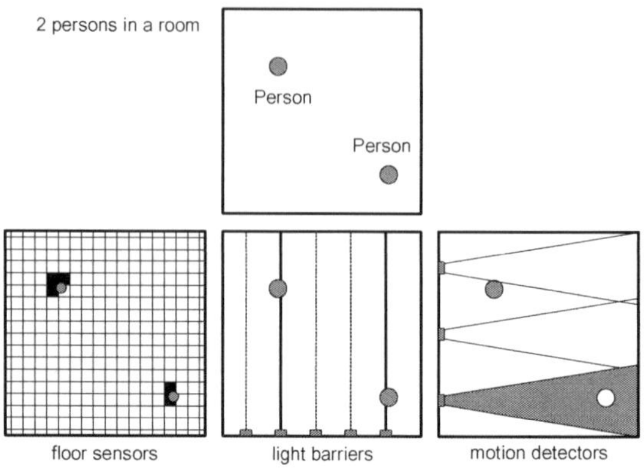

Figure 1: Redundant information is used to determine the position of persons

A simplified example is shown in Figure 1: the top square shows a room with two persons. Pressure sensors in the floor provide information about the position of the two persons. This information is also available from the light barriers and the motion detectors. By using all this redundant information, the ARS system can deduce the current position of each person, even if some sensors cannot contribute to the position, because they are obscured by other objects or they are faulty.
This person surveillance is the basic application that enables other applications built on top of it. These applications are, for example, a geriatric care system that supervises elderly people and recognizes possibly dangerous situations.
Although we expect the costs for sensors to decrease significantly in the next years, we are today unable to provide the amount of sensors that are envisioned in the ARS project. We estimate that it will be necessary to have some ten thousand sensors of different domains in each building. This is the point where we expect classic building automation to be unable to exploit the full potential of the available installation. Therefore we look at such a setup today, however without having a real installation available.
The solution is to use the simulator to apply the bionic model of the ARS project on the simulated environment and the available sensor data.

The second example where a large quantity of sensory input is used is the in-house project Building Assistance System for Safety and Energy Efficiency (BASE). BASE makes use of numerous sensory input devices to generate a statistical model of what happens "normally" in a building. The knowledge of normal sensor behaviour is used to predict system behaviour in the near future and to recognise sensor malfunction.

In order to model the building's behaviour, it is necessary to have huge numbers of sensors. For building up models that are capable of representing useful situations in, for example, an office building, pressure sensors in the ground, light barriers, motion sensors, door contacts and sensors for temperature, light, humidity as well as measures of air-quality and particles in the air are necessary. All factors that cause humans to change their behaviour have to be observed in order to have the chance to predict their behaviour.

An example scenario in a kitchen in a dwelling house with unlikely sensor behaviour could be as follows: The system detects the furniture, a person and the stove because of the pressure sensors in the ground and the behaviour in the past by comparing it with templates for tables, chairs, cupboards and so on. This means the system's representation of the room is initialized and stable.

In the current example a person lingers in front of the stove for some time and the temperature in that area rises. From previous observations the system knows that in case of the detection of a lot of steam and temperature around the stove, the person turns down the stove. If, for some reason like in Figure 2, the person leaves the room while cooking and a pot overflows, the system computes very unlikely probabilities for the combination of temperature, steam, stove control and presence and can therefore launch an alarm.

Another possibility is that one of these values is incorrect due to sensor faulty. In this case the system expects different values than it gets and launches an alarm.

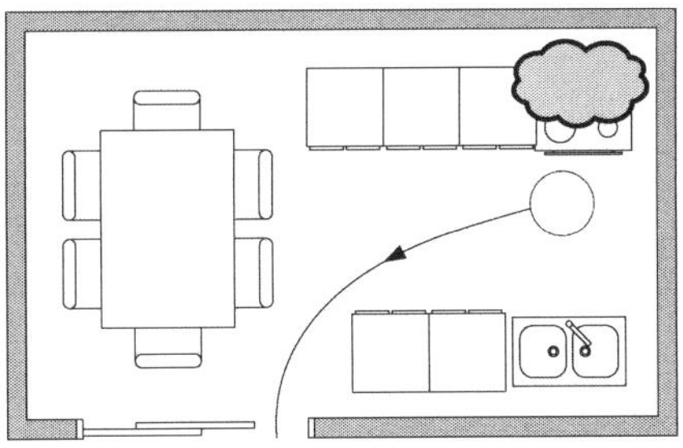

Figure 2: An alarm generating scenario: After some cooking the person leaves the kitchen. One plate is heated too much and therefore steam appeared. The system computes very low probability for the combination of temperature, steam, stove control and presence of a cook and launches an alarm.

5. FIRST RESULTS

The test object is an office building with 6 floors. The floor space on each level is approximately 3200 m^2 with gives a total area of 19200 m^2. Each floor consists of 53 to 78 rooms including halls, giving a total of 395 rooms. The mean amount of sensors per room is nearly 1200, the most of them being pressure sensors on the floor. The total amount of sensors is almost 500000.

Each room is equipped with office furniture like desks and cupboards. The scenario consists of 1500 persons moving on random paths through the building. Additionally, the temperature level in each room is different on start-up and the temperature equalization is calculated.

The test system consists of a standard office PC which includes a 3 GHz Pentium 4 Processor,1 GB of Ram and a NVIDIA Quadro PCI-E video card.

Running the visualization interface on the same computer as the simulation tool make the visualization a bit slow in solid mode. This is due to overload of the graphic hardware which has to calculate the shading for each lightning source in every visualization cycle. Switching to wire frame mode let the visualization run more smoothly.

The time basis for fast sensors was fixed to 50 ms which demands 20 recalculations per second. The mean amount of sensors recalculated on each calculation step was nearly 600 with gives a total of 12000 calculations per second. Here the transmission of the data into the database shows a possible bottle neck. The used MySQL-Server was overstrained.

Another weak point of the simulation tool is the memory usage. On runtime the simulation uses up to 520 Mb. A closer look on the memory management is therefore necessary.

6. CONCLUSION AND OUTLOOK

We have presented the concept of a simulation tool, which is able to generate the output of a large number of sensors that are triggered by simulating different scenarios. These sensor data are the basis for the evaluation of new models for future automation systems based on statistical models or biological archetypes. These systems should be able to recognize the simulated scenarios and react in a given way.

Most of the presented concepts were integrated in the simulation so far. Some of them have to be improved or replaced. Some are still concepts and will be implemented in near future. The first steps have been done, and there are some to do to be ready for future requirements.

REFERENCES

Bratley, P., B.L. Fox, L.E. Schrage (1987). *A guide to simulation,*. (Edition 2) Springer, New York

Dhillon, S.S., K. Chakrabarty (2003). Sensor placement for effective coverage and surveillance in distributed sensor networks. *Wireless Communications and Networking, **Vol. 3,*** pp. 1609 - 1614

Kastner, W., P. Palensky, T. Rausch, CH. Roesener (2004). A Closer Look on Today's Home and Building Networks, *7th AFRICON Conference in Africa, **Vol. 2,*** pp. 1239 - 1244

Lin, F.Y.S., P.L. Chiu (2005). A near-optimal sensor placement algorithm to achieve complete coverage-discrimination in sensor networks. *Communications Letters, **Vol. 9, Issue 1**.* pp. 43 - 45

Microsoft (2004). X File Reference. http://msdn.microsoft.com/archive/default.asp?url=/archive/en-us/directx9_c_Dec_2004/directx/graphics/reference/d3dxfile/xfilereference.asp

Pratl, G (2005). A Bionic Approach to Artificial Perception and Representation in Factory Automation. EFTA 2005

Roesener, Ch. Hareter, H. Burgstaller, W. Pratl, G. (2004). Environment Simulation for Scenario Perception Models, *5th IEEE International Workshop on Factory Communication Systems*, pp. 349 – 352

Russ, G. (2003). Situation-dependent Behavior in Building Automation, PhD Thesis, pp 140

Vorwerk (2005). Intelligent carpets as a navigation system for robots. http://www.vorwerk-teppich.de/sc/vorwerk/rfid_en.html

ELSEVIER
IFAC
PUBLICATIONS

ANALYSIS OF NETWORKED CONTROL SYSTEM WITH PACKET DROPS GOVERNED BY (*m,k*)-FIRM CONSTRAINT

Ning JIA[1], Ye-Qiong Song[1] and Rui-Zhong LIN[2]

1. LORIA – INPL
Campus Scientifique - B.P. 239
54506 VANDOEUVRE-lès-NANCY, France
{Ning.Jia; song}@loria.fr

2. ZHEJIANG UNIVERSITY
HANG ZHOU
310027, Chine
rzlin@iipc.zju.edu.cn

Abstract: In this paper we study the effect of the packet drop process governed by (*m,k*)-firm constraint in the feedback loop of a control system. Our approach is to first analyse the stability and the optimality of such a system, then by considering the amplitude variance of the system state as the system performance, we propose a method to optimally distribute the packets to offer a better system performance. *Copyright © 2005 IFAC*

Keywords: Networked control system, real-time, (*m,k*)-firm, packet loss, stability, optimization

1. INTRODUCTION

Much attention has been paid recently towards the networked control systems (NCS). In such systems, the feedback loops are implemented over a network link. The measurements quantized by the sensors are sent to the controller over the network link. After the computation of the controller based on these measurements, the control output is then sent to the actuators via the network. As the network link is often shared among different applications and control loops, the network overload may occur and must be managed. During network overload, some proposed solutions discard data packets in order to reduce the effective utilization of the system. Among these solutions, the technique often mentioned is (*m,k*)*-firm* policy (Ramanathan99) (Quan00). The (*m,k*)*-firm* guarantee model requires that at least *m* out of any *k* consecutive packets must be delivered by the network. An interesting question is: How to choose the rate of *m* over *k* so that the system controlled remains stable and how to distribute the packet drops in the network so that the performance of system can be improved. This paper provides the answers to these questions.

There has, however, been relatively little work studying the impact of packet drop on the control system performance. Most of the previous work has only focused on the impact of packet drops on overall system stability by assuming certain statistical drop models. In (Walsh02), a protocol to schedule the use of shared resources is proposed and the stability of such a networked control system for a continuous plant and a continuous controller is studied. In (Zhang01), the stability of the control system that is modelled as asynchronous switching system is studied under the network induced delay and dropped packets. An approach for data loss compensation in the network link was proposed in (Nilsson98), where the dropouts are governed by a Markov chain and are treated as vacant sampling. The author then presented an analysis of system stability. In (Seiler01), author investigated the

This work has been partially supported by Franco-Chinese research program PRA SI03-02 stability of a linear networked control system in which the packet drop process is i.i.d (independently and identically distributed). Analysis of the performance with data losses replaced by zeros has been made in (Hadjicostis02). In (Ling02) and (Ling03), performance of the networked control system is characterized by the output signal power. By assuming a certain statistical drop model, the performance of system that is defined as a function of packet loss rate is investigated. In (Liu03), to provide a better system performance that is measured by the output signal power, the author introduced a Markov chain constraint (MC constraint) to represent the optimal dropout process and designed scheduling approaches with respect to this MC constraint. Unfortunately, none of the above solutions answers to our question.

In this work, the system setting is much like the setting in (Hadjicostis02), but the packet drop process is quite different. We are specifically interested in control systems communicating over network link which selectively drop packets according to the (m,k)-*firm* constraint. This selective packet dropping mechanism could be implemented at the communication nodes (switch or router) when a packet switching network is used (e.g. switched industrial Ethernet or IP network). The variance of the system state is used as a criteria in the first part of this paper to investigate the stability and optimality of control system, then we will introduce a metric that relates the system performance to the packet drop sequence, and propose an approach that uses the method of Lagrange multiplier to improve the system performance by rationally organizing the drop distribution.

The rest of this paper is organized as follows. In section 2 we describe and formalize the system under study. Section 3 states the necessary and sufficient condition of the stability and gives the interval of the gain of controller and the parameter of network (represented by (m,k)-firm constraint) in order that the variance of the system state remains finite in long run. Section 4 presents a method for minimizing the variance of the system state. In section 5, we will consider the amplitude variance of the system state as the system performance and give a method based on Lagrange multiplier to improve the system performance. In section 6, we will give some numerical examples. We conclude our work and show the perspective in section 7.

2. SYSTEM MODEL

In this paper, we will consider one-dimensional discrete linear time-invariant system with a communication network in the feedback loop (see Figure 1). The system we consider are of the form:

$$x[t+1] = \alpha x[t] + \beta v[t]$$

where α satisfies $|\alpha| > 1$ and β is a nonzero constant.

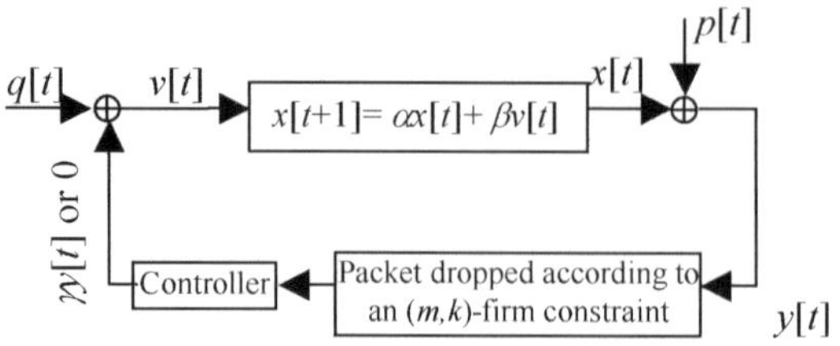

Figure 1: Feedback loop with a network

The measurement of the system state $x[t]$ is quantized as $y[t] = x[t] + p[t]$, where $p[t]$ is the quantization noise. We will assume that this quantized measurement $y[t]$ is sent in a data packet through the network, and the controller has access to the communicated data packet. Specifically, in our work, the data packets are dropped selectively in the communication network during the overload period, and the packet drop process is governed a (m,k)-*firm* constraint, i.e., at least m out of any k consecutive packets sent from the plant must be received by the controller. For analyzing the system, we placed our selves in the worst-case by assuming a long network overload period and the network drops systematically k-m packets every k. So the packet delivery sequence is periodic, i.e., if the nth packet is dropped, the $(n+k)^{th}$ packet will be also dropped; otherwise, if the nth packet is received by the controller, the $(n+k)^{th}$ packet will be also received. If the packet containing $y[t]$ is available to the controller, $\gamma y[t]$ for some gain γ will be determined and be sent to the actuator. Otherwise, the controller outputs zero. We assume that there is no communication link from the controller to the actuator. Therefore, the command sent from the controller is received directly by the actuator with a driving noise $q[t]$, i.e., $v[t] = \gamma y[t] + q[t]$.

The system then can be evaluated as following:
If a packet is received by the controller, the system state evolution can be expressed as:

$$\begin{aligned} x[t+1] &= \alpha x[t] + \beta[q[t] + \gamma(x[t] + p[t])] \\ &= (\alpha + \beta\gamma)x[t] + \beta(q[t] + \gamma p[t]) \qquad (1) \end{aligned}$$

When a packet is dropped in the network link, the controller's output is zero, the state evolution becomes:

$$x[t+1] = \alpha x[t] + \beta q[t] \qquad (2)$$

For simplicity, the noises $p[t]$ and $q[t]$ are assumed to be the white noise with zero mean, and have the variance σ_P^2 and σ_Q^2 respectively.

3. CONDITION OF STABILITY

In this section, we will show that the system state $x[t]$ is a zero-mean random variable, and we will discuss the conditions under which the variance of the system state remains finite in long run. We suppose that $p[t]$ and $q[t]$ are uncorrelated (i.e, $E[p[t]q[t+\tau]] = 0$ for all t, τ), so the theorem 1 can be given to describe the property of the system state.

Theorem 1. *Under the assumption that $p[t]$ and $q[t]$ are uncorrelated white noise processes with zero mean, and the system described in the previous section with $x[0] = 0$, the state variable $x[T]$ is a zero-mean random variable with variance:*

$$\sigma_{x[T]}^2 = \beta^2\Bigg(\sigma_Q^2\left(1+A[T]\right)+\gamma^2\sigma_P^2\left(1_{[\text{if the } T^{\text{th}} \text{ packet is received}]}+A^{'}[T]\right)+\left(\sigma_Q^2B[T] +\gamma^2\sigma_P^2B^{'}[T]\right)\frac{1-\left((a^mb^{k-m})^2\right)^{\left\lfloor \frac{T}{k}\right\rfloor}}{1-(a^mb^{k-m})^2}-\left(\sigma_Q^2C[T]+\gamma^2\sigma_P^2C^{'}[T]\right)(a^mb^{k-m})^{2\left\lfloor \frac{T}{k}\right\rfloor}\Bigg) \quad (3)$$

where the integer variables m and k represent (m,k)-firm constraint, $a = \alpha+\gamma\beta$, $b = \alpha$ and

$$A[T]=\sum_{i=1}^{T-k\left\lfloor \frac{T}{k}\right\rfloor}\prod_{j=i}^{T-k\left\lfloor \frac{T}{k}\right\rfloor}K_j \qquad A^{'}[T]=\sum_{i=1}^{T-k\left\lfloor \frac{T}{k}\right\rfloor}\left(\left(\prod_{j=i}^{T-k\left\lfloor \frac{T}{k}\right\rfloor}K_j\right)0_{\left[\substack{\text{if the } (i-1)^{th} \text{ packet} \\ \text{in a period is dropped}}\right]}\right)$$

$$B[T]=\sum_{i=1}^{k}\left(\prod_{j=i}^{k}L_j\prod_{l=1}^{T-k\left\lfloor \frac{T}{k}\right\rfloor}K_l\right) \qquad B^{'}[T]=\sum_{i=1}^{k}\left(\left(\prod_{j=i}^{k}L_j\prod_{l=1}^{T-k\left\lfloor \frac{T}{k}\right\rfloor}K_l\right)0_{\left[\substack{\text{if the } (i-1)^{th} \text{ packet} \\ \text{in a period is dropped}}\right]}\right)$$

$$C[T]=\prod_{i=1}^{T-k\left\lfloor \frac{T}{k}\right\rfloor}K_i \qquad C^{'}[T]=\prod_{i=1}^{T-k\left\lfloor \frac{T}{k}\right\rfloor}K_i0_{\left[\substack{\text{if the last packet} \\ \text{in a period is dropped}}\right]}$$

with $K_i = (\alpha+\gamma\beta)^2$, if the $\left(k\left\lfloor \frac{T}{k}\right\rfloor+i\right)^{th}$ packet is received by the controller, if not, $K_i = \alpha^2$; $L_i = (\alpha+\gamma\beta)^2$, if the ith packet is received by the controller, if not, $L_i = \alpha^2$; $B[T]$ and $B^{'}[T]$ are equal to 0 while $\left\lfloor \frac{T}{k}\right\rfloor<1$.

Proof: The proof is inspired from that in (Hadjicostis02) for the calculation of variance of the system state.

Since the initial state $x[0]$ is equal to 0, from (1) and (2), the state $x[T]$ of the system at time step T is given by

$$x[T]=\beta\left(\sum_{i=0}^{T-1}C_{i+1}n[i]+\gamma\sum_{i=0}^{T-1}C_{i+1}I_im[i]\right)$$

where $C_i = A_{T-1}A_{T-2}\ldots A_{i-2}A_{i-1}A_i$ (specifically $C_T = 1$) with A_i and I_i (for $0 \le i \le T-1$) satisfying:

$I_i = 1$ and $A_i = \alpha+\gamma\beta$, if the $(i+1)^{\text{th}}$ packet is received by the controller;

$I_i = 0$ and $A_i = \alpha$, if the $(i+1)^{\text{th}}$ packet is lost.

We can find that the expected value of $x[T]$ is zero, because the variables $p[t]$ and $q[t]$ are zero-mean random processes. Since $p[t]$ and $q[t]$ are uncorrelated, the variance of $x[T]$ can be calculated as follow.

$$\begin{aligned}\sigma_{x[T]}^2 &= E\left[\beta\left(\sum_{i=0}^{T-1}C_{i+1}q[i]+\gamma\sum_{i=0}^{T-1}C_{i+1}I_ip[i]\right)\beta\left(\sum_{j=0}^{T-1}C_{j+1}q[j]+\gamma\sum_{j=0}^{T-1}C_{j+1}I_jp[j]\right)\right] \\ &= E\left[\beta^2\left(\sum_{i=0}^{T-1}C_{i+1}^2q^2[i]+\gamma^2\sum_{i=0}^{T-1}C_{i+1}^2I_i^2p^2[i]\right)\right] \\ &= \beta^2\left(\sigma_Q^2\sum_{i=0}^{T-1}C_{i+1}^2+\gamma^2\sigma_P^2\sum_{i=0}^{T-1}C_{i+1}^2I_i^2\right) \end{aligned} \quad (4)$$

In our system, exactly m out of k consecutive packets are received by the controller, therefore, we get

$$\begin{aligned}\sum_{i=0}^{T-1}C_{i+1}^2 &= 1+A[T]+B[T]\sum_{i=0}^{\left\lfloor \frac{T}{k}\right\rfloor-1}(a^mb^{k-m})^{2i}-C[T](a^mb^{k-m})^{2\left\lfloor \frac{T}{k}\right\rfloor} \\ &= 1+A[T]+B[T]\frac{1-\left((a^mb^{k-m})^2\right)^{\left\lfloor \frac{T}{k}\right\rfloor}}{1-(a^mb^{k-m})^2} -C[T](a^mb^{k-m})^{2\left\lfloor \frac{T}{k}\right\rfloor}\end{aligned} \quad (5)$$

Similarly,

$$\begin{aligned}\sum_{i=0}^{T-1}C_{i+1}^2I_i^2 &= 1_{[\text{if the T}^{\text{th}} \text{ packet is received}]}+A^{'}[T]+B^{'}[T]\sum_{i=0}^{\left\lfloor \frac{T}{k}\right\rfloor-1}(a^mb^{k-m})^{2i} -C^{'}[T](a^mb^{k-m})^{2\left\lfloor \frac{T}{k}\right\rfloor} \\ &= 1_{[\text{if the T}^{\text{th}} \text{ packet is received}]}+A^{'}[T] +B^{'}[T]\frac{1-\left((a^mb^{k-m})^2\right)^{\left\lfloor \frac{T}{k}\right\rfloor}}{1-(a^mb^{k-m})^2}-C^{'}[T](a^mb^{k-m})^{2\left\lfloor \frac{T}{k}\right\rfloor}\end{aligned} \quad (6)$$

Finally, if we plug (5) and (6) into (4), we get

$$\begin{aligned}\sigma_{x[T]}^2 &= \beta^2\Bigg(\sigma_Q^2\left(1+A[T]+B[T]\frac{1-\left((a^mb^{k-m})^2\right)^{\left\lfloor \frac{T}{k}\right\rfloor}}{1-(a^mb^{k-m})^2}-C[T](a^mb^{k-m})^{2\left\lfloor \frac{T}{k}\right\rfloor}\right) \\ &\quad +\gamma^2\sigma_P^2\left(1_{[\text{if the T}^{\text{th}} \text{ packet is received}]}+A^{'}[T]+B^{'}[T]\frac{1-\left((a^mb^{k-m})^2\right)^{\left\lfloor \frac{T}{k}\right\rfloor}}{1-(a^mb^{k-m})^2} -C^{'}[T](a^mb^{k-m})^{2\left\lfloor \frac{T}{k}\right\rfloor}\right)\Bigg) \\ &= \beta^2\Bigg(\sigma_Q^2\left(1+A[T]\right)+\gamma^2\sigma_P^2\left(1_{[\text{if the T}^{\text{th}} \text{ packet is received}]}+A^{'}[T]\right) \\ &\quad +\left(\sigma_Q^2B[T]+\gamma^2\sigma_P^2B^{'}[T]\right)\frac{1-\left((a^mb^{k-m})^2\right)^{\left\lfloor \frac{T}{k}\right\rfloor}}{1-(a^mb^{k-m})^2} \\ &\quad -\left(\sigma_Q^2C[T]+\gamma^2\sigma_P^2C^{'}[T]\right)\left(a^mb^{k-m}\right)^{2\left\lfloor \frac{T}{k}\right\rfloor}\Bigg)\end{aligned}$$

The variables $A[t]$, $B[t]$, $A^{'}[t]$ and $B^{'}[t]$ take their values periodically in a finite set of length k in long run, thus γ can be chosen so that the variance remains finite as T goes to infinity, i.e., $\left((\alpha+\gamma\beta)^m \alpha^{k-m}\right)^2 < 1$.

We have therefore the following lemma:

Lemma 1. *Given a system described in the previous section with $x[0] = 0$, under the same assumptions in the theorem 1, the gain γ can be chosen so that the variance $\sigma^2_{x[T]}$ remains finite as T goes to infinity if and only if*

$$\left((\alpha+\gamma\beta)^m \alpha^{k-m}\right)^2 < 1 \qquad (7)$$

Note that the above condition for system stability (in the mean square sense) is necessary and sufficient for our system. By equation (1), we can find that the system stability depends on not only the choice of the gain γ but also the (m,k)-firm constraint. I.e., the variance of the system state can be bounded in long run by adjusting the gain γ for a fixed (m,k)-firm constraint, or by adjusting the rate of m over k for a fixed gain γ. Furthermore, note that the distribution of packet drops in a packet delivery sequence has no influence upon the system stability.

Given an (m,k)-firm constraint, the gain γ should be chosen by considering the parity of m, the sign of β and α^{k-m} for (7) being justified. If we suppose that β is positive, by a bit of algebraic manipulation, the following intervals of γ can be obtained for each corresponding condition.

$$\frac{\left(-\sqrt[m]{\frac{1}{\alpha^{k-m}}}-\alpha\right)}{\beta} < \gamma < \frac{\left(\sqrt[m]{\frac{1}{\alpha^{k-m}}}-\alpha\right)}{\beta}$$

if $\alpha^{k-m} > 0$ and m is a even number

(8.1)

$$\frac{\left(\sqrt[m]{-\frac{1}{\alpha^{k-m}}}-\alpha\right)}{\beta} < \gamma < \frac{\left(\sqrt[m]{\frac{1}{\alpha^{k-m}}}-\alpha\right)}{\beta}$$

if $\alpha^{k-m} > 0$ and m is a odd number

(8.2)

$$\frac{\left(-\sqrt[m]{-\frac{1}{\alpha^{k-m}}}-\alpha\right)}{\beta} < \gamma < \frac{\left(\sqrt[m]{-\frac{1}{\alpha^{k-m}}}-\alpha\right)}{\beta}$$

if $\alpha^{k-m} < 0$ and m is a even number

(8.3)

$$\frac{\left(\sqrt[m]{\frac{1}{\alpha^{k-m}}}-\alpha\right)}{\beta} < \gamma < \frac{\left(\sqrt[m]{-\frac{1}{\alpha^{k-m}}}-\alpha\right)}{\beta}$$

if $\alpha^{k-m} < 0$ and m is a odd number

(8.4)

If β is negative number, it is enough to replace the sign 'less than' by the sign 'greater than' for each above obtained interval.

If the gain γ and the parameter k of the (m,k)-firm constraint are given, we can derive the following intervals from (7) for the choice of m that justify the inequality (7): if $(\alpha+\gamma\beta)\alpha^{-1} > 0$ and k is a even number; or if $(\alpha+\gamma\beta)\alpha^{-1} < 0$, m is a even number and k is a even number; or if $(\alpha+\gamma\beta)\alpha^{-1} > 0$, $\alpha > 0$ and k is a odd number; or if $(\alpha+\gamma\beta)\alpha^{-1} < 0$, m is a even number, $\alpha > 0$ and k is a odd number; or if $(\alpha+\gamma\beta)\alpha^{-1} < 0$, m is a odd number, $\alpha < 0$ and k is a odd number, then

$$m > \log_{(\alpha+\gamma\beta)\alpha^{-1}} 1/\alpha^k$$

else

$$m > \log_{(\alpha+\gamma\beta)\alpha^{-1}} \left(-1/\alpha^k\right)$$

Obviously, the adjustable range of the rate of m over k for guaranteeing the system stability is not as great as that of the gain γ because the variable m must be integer and be located between 1 and k.

4. OPTIMAL CHOICE OF THE GAIN CONSTANT

By equation (3), we can find that the more the quantity of m is, the more the quantities of $A[t]$, $B[t]$, $A'[t]$, $B'[t]$ and $\left((\alpha+\gamma\beta)^m \alpha^{k-m}\right)^2$ decrease. Therefore, given a gain γ, the variance of the system state is minimized when m is equal to k. Obviously, the quantity of m must be chosen by considering the parameters of network link. In this section, we will discuss how to choose γ so that the variance of the system state is minimized as T goes to infinity.

If $\left((\alpha+\gamma\beta)^m \alpha^{k-m}\right)^2 < 1$, when T goes to infinity, the expression of the variance of system state is transformed to:

$$\sigma^2_{x[T]} = \beta^2 \Bigg(\sigma^2_Q \left(1 + A[T]\right) + \gamma^2 \sigma^2_P \left(1_{[\text{if the T}^{\text{th}}\text{ packet is received}]} + A'[T]\right) + \left(\sigma^2_Q B[T] + \gamma^2 \sigma^2_P B'[T]\right) \frac{1}{1-(a^m b^{k-m})^2} \Bigg) \qquad (9)$$

Clearly, when $\gamma = \pm\infty$, the quantity of the term $\left(\sigma^2_Q B[T] + \gamma^2 \sigma^2_P B'[T]\right) \frac{1}{1-(a^m b^{k-m})^2}$ in equation (9) is minimized and is equal to zero (since $1-(a^m b^{k-m})^2$ becomes infinite). Within one interval S of γ defined in (8), the quantity of $\left(\sigma^2_Q B[T] + \gamma^2 \sigma^2_P B'[T]\right) \frac{1}{1-(a^m b^{k-m})^2}$ is finite and tends to $+\infty$ as we reach the limits of S from within S (since the quantity $\sigma^2_Q B[T] + \gamma^2 \sigma^2_P B'[T]$ is always positive and that $1-(a^m b^{k-m})^2$ tends to zero). On the other hand, the term $\gamma^2 \sigma^2_P \left(1_{[\text{if the T}^{\text{th}}\text{ packet is received}]} + A'[T]\right)$ in (9) forms a symmetrical concave curve by considering γ as a variable. Therefore, the value of γ that minimizes the quantity of the sum of these two terms $\left(\sigma^2_Q B[T] + \gamma^2 \sigma^2_P B'[T]\right) \frac{1}{1-(a^m b^{k-m})^2}$ and $\gamma^2 \sigma^2_M \left(1_{[\text{if the T}^{\text{th}}\text{ packet is received}]} + A'[T]\right)$ is not certainly located in a interval defined in (8). I.e., one can not simply take the derivative of the expression in equation (9) with respect to γ to derive the optimal gain. So to find a reasonable value of the gain γ, a feasible method is to examine all the values in a interval defined in (8) by fixing a acceptable search precision and select the best one that minimizes the quantity of equation (9). Obviously, the value of the gain γ is not unique for the different values of $A[t]$, $B[t]$, $A'[t]$ and $B'[t]$. The variables $A[t]$, $B[t]$, $A'[t]$ and $B'[t]$ take their values periodically in a finite set of length k, therefore there are k gains that need to be switched by the controller to generate the control data.

Notice that when σ_P^2 is equal to zero, the value of gain γ that minimizes the quantity of equation (9) is the one that makes $(a^m b^{k-m})^2$ zero, i.e., $\gamma = -\frac{\alpha}{\beta}$. On the other hand, when σ_Q^2 is equal to zero, equation (9) is equal to zero if the value of gain γ is zero.

5. OPTIMAL PACKET DELIVERY SEQUENCE

In section 3, we have discussed how to guarantee that the variance of the system state $x[t]$ remains finite as t goes to infinity, but in many actual situation, the bounded variance of control system is not enough to guarantee a stable operation of system. The high amplitude variation of the system state may be very troubling for it indicates that even though the variance of the system state is bounded, the large signal amplitudes may be unacceptable. Therefore the variation level of control system state is often considered as an important metric of system performance. A way to optimize this system performance is to minimize the variation of the system state as we have discussed in the previous section. Furthermore, in our work, the fact that the packet dropout sequence is governed by the (m,k)-firm constraint makes a further optimization possible for the positions of the packet drops in a packet delivery sequence has a important influence to the amplitude variation of system state. In this section, we will present a metric that directly relates control system performance to the packet drop sequence, then an approach based on this metric to deduce an optimal packet delivery sequence.

In the rest of the paper, we will use the number 0 and 1 to facilitate the presentation of the packet delivery sequence. I.e. 1 represents that the packet is received by the controller, and zero represents a packet drop. E.g., the sequence 10 represents that the first packet is received by the controller and the second is dropped.

To focus on the relationships between the packets, the effects of the noises will be eliminated from the system. Therefore the system is transformed as following.

If a packet is received by the controller, the system state evolution can be expressed as:

$$x[t+1] = \alpha x[t] + \gamma x[t]$$
$$= (\alpha + \gamma) x[t] \qquad (10)$$

When a packet is dropped in the communication link, the state evolution becomes:

$$x[t+1] = \alpha x[t]. \qquad (11)$$

The square value of the difference between two peak values of system state determine a level of amplitude variation of these two peaks: the greater this square value is, the greater the distance between these two peaks is. Therefore the sum of the square values of the difference between any two peak values can be used as a metric to measure the control system performance. Since the packet delivery sequence is periodic (i.e. the same situation will be repeated in each period of the packet delivery sequence), the measurement of performance will be calculated in a single period of the packet delivery sequence. Obviously, by increasing the dimension of this calculation (i.e. the squares of the difference between two peak values in two different period is also taken into account), a more precise measurement of system performance can be obtained. However, such an increase of dimension can also augment the calculation complexity. So in this work, the measurement of system performance is taken in a period of the packet delivery sequence. The control system performance is therefore characterized by the following equation.

$$\sum_{i=1}^{2m-1} \sum_{j=i+1}^{2m} \left(x[t_i] \text{-} x[t_j] \right)^2 \qquad (12)$$

where t_i is a peak moment (called peak moment in the following) that the last packet drop in the consecutive packet drops occurs or the moment that a packet is received by the controller.

Example 1. Given a (3,9)-firm constraint, if the system states in a period of packet delivery sequence are $x[1]$,$x[2]$,...,$x[9]$, and a period of packet delivery sequence can be described as 100100010, then the control system performance is represented by $\sum_{i=1}^{5} \sum_{j=i+1}^{6} \left(x[t_i] \text{-} x[t_j] \right)^2$ with t_1=1, t_2=3, t_3=4 t_4=7 t_5=8 t_6=9.

In our work, we suppose that the first packet in a period of the delivery sequence is received (the reason will be detailed in the following). So (12) can be transformed to:

$$x[t_1 - 1] \sum_{i=1}^{2m-1} \sum_{j=i}^{2m} \left(a^{\left\lceil \frac{i}{2} \right\rceil} b^{\sum_{l=1}^{\left\lceil \frac{i-1}{2} \right\rceil} P_l} \text{-} a^{\left\lceil \frac{j}{2} \right\rceil} b^{\sum_{l=1}^{\left\lceil \frac{j-1}{2} \right\rceil} P_l} \right)^2 \qquad (13)$$

where $a = \alpha+\gamma$, $b = \alpha$, P_j represents the number of the packets dropped between two consecutive packets received by the controller:

$$\sum_{i=1}^{m} P_i = k - m \tag{14}$$

The problem now becomes to determining the value of P_i so that the quantity of expression (13) is minimized. I.e. we will try to find the extreme value of the following expression (15) subject to the constraint (14).

$$\sum_{i=1}^{2m-1} \sum_{j=i}^{2m} \left(a^{\left\lceil \frac{i}{2} \right\rceil} b^{\sum_{l=1}^{\left\lceil \frac{i-1}{2} \right\rceil} P_l} - a^{\left\lceil \frac{j}{2} \right\rceil} b^{\sum_{l=1}^{\left\lceil \frac{j-1}{2} \right\rceil} P_l} \right)^2 \tag{15}$$

One way to accomplish this objective is to use the method of Lagrange multiplier. We will show an example to illustrate this.

Example 2. Given a system described by (10) and (11), the packet drop process is governed by (2, k)-firm constraint ($k > 2$ and $k \in Z$). To find an optimal distribution of packet drops, we set up firstly the metric describing the system performance. According to the formulas (14) and (15), we get:

$$\begin{aligned} f(p_1, p_2) = {} & 3a^2 - 2a^2 b^{p_1} + 3a^2 b^{2p_1} - 2a^3 b^{p_1} + 3a^4 b^{2p_1} \\ & - 2a^3 b^{p_1+p_2} + 3a^4 b^{2(p_1+p_2)} - 2a^3 b^{2p_1} \\ & - 2a^3 b^{(p_1+p_2)} - 2a^4 b^{(2p_1+p_2)} \end{aligned} \tag{16}$$

$$g(p_1, p_2) = p_1 + p_2 = k - 2 \tag{17}$$

Therefore, equation (16) is the formula that we try to minimize subject to (17).

According to the method of Lagrange multipliers, we introduce a variable λ (the Lagrange multiplier) and set up the equations $\frac{\partial f}{\partial p_1} = \lambda \frac{\partial g}{\partial p_1}$ and $\frac{\partial f}{\partial p_2} = \lambda \frac{\partial g}{\partial p_2}$, we get respectively

$$\begin{aligned} & -2a^2 b^{p_1} \ln(b) + 6a^2 b^{2p_1} \ln(b) - 2a^3 b^{p_1} \ln(b) + 6a^4 b^{2p_1} \ln(b) \\ & -2a^3 b^{(p_1+p_2)} \ln(b) + 6a^4 b^{2(p_1+p_2)} \ln(b) - 4a^3 b^{2p_1} \ln(b) \\ & -4a^3 b^{(2p_1+p_2)} \ln(b) - 4a^4 b^{(2p_1+p_2)} = \lambda \end{aligned}$$

and

$$\begin{aligned} & -2a^3 b^{(p_1+p_2)} \ln(b) + 6a^4 b^{2(p_1+p_2)} \ln(b) - 2a^3 b^{(2p_1+p_2)} \ln(b) \\ & -2a^4 b^{(2p_1+p_2)} \ln(b) = \lambda \end{aligned}$$

By solving these equations with the constraint (17), we get:

$$p_1 = \frac{\ln\left(\frac{1 + a + a e^{(\ln(b)(k-2))} + a^2 e^{(\ln(b)(k-2))}}{3 + 3a^2 - 2a} \right)}{\ln(b)} \tag{18}$$

$$p_2 = -\frac{\ln\left(\frac{1 + a + a e^{(\ln(b)(k-2))} + a^2 e^{(\ln(b)(k-2))}}{3 + 3a^2 - 2a} \right) - k\ln(b) + 2\ln(b)}{\ln(b)} \tag{19}$$

Obviously, the number of the dropped packets must be a positive integer. From the method of Lagrange multiplier, it's difficult to obtain the integer solution, and sometimes there are some negative numbers in the results. When this happens, we will carry out an approximate calculation as following:

1. replace all the negative numbers by 0, and divide the sum of these negative numbers by the number of the positive numbers in the results;
2. add the result of the division to all the positive numbers;
3. delete all the decimal parts of the positive numbers and add 1 to the number whose decimal part was the largest after step 2.

For example, given $a = 0.03$, $b = 3$ and $k = 9$ from equations (18) and (19), we get $p_1 = 2.866$ and $p_2 = 4.134$. By the method of approximation, we get $p_1 = 3$ and $p_2 = 4$. Therefore, a period of the packet delivery sequence is described as 100010000. If $k = 3$ and the other parameters remain same, we get $p_1 = -0.8773$ and $p_2 = 1.877$. By the method of approximation, we get $p_1 = 0$ and $p_2 = 1$. A period of the packet delivery sequence is thus described as 110.

Before proceeding further, we make several comments about this section. If the number of the packets that must be received by the controller in a period is different from 2, then example 2 gives a procedure for finding the optimal distribution of packet drops. We force the first packet in a period of packet delivery sequence to be delivered, as by doing this, the general level of amplitude variation of system state in a period can be depressed (the reason being that the last peak value in a period is unchangeable). Also note that the effect of the noises is not considered in this section. But if the initial disturbance in a control system described by (1) and (2) is considerable, the packet delivery sequence obtained by our approach can rapidly eliminate the high amplitude bursts caused by this initial disturbance. We will illustrate this in the next section.

6. NUMERICAL EXAMPLES

In this section, we present two examples which exploit the research results obtained in the previous sections. The first example is based on an instance of the system described by (1) and (2). We will firstly calculate the value interval of the gain so that the variance of system state is bounded in long run, then we will give the optimal gain that minimizes the variance of system state. In the next example, we will apply our approach to find an optimal packet drop distribution, then we will compare the performance of this packet delivery sequence with that of the packet delivery sequence obtained by the algorithm in (Ramanathan99) under two different system settings.

Given a system described by (1) and (2) with $\alpha=3$, $\beta=1$, $\sigma_M^2=1$, $\sigma_N^2=4$, and the packet drop process is governed by (4,7)-firm constraint, we assume that a period of packet delivery sequence is given by 1101010. By inequality (8.1), we find that the value of the gain γ must satisfy $-3.333 < \gamma < -2.667$ so that the variance of the system state remains finite in long run. To calculate a optimal gain γ that minimizes the variance of the system state, we examine all the values in the interval of the gain γ that we have obtained by fixing the search precision as 0.001. Therefore, we get $\gamma_{op}[t_1]=-2.974$, $\gamma_{op}[t_2]=-2.832$, $\gamma_{op}[t_3]=-2.832$, $\gamma_{op}[t_4]=-2.962$, $\gamma_{op}[t_5]=-2.962$, $\gamma_{op}[t_6]=-2.963$, $\gamma_{op}[t_7]=-2.963$ where t_i represents a moment that the ith packet delivery occurs in a period of the packet delivery sequence. Figure 2 shows the system state trace with the obtained optimal gain constants.

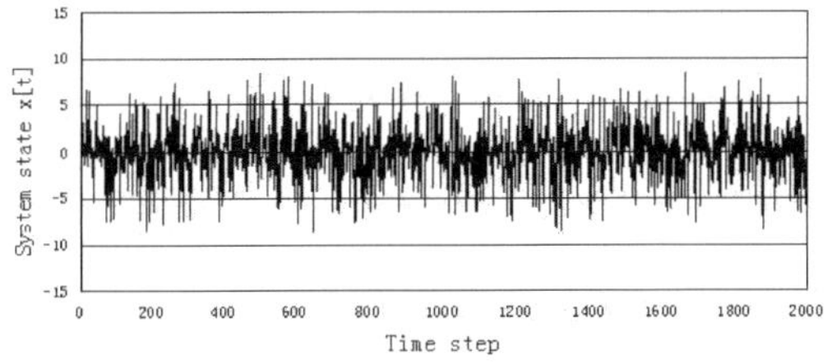

Figure 2. The system state trace under the optimal gain constants

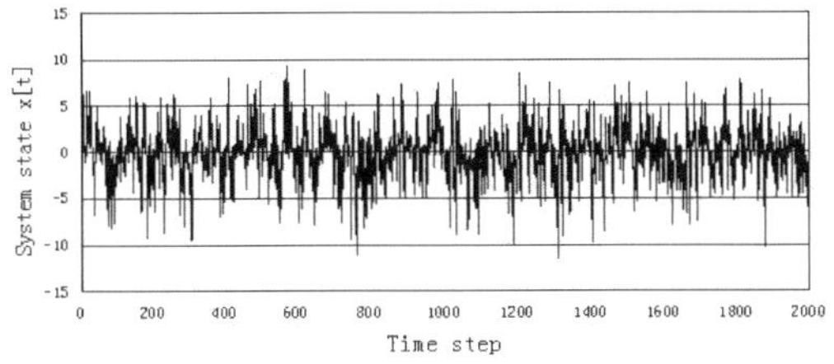

Figure 3. The system state trace under a fixed gain constants

In figure 3, we give the state trace of the system whose gain γ is fixed to -2.8. We can see that the variance is slightly higher than the first one.

In the next experiment, we will take a instance of the system described by (10) and (11) with $\alpha=3$, $\gamma=-2.97$, and the system is under (3,10)-firm constraint (the system is stable in mean square sense, because $(\alpha+\gamma)^3\alpha^7<1$). By the procedure introduced in the section 5, we get the optimal packet delivery sequence that is 1100010000. The initial system state is set to be 5000, the system state traces under the packet delivery sequence obtained by our approach and that obtained by the algorithm in (Ramanathan99) are illustrated in figure 4 and 5.

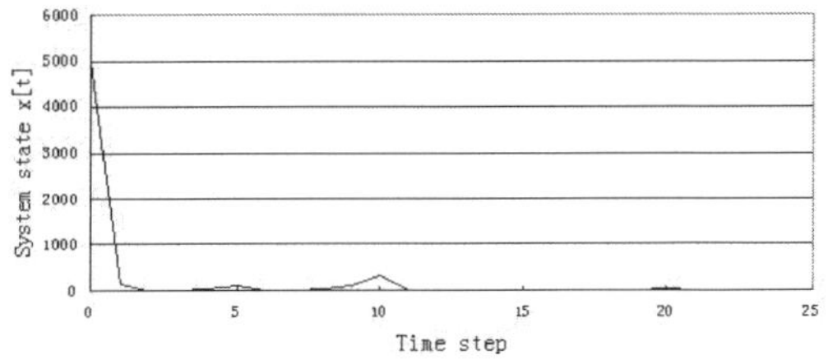

Figure 4. The system state trace with the packet delivery sequence 1100010000

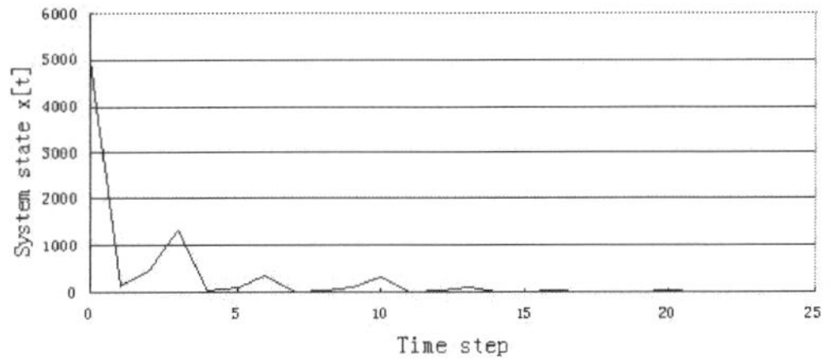

Figure 5. The system state trace with the packet delivery sequence 1001001000

Comparing the figure (4) and (5), we see that the amplitude variance of the system state in figure (5) is much more higher than that in the figure (4), and the system state converges more rapidly to the steady state in figure (4).

Now we introduce a quantization noise with $\sigma_M^2=3$ to this control system, the system state traces with these two packet delivery sequence are given in figure 6 and 7.

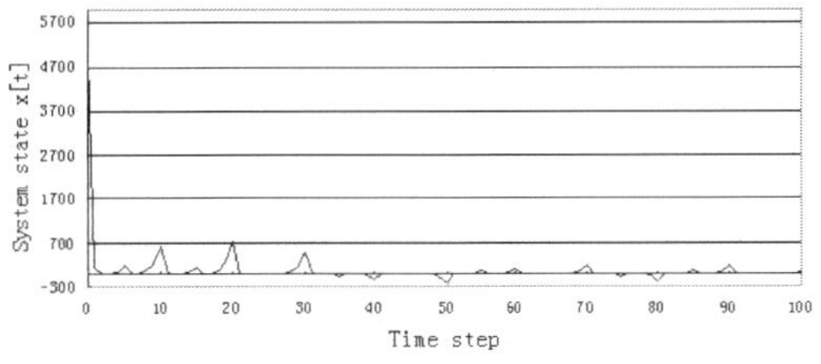

Figure 6. The state trace of the system under the quantization noise with the packet delivery sequence 1100010000

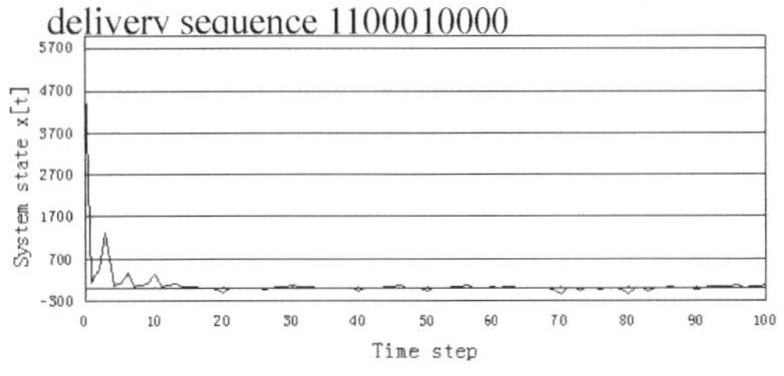

Figure 7. The state trace of the system under the quantization noise with the packet delivery sequence 1001001000

In figure 6, the influence of the initial disturbance can be rapidly eliminated by using the packet delivery sequence obtained by our algorithm, but there still exist occasional high amplitude bursts in long run. In figure 7, the system state slightly changes around zero after a high amplitude burst at the first period.
By the same procedures for the first and second experimentation, we have tested many others independent trials by changing the setup of the system, and all results have confirmed our research results.

7. CONCLUSION AND PERSPECTIVE

This paper studied the networked control system in which packet drop process is governed by the (m,k)-firm constraint. We first analyzed the problem of stability of such a system and identified the interval of the gain constant and the network parameter (that is identified by (m,k)-firm constraint) over which the variance of the system state is bounded in long run, and a method for minimizing the variance of the system state was proposed. These research results allow to optimally choose the gain of the controller so that the system stability (in mean square sense) can be guaranteed and to specify bounds on the network's QoS (packet drop rate) that improve control system performance. We then considered the amplitude variance of the system state as an important system performance, and associated the system performance with the packet drop strategy. For improving the system performance (i.e. minimizing the amplitude variance of the system state), we proposed a method based on the Lagrange multiplier to optimally lay out the packet drops in the packet delivery sequence.

In our work, the packet drop process is assumed to be governed by (m,k)-firm constraint, and the packet delivery sequence is periodic. A future work intends to explore the networked control system possessing a more flexible packet drop process model. This flexible drop process is realized by a judicious choice for the real-time scheduler. One such candidate is a scheduler implementing the Markov chain constraint (Liu03).

REFERENCES

Hadjicostis, C. N., and R. Touri. «Feedback control utilizing packet dropping network links», *Proc. Of the IEEE Conference on Decision and Control,* 2002.

Ling Q and M.D. Lemmon, «Robust performance of soft real-time networked control systems with data dropouts», *Proc. Of the IEEE Conference on Decision and Control,* 2002.

Ling, Q., and M.D. Lemmon, «Soft real-time scheduling of networked systems with dropouts governed by a Markov chain», *Proc. Of the American Conference on Control,* 2003.

Liu, Donglin., and Xiaobo. Sharon Hu, Michael. D. Lemmon, and Qiang. Ling, «Firm Real-Time System Scheduling Based on a Novel QoS Constraint», *Proc. Of the 24th IEEE International Real-Time Systems Symposium,* 2003.

Nilsson, J., «Real-time control systems with delays», PhD thesis, Lund Institute of Technology, 1998.

Quan, G., and X. Hu, « Enhanced Fixed-priority Scheduling with *(m,k)-firm* Guarantee », *Proc. Of 21st IEEE Real-Time Systems Symposium*, pp.79-88, Orlando, Florida, (USA), November 27-30, 2000.

Ramanathan P., «Overload management in Real-Time control applications using *(m,k)-firm* guarantee », *IEEE Transactions on Parallel and Distributed Systems*, 10(6) :549-559, Jun 1999.

Seiler, P., and R. Sengupta, «Analysis of communication losses in vehicle control problems», *Proc. Of ACC 2001, the 2001 American Control Conference*, Volume 2, pages 1491-1496, Arlington, VA, 2001.

Walsh, G., C. and Y. Hong and L. G. Bushnell, «Stability Analysis of Networked Control Systems», *IEEE Transactions on Control Systems Technology*, 10(3):438-46, 2002.

Zhang, W., M. S. Brannicky, and S. M. Philips, «Stability analysis of networked control systems», *IEEE Control System Magazine*, 21(1):84-89, Feb 2001.

ELSEVIER
IFAC
PUBLICATIONS

A SCHEDULABILITY ANALYSIS OF AN IEC-61499 CONTROL APPLICATION

Mohamed Khalgui, Xavier Rebeuf, Françoise Simonot-Lion

LORIA - INPL
Campus Scientifique, BP 239
54506 Vandoeuvre-lès-Nancy - France
{khalgui, rebeuf, simonot}@loria.fr
tel: +33 3 83 58 17 28, fax: +33 3 83 58 17 01

Abstract: This paper deals with the temporal correctness of control applications designed using the component-based standard IEC 61499. In this standard, a function block is defined as an event trigger component containing its own data. To validate the temporal behavior of an application, we have to take into account its scheduling on the execution support. We propose an abstraction of the component behavior taking into account all its possible executions. Moreover, we propose to verify the temporal correctness of the application with regard to global temporal properties (end to end delays). Thanks to such characterization, we show that it is possible to check deadlines for the application to ensure its correctness. To reach this goal, we transform the application into a dependant tasks model(Copyright © 2005 IFAC).

Keywords: Function Blocks, IEC 61499, Real Time, schedulability analysis, offline scheduling.

1. INTRODUCTION

Industrial control applications need to satisfy not only functional properties but also temporal ones. To validate a priori the correctness of such applications, one of the key issues is to model their temporal behaviors. To manage the design complexity, several methods based on components have been proposed (Pecos 2005, holobloc 2005, Stewart *et al.* 1997, Articus 1996). In such approaches, evaluating temporal behavior requires also a model of the execution support.

In the control systems field, the IEC 61499 standard (WG6 2003, WG6 2004) is a component-based methodology allowing to design applications as well as the execution support (Crnkovic and Larsson 2002). In the standard, the Function Block is defined as an event trigger component (Lewis n.d.). It is a reusable functional unit of software owning data. A control application is specified by a "function blocks network" which can be distributed on one or more devices. Several complex applications have been specified using function blocks (Ifak 2005). Moreover, reusable function blocks libraries already exist (holobloc 2005).

The standard allows to validate static interoperability between blocks. Nevertheless, temporal behavior depends on dynamic data. Therefore, it is difficult to a priori validate temporal interoperability. We propose an abstraction of each block behavior taking into account all its possible executions. Such abstraction allows to compute an upper bound of the application execution time. We show that it is possible to verify the temporal correctness of the application with regard to global temporal properties (end to end delays).

To validate temporal interoperability, we propose a schedulability analysis of a function blocks network distributed on one device. To perform such analysis, we propose to transform the function blocks application into a particular dependent tasks model in the order to take advantages of the results in this field.

Some works are proposed about schedulability analysis of systems in one processor. In the non-preemptive case, (Cucu *et al.* 2002) presents an optimal algorithm to schedule systems with precedence, periodicity and latency constraints (end to end deadline). In addition, (Cucu and Sorel 2003) proves the existence of a hyper-period T for synchronous systems allowing the proposed algorithm to find a schedule applying it from 0 to T, instead of from 0 to ∞. On another side, (Jeffay *et al.* 1991) gives a schedulability condition for a set of periodic and sporadic operations with arbitrary release times. (Howell and k. Venkatrao 1995) studies the complexity for the problem of non-preemptively scheduling of periodic and sporadic tasks on one processor using inserted idle times. Using part of this result, we show that it is possible to generate a safe off-line scheduling for a function blocks network. Moreover, the proposed schedulability analysis allows to define a strategy for a function block to adapt its internal behavior according to the scheduling.

In the next section 2, we briefly present the IEC 61499 standard. Then we present our behavior characterization of an application. The section 4 deals with its transformation into a tasks model. In the section 5, we propose a generation of tasks deadlines according to the application delays. Finally, we present a schedulability analysis based on the generated deadlines.

2. THE IEC 61499 STANDARD

We present the main concepts of the IEC 61499 Function Blocks standard (WG6 2003, WG6 2004). This standard is an extension of the IEC 61131.3 (IEC61131-3 1993) for the Programmable Logic Controllers. We can divide its description into two parts: the architecture description and the block behavior through the events selection mechanism.

2.1 Architecture description

An application function block (FB) (figure 1) is a functional unit of software that supports some functionalities of an application. It is composed by an interface and an implementation. The interface contains data/event inputs and outputs supporting the interaction with the environment. Events are responsible for the activation of the function block while data contain valued information.

The implementation consists of a body and a head. The body is composed of internal data and algorithms implementing the block functionalities. Each algorithm gets values in the input data channel and produces values in the output data ones. They are programmed in structured text (ST) language (IEC61131-3 1993).

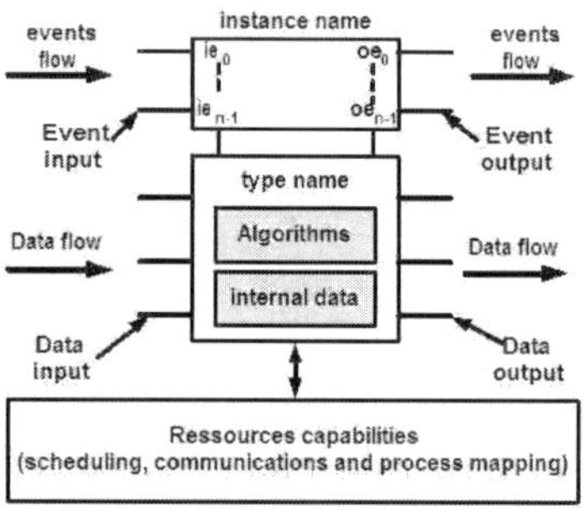

Figure 1. An IEC 61499 function block

The block head is connected to event flows. It selects the sequence of algorithms to execute with regard to an input event occurrence. The selection mechanism of an event occurrence is encoded in a state machine called the Execution Control Chart (ECC). At the end of the algorithms execution, the ECC sends the corresponding output event occurrences.

In the standard, a function blocks network defines the functional architecture of a control application. Each function block event input (resp. output) is linked to an event output (resp. input) by a channel. Otherwise, it corresponds to a global application input (resp. output). Data inputs and outputs follow the same rules (figure 2).

The execution support architecture (i.e. the industrial control system) is defined by a devices network. A device is composed of one processing unit, sensors, actuators and network interfaces. In order to manage interactions with a given physical processes, a device contains several resources. A resource contains FBs networks interacting with a physical process. Considered as a logic execution unit corresponding to time slots of the processing unit, the resource provides scheduling function for its local FBs networks.

The operational architecture corresponds to a distribution of the application function blocks over the different resources of the execution support architecture. The advantage of such approach is to take into account hardware as well as software components. For sake of simplicity, we consider in this paper only one function blocks network distributed on a single resource of a device.

Running Example. *For all the continuation, we consider the following running example of a control application located in a resource of a device. This control application is composed by four FBs supporting its functionalities (figure 2).*

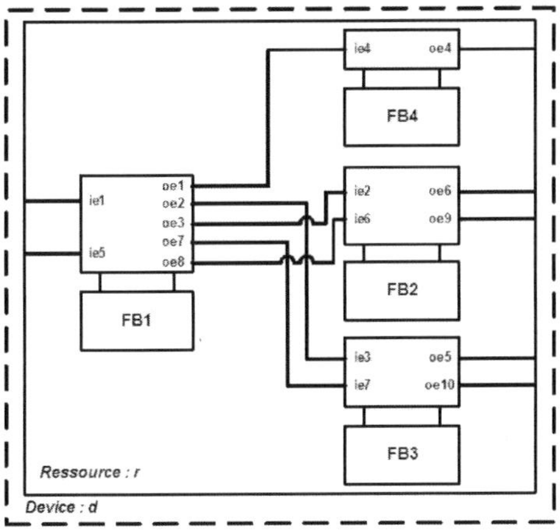

Figure 2. A control application *fbn*.

2.2 Events selection mechanism

Let turn to the internal description of a function block. Note that only algorithms execution spends time. In a given function block, the ECC is said idle if there is no algorithm to execute. Otherwise, the ECC is busy (figure 4).

According to the standard (WG6 2004), the FB contains an internal buffer for input occurrences. The ECC behavior is devised into three steps:

* First, it selects one input event occurrence according to priority rules defined in the resource.

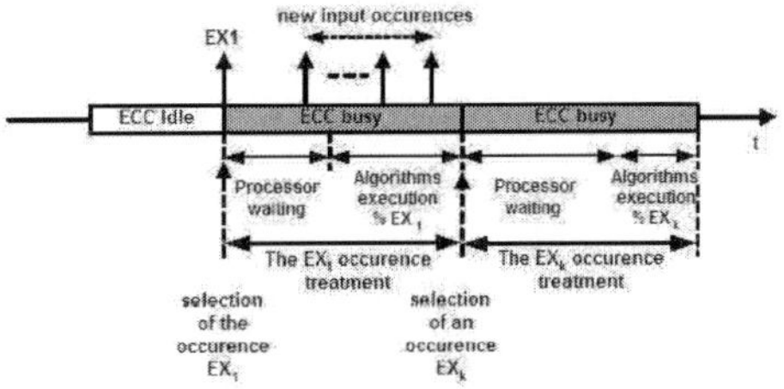

Figure 3. The ECC behavior

* It activates the algorithms sequence corresponding to the selected event. Then, it waits for the resource scheduler to execute this sequence.

* When the execution ends, it emits corresponding output events occurrences.

We note that an algorithms sequence is atomic. Moreover, the resource scheduling function applies only non preemptive policies. On the other hand, the policy of events priorities in the ECC is not specified in the standard. Therefore, it is up to the designer to fix such priority for each function block (Khalgui *et al.* 2004). Note that the ECC is specified as a state machine where each trace is composed by a waiting of an input event, invocations of algorithms, sending of output events.

Running Example. *we present the ECC behavior of the function block $FB1$ (figure 4). The selection mechanism is performed thanks to a state variable 'a' of $FB1$.*

When the ECC selects an ie_1 occurrence, it asks (!ex_fb) the processor to perform the corresponding algorithms sequence. When the execution is finished (?end_ex) and depending on FB_1 state variables, the ECC sends oe_1 to FB_4 or simultaneously oe_2 and oe_3 to respectively FB_3 and FB_2.

When the ECC selects an ie_5 occurrence, it waits the processor to execute the corresponding algorithms sequence. When it is finished, it sends oe_7 to FB_3 or oe_8 to FB_2 depending on the internal variables.

3. BEHAVIOR CHARACTERIZATION

To characterize temporal behavior of a function block, we have to take into account the execution of the ECC. Indeed, the ECC decides not only algorithms to execute but also the output events to send. Nevertheless, the selection of transitions inside the ECC may depend on internal state variables. Then we propose to define sets of output events corresponding to all possible executions.

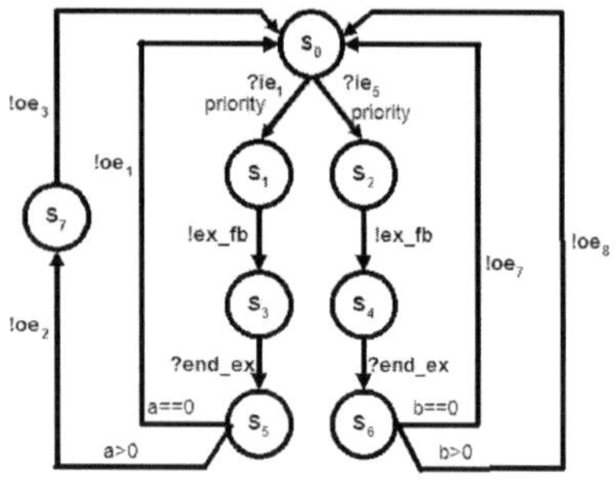

Figure 4. The ECC behavior of FB_1

To validate the temporal correctness of the application, we propose to define end to end delays corresponding to time constraints imposed by the problem specification. In this section, we first present the abstraction of the function block behavior. Then we propose a formalization for end to end delays. For all the continuation, we denote by fbn a function blocks network.

3.1 Function block behavior

We propose to compute an abstraction of the function block behavior. The difficulty is to identify the output events sent corresponding to an input event occurrence. Such association is specified in the ECC state machine. Nevertheless, firing a transition in the ECC can depend on internal variables of the block. Therefore, we propose to identify the supersets of output events possibly occurring simultaneously.

For each trace of the ECC automaton (i.e. each possible execution), we associate a superset gathering all the output event occurring in this trace.

Let consider IE (resp OE) the set of input (output) events of fbn. Let consider IE_{FB} (resp OE_{FB}) the set of input (resp output) events of a function block FB.

Let tr be a trace in ECC_{FB}, we denote by

- $IE(tr)$ the input event occurring in tr
- $OE(tr)$ the set of output events occurring in tr

We propose a function *follow* associating to an input event $ie \in IE_{FB}$, the sets of simultaneous output events.

$$follow(fb, ie) = \{OE(tr)/ie = IE(tr),\ tr \in ECC_{FB}\}$$

Running Example. *For example, we associate to ie_5 two sets of output events corresponding to the two traces in the ECC starting from the transition triggered by $?ie_5$.*

$$follow(fb_1, ie_1) = \{\{oe_1\}, \{oe_2, oe_3\}\}$$

$$follow(fb_1, ie_5) = \{\{oe_7\}, \{oe_8\}\}$$

$$follow(fb_4, ie_4) = \{\{oe_4\}\}$$

3.2 fbn temporal constraint

A real time application must often respect temporal constraints as end to end delays. To associate such delay to the function blocks network, we need to first formalize the composition between function blocks.

We propose a function *cause* that specifies causalities between an event input of a function block and an output of another one according to the function blocks network. Note that *effect* specifies the opposite function that associates to an output event the input event target of the sent occurrences.

Running example. *From the link between $FB1$ and $FB4$, one can deduce :*

$$cause(ie_4) = oe_1$$

$$effect(oe_1) = ie_4$$

We define *inputs* (respectively *outputs*) the set of input (respectively output) events in fbn which is not linked to another event.

$$inputs = \{ie \in IE/cause(ie) \notin OE\}$$

$$outputs = \{oe \in OE/effect(oe) \notin IE\}$$

Running example. *We have the following sets :*

$$inputs = \{ie_1, ie_5\}$$

$$outputs = \{oe_4, oe_5, oe_6, oe_9, oe_{10}\}$$

We propose the function *delay* that encodes all the end to end delays for fbn. $delay(ie, oe)$ denotes the maximum duration that can take the execution between the receive of an $ie \in inputs$ occurrence and the sent of an $oe \in outputs$ one.

Running Example. *$delay(ie_1, oe_4)$ specifies the maximum duration for the treatment of the ie_1 and ie_4 algorithms sequence. Let suppose the following constraints : $delay(ie_1, oe_4) = delay(ie_1, oe_5) = delay(ie_1, oe_6) = 10ms$ and $delay(ie_5, oe_9) = delay(ie_5, oe_{10}) = 17$.*

4. TRANSFORMATION INTO A TASKS MODEL

In this part, we propose to transform fbn to a tasks system S with precedence constraints (Babanov *et al.* 2003). The purpose is to exploit the results on this topic to perform the schedulability analysis of fbn.

In this section, we define the task characterization in the system S. A task corresponds to one execution of a function block. Then, we define a trace as a causality sequence of tasks.

4.1 Task definition starting from a FB network

An application task T corresponds to the execution of a FB activated by an input event occurrence ie. We define the function $generate(ie)$ that constructs for an input event ie the corresponding task T. Note that $is\_generated\_by(T)$ is the opposite function of $generate$. We denote by $Task$ the tasks set of S.

Let set_{OE} be a set of output events. We define the function $target(set_{OE})$ that associates for set_{OE} the following set of tasks,

$$target(set_{OE}) = \{T \in Task / \exists oe \in set_{OE},\ oe = cause(is\_generated\_by(T))\}$$

A task T is characterized as follows,

$$T = \{WCET,\ pred,\ succ\}$$

such as,

- $WCET$: the worst case execution time of the algorithms sequence corresponding to ie. It can be evaluated using the code and the characteristics of the execution support.
- $pred$: The task that must be executed before the execution of T. It corresponds to the execution of the FB producing $cause(ie)$.
- $succ$: a set of tasks sets. All the elements of a set correspond to tasks to execute once the execution of T is finished. Note that each set corresponds to a possible execution scenario. (i.e. one trace in the ECC)

$$T.succ = \{set_T / set_T \subset Task,\ \exists set_{OE} \in follow(fb, is\_generated\_by(T)),\ set_T = target(set_{OE})\}$$

Running example. *We generate the predecessor of* T_4 *and the successors of* T_1 *as follows,*

$$T_4.pred = T_1$$

$$T_1.succ = \{\{T_2, T_3\}; \{T_4\}\}$$

When the T_1 *execution is finished, two scenarios are possible : Either we execute* T_2 *and* T_3 *or we execute* T_4.

We define $first$ (resp $last$) the tasks set that they have not a predecessor (resp successor). These sets correspond to the $inputs$ and $outputs$ sets,

$$first = \{T \in Task / T.pred \notin Task\}$$

$$last = \{T \in Task / \forall s \in T.succ,\ \forall T_j \in s,\ T_j \notin Task\}$$

4.2 Trace definition

To specify a causality sequence of tasks, we define in S a trace tr as a Tasks sequence,

$$tr = T_0, T_1, ..., T_{n-1}$$

such as,

- $T_0 \in first$
- $T_{n-1} \in last$
- $\forall i \in]1, n-1],\ T_{i-1} = T_i.pred$

Thanks to such traces principle, it is possible to check the application end to end delays. Let $Trace$ be the traces set in S. We denote by $first(tr)$ the first task of the trace tr. In this paper, we consider the case of not reentry traces (Klein *et al.* 1993, Liu 2000) : the execution of the $k-th$ instance of the first task must not start before the execution end of the $(k-1)-th$ instance of the last task. Otherwise the system is not feasible.

Running Example. *We distinguish five traces in this control application :* $tr_1 = T_1, T_2$ *;* $tr_2 = T_1, T_3$ *;* $tr_3 = T_1, T_4$ *;* $tr_4 = T_5, T_6$ *and* $tr_5 = T_5, T_7$. *Each trace specifies a tasks sequence in* fbn.

Note that we can classically define an operation $op(T)$ in S as the set of traces having the same root. It corresponds to all possible executions of fbn when T is activated (the corresponding input event occurs)

$$op(T) = \{tr \in Trace / first(tr) = T\}$$

5. TASKS DEADLINES GENERATION

We classically define $delay(tr)$ as the end to end deadline of a trace $tr \in Trace$. This temporal constraint corresponds to the $delay(ie, oe)$ where ie corresponds to the $first(tr)$ and oe corresponds to an event producing by the last task of tr.

We define $d(T)$ the deadline of the task $T \in tr$. $d(T)$ has to take into account the time for executing all the successors belonging to a same set of $T.succ$ before their respective deadlines.

We characterize $d(T)$as follows,

If $T \in last$,

$$d(T) = delay(tr)$$

Otherwise,

$$d(T) = min_{s \in T.succ}\{min_{T_i \in s}\{d(T_i) - \sum_{T_j \in s}^{d(T_j) \leq d(T_i)} T_j.WCET\}\}$$

Running Example. *We construct the different deadlines of the different tasks. For the operation* op_1*, we compute first of all the deadlines of the tasks* T_2, T_3 *and* T_4.

- $d(T_2) = delay(tr_1) = 10$
- $d(T_3) = delay(tr_2) = 10$
- $d(T_4) = delay(tr_3) = 10$

We compute then the deadline of T_1.

$$d(T_1) =$$

$$min\{d(T_2) - T_2.WCET - T_3.WCET;$$
$$d(T_4) - T_4.WCET\}$$

Therefore,

$$d(T_1) = min\{10 - 3 - 4, 10 - 1\} = 3$$

6. SCHEDULABILITY ANALYSIS

In this part, we propose a schedulability analysis of an IEC 61499 control application. Let S be the tasks system.

The system S is *schedulable* if its temporal behavior is correct. The schedulability criterion is the respect of the deadline for each task. To perform the schedulability analysis of S, we propose to construct an accessibility graph. The accessibility graph is a set of scheduling trajectories. Each trajectory represents a possible execution scenario of the application. Therefore, several traces are used to construct a trajectory. We apply the EDF algorithm during each trajectory construction to verify end to end delays of the corresponding traces.

We present a temporal characterization of S. Since all input events of fbn are periodic, then each task T belonging to $first$ is activated periodically. We characterize such task by a release time r, a period p and a jitter j (the maximum deviation of the period).

$$T = \{r, p, j\}$$

Let lcm be the least common multiple of the tasks periods.

Let $T_{max} = \{r_{max}, p_{max}, j_{max}\}$ and $T_{min} = \{r_{min}, p_{min}, j_{min}\}$ be two tasks of $first$ such as,

$$\forall T_i \in first,\ r_{min} + j_{min} \leq r_i + j_i \leq r_{max} + j_{max}$$

As we treat only not reentry traces, we can exploit the result on the hyper period proposed for the schedulability analysis of asynchronous systems (Leung and Whitehead 1982). By analogy with our case, the analysis may be done in $[r_{min} + j_{min}, r_{max} + j_{max} + 2.lcm]$.

6.1 Accessibility graph generation

Let G be the accessibility graph corresponding to the schedulability analysis of S. We define a tasks state C of G as follows :

$$C = \{S, T, t\}$$

Where,

- S a tasks set to execute,
- T a selected task to execute between all the activated tasks of S. We apply the EDF policy to perform such selection.
- t the start time of the T execution.

We present the following algorithm to construct G. We apply such algorithm to analyze all the executions scenarios in the interval $[r_{min} + j_{min}, r_{max} + j_{max} + 2.lcm]$. We present the following steps to construct G. The first step lets to construct the first tasks state C_0.

First step: Let $C_0 = \{S_0, T_{min}, t_{min} = r_{min} + j_{min}\}$ be the first state in G. S_0 contains the tasks belonging to first.

$$S_0 = first$$

We generate then step by step the different tasks states in the different G trajectories as follows,

Step: Let $C_i = \{T_i, S_i, t_i\}$ be a state in the graph G where $T_i = \{WCET_i, pred_i, succ_i\}$

if $T_i \notin last$

Let suppose that $succ_i$ contains k tasks sets.

$$T_i.succ = \{ts_1, ..., ts_k\}$$

We construct k tasks states $C_0, ..., C_{k-1}$ target of C_i as follows,

$$\forall j \in [0, k-1],\ S_j = (S_i \backslash \{T_i\}) \cup ts_j$$

if $T_i \in last$

Let tr be the trace containing T_i.

We construct C_j target of C_i as follows

$$S_j = (S_i \backslash \{T_i\}) \cup \{first(tr)\}$$

Thanks to this graph construction, we deduce that the application is not schedulable if one deadline of a task belonging to a tasks state is not satisfied.

Running the example. *We perform the proposed algorithm to analyze the schedulability of the control application fbn. We propose the following temporal characteristics of T_1 and T_5.*

- $r_1 = 2$; $p_1 = 50$; $j_1 = 1$.
- $r_5 = 1$; $p_5 = 50$; $j_5 = 1$.

We suppose the following worst case execution times of the different tasks :

Task	T1	T2	T3	T4	T5	T6	T7
WCET	2	3	4	1	2	1	3

By constructing the accessibility graph in the hyper period $[2, 103]$, the algorithm verifies all the temporal constraints and proves that the application is schedulable. We present a part of such graph (figure 5). Applying the "first step", we construct the first tasks state $C_0 = \{\{T_1, T_5\}, T_5, t = 2\}$. Then, we apply "step" to construct the remainder states.

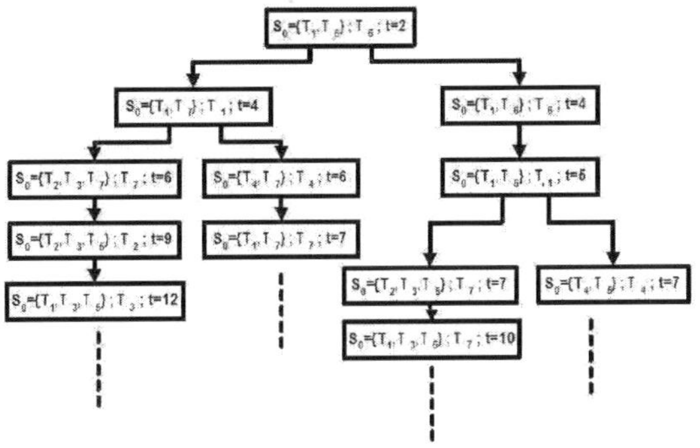

Figure 5. *The accessibility graph*

We deduce from the accessibility graph an off-line scheduling of the application (figure 6). We show in each state of this graph the task and the start time of its execution. Note that for sake of conciseness, identical branches parts of the tree are merged.

6.2 Events priorities policy

In a function block fb, the selection of input occurrences must be based on the corresponding tasks deadlines. Indeed, the occurrence to select corresponds to the task that has the earliest deadline. But the ECC is unaware of these temporal properties. We propose therefore to exploit the previous schedulability analysis to generate events priorities for each block.

As the schedulability analysis is performed in the hyper period $[r_{min} + j_{min}, r_{max} + j_{max} + 2.lcm]$, we propose to generate for each ECC the order

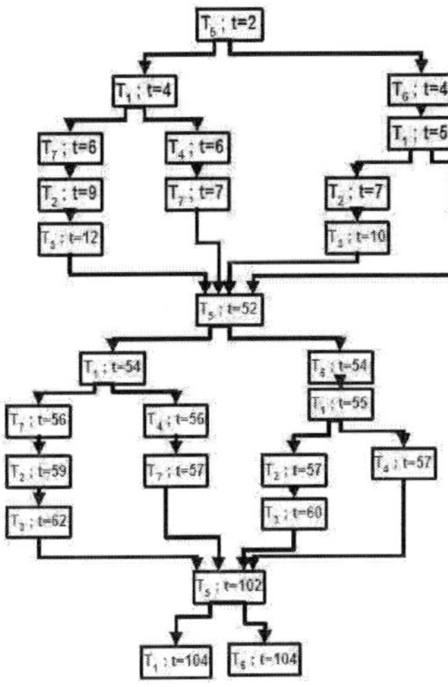

Figure 6. The schedulability analysis of the FBs network

of occurrences to select in such hyper period. The selection mechanism must be compatible with the generated schedulability analysis. Let $T_{i,m}$ and $T_{j,n}$ be two occurrences to select, we note that $T_{i,m} \ll T_{j,n}$ if the ECC has to select $T_{i,m}$ before $T_{j,n}$.

Therefore, the scheduler will receive at a time t from an ECC the adequate task that it is dedicated to process according to the schedulability analysis. Such policy lets to guarantee the conformity between the internal behavior of a FB and the scheduler behavior inside the resource.

Running Example. *We deduce the following events priorities for ECC_1, ECC_2 and ECC_3.*

- $FB_1 : T_{51} \ll T_{11} \ll T_{52} \ll T_{12} \ll T_{53} \ll T_{13}$
- $FB_2 : T_{61} \ll T_{21} \ll T_{62} \ll T_{22} \ll T_{63}$
- $FB_3 : T_{71} \ll T_{31} \ll T_{72} \ll T_{32}$

In each block, the ECC must select always the input occurrences according to such order to guarantee the correct composition with the scheduler.

7. CONCLUSION AND FUTURE WORKS

This paper presents a contribution to develop an industrial control application according to the IEC 61499 standard. This contribution lets to validate the temporal interoperability of such application. First of all, we propose an abstraction of the function block behavior to take into account all its possible executions. Moreover, we propose to verify the correctness of the application with regard to end to end delays. To perform such validation, we propose to transform the application to a particular dependant tasks model. We propose then the generation of tasks deadlines to guarantee the temporal correctness of the application. Finally, we propose a schedulability analysis and an off-line scheduling generation. On the other hand, to guarantee the correct composition between blocks and the scheduler, we propose to take into account the off-line scheduling in the definition of the events selection policy.

We are currently working to find heuristics reducing the number of states in the accessibility graph in order to reduce the combinatory explosion. Moreover, we are plan to perform the schedulability analysis of an application distributed over several resources of a device. We also will extend our work to the distribution over resources of several devices. Such extension imposes to take into account the communication interface inside each device and the networks delays.

On another side, we will extend our researches by exploiting the fault tolerant scheduling approach to reduce temporal constraints in the case of non critical systems. Therefore, we authorize the application to not respect some of its end to end delays.

REFERENCES

Articus (1996). *Rubus OS Reference Manual.* Articus Systems.

Babanov, A., J. Collins and M. Gini (2003). Scheduling tasks with precedence constraints to solicit desirable bid combinations. In: *International Conference on Autonomous Agents. Melbourne. Australia.*

Crnkovic, I. and M. Larsson (2002). *Building reliable component-based software systems.* Artech House. London.

Cucu, L. and Y. Sorel (2003). Schedulability condition for real-time systems with precedence and periodicity constraints, without preemption. In: *11th Real-Time Systems Conference. France.*

Cucu, L., R. Kocik and Y. Sorel (2002). Real-time scheduling for systems with precedence, periodicity and latency constraints. In: *10th Real-Time Systems Conference. France.*

holobloc (2005). http://www.holobloc.com.

Howell, R R. and M k. Venkatrao (1995). On non-preemptive scheduling of recurring tasks using inserted idle time. In: *Information and computation Journal.*

IEC61131-3, International Standard (1993). *Programmable Controllers Part3.* Bureau Central de la commission Electrotechnique Internationale. Switzerland.

Ifak (2005). http://www.ifak-md.de/wg7/.

Jeffay, K., D. Stanat and C. Martel (1991). On non-preemptive scheduling of periodic and sporadic tasks. In: *IEEE Real-Time Systems Symposium.*

Khalgui, M., X. Rebeuf and F. Simonot-Lion (2004). A behavior model for iec 61499 function blocks. In: *Third Workshop on modelling of Objects, Components and Agents. Denmark.*

Klein, M H., T. Ralya, B. Pollack, R. Obenza and MG. Harbour (1993). *A practioner's handbook for Real-Time analysis. Guide to Rate monotonic Analysis for Real-Time Systems.* Kluwer Academic Publisher.

Leung, J. and J. Whitehead (1982). On the complexity of fixed-priority scheduling of periodic real-time tasks. In: *performance Evaluation 2(1982). 237250.*

Lewis, R. (n.d.). *Modelling control systems using IEC 61499.* The institution of Electrical Engineers.

Liu, JWS. (2000). Real-time systems. In: *Prentice Hall.*

Pecos (2005). http://www.pecos-project.org.

Stewart, D B., R A. Volpe and P K. Khosla (1997). Design of dynamically reconfigurable real-time software using port-based objects. In: *IEEE Transactions on Software Engineering.*

WG6, International Standard IEC TC65 (2003). Function blocks for industrial process measurements and control systems. Committee Draft.

WG6, International Standard IEC TC65 (2004). *Industrial Process Measurements and Control Systems.* Committee Draft.

ELSEVIER
IFAC
PUBLICATIONS

FUNDAMENTAL CONSIDERATIONS FOR IMPLEMENTING CONTROL SYSTEMS ON A CAN NETWORK

Guy Juanole * **Gérard Mouney** **
Christophe Calmettes *** **Marek Peca** ****

** LAAS-CNRS, 7 Avenue du colonel Roche*
31077 TOULOUSE (FRANCE); juanole@laas.fr
*** LAAS-CNRS, 7 Avenue du colonel Roche*
31077 TOULOUSE (FRANCE); mouney@laas.fr
**** LAAS-CNRS and Centre Universitaire*
J-F Champollion 81000 ALBI (FRANCE);
christophe.calmettes@univ-jfc.fr
***** LAAS-CNRS, 7 Avenue du colonel Roche*
31077 TOULOUSE (FRANCE); marek@laas.fr

Abstract: Process control applications implemented through communication networks require, for the transfer of data between the remote tasks, a timeliness Quality of Service in order to guarantee performances (stability for example). The goal of this paper is by considering an implementation, on a saturated network CAN, of a regulation system, to study the influence, on the performances of the regulation system, of the priority scheme on CAN. *Copyright ©2005 IFAC*

Keywords: Automatic Control, Stability, Communication, Networks, Simulation

1. INTRODUCTION

In the technological context of today, distributed systems (computers connected through a communication network) are more and more used, particularly, for implementing industrial applications like, for example, process control based on a closed loop structure (Astrom and Wittenmark, 1997), (Halevi and Ray, 1988). The study and the design of such systems, called Networked Control Systems (Zhang *et al.*, 2001), represent an area of research very important because of its pluridisciplinary aspect. This aspect involves a deep knowledge in Automatic Control, Control Science (mainly task scheduling) and Communication Networks (mainly message scheduling and communication protocols).

Two kinds of works have emerged these last years:
(i) one which wants to link the Quality of Service of the distributed systems to the performances (stability for example) of the closed loop process control systems (Juanole, 2002); but these works consider the design of the distributed systems mechanisms and of the controller for the process control system in a separate way;
(ii) another one (more recent) wants to consider the view of an integrated system (i.e consideration of an approach of codesign for the controller and the scheduling in particular (Cervin, 2003), (Marti Colom, 2002), (Sename *et al.*, 2003)). The work presented in this paper stands in the first kind and is a continuation of a first work submitted to the ETFA05 conference (Juanole *et al.*, 2005).

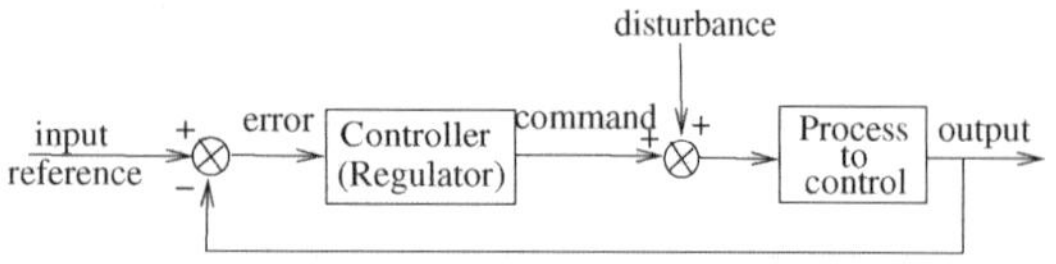

Fig. 1. Regulation system

We want to link in a very intimate way the parameters and performances related to the process control (Automatic Control view) with parameters and mechanisms related to the network and more precisely the message scheduling (Communication Network view). Control problems can be classified into regulation problems and servo problems. The major issue on the regulation problem is to control (reject) disturbances. The objective of a servo problem is to make the output of the system respond to command signals on a specified way (model to follow which gives the desired response of the output). Practical control problems involve both servo and regulation properties. *Here we restrict our work to regulation problems.*

This paper includes three parts. The first part presents the context of the study (regulation systems, CAN network, the tool for the study). The second part presents fundamental considerations concerning the influence of the priority scheme used in CAN on a regulation problem. The third part concerns a study of a regulation example (control of a continuous power motor).

2. CONTEXT STUDY

2.1 Regulation systems

The structure of a regulation system is given on the figure 1. The output of the process to control is enslaved to a reference which represents the input of the system. The role of the controller is to provide performances in terms of stability and accuracy in steady state (accuracy with respect to the input and with the disturbances). Here we do not consider the accuracy aspect and we focus on the stability. The process to control and the controller are represented by a transfer function (here we will consider transfer functions based on a Laplace transform (Franklin *et al.*, 2002)). The process to control that we consider has a transfer function of the type :

$$G(s) = \frac{K}{s(1+\tau s)}$$

This type of transfer function is characteristic of a lot of processes (for the closed loop system, no position error and speed error control).

The regulation system implemented through a communication network is represented in figure

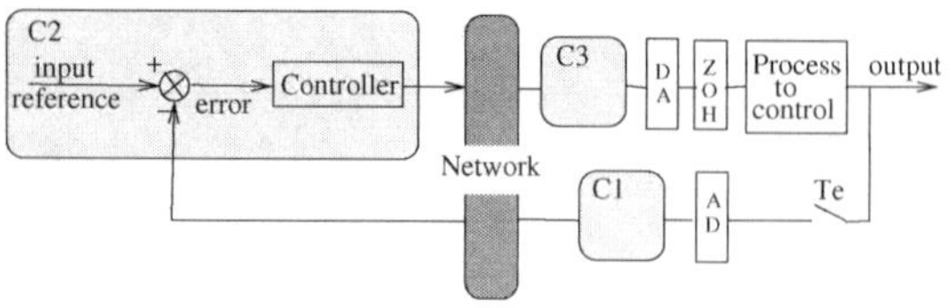

Fig. 2. Implementation of a regulation system on a network

2. The network operates in the loop both between the computer 1 (C1) associated to the numerical information provided by the AD conversion (this computer includes a task that we call the sensor task and which generates the sensor flow) and the computer 2 (C2) where we have the reference and the controller (in C2 we have a task called controller task which generates the controller flow) and between C2 and the computer 3 (C3) which provides numerical information to the DA conversion in front of the Zero Order Hold (ZOH) (figure 2). The sensor flow goes from C1 to C2 and the controller flow from C2 to C3.

The task which generates the sensor flow is time-triggered (the sampling is based on a clock) whereas the task which generates the controller flow is event triggered (the controller waits for the sensor sample receptions before computing and generating its flow). The implementation through the network then induces two delays (the delays of the sensor flow and the controller flow) and requires a ZOH module (such a module can be seen, in conditions that we precise nextly, as introducing also a pure delay in the loop). We neglect here the task computation delays.

2.2 On the network CAN : main ideas

CAN network (CiA, 2002), (Tindell *et al.*, 1994) is based on a serial bus shared by stations by means (in the MAC sublayer) of a CSMA/CD scheme with a deterministic collision resolution based on priorities associated to the identifiers (adresses) of the frames which carry data (n bytes with $n \leq 8$) of application tasks. Smaller is the adress (*i.e.* it has many dominant bits (bits 0)), higher is the priority. A station, which does not recognize the adress of the frame it is emitting, detects a collision and stops till the bus becomes free. The only station which does not detect a collision (it is the station which, among the stations trying to transmit, has the longer sequence of dominants bits) continues and then is the winner. The priorities associated to the frames identifiers reflect the priorities of the data included in the frames.

The length of a frame is given by the following formula (Navet, 1999) wich results from the frame format :

$$L = 47 + 8n + \lfloor \frac{34 + 8n - 1}{4} \rfloor$$

The bit rate which is considerated here is 1Mbit/s.

We only consider periodic traffic and we characterize the use of CAN by two rates :

- the Use Request Factor (URF) which represents the frame rate requested to the MAC sublayer by the data requests coming from the application layer (by considering i periodic flows with L_k and T_k respectively the frame duration and the period of the flow, we have URF= $\sum_{k=1}^{i} \frac{L_k}{T_k}$),
- the Use Network Factor (UNF) which represents the frame transmission rate on the bus (obviously UNF ≤ 1)

URF can be ≤ 1 or > 1. This last case traducts application tasks requests rate higher than the possibilities of frame transmission on the bus.
It is also important to specify the buffering technique at the MAC layer : FIFO queue (which implies waiting delays which can be very long) or one buffer with overwritting (which can induce erasures and then losses).

2.3 The tool for the analysis

We use the tool TrueTime (Henriksson and Cervin, 2004), a toolbox for simulation of distributed real-time control systems. TrueTime can simulate the timely behaviour of real-time kernels executing controller tasks and also simple models of network protocols and their influence on networked control loops (the MAC model protocol of CAN is supported).

In TrueTime the buffering mechanisms in the communication entity is only FIFO queuing (we have added the "one buffer with erasure" technique). *We will only consider this last technique in this paper.*

3. FUNDAMENTAL POINTS ON A REGULATION SYSTEM : INFLUENCE OF THE PRIORITY SCHEME USED FOR THE NETWORK ACCESS

3.1 Objective

The objective is to make a complete study of the influence, on the behaviour of the control application (response to a position step input), of the priority scheme (priorities between the control application flows and the external flow, priorities inside the control application between the sensor flow and the controller flow).

In order to well see this influence, we consider a *saturated network* (the global User Request Factor URF, *i.e* the sum of the URF of the control application flows and the external flow, is higher than one). We will have the following priorities : Priority sensor flow (Psf), Priority controller flow (Pcf) and Priority external flow (Pef). When talking about the process control application, we will also use the notation (Psf, Pcf) to characterize the couple of the two flows.

3.2 Hypothesis

The considered system We consider a regulation system (figure 1) where the process to control has a transfer function of the type $\frac{K}{s(1+\tau s)}$, the controller is a proportional controller (gain 1) and the feedback is an unity feedback. So the open loop transfer function is :

$$G(s) = \frac{K}{s(1 + \tau s)}$$

Such system is representative of a lot of regulation systems, which (due to the integration $\frac{1}{s}$) have no position error, which control the speed error and which have a second order closed loop transfer function :

$$F(s) = \frac{G(s)}{1 + G(s)} = \frac{\frac{K}{\tau}}{s^2 + \frac{1}{\tau}s + \frac{K}{\tau}}$$

$$F(s) = \frac{\omega_n^2}{s^2 + 2\zeta\omega_n s + \omega_n^2}$$

with ω_n : natural pulsation and ζ: damping. We have the relation $\frac{K}{\tau} = \omega_n^2$ and $\frac{1}{\tau} = 2\zeta\omega_n$.

A fundamental criteria which is considered here is the stability criteria which is characterized by the phase margin

$$\phi_m = \pi + \arg\{G(j\omega)\}$$

$$\phi_m = \pi - \frac{\pi}{2} - \arctan(\omega\tau)$$

for $\omega = \omega_0$ such that

$$|G(j\omega)| = \frac{K}{\omega\sqrt{1 + \omega^2\tau^2}} = 1$$

The phase margin gives, in particular, informations on the damping ζ and the overtaking $D\% = e^{-\frac{\pi\zeta}{\sqrt{1-\zeta^2}}}$ (Franklin *et al.*, 2002) of the response of the closed loop transfer function. Note still the following performances : t_r (rise time) $\approx \frac{1.8}{\omega_n}$ and t_{resp} (response time at 5%) $\approx \frac{3}{\zeta\omega_n}$.

Quantitatives constraints We specify now the three types of constraints which are considered in our study : the constraint of the type of the considered control applications, the constraint of the service requested to the network and the constraint of the network service delay.

(i) Constraint of the type of the considered control application We consider control applications which, without the network, have a phase margin of 65° which implies $\arctan(\omega_0\tau) = 25°$ (and then $\omega_0\tau = 0.47$), and $K\tau = 0.5145$. The closed loop system has then good performances i.e $\zeta = 0.7$ and $D\% = 5\%$. Note that the two relations

$$\omega_0\tau = 0.47 \quad \text{and} \quad K\tau = 0.5145$$

are representative of a set of open loop transfer functions (different values of K and τ) which have a margin phase of 65° for a pulsation $\omega_0 = \frac{0.47}{\tau}$.

(ii) Service requested to the network. We consider that the data included in the sensor flow and controller flow frames are short informations of two bytes ($n = 2$) which give frames of 75 bits. As we use the simulator TrueTime for the analysis and we want to use it in the simplest way, we only consider frames which are multiples of one octet and then we choose frames of 72 bits which gives frames of duration $72\mu s$. We furthermore want to make the study with a URF, relative to the control application, close to 1 (but < 1) : this will allow, at first, to study normally the behaviour of the network dedicated to the control application, and second, to easily bring the network at the saturation (by considering one flow external to the control application and making the global URF > 1) in order to make the study of the control application in a network shared in hard conditions. That is why we choose the sampling period of the control application (call Te this period) equal to $150\mu s$. So, as the transmission in sequence of the frame of the sensor flow and the frame of the controller flow takes $144\mu s$, it remains $6\mu s$ during which the external flow can take the bus. In this way we are able to illustrate, in particular, the importance of the choice of the priority scheme associated to the frames of the control application and to the external frames. The frames of the external flow carry data of 8 bytes (this would give frames of length 135 bits but as we consider in TrueTime simulations multiples of an octet, we will then takes frames of length 128 bits (duration $128\mu s$)). Concerning still the external flow we will consider several values of the period (in order to make to vary the global URF).

(iii) Network service delay. The network imposes to the regulation system a delay $\tau_D = \tau_S$ (delay of a sensor flow frame) $+\tau_C$ (delay of a controller flow frame) $+\frac{Te}{2}$ (delay introduced by the ZOH module). This last delay is an approximation of the ZOH behaviour if we have $\frac{1}{Te} \gg$ higher frequency in the spectrum of the loop signal (this frequency is $f_0 = \frac{\omega_0}{2\pi} = \frac{0.47}{2\pi\tau}$), then we find $\tau \gg 10\mu s$. The open loop transfer function of the regulation application implemented through the network is then :

$$G(s) = e^{-\tau_D s}\frac{K}{s(1+\tau s)}$$

Flows characteristics We have two flows which are time-triggered (sensor flow, external flow) and one flow which is event-triggered (controller flow which results from the reception of the sensor flow). The two time-triggered flows start at the same time (the waking time is the time 0).

The function $G(s) = \frac{K}{s(1+\tau s)}$ which is studied We take a function which belongs to the functions satisfying our hypothesis *i.e.* the phase margin of 65° of which gives for the closed loop system $\zeta = 0.7$ and $D\% = 5\%$. Furthermore as we consider a sampling period T_e of $150\mu s$ we can define a rise time t_r which can be choosen (we know the following relation : $4 \leq \frac{t_r}{T_e} \leq 10$ (Astrom and Wittenmark, 1997)); we choose $t_r = 4T_e = 600\mu s$. From t_r we have the relation $tr \approx \frac{1.8}{\omega_n}$, and we get $\omega_n = 3.10^3$ rd/s. We know that $\frac{1}{\tau} = 2\zeta\omega_n$ (then $\tau \approx 0.25$ms) and $\frac{K}{\tau} = \omega_n^2$ (then $K = 2020$rad/s). The function $G(s) = \frac{K}{s(1+\tau s)}$ is then defined.

3.3 Previous results

We summarize the results which have been given in (Juanole *et al.*, 2005).

(Psf,Pcf) > Pef The period of the requests of the external flow T_{ef} is $128\mu s$ *i.e.* the URF of this flow is already one. The case 1 (Psf > Pcf) and the case 2 (Pcf > Psf) are considered. Though the external flow has an URF of 1, that does not prevent the process control application to be implemented. It works however less well than with a dedicated network (where there is only the delay for the transmission of the frames) : we have now in more the priority inversion phenomenon (common to the cases 1 and 2) but the case 2 (Pcf > Psf) is better because the controller (which has the higher priority) can act faster after the sensor flow frame reception.

Pef > (Psf,Pcf) We cannot here consider that the external flow saturates the network because, as it has the highest priority, the process control application could not be implemented. We give

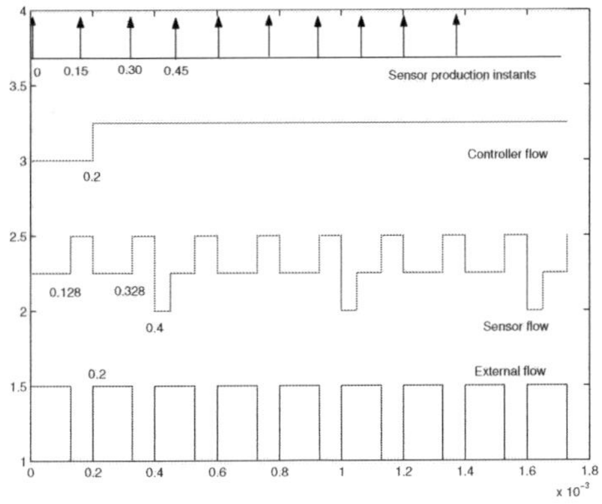

Fig. 3. Flows (case 1)

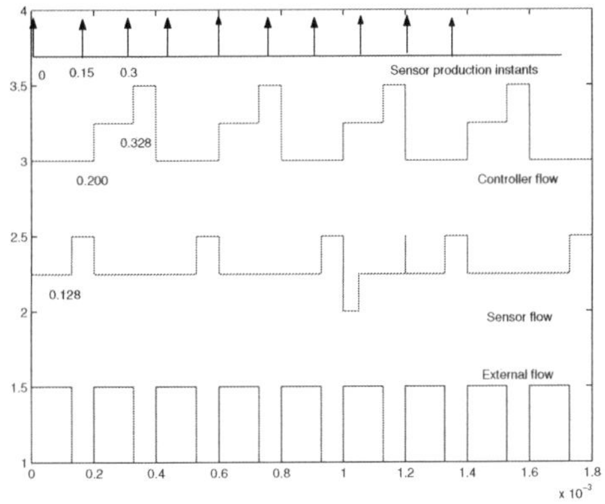

Fig. 4. Flows (case 2 and Pef>Pcf > Psf)

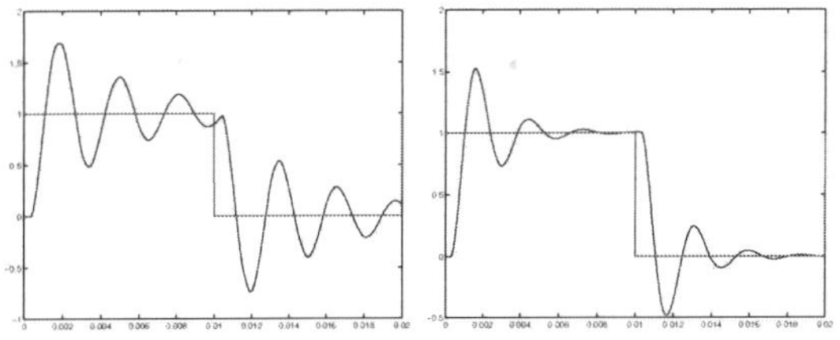

Fig. 5. Output (case 1, left; case 2, right)

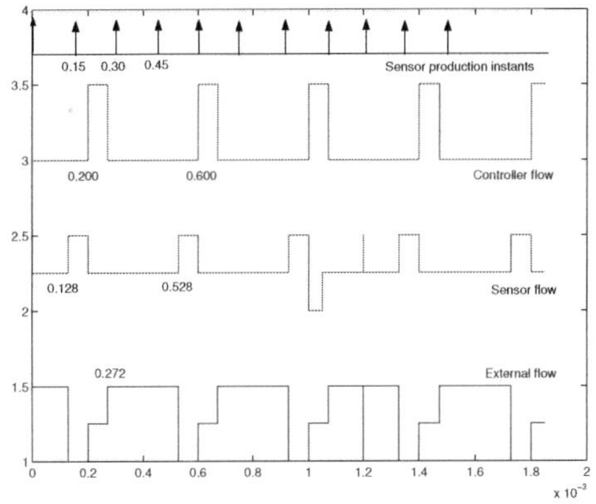

Fig. 6. Flows : Pcf > Pef > Psf

the results of two studies : we consider, for the external flow, successively a period of $200\mu s$ (as $200\mu s = 128\mu s + 72\mu s$, during the external flow period, only one frame of the control application can be sent) and a period of $500\mu s$ (several frames of the control application can be sent during this period).

For each study, we consider the case 1 (Psf > Pcf) and the case 2 (Pcf > Psf). The simulator TrueTime provides diagrams representing the flow shapes with three levels : the low level represents "nothing to send", the middle level represents "network service request" and the high level represents "sending a frame in the network". In the figures 3 and 4 we represent the external flow, the sensor flow, the controller flow and the sensor production instants; the time is given in ms.

- study 1 (period of $200\mu s$)
  In the case 1, the controller can never send a frame as it has the lowest priority (see the flows exchanged figure 3) and then the process cannot be implemented; in the case 2 the controller can now send frames (see the flows exchanged figure 4) and then now the process control can be implemented.

- study 2 (period of $500\mu s$)
  As we have now more room during the period of the external flows, the process control application can be implemented (but the case 2 (priority of the controller flow higher than the priority of the sensor flow) is better than the case 1: see the response to an unit position step input on the figure 5, where in the case 2 the response has a lower damping and is less oscillatory).

3.4 New studies

The objective is, at first, to analyse the case where Pef is comprised between the priorities of the couple (Psf, Pcf) and, second to summarize all the different cases resulting of the combinatorial of the priorities.

Pef comprised between the priorities of the couple (Psf, Pcf) We still consider the period of $200\mu s$ for the external flow.

- Psf >Pef> Pcf
  We have the result (no possibility of implementing the process control application) identical to the one in the case Pef >Psf> Pcf (the controller can never send a frame). We will refine the comparison between the two cases (Pef >Psf> Pcf and Psf >Pef> Pcf) in the summary of all differents cases by considering external flow periods higher than $200\mu s$ and then by lowering the URF.

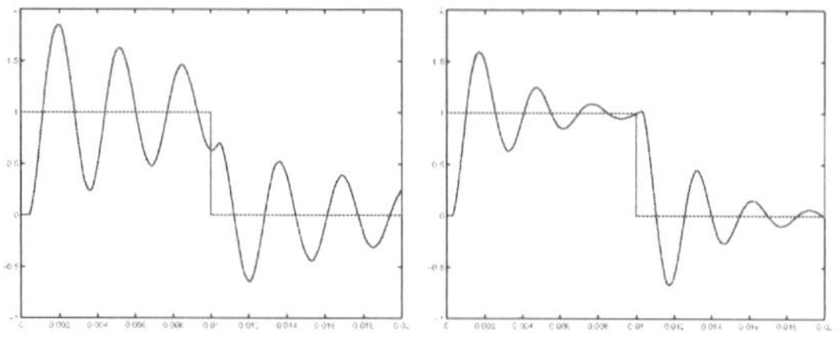

Fig. 7. Position step answer : left (Pef > Pcf > Psf); right (Pcf > Pef > Psf)

- Pcf >Pef> Psf
  We have a result better than in the case where we have Pef >Pcf> Psf. The diagram of the figure 6 (compared to the diagram of the figure 4) shows that now the controller flow (as it has the highest priority) sends a frame immediately after the reception of the sensor flow. The consequence of this difference appears on the figure 7 representing the performance of the process control application (response to a position step input).

Summarizing all the cases
(a) Context of the study We observe the behaviour of the flows (*ef, sf, cf*) on the network during a time of 5000μs (as the rise time ($t_r = \frac{1}{\zeta\omega_n}$) of the process control application is 1.4 ms, we then have informations on the steady state behaviour of this application). The number of requests of the flow *sf* is $\lfloor \frac{5000}{150} \rfloor = 33$. That means that the optimal behaviour of the process control application is when we have 33 frames *sf* and 33 frames *cf* which are exchanged through the network. In order to analyse the behaviour of this application, we consider that the period of *ef* changes between 200μs and 3000μs (so the global URF (*sf, cf, ef*) changes from 1.6 to 1.0027) and we look at the number of frame exchanges on the network, in function of the combinatorial of the priority scheme.

(b) Results The results are given on the table 1. The main comments are the following :
- the case where Pcf > Psf gives an identical distribution of the frames *cf* and *sf* (which is the normal behaviour); obviously if the flow *ef* has an important URF (less room for the frames *cf* and *sf*, we have not the optimal behaviour (33 frames *sf* and 33 frames *cf*) which only appears when the global URF arrives to the value 1. An aspect (which does not appear on the table but which was shown on the figure 7) is that it is better to have Pef between Pcf and Psf than higher than Pcf and Psf (the delay of the controller frame (which follows the reception of the sensor frame) is shorter because the controller has the highest priority).
- the case where Psf > Pcf is less good because we have not (except when the global URF arrives to 1) the same number of frames *sf* and *cf*; if Psf > Pef the number of frames *sf* is always 33 (it is normal as Psf has the highest priority in the network; on the other hand, if Pef > Psf, we have not always 33 frames *sf* (we have erasures of these flows because the flow *ef* (highest priority) is always transmitted when it needs and so delays the flows *sf*, that can be erased); however (in strong saturated condition like URF =1.6 or 1.472 or 1.216) the case Pef > Psf has better performance (damping for exemple) than the case Psf > Pef because we have less differences between the number of frames *sf* and *cf* (a waited solution being the equality and the optimal being the number 33).

3.5 Important conclusion

In saturated conditions, it is recommended to choose Pcf > Psf and Pcf > Pef > Psf (instead of Pef > Pcf > Psf). However, this conclusion makes to appear the question of the choice of the scheduling algorithm : for example, if we use the Rate Monotonic algorithm, the second relation cannot be satisfied in some cases (*ef* period < regulator sampling period); on the contrary if we associate the priorities to importance degrees of the different applications, we can choose these priorities in such a way that the second relation is satisfied.

4. EXAMPLE OF AN IMPLEMENTATION ON CAN OF A CONTINUOUS POWER MOTOR REGULATION

This is a simple example of implementation of a regulation of a real system (motor control) in a CAN context. The time constant of the motor is such that the URF of the regulation application is very low ($\ll 1$) and then the analysis which is made is just a particular use of the general analysis done in 3.

4.1 Hypothesis

We consider a motor with the Laplace transfer function $G(s) = \frac{8}{s(1+0.25s)}$ and here is a regulator (proportional corrector) with a gain $k = 0.257$ which allows a phase margin of 65°, then a damping $\zeta = 0.7$ and an overtaking of 5%. As $\omega_n = 2.85$rd/s (see the general equation in 3.2) we have t_r (rise time) $\approx 630ms$ and the we choose, as sampling period (see 3.2) $Te = \frac{t_r}{4} = 150$ms. Concerning the frames of the sensor flow and the controller flow, we consider that they are like in the study of the section 3.3 (duration 72μs).

Table 1. Frame exchanges

	T_{ef} (μs)	200	250	500	1000	1500	3000
	URF	1.6	1.4720	1.216	1.08	1.045	1.0027
	Application flows	(sf, cf)	(sf, cf)	(sf, cf)	(sf, cf)	(sf, cf)	(sf, cf)
Psf > Pcf							
	Psf > Pef > Pcf	(33, 0)	(33, 1)	(33, 18)	(33, 27)	(33, 29)	(33, 33)
	Pef > Psf > Pcf	(25, 0)	(25, 9)	(32, 20)	(32, 28)	(31, 31)	(33, 33)
Pcf > Psf							
	Pcf > Pef > Psf	(13, 12)	(17, 17)	(26, 26)	(30, 30)	(31, 31)	(33, 33)
	Pef > Pcf > Psf	(13, 12)	(17, 17)	(26, 26)	(30, 30)	(31, 31)	(33, 33)

4.2 Implementation on a dedicated network

Taking into account for the sampling and the use of CAN, we have (see section 3.2) a delay (in seconds) $\tau_D = \frac{Te}{2} + \tau_s + \tau_c = 75 \times 10^{-3} + 72 \times 10^{-6} + 72 \times 10^{-6} \approx 75 \times 10^{-3}$. It results mainly of the delay introduced by the ZOH module (the delays resulting of the frames *sf* and *cf* being insignificant). Then we have a decrease of the phase margin of 9° which gives an overtaking of 10%.

4.3 Implementation on a shared network

Hypothesis In more of *sf* and *cf* flows (URF$= \frac{72\times10^{-6}}{150\times10^{-3}} = 0.99 \times 10^{-3}$) we consider an external flow *ef* which has the characteristics of the *ef* of the study in section 3.3 (frame duration 128 μs, period $\geq 128\mu$s) and which allows to simulate changes of the global URF.

General results We consider two cases : *ef* period = 128 μs and *ef* period $\geq 128\mu$s.

(i) ef period=128μs (global URF =1.00099) : If Pef > (Psf, Pcf), the regulation application cannot be obviously implemented (*ef* is requesting the bus as soon as it has finished to transmit and it has the highest priority). On the contrary, if (Psf, Pcf)> Pef the regulation application can be implemented (when an *sf* or *cf* flow is waiting for accessing the bus and if it is in competition with the *ef* flow, it wins) and the performances do not depend on the position of Psf with respect to Pcf (as Te = 150 ms, (which is a large value with respect to the length of the frames *sf* and *cf* and of the frame *ef*), we cannot have competition between a request to send by the controller (after the reception of a sensor frame) and a new request to send by the sensor; the competition appears when T_e becomes smaller; see the study in section (3)).

(ii) ef period > 128μs (global URF $\leq$ 0.9931) : The regulation application can always be implemented and also the performances depend
- neither on the position of Pef with respect to (Psf, Pcf) : as URF of *ef* is less than 1, *ef* (if it has the highest priority) let room for a flow *sf* or *cf*, waiting for the access of the bus, to access the bus,
- and nor on the position of Psf with respect to Pcf : for the same reason explained in the case "*ef* period=128μs", relatively to the large value of T_e with respect to the length of the frames *ef*, *sf* and *cf*, the position of Psf with respect to Pcf has no influence.

Remark : the case where *ef* period > 128 μs (*i.e.* URF(ef)< 1) is the normal condition for implementing the regulation application. It would be interesting, in this case, to make a study where the regulation application can saturate the network (by decreasing T_e) and analyse, in this case, the influence of the position of Psf with respect to Pcf and to link this study to the results given on the table 1. This work will be done nextly.

5. CONCLUSION

This study has demonstrated two important results for the implementation of a regulation system including two flows in sequence (the sensor flow from the sensor to the controller and the controller flow, from the controller to the actuator) in a context of a saturated network CAN (the saturation comes from the presence of other flows (external flows)).

The first result is : we must have priority of the controller flow > priority of the sensor flow; in this case we have an identical distribution of the two flows. The second result is : it is better to have the priority of the external flow between the priority of the controller flow and the priority of the sensor flow than above the priority of the controller flow (in this way the delay of the response of the controller flow to the sensor flow is shorter).

This study will be continued in two directions : the first one concerns the implementation of control system consisting into regulation and servo problems; the second one will consider the implementation of several control applications.

REFERENCES

Astrom, Karl J. and Bjorn Wittenmark (1997). *Computer-controlled systems – Theory and design*. International Edition – Prentice Hall.

Cervin, Anton (2003). *Integrated Control and Real-Time Scheduling*. Department of Automatic Control, Lund Institute of Technology.

CiA, CAN in Automation group (2002). *CAN, CANopen, DeviceNet*. URL : www.CAN-CiA.de.

Franklin, Powell and Emami-Naeini (2002). *Feedback Control of Dynamic Systems*. Prentice Hall.

Halevi, Yoram and Asok Ray (1988). Integrated communication and control systems: part 1 - analysis. *Journal of Dynamic Systems, Measurement, and Control* **110**, 367–373.

Henriksson, Dan and Anton Cervin (2004). Truetime 1.2 - reference manual. *Deparment of Automatic Control, Lund Institute of Technology*.

Juanole, Guy (2002). Quality of service of communication networks and distributed automation: models and performances. $15^t h$ *Triennial world Congress of the IFAC, Barcelone*.

Juanole, Guy, Christophe Calmettes, Gérard Mouney and Marek Peca (2005). On the implementation of a process control system on a can network : linking the process control parametres to the network parameters.. *ETFA'2005*, 10th *IEEE International conference on Emerging Technologies and Factory Automation, Catania, Italy*.

Marti Colom, Pau (2002). *Analysis and design of Real-Time Control Systems with Flexible Timing Constraints*. Universitat Politècnica de Catalunya, Departament d'Enginyera de sistemes, Automàtica i Informàtica Industrial.

Navet, Nicolas (1999). *Evaluation de performances temporelles et optimisation de l'ordonnancement de tâches et messages*. Doctorat de l'Institut National Polytechnique de Lorraine.

Sename, Olivier, Daniel Simon and David Robert (2003). Feedback scheduling for real-time control of systems with communication delays. *IEEE International Conference on Emerging Technologies and Factory Automation, Lisbon*.

Tindell, K., H Hansson and A. Wellings (1994). Analysing real-time communications : Controller area network (can). *IEEE RTSS, Puerto-Rico*.

Zhang, W, M. S. Branicky and S. M. Philipps (2001). Stability of networked control systems. *IEEE Control Systems Magazine* **21**(1), 84–99.

ELSEVIER
IFAC
PUBLICATIONS

DEVELOPMENT OF A STANDARDIZED FIELDBUS-BASED GREENHOUSE CLIMATE CONTROL

Olga Plaksina, Thomas Rausch

Institute of Computer Technology, Vienna University of Technologies
Gusshausstrasse 27-29, A-1040 Vienna, Austria
{plaxina, rausch}@ict.tuwien.ac.at

Abstract: The paper investigates a new application area for standardized control networks which have been primary developed for the use in home and building automation. The work concentrates on the feasibility of building automation technologies for the climate control in growing environments. This includes, but is not limited to, the acquisition and processing of environmental data. The paper analyzes the requirements for greenhouse climate control and compares these demands to those of residential building automation. The authors provide a concept of a KNX-based control system for this application domain and give an outlook on the future phases of the project. *Copyright © 2005 IFAC*

Keywords: Agriculture, data handling systems, distributed control, enterprise integration, environment control, fieldbus, knowledge representation.

1. INTRODUCTION

Growing crop in greenhouses is an important branch of agriculture and it is a labour intensive and technically challenging business. Optimized control helps to increase production despite saving precious resources. In recent years, growers have invested in a broad range of new technologies, including more efficient growing media, reconstruction of the heating or irrigation equipment itself. In many cases these modifications also involve improvements of associated climate control systems.

The complexity of horticultural production and a growing demand for effective and economic resource management leads to an immediate need for increasing the volume of monitoring and automated control. The number of parameters in greenhouses has been constantly increasing over the last decade (Gieling, 1998). Horticultural structures like greenhouses are very often characterized with significant spreading and decentralization. Therefore the implementation of centralized systems is cumbersome and decreases reliability. A distributed control system exhibits better performance and can be used to improve transparency of the installation and to ease the maintenance. But, in spite of superficial similarities of applications framing climate control in buildings and greenhouses, fieldbus-based building automation/control networks and computerized greenhouse control systems (presented mostly by proprietary solutions) show major differences. That complexity of greenhouse control outperforms regular building automation. Plants grown in covered structures react very sensitive to climatic variances. Therefore the control must be very smooth and constant.

2. GREENHOUSE CONTROLS vs. BUILDING CONTROLS

In recent years research on greenhouse climate control has been dealing more with optimization of climate models or controlling algorithms (Seginer and McClendon, 1992; Langang, *et. al.* 2000) rather then with communication techniques. Technical implementations (e.g., Stipanicev and Marasovic, 2003) present applications with newest technology achievements but did not concern on the variety of environmental data and climate models and specific greenhouse requirements. This paper combines both areas, together with integration them into global enterprise information system.

2.1 A modern greenhouse as a control object.

A highly dynamic climate is typical for greenhouses (Bailey, 1995). For the comfort in residential buildings, the thermal environment is playing the leading role. An economically and environmentally reasonable system for a modern room climate control concerns a few parameters such as air temperature and velocity, wall and heater surfaces temparature, CO_2 concentration, as well as exterior weather conditions (D. Jelondz, 2001). Due to much more intensive response of plants to their environments and the dependance on the outside weather, greenhouse climate conditions are more dynamic comparing to domotics application domain. Moreover, there could be a necessity to change internal climate several times within 24 hours upon diurnal or multi-day schedule depending on the crop and the growing phase (Seginer and McClendon, 1992).

In (Hanan, 1998a) 33 different subset of different factors for greenhouses and plant production can be distinguished. Five of the most important parameters are listed in Table 1.

Table 1 Greenhouse environmental criteria under measurement and subject to control

Main factor	Subsets	Actuators
Radiation	Global radiation outside, net radiation, etc.	Shade screens, supplemental irradiation
Temperature	Outside, inside, crop and root zones. Water and pipe temperature, etc.	Boilers, ventilators, evaporative pads, thermal screens, etc.
Humidity	Inside and outside	Misting systems, ventilators, heating systems
Carbon dioxide	Inside and outside	CO_2 injection
Wind	Outside, velocity	Ventilator systems

These characteristic values are needed for precise description of interior and exterior climatic conditions of a greenhouse. Building upon these information a control system can influence the internal enviromental conditions. All parameters must be continuously measured and analyzed. Beside the active climate control, the recorded data is of great importance for plant growth related research. Out of this historical information, the grower can apply further improvements to increase the quality and the amount of the crop. This data can be used in the long planning horizon.

Another very important aspect to keep in mind is the tolerance of greenhouse "inhabitants": In residential buildings, temperature fluctuation within some degrees over a particular time interval would not cause serious harm to the tenants. Naturally these fluctuations won't please any inhabitant however if not occurring constantly, these events can be regarded as tolerable inconveniences. For industrial farming failures in the temperature profiles could be vital for the some types of crop and thus lead to serious losses. Therefore even solitary failures in the climatic control must be utterly avoided.

The greenhouse control strategies were developed through years and represent sets of rules defining procedures to bring the interior climatic system to a desired safe state. The operation of the affected equipment is of the major importance since the most climatic parameters rely on the correct functionality of the installed devices. Examples for rules of operation are particular starting sequences for exhaust fans and evaporative pads in order to provide an optimal and efficient cooling process. Proportional, integral and derivative algorithms as well as their combinations are usual for calculating of controlling outputs in greenhouses, though a problem of conflicting requirements and interacting control loops for systems of heating, ventilation, etc. still remains a subject of research (Hanan, 1998). Also physical conditions of operation shall be taken into account to prevent damages of the equipment. In case of snowfall, a shade curtain should be fully open and stay in this position until the snowfall ends; otherwise there is a risk of damage caused by the accumulated snow. Some types of greenhouse-specific equipment (e.g. high-intensity lamps used for supplementary lighting, which are not common for residential buildings) require certain delay after switching off, in order to prohibit rapid reversals.

Similar to buildings, optimization of energy consumption and reduced operation costs are important in a horticultural industry. Additionally, when a greenhouse is operated as business enterprise, the planning horizon usually goes beyond just 24-hours continuous operation to solve short-term problems (Hanan, 1998). It spreads over certain longer periods, and provokes development of prospective policies, which could depend on fuel costs, taxes, marketing course, quality of products, etc. In this case, the environmental data and information of current expenditures or energy loads, recorded by the greenhouse control system, could be used for long-term company management, on the one hand, and the set-points for the current crop production could be defined by the company's strategy, on the other (e.g. to produce more flowers by exact specific date, and not a week earlier, Fig. 1).

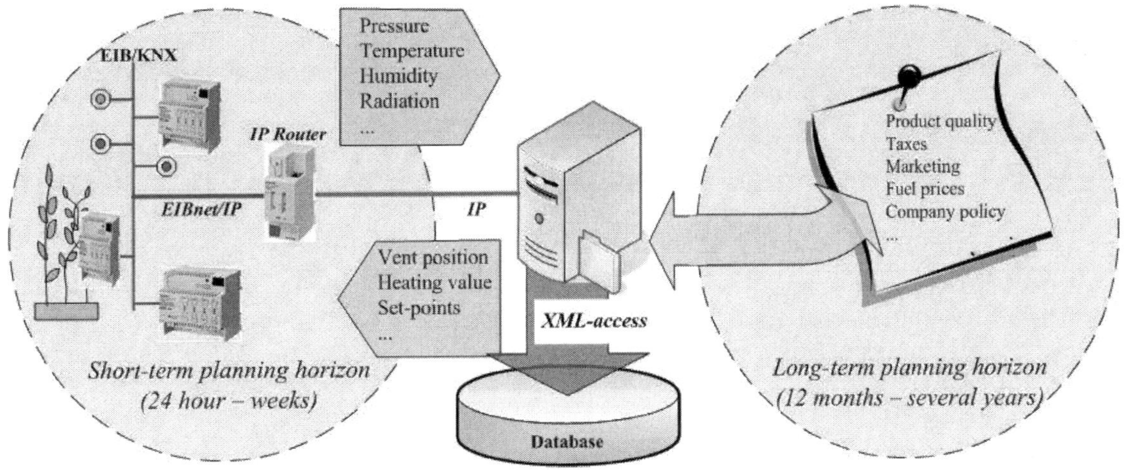

Fig. 1. System architecture from the operational point of view and planning horizons.

2.2 The importance of standardized control networks.

KNX is a standardized European technology for control networks. It has evolved mainly from the well-known European Installation Bus (EIB), which is supported by the leading European companies in electrical industry (Dietrich, *et al.*, 2000), but it also bases upon two other field bus technologies: Batibus and the European Home System (EHS).

There are many arguments for using a standardized control network like KNX in comparison to some proprietary technology. Proprietary solutions are not open, in terms of accessible to the public. Only the company or consortium which has developed the particular technology has the legal rights and the technical know-how to use it properly. Standardized network technologies on contrary can be implemented by anyone. As a result, a broad spectrum of different products is available for these technologies (Kastner, 2004). For the customers this also means more safety on their investments. They do not rely on a single company for future extensions, but are able to choose from various vendors. These and other reasons explain why today the majority of projects in building automation are implemented with standardized technologies.

The use of a standardized communication subsystem provides a proven technology which mainly consists of standard components. Due to the complexity of climate control, some special functions will however not be implemented in regular devices found for home and building automation. In this case additional features have to be implemented to fulfil the desired tasks.

2.3 Requirements to the system

Summarizing the features described in the previous sections, it is now possible to formulate the requirements for a KNX-based system for greenhouse climate control:

- Highly efficient data acquisition and analysis to cope with continuous measurements.
- Accuracy and stability meaning the ability of the system to achieve the desired set-points and control the output with a permissible error.
- Ability to follow the rule sets, which prescribe the sequence of the operation, designed for safe and efficient running of equipment and choosing proper algorithms to avoid energy load peaks.
- Interface for information exchange with enterprise resource planning systems or other company management processes.

These requirements build the basement for the conceptual model of the climatic control system.

3. CONCEPTUAL MODEL

This chapter describes a general scheme for the communication between different components of the climate control system. It also illustrates how data is represented in order to achieve the required performance and integrability.

The definition of clean interfaces for each component allows a modular architecture of the system. On the field level (Fig. 2), the KNX bus provides communication between the sensors and the controlling units, which have internal "intelligence". Once the set-points and reaction algorithms are loaded into system components, they do not need a central supervision station and provide with climate maintenance by fulfilling their local tasks. This distributed approach allows more scalability and reliability than traditional centralized systems.

Though the computing capacity of field controllers is sufficient for local control loops, the system-wide supervision might require complex calculations of new setpoints. These complicated processes run on dedicated servers, which constantly acquire environmental data. The interconnection between the distributed control network and the servers used for additional global control can be established over any internet work.

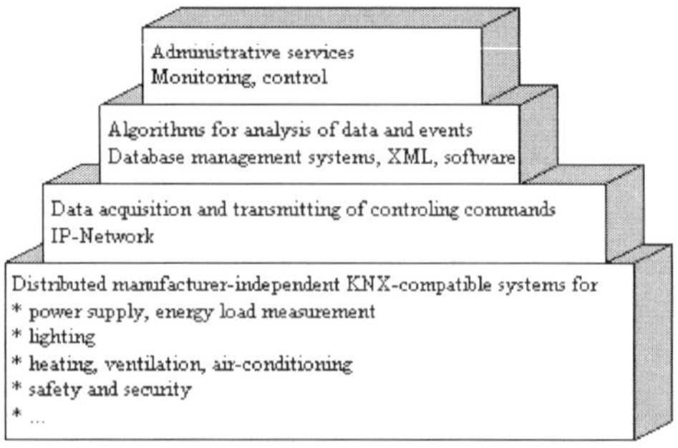

Fig. 2. Modular system architecture.

All data generated in the control network is stored in a database. The server calculates globally optimized controlling commands (e.g. to avoid energy peaks) and set-points (when necessary) and transmits them back to the bus to update the settings of the system components.

The database records should be available for remote data retrieval and greenhouse monitoring, as well as for information exchange with a company management system. The XML technology allows integration of different data sources (Vakali, *et. al.*, 2005) and provides with web-services (e.g. using Internet-resources like weather forecasts).

One of the main advantages of XML data model consists of its ability to work with unstructured data. Structured data is for instance sensory data retrieved from the KNX-telegrams, which can be partition into records with similar fields and therefore stored in a relational database. Unstructured information such as device parameters or greenhouse climate model could have different attributes and needs some other mechanisms to work with. For solving a problem of integration of heterogeneous information sources, a virtual approach based on XML data model seems to be a promising perspective. Data is not physically stored in a centralized unified database but the end-user queries are translated into sub-queries and data is interpreted as if it had unitary logical representation.

4. CONCLUSION

The paper comprehends application requirements and conceptual model for a greenhouse control system, which deals with climate control and acquiring of environmental data used for facilitating plant growth activity as well as for horticultural research and greenhouse business management. The model presents a balanced combination of complex distributed control network and information system coverage.

The current work is being focused on investigation and definition of proper knowledge representation formalisms to provide the most efficient processing of the environmental data and development of corresponding XML-enabled data stores. Analysis of control strategies and mechanisms for choosing an optimized one is also a subject of investigation.

REFERENCES

Bailey, B. J. (1995) Greenhouse Climate Control - New Challenges In *Greenhouse Environment Control and Automation* (Ed. Kano, A.). Acta Hort. Kyoto, Japan.

Cunha, J.B., de Moura Oliveira, J.P. (2003). Optimal management of greenhouse environments In. *Proceedings of EFITA 2003 Conference, 5-9 July 2003*, pp. 559 – 564. Debrecen, Hungary.

Dietrich, D., Kastner, W., Sauter, T. (2000). *Gebäudebussystem*, p. 41. Hüthig, Heidelberg.

Gieling, Th. H. (1998) Sensor and Measurement, a Review In *II Intrenational Symposium On Sensors in Horticulture* (Ed. van Meurs, W. Th. M., Gieling, T. H., Bennedsen, B. S.). Acta Hort. Tune Landboskole, Greve, Denmark.

Hanan, J.J. (1998). *Greenhouses: advanced technology for protected horticulture.* CRC Press, Boca Raton.

Kastner W. et al. (2004). A Closer Look on Today's Home and Building Networks In *IEEE Africon, 7th Africon Conference in Africa, Technology Innovation*, **vol. 2** pp. 1239 – 1245.

Jelondz, D., Spasokukotskiy, K., Ruser, H. (2001). Concept and realisation of an EIB based automated room climate control In *Proceedings of EIB Conference*. Technical University Munich.

Langang, P., Wanliang, W., Qidi, W. (2000). Application of Adaptive Fuzzy Logic System to Model for Greenhouse Climate In *3rd World Congress on Intelligent Control and Automation*, pp. 1687 - 1691. Hefei, P.R. China.

Seginer, I., R.W. McClendon (1992). Methods for optimal control of the greenhouse environment In *Transactions of the ASAE 35(4)1299-1307.*

Stipanicev, D., Marasovic, J. (2003). Networked embedded greenhouse monitoring and control In *IEEE Conference on Control Applications*, **vol. 1**, pp. 1350 – 1355.

Vakali, A., Catania, B., Maddalena, A. (2005) XML Data Stores: Emerging Practices In *Internet Computing* (Ed. Clarke Siobhan), **vol. 9**, issue 2, pp. 62 - 69. IEEE Computer society.

ELSEVIER
IFAC
PUBLICATIONS

COMMUNICATIONS REQUIREMENTS FOR AUTONOMOUS MOBILE ROBOTS: ANALYSIS AND EXAMPLES

Valter Silva[1),2)], José A. Fonseca[2)], Urbano Nunes[3)], Rodrigo Maia[3)]

[1)] *Escola Superior de Tecnologia e Gestão de Águeda*
Universidade de Aveiro, Portugal
vfs@estga.ua.pt

[2)] *Departamento de Electrónica e Telecomunicações*
Universidade de Aveiro
jaf@det.ua.pt

[3)] *Institute for systems and Robotics*
University of Coimbra, Portugal
{urbano,rmaia}@isr.uc.pt

Abstract: Autonomous Mobile Robots (AMRs) are becoming pervasive and, in consequence, the optimization of their electronics is also becoming a current research topic. In particular, the use of distributed architectures is an issue. In the presence of distribution, a network must be used to interconnect the several modules of an AMR. This network must be able to support the traffic requirements to carry on the data flows produced by the nodes. In this paper a characterization of the subsystems required in AMRs is made, analyzing typical AMRs. Then, the data flows produced by them are quantified and the adequacy of popular fieldbuses, in particular CAN - Controller Area Network, to accommodate the required traffic in two robots is analyzed. *Copyright © 2005 IFAC*

Keywords: Data Flow Analysis, FieldBuses, Mobile robots, Sensors

1. INTRODUCTION

Nowadays, Autonomous Mobile Robots (AMRs) are becoming widely used in several fields, for example factory automation systems, rescue, research, science and technology contests and others.

AMR architectures are almost always proprietary. Currently, fieldbus (Thomesse, 1998) based distributed architectures are starting to be used. Processing tasks are often distributed by several nodes, some of them based on microprocessors or microcontrollers and other based on PC computers (or even PC computers themselves).
An AMR is, often, a safety critical device as, during operation, a defective behavior can put in risk humans or property. When AMRs are based on a distributed architecture, the communication network plays a very important role in what concerns fulfilling the requirements that permit the system to show properties such as dependability.

The use of communication networks in safety critical applications is a current research topic led by the automotive industry and distributed real-time systems research community. A recent overview of this field in what concerns automotive applications can be obtained in (Koopman, 2002). Many of the theoretical studies in this field use data coming from the automotive industry itself. The data characterizes the traffic requirements for the operation of a car. Examples of such data are the so-called PSA benchmark (Navet, 1997) which describes the most important data streams in the distributed system of a car and the SAE benchmark used in (Tindell, 1994) which describes the same type of data for an electrical vehicle.

In what concerns AMRs, a first study of the use of fieldbuses in these devices was done in (Nunes, 2003). However, the study is just qualitative. So, if an adequate assessment of the possibility to use distributed embedded systems in AMRs is to be done, then quantitative data is required. The quantification of the data flows in a robot distributed architecture is important because, in some cases, the data generated by one sensor can vary according to the environment conditions and thus it can leave or restrict the bandwidth used by data from other sensors. This means that the communication fieldbus would benefit from QoS management to achieve a better usage of the available bandwidth. For example, if the robot has different speeds, the sampling of perceptional sensors systems and consequently the data generated by them, can change significantly. Thus an important issue with strong impact in the communications is the use of a vision system. In fact, the amount of data produced can be unsuitable for the most usual fieldbuses.

In this paper the problem of using distributed embedded systems in AMRs starts to be addressed. To do this, the paper makes an initial characterization of the distributed architecture, identifying the subsystems that will generate data streams that must be delivered for the different system elements. The characteristics of these data streams are identified, leading to figures like the data length and period of transmission. With these figures, an insight of the communication requirements of the AMR distributed system is obtained. This study will enable in the future the evaluation of the adequacy of the most relevant fieldbuses to be used in this specific application.

In particular, the use of CAN fieldbuses (Bosch, 1991) and their implications, advantages and disadvantages in a robot agent for robotic soccer and in a robotic wheelchair is studied. In both cases a time triggered protocol has been applied to increase the performance of the distributed system. The end to end delay and the jitter is reduced when using this time-triggered protocol (Silva, 2005).

2. FIELDBUS BASED AMRS

AMRs can be applied in very diverse fields. In all applications fields one can find distributed architectures even if those are not the most common. Below we identify some AMRs that rely on fieldbuses to support distributed architectures.

In (Mock, 1999) a snake like robot is presented for test purposes of the on-board real time communication system. The robot is divided in sections, each one has 4 servo motors and one sensor. The servos are responsible for the section motion and the sensor is responsible for the servos' state. Each section communicates with the others via a CAN network where 4 nodes are connected to interface the motors and sensors.

The robot presented in (Moore, 2000) is an unmanned ground vehicle like the previous one. This robot is a six-wheeled omnidirectional autonomous mobile robot. Each wheel has three degrees of freedom, drive, steering and height. In each wheel there is a microcontroller to take care of the tasks related with speed control. This leads to a distributed architecture where a CAN fieldbus interconnects some sensors and a local area network interconnects some subsystems.

In (Cavalieri, 1997) the authors present an orange picking robot. The robot is a mobile vehicle that moves between the rows of trees. The robot is composed by four platforms, each one with two picking arms. A monochrome camera mounted in the vehicle provides the basic information for movement. An SP50 (later Foundation Fieldbus FF-H1) fieldbus is used to provide connectivity between the four platforms and support the required data exchanges.

Other application of AMRs is in contests intended to promote research and training in science and technology. In the authors' research team AMRs for robotic soccer are being develloped (Almeida, 2004). In those devices a laptop computer controls each robot and performs the upper level control behavior. The laptop computer communicates with several PIC microcontrollers through a CAN bus. A real time protocol, called FTT CAN (Almeida, 2002) has been implemented, which can accommodate synchronous and asynchronous traffic in the same fieldbus. The microcontrollers are responsible for the sensors and actuators and for the low level processing. Three motors with a holonomic steering are used to move the robots. These robots use a Web Cam for vision and localization and act like a team, communicating one with the other and with a base station through a wireless LAN.

The AMRs briefly described above, although targeted for very different applications, present similar architectures with sensors, processing units and actuators interconnected by fieldbuses. In the literature we could not find analysis of the traffic requirements and the correspondent relation with sensors for mobile robots. In order to be able to do that analysis, we start, in the next section, to identify and briefly describe the typical electronic elements which can be found in such robots.

3. ELECTRONIC ELEMENTS AND ARCHITECTURES

To be autonomous, AMRs must take decisions by themselves. One or more controllers installed onboard take decisions and compute all the sensor data, sending commands to actuators.

In what concerns sensors, many types can be found. The tasks associated with these sensors are the following:

- Proximity detection (infrared sensors, ultrasonic sensors, hall effect detectors)
- Range determination (radars, sonars, lasers, microphones, infrared sensors)
- Imaging (cameras, photocell arrays)
- Orientation (compasses, accelerometers, gyroscopes)
- Positioning (GPS)
- Contact detection (bumpers, limit switches, push buttons)
- Internal (rotary encoders, magnetic field, battery level sensors)
- Other detection mechanisms (for example gas detectors, humidity detectors, ...)

For the purpose of this paper an important issue is to determine the bandwidth required for each type of sensor which depends on the data generated by the sensor. The discussion is then centered on this topic. However, there is not a consensus in this issue. Some authors claim that sensors must be sampled as often as possible (Greenwald, 2003), but, if restricted resources are used, the sampling rate must be carefully chosen. Others claim that, in most of the cases, the sampling rate for reading or writing I/O devices is determined in an ad-hoc manner (Stewart, 2000). A common approach that takes into consideration the limits of the resources available is to use different timings at different control levels. If we consider a global navigation system like the one described in (Nunes, 2003), sampling rates of the order of 100 Hz are used in the motion control and motion tracking levels, and 10 Hz for the local motion planning level. The following paragraphs briefly analyze the data flow requirements for each type of sensor.

In proximity detection the sensors mostly used are infrared or ultrasonic. These sensors can output digital or analog data. If a digital sensor is used, the traffic generated by the sensor is one bit length and the sensor behaves like a bumper. On the other hand, if a sensor with analog output is used, the data generated depends of the used ADC. Typically an 8 or 10 bits ADC is sufficient for most applications. Due to its limited range of detection, this kind of sensors is more suitable to navigate near obstacles. That means that, in the absence of obstacles, the acquisition rate of these sensors can be reduced or even switched off. In that way, the amount of traffic generated by these sensors can be important if they are connected to a fieldbus. In some contexts, e.g., robotic soccer, an efficient management of bandwidth usage can be done if the sampling rate of these sensors is properly chosen, and, if possible, changed during the robot operation. If the sensors have a large sampling period the used bandwidth is lower but the robots can collide. If the sampling period of the sensors is decreased, the robot can have a larger speed but the usage of the fieldbus bandwidth is also larger. For example, if the range of the sensor is 50 cm and the robot has a maximum speed of 50 cm/s the sampling rate must be greater than 1 Hz. In other words, the maximum blind space must be less than the range of the sensor, otherwise the robot will be always blind.

Bumpers are probably the simplest sensors for a mobile robot and change state (one bit) if the robot collides with an obstacle. Usually bumpers work in an interrupt driven fashion. But if not, all the considerations made for the infrared or ultrasonic sensors are valid.

Laser sensors are also range sensors used in mobile robots. This sensor outputs the obstacle distance in an angular range. For example the SICK LMS laser scanner (Sick, 2002) can work with an angular range of 100° or 180°, with 0.5° or 1° of resolution. The laser has ranges of 8.1 m or 81 m with resolutions of 1 mm and 1 cm respectively. This laser performs a complete scanning in 52, 26 or 13 ms (depending on the resolution) and outputs the data using a RS232 or RS485 connection. In the coarser operation mode (100° with 1° of resolution) the laser scanner outputs 212 bytes per frame (202 of usable data), and performs about 19 scans per second. This originates 110704 bps (bits per second) of usable data.

Video cameras can be used in robot tasks such as obstacle detection/avoidance and target detection/following. The generated traffic is dependent of the desired video resolution and video rate. For example, the manufacturer OmniVision (OmniVision, 1999) has many types of sensors with different resolutions and video rates. For many applications a QCIF resolution (356x292), 8 bits per pixel, with 15 frames per second is appropriate. With this resolution the generated traffic is 1.55 Mbytes per second. This can be reduced if the resolution is reduced and some compression is made.

GPS are widely used in outdoor environments positioning devices. These use commonly the NMEA-0183 protocol (NMEA, 2002) where the communications is done over a RS232 link. The frames are variable in size up to 82 characters (including header and footer). The frame rate is less than one message per second. The total maximum amount of traffic is 80 bytes per second (640 bps).

One of the most used compasses is the Vector2X. This compass outputs 10 bits (2 bytes with bit stuffing) 5 times per second. This originates 50 bps.

The rotary incremental encoders are used for dead reckoning in a great number of robots (Everett, 1995). Usually encoders have a pre-processing circuit which integrates the counting pulses produced. This circuit has often a 16 bits counter. Whenever a reading is required, two bytes of data must be acquired. In the soccer AMR described later in this paper, readings of the encoders are made each 5 ms. However, if the motion control level uses a 10 ms cycle, as referred above, then the traffic generated by such an encoder is 200 bytes per second.

The data for motor set points can vary in length, but 1 byte is sufficient for many applications such as the ones described below. For example, if the actualization frequency is 33 Hz it will generate 33 bytes per second of data.

In the next two sections, the characterization of the hardware and application requirements of the CAMBADA team and RobChair is made. These two robots have a distributed system based on the CAN network.

4. CAMBADA ROBOTIC AGENT CHARACTERIZATION

The general architecture of the CAMBADA robots has been described in (Almeida, 2004). Basically, the robots follow a biomorphic paradigm, each being centered on a main processing unit, the brain, which is responsible for the higher-level behavior coordination, i.e. the coordination layer. This main processing unit handles external communication with the other robots and has high bandwidth sensors, typically vision, directly attached to it. Finally, this unit receives low bandwidth sensing information and sends actuating commands to control the robot attitude by means of a distributed low-level sensing/actuating system, the nervous system (Figure 1).

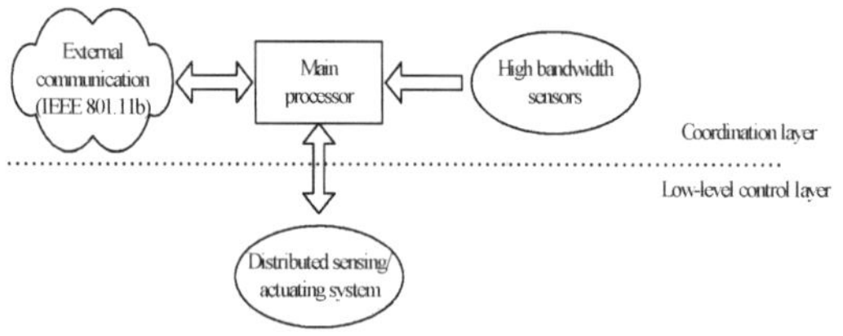

Fig. 1. The biomorphic architecture of the CAMBADA robots

At the heart of the coordination layer is the Real-Time Database (RTDB) that contains both the robot local state information as well as local images of a subset of the states of the other robots. A set of processes update the local state information with the data coming from the vision sensors as well as from the low-level control layer. The remote state information is updated by a process that handles the communication with the other robots via an IEEE 802.11b wireless connection. The RTDB is then used by another set of processes that define the specific robot behavior for each instant, generating commands that are passed down to the low-level control layer (Figure 2).

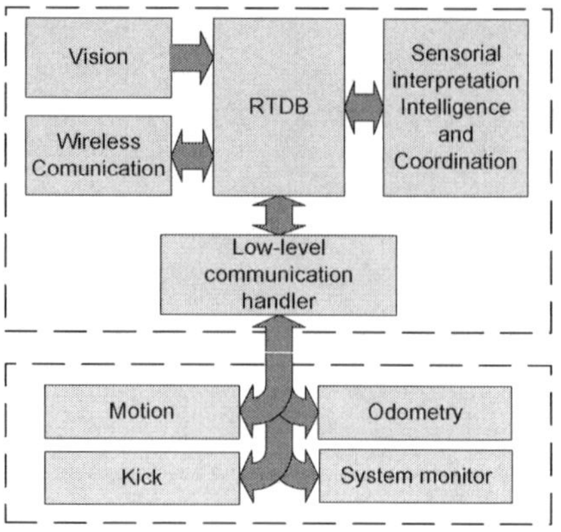

Fig. 2 - The robots functional architecture built around the RTDB

The low-level sensing/actuating system follows the fine-grain distributed model (Silva, 2005) where most of the elementary functions, e.g. basic reactive behaviors and closed-loop control of complex actuators, are encapsulated in small microcontroller-based nodes interconnected by means of a network. The nodes are based on the PIC microcontroller 18Fx58 operating at 40MHz while the network uses the CAN protocol with a bit rate of 250Kbps.

At this level there are 3 DC motors with their respective controllers plus an extra controller that, altogether, provide holonomic motion to the robot. Each motor has an incremental encoder that is used to obtain speed and displacement information. Another node is responsible for combining the encoder readings from the 3 motors and building a coherent displacement information that is then sent to the coordination layer. Moreover, there is a node responsible for the kicking system that consists of a couple of sensors to detect the ball in position and trigger the kicker. This node also carries out battery voltage monitoring. Finally, the low-level control layer is interconnected to the coordination layer by means of a gateway attached to the serial port of the PC, configured to operate at 115Kbaud. From the perspective of the low-level control layer, the higher coordination layer is hidden behind the gateway and thus, we will refer to the gateway as the source or destination of all transactions arriving from or sent to that layer.

5. ROBCHAIR CHARACTERIZATION

Figure 3 depicts the robotic wheelchair RobChair (Pires, 2002). A brief description of its sensory and actuation system is presented in order to permit an analysis of its data communications requirements. The RobChair has a mechanical structure built by Vector Mobility, equipped with two driving wheels powered by two 24V Permanent Magnet DC Motors with a nominal torque of 29.3 Nm, and three caster wheels to assure stability. The motors are driven by two power amplifiers 80A8T from Advanced Motion Controls, which allow voltage and current control modes with resolutions above 12bits. Each motor has an optical encoder with a resolution of 20000 pulses per wheel revolution that is used for odometry calculations. A 2-axis inductive analog joystick allows HMI between user and RobChair. To obtain surrounding obstacle information 12 IR proximity sensors (On/Off), 12 Sharp GP2D12 analog IR range sensors and an ultrasonic based range finder system ME-EERUF (Moita, 201) are used. Additionally a SICK LMS 200 laser scanner and low resolution firewire cameras are being integrated. To provide absolute positioning a magnet sensor ruler developed at ISR (Bento, 2005) is also being installed.

Fig. 3 – The wheelchair robot

In (Nunes, 2003) is presented a Global Navigation Architecture for this type of autonomous mobile robots. This layered architecture enables a distributed architecture where one or more functional units provide one ore more functionalities. In the RobChair project, the first functionalities implemented were the interfaces with the sensory and actuation systems, the low-level motion control and a reactive shared-control behavior (Pires, 2002). In order to increase robustness, modularity and flexibility, each functionality is implemented by one processing unit. In Figure 4, the distribution of the functionalities by several processing units is presented. The sensory and actuation systems interface units, which do not required large processing power, are based in a Microchip 18F258 PIC microcontroller. On the other hand, tasks that involve more processing power like motion tracking (Maia, 2003), path-following (Coelho, 2005), laser-based perception (Mendes, 2004) are implemented in embedded PCs (currently Advantech PCM-9577).

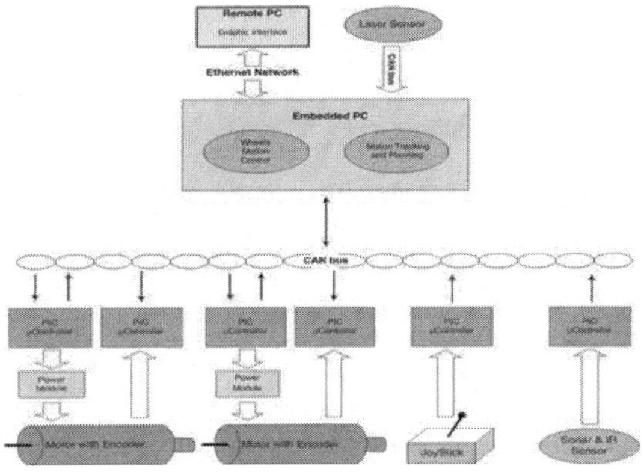

Fig. 4 – wheelchair architecture

6. BEST CASE COMMUNICATION REQUIREMENTS USING A CAN BUS

When a fieldbus is used, all the traffic can be transmitted in a single network or a division between heterogeneous networks can be made. Currently, in the automotive industry, fieldbuses like CAN (Bosch, 1991), LIN (Lin, 2000), Flexray (Flexray, 2005) and MOST (Most, 1999) are the most widely used (Axelsson, 2003). Cars will use some or most of these networks simultaneously, each transmitting the traffic for which it is adequate. A prospective view of the use of networks in next generation cars can be obtained in (Leohold, 2004).

Concerning the case of robots, the research is yet in an early phase, so the issue of using a single network or heterogeneous networks must still be addressed. However, a distributed system with CAN fieldbus is installed in our robotic soccer team low level hardware and in RobChair. In the early versions, a CAN network without any protocol has used in the robotic soccer robots. A time trigger protocol, called Flexible Time Triggered CAN (Almeida, 2002), has been installed in that hardware architecture to support the data exchange. In this paper the use of the fieldbus without any protocol is studied. In the RobChair, a CAN network without any protocol is used.

For the robotic soccer, each robot has 8 PIC nodes, each one responsible for a specific task. The robot has 3 motors with respective encoders and an actuator to kick the ball (Kicker). Also two infrared sensors for ball detection for kicking are used. The motion orders come from a laptop computer connected to the CAN network via a gateway and are computed by a separated node which sends commands to each one of the three motor nodes. The encoders data is computed by one node which, after some calculations, transmits the position information using the CAN network (normally to be read by the high level software via the referred gateway). The infrared sensors are used for ball detection. If the ball is near the robot, the sensors are sampled at 50Hz leading to 100 bytes of usage CAN bandwidth for both infrared sensors.

The traffic generated to command the kicker is very small (due to the asynchronous and sporadic usage) so we will not take it into account in the Table 1. The values shown in Table 1 are for a encoders acquisition period of 10 ms and a 8 bit encoder counter. The motor setpoints has a resolution of 1 byte and are done every 30 ms.

Like presented in the previous chapter, the RobChair has 2 motors with encoders. The data provided by the encoders is sent to the central computer via the CAN bus. The central computer is also in charge of sending commands to the microcontrollers which control the power drive. The microcontrollers connected to the power drivers are also responsible for sending a message with voltage and current values that are used in each instant. This is important to the higher layer software to compute the speed and torque algorithms. In the table 1 this traffic will be considered as "other" and its period is the same as the actuation period of the motors.

The wheelchair can also be controlled by a human through a joystick. This joystick is connected to a PIC microcontroller that communicates with the CAN bus to send information to the motion coordinator. The infrared sensors and the sonar are also connected to a PIC microcontroller to communicate with the CAN bus. Table 1 also presents data flow of the RobChair, not including, however, the traffic generated by the joystick and by the infrared sensors. In that table is also included the traffic for a sequence number to identify any loss of sampling of the encoders. The encoders are sampled at 100 Hz outputting a byte each time and the motors are actuating with one byte every 10 ms.

In Table 1 the sensors and actuators of the two robots presented bellow are summarized. In that table the traffic generated by the robots sensors or to command the actuators is presented. The accommodation of the traffic in the bus is dependent on the hardware and on the protocol used.

In the referred CAN fieldbus, the information is organized into messages. Each message can include till 8 data bytes. An overhead of 34 bits for message identification (CAN 1.1), control and other features must be added as well as a maximum of 25 bits due to the bit stuffing (for a message with 8 bytes) (Nolte, 2003).

In Table 2 we summarize the bandwidth utilization for the robot 1 using two scenarios of traffic accommodations of the CAN messages. Scenarios 1 use piggybacking of the sensor data, i.e., several sensors share the data field of just one message. In scenario 2 each sensor produces its own message, independently of the length required for the data field. In Scenario 2, each sensor produces one CAN message leading to a larger bandwidth usage. However, each sensor becomes independent of the others. In this case, the addition or substitution of a sensor is easy and can be done rapidly.

In scenario 1, all the data generated by sensors of the same type are joined together in the same CAN message. In that case, some of the flexibility of using a distributed architecture is maintained, but the sensors of the same type must be connected to the same node. This can be difficult and lead to a significant increase in the wiring complexity when compared with scenario 2. For example, the infrared normally are dispersed for the entire robot and in this case must be connected to the same node.

In Table 2, it is shown that the use of different traffic accommodation scenarios can lead to significant differences in the produced traffic. Also, the system architecture must be different because the traffic accommodation is also different. For a larger flexibility of the system (scenario 2) more bandwidth is used, and also, for less usage of bandwidth, the system becomes more concentrate, with several sensors connected to the same node. In that scenario the connection of a new sensor, for example an infrared sensor, implies to connect it to an existing node and to change the software of this node. On the other hand, in scenario 2, the connection is easy, and, the same software that is used for other nodes can be used.

In table 1, the two traffic patterns generated by the two robots can be evaluated. The soccer robot uses about 3.2% of the available bandwidth (in 125KBps of bit rate) while the RobChair uses 26% of the bandwidth. In that way, more sensors can be added to the system if it is necessary. In spite of the low usage of bandwidth, collisions can occur in the bus increasing the end-to-end delay and, in consequence, the jitter. To avoid this, an implementation of the system architecture for the soccer team using a time-trigger protocol (Silva, 2005) was decided.

In (Silva, 2005) some practical measurements where made in order to verify the capability of the time-trigger protocol to reduce the end-to-end delay. In fact, without FTT, for an average end-to-end delay of 51ms the jitter is 30 ms while, with FTT the jitter is 1ms for a average end-to-end delay of 27ms. On the other hand, the time-trigger protocol adds an additional communication overhead of about 10% to the system.

This preliminary study shows that a CAN fieldbus with a reduced bit rate, e.g., 125Kbps which is typical for some applications, can accommodate the traffic requirements for an AMR if no imaging sensors such as video cameras are used. This example does not include also laser data but it seems preliminarily that it can be also accommodated using a larger bit rate.

However, this conclusion is not definitive as the implications of the mutual influence of traffic flows in the control performance are not taken into consideration. In fact, sampling periods obtained will be affected by network induced jitter. These systems can be considered NCSs, i.e. networked control systems which are a current topic of research not only in the communications research community but also in the control community. See (IEEE, 2004) for an overview of this field from the control perspective.

Table 1 – Fieldbus traffic for Soccer Team and RobChair

	Robot 1 (Soccer team)		Robot 2 (RobChair)	
Sensor	Qty	Traffic (Bytes/s)	Qty	Traffic(Bytes/s)
Infrared	2	100		
Motors	3	100	2	1200
Encoders	3	300	2	1200
Kicker	1	0	0	
Other			2	1600
Traffic (bytes per second)		500		4000

Table 2 – Communication scenario for Robot 1

	Sensors		Scenario 1				Scenario 2			
Sensor	Qty	Traffic	Messages	Bytes	Period	bps	Messages	Bytes	Period	bps
Infrared	2	100	1	2	20ms	3750	2	1	20ms	6500
Motors	3	100	1	3	30ms	2805	3	1	30ms	6435
Encoders	3	300	1	3	10ms	8500	3	1	10ms	19500
Total						15055				32435

7. CONCLUSION

In this paper a first study to incorporate fieldbuses in AMRs is presented. Using a centralized architecture the generated data and sampling rates of the sensors does not have a strong influence in the system performance. On the other hand, if a distributed architecture is used, the sampling rate and generated data affects directly the network utilization. In this case it may be necessary to optimize the sampling rate and the accuracy (for example changing the ADCs in analog sensors) of the sensors to reduce the bandwidth used.

In the paper preliminary values of the traffic requirements in AMRs were presented. First, a general view was obtained. Then, particular cases using a well known and popular fieldbus coming from automotive industry, CAN – Controller Area Network, were identified and quantified. It is shown that CAN can be an option if some sensors that are greedy in bandwidth are not used. However, the paper does not take in consideration the impact of the mutual influence of traffic flows in the controllers' performance.

In the samples robots presented, some practical advantages of the distributed architecture are visible. The wiring complexity has been reduced compared with the centralized architecture and the software has been split across the nodes and so, each node is responsible by a specific task associated with the hardware it controls. If a time triggered protocol is used like in the new version of the robotic soccer, the end to end delay and the jitter of the data flow can be both reduced.

ACKNOWLEDGMENTS

This work was supported by Fundação para a Ciência e Tecnologia under grant PRODEP 2001 – Formação Avançada de Docentes do Ensino Superior Nº 200.019, by project ARTIST2: Embedded Systems Design Proposal, Contract no.: IST-004527 and by project POSC/EEA/SRI/58016/2004.

REFERENCES

Almeida, L., Pedreiras, P., Fonseca, J.A.G. (2002), The FTT-CAN protocol: Why and how, IEEE Transactions on Industrial Electronics, Volume 49, Issue 6, pp. 1189-1201

Almeida, L., Santos, F., Facchinetti, T., Pedreiras, P., Silva, V., Lopes, L. (2004), Coordinating distributed autonomous agents with a real-time database: The CAMBADA project, Computer and Information Sciences - ISCIS 2004:19th Int. Symposium, Turkey.

Axelsson, J., Froberg, J., Hansson, H., Norstrom, C., Sandstrom, K., Villing, B. (2003), Correlating Business and Network Architectures in Automotive Applications – A comparative case study, Proceedings of the 5th IFAC International Conference on Fieldbus Systems and their Applications, July 7-8.

Bento, L., Nunes, U., Moita, F., Surrecio, A. (2005), Sensor fusion for precise autonomous vehicle navigation in outdoor semi-structured environments. IEEE International Conference on Intelligent Transportation Systems (ITSC'05).

Bosch GmbH (1991), CAN Specifications Version 2.0 – Technical Report, Stuttgart, Germany.

Cavalieri, S., Stefano, A., Mirabella, O. (1997), Impact of Fieldbus on Communication in Robotic Systems, IEEE Transactions on Robotics and Automations, Vol 13, No 1.

Coelho, P., Nunes, U. (2005), Path-following control of mobile robots in presence of uncertainties. IEEE Transactions on Robotics, vol. 21, n. 2, 252-265.

Everett, H. R. (1995), Sensors for Mobile Robots – Theory and Application, A K Peters.

FlexRay (2005), FlexRay requirements Specifications, Version 2.0.2/ April 2002 [online].

Greenwald, L., Kopena, J. (2003), Mobile Robot Labs, IEEE Robotics & Automation Magazine, pp. 25 – 32.

IEEE (2004), Special Issue on Networked Control Systems, IEEE Transactions on Automatic Control, Vol. 49, No. 9.

Koopman (2002), P., Critical Embedded Automotive Network", IEEE Micro Special issue on Critical Embedded Automotive Network.

Leohold, J. (2004), Communications requirements for Automotive Systems, Keynote Automotive Communication, 5th IEEE Workshop on Factory Communication Systems.

LIN-Protocol (2000), Development TOOLS, and Software Interfaces for Local Interconnect Networks in Vehicles, 9th Conference on Electronic Systems for Vehicles, Baden-Baden.

Maia, R., Cortesão, R., Nunes, U., Silva, V., Fonseca, J. (2003), Robust low-level motion-control of WMR with active observers. IEEE Int. Conference on Advanced Robotics (ICAR03) vol. 2, 876-882.

Mendes, A., Nunes, U. (2004), Situation-based multi-target detection and tracking with laserscanner in outdoor semi-structured environment. IEEE/RSI Int. Conference on Intelligent Robots and Systems (IROS 2004), vol. 1, 88 - 93.

Mock, M., Nett, E. (1999), Real-Time Communication in Autonomous Robot Systems, Proceedings of the 4th International Symposium on Autonomous Decentralized Systems, Integration of Heterogeneous Systems, pp. 34-41

Moita, F.; Nunes, U. (2001), Multi-echo technique for feature detection and identification using simple sonar configurations. IEEE/ASME International Conference on Advanced Intelligent Mechatronics (AIM01), vol. 1, 389-394.

Moore, K, Flann, N. (2000), A Six-Wheeled Omnidirectional Autonomous Mobile Robot, IEEE Control Systems Magazine, pp. 53-66.

MOST Coorperation (1999), MOST specification framework Rev 1.1, available at www.mostcooperation.com

Navet, N., Song, Y. (1997), Performance and Fault Tolerance of Real-Time Applications Distributed over CAN (Controller Area Network), CiA – CAN in Automation Research Award.

NMEA 0183 (2002), Standard for Interfacing Marine Electronic Devices, Version 3.01, National Marine Electronics Association.

Nolte, T., Hansson, H., Norstron, C. (2003), Probabilistic Worst-Case Response-Time Analysis for the Controller Area Network, Proceedings of the 9th IEEE Real-Time and Embedded Technology and Applications Symposium.

Nunes, U., Fonseca, J.A., Almeida, L., Araújo,R., Maia, R. (2003), Using distributed systems in real-time control of Autonomous Vehicles, ROBOTICA, Cambridge Univ. press, vol. 21, pp. 271-281.

OmniVision (1999), OV6620 Single-chip CMOS CIF Color Digital Camera.

Pires, G., Nunes, U. (2002), A wheelchair steered through voice commands and assisted by a reactive fuzzy-logic controller. International Journal of Intelligent and Robotic Systems, vol. 34, n. 3, 301-314.

Sick AG Auto Ident (2002), Quick Manual for LMS communication Setup.

Silva, V., Marau, R., Almeida, L., Ferreira, J., Calha, M., Pedreiras, P., Fonseca, J. (2005), Implementing a Distributed sensing system: The CAMBADA Robots case study, in 10th IEEE Int. Conference on Emerging Technologies and Factory Automation (ETFA'05).

Stewart, D., Moy, M. (2000), An Engineering Approach to Determining Sampling Rates for Switches and Sensors in Real-Time Systems, 6th IEEE Real-Time Technology and Applications Symposium, pp. 34-45.

Thomesse, J.P. (1998), A Review of the Fieldbuses, Annual Reviews in control, 22 pp. 35-45.

Tindell, K., Burns, A. (1994), "Guaranteeing Message Latencies on Controller Area Network (CAN)", Proceedings of the ICC'94 (1st International CAN Conference), Mainz, Germany.

Copyright © Fieldbus Systems and Their Applications
Puebla, Mexico, 2005

ELSEVIER
IFAC
PUBLICATIONS

CRITICAL DEMANDS OF DATA TRANSMISSION BETWEEN TRAINS AND TRACKSIDE INFRASTRUCTURE

Rainer Hornstein[1], Martin Pottendorfer[2], Herbert Schweinzer[1]

[1] *Technical University of Vienna, Institute of Electrical Measurements and Circuit Design, Gusshausstrasse 25/354, 1040 Vienna*
[2] *Alcatel Austria AG/TSD, Scheydgasse 41, 1211 Vienna*
e-mail: rainer.hornstein@tuwien.ac.at

Abstract: This article focuses on problems and implementation details of spot transmission and distributed communication systems in railway applications on the basis of the European Train Control System (ETCS). For spot transmission the "Eurobalise Subsystem" uses a passive transponder on the track called "Eurobalise" and a receiver unit at the train. Being still in the standardization phase, the "Euroloop Subsystem" uses a leaky coaxial cable as trackside antenna and a receiver unit at the train for distributed communication. In this work a system design is presented that comprises the authors experience with experimental implementations of the communication concept of both subsystems. *Copyright © 2005 IFAC*

Keywords: Communication System, Rail Traffic, Telecommunication, Train Control, Transponder, Vehicles

1 AN INTRODUCTION TO THE EUROPEAN TRAIN CONTROL SYSTEM

As all over Europe different railway signaling systems are in use, a train applied in cross-border railway traffic has to be equipped with communication facilities for all necessary different systems. To achieve a standardization of the signaling system, the European Union and European railway and signaling industries have cooperated closely over the past ten years to specify a standard for future railway signaling. The result of this cooperation is the European Train Control System (ETCS) standard. A part of this standard arranges the communication between track side equipment and the train what should be the focus of this article. According to the system performance three implementation levels of ETCS are defined. Most interesting for this article is level one since it introduces the communication methods of interest. Levels two and three use them too but they extend the system performance by other functionalities.

1.1 Level 1

Conventional railway signaling systems use light and drop signals for displaying the track conditions to the train driver. Applying ETCS level one the signal conditions are additionally transferred to the train using communication via the Eurobalise Subsystem that transmits information telegrams to the train. The transmitted telegram is received at the train by the Balise Transmission Module (BTM). After signal processing, the telegram is passed to the train control computer, called the European Vital Computer (EVC). The Euroloop Subsystem is an extension to the Eurobalise Subsystem and also a part of ETCS level one. The data handling works as with the Eurobalise Subsystem, what means, the trackside loop transmits the data telegram and the onboard Loop Transmission Module (LTM) receives and processes it.

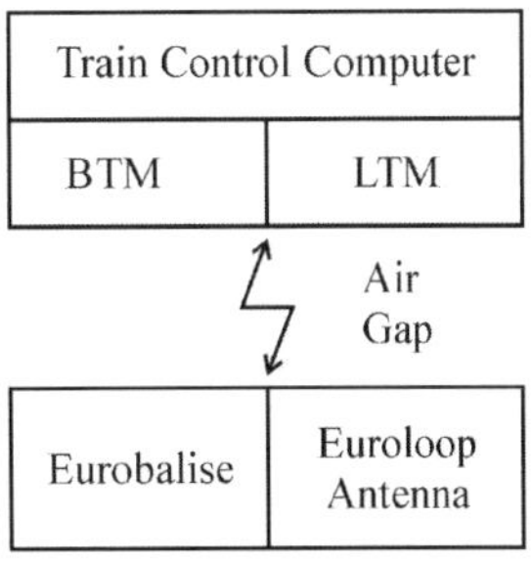

Fig. 1 – System Overview

On the one hand the EVC displays the information to the train driver and on the other hand it processes the information in an adequate way to supervise the actions of the driver. If the driver wants to operate the train in a way that is not allowed by the control rules of the EVC, it can interrupt the operation and take the train to a safe state.
Fig. 1 shows the modules of interest of a level 1 ETCS system. Being a passive transponder, the Eurobalise has to be powered by the BTM using magnetic coupling. If powered the Eurobalise transmits the stored telegram that is received by the BTM. The trackside Euroloop components are not passive but they have to be supplied by additional trackside equipment.
Because this article is concentrated on data transmission and data processing of the Eurobalise and Euroloop Subsystem, levels 2 and 3 of ETCS will just be mentioned for completeness.

1.2 Level 2 and Level 3

Those levels change the railway signaling by the use of an additional communication method and another kind of traffic management. Additional information can be found in (AEIF, 2005).

2 PHYSICAL INTERACTION OF THE SYSTEM COMPONENTS

In this chapter the main components of the Eurobalise and Euroloop subsystem will be presented and the communication link characteristics will be described basically. Generally a transmission from track to train is referred as uplink and from train to track as downlink. Since the downlink function is still in specification phase, this article will concentrate on the uplink and some details to the downlink will be mentioned in later chapters.

2.1 Eurobalise

In a simple case, a Eurobalise is a passive transponder that stores a 1023 bit data telegram and transmits it if powered by a magnetic field sent by the BTM. As explained in Finkenzeller (2003), a Eurobalise can be seen as an inductively coupled RFID system. However, Eurobalises are quite different to normal RFID-tags because of several critical demands: a short contact time because of the high traveling speed of passing trains, the large dynamic of the received signal due to environmental conditions and the high demands to the overall system safety. The minimum configuration of the Balise subsystem is shown in the solid line part of Fig. 2. The moving train transmits a 27.095MHz magnetic field called the Tele-Powering signal. The transmitted energy is used to power the stand alone Balise via the air gap. If the Balise is powered, it transmits the stored telegram using a binary frequency shift keying (FSK). The two frequencies used are 3.951MHz and 4.516MHz with a data rate of 564.48kbits/s. In a Balise one telegram is stored with an overall length of 1023 bits. This telegram is transmitted cyclically as long as the Balise is powered by the BTM.
If the traffic management needs the possibility to transmit different telegrams, a so called "switched Balise" has to be used. That kind of Balise can not be used stand alone. It has to be connected to a Line Electronic Unit (LEU) which acts as an interface to the interlocking machine. This kind of system configuration is shown in Fig. 2 by the solid and dashed line parts. Using a switched Balise system, the telegram is configured according to the settings of the interlocking and sent to the LEU that passes it to the Balise by a cable link. If the Balise is powered by the train, it does not transmit a locally stored telegram but the telegram from the LEU connected at the so called interface "C". At a system malfunction, a standard telegram locally stored into the Balise reports this circumstance to the train.
Depending on the physical size, two kinds of Eurobalises can be differed: standard size Eurobalises with a size of 490mm x 360mm and reduced size Eurobalises with a size of 390mm x 200mm. Applying a standard size Eurobalise, a 1023 bit telegram can be transmitted up to a line speed of 500km/h according to the Eurobalise specification (ERTMS/ETCS, 2003). Reduced size Eurobalises can transmit a 1023 bit telegram up to a speed of 300km/h. For a track speed up to 500km/h, reduced size Eurobalises can only transmit short telegrams containing 341 bits. Which kind of Eurobalises are used on the track depends strongly on the track configuration.

2.2 Euroloop

The Euroloop Subsystem is a distributed transmission system that is used in those track ranges where the system performance can be improved by continuous data transmission. Such ranges are for example railways stations. A Eurobalise is not the best solution in this case, because if the train stops it is unlikely that it will stand over a Eurobalise. Hence not being in the range of the Eurobalise transmission,

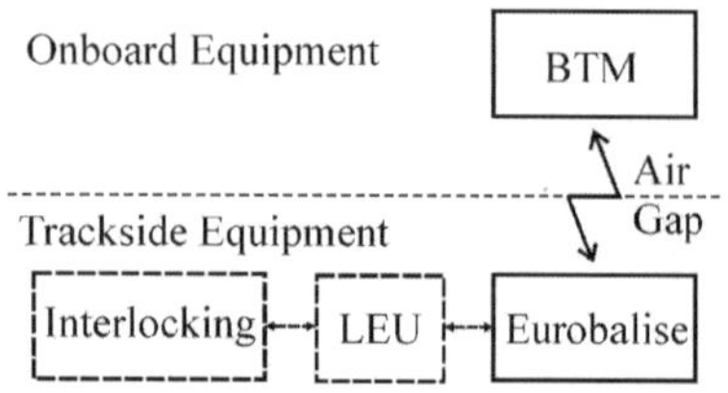

Fig. 2 - Balise Subsystem

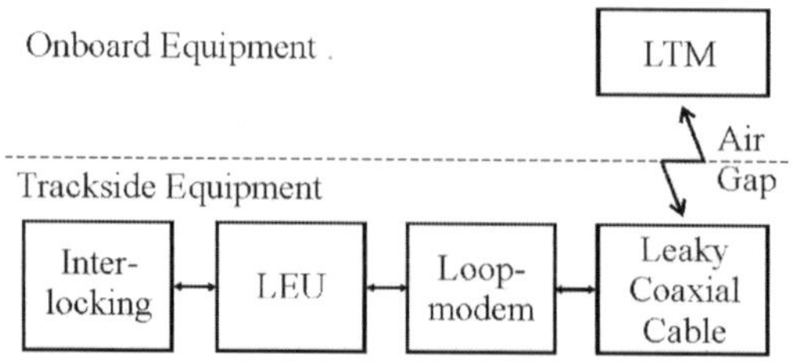

Fig. 3 - Loop Subsystem

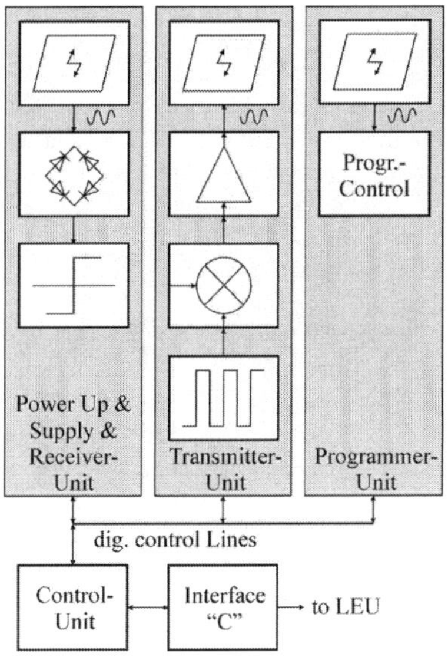

Fig. 4 - Block Diagram of a Eurobalise

information cannot be transmitted to the train. For this application a Euroloop Subsystem as shown in Fig. 3 is applied. It consists of a Loop Modem that generates the output signal and supplies the antenna. The Loop modem is connected to the LEU which acts as an interface to the interlocking machine and generates and stores the telegrams for transmission according to the settings of the interlocking. As an antenna a leaky coaxial cable is used that is mounted next to the rail. According to the Euroloop specification (ERTMS/ETCS, 2005 and ERTMS/ETCS, 2000) the length of this antenna is ranging from 50m up to 1000m that covers all application ranges. At the beginning and the end of a Euroloop covered track range, Eurobalises are mounted called "end of loop marker" that store LTM settings for loop transmission.

The Euroloop system can not work in a stand alone mode like Eurobalises. It always needs the loop modem which is powered by the interlocking and equipped to generate the transmission signal according to the interlocking settings.

If a train reaches the Euroloop range, it is informed about this fact by a Eurobalise and turns on the onboard Loop Transmission Module (LTM) that

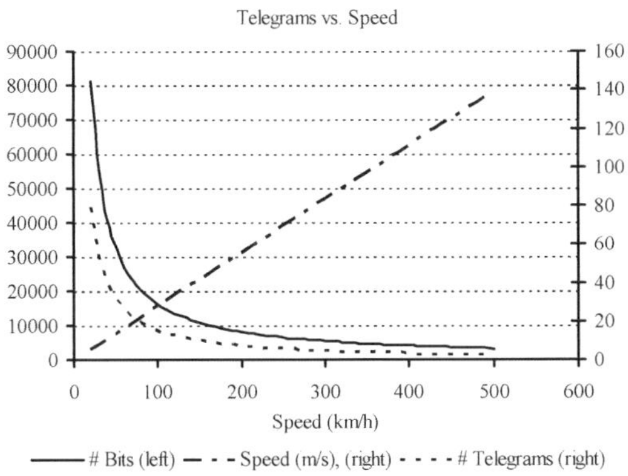

Fig. 5 - Transmitted Bits vs. Traveling Speed

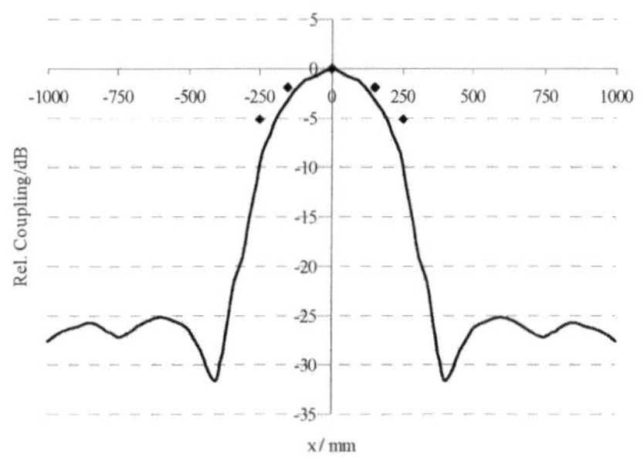

Fig. 6 - Relative Coupling vs. Distance

enables reading of Euroloop telegrams. If the Loop Modem receives a 27.095MHz signal from the train that is just used as a recognition signal and not for Tele-Powering, it starts the telegram transmission. As well as with Eurobalises, a telegram of the Euroloop has a length of 1023 bits and is transmitted cyclically. The difference to the Eurobalise subsystem is the modulation format that is a binary phase shift keying (BPSK) with a carrier frequency of 13.548MHz. Additionally the data is modulated by direct sequence spread spectrum with a chip rate of 4.516Mchips/s. Calculating the bit rate leads to 9567.4kbit/s if the specified spread spectrum sequence length of 472 chips is taken into consideration. Additionally similar to the Eurobalise subsystem, a short telegram with 341 bits can be used for some track characteristics.

3 EUROBALISES – A POINT ORIENTED COMMUNICATION METHOD

The following detailed system description will concentrate on the Eurobalise that acts as trackside telegram transponder. Basing on an example block diagram the critical aspects of the standard will be worked out.

3.1 Block Diagram

A block diagram that concentrates on the main functions of a Eurobalise is shown in Fig. 4. The main components are the power supply and receiver unit, the transmitter unit, programmer unit and a control and interface unit.

3.2 Power Up, Power Supply and Receiver Unit

If a 27.095MHz signal is applied to a Eurobalise the signal is received by the powering antenna. After rectification, the signal is used for powering the Balise. Power management is the most critical part of the Eurobalise where two points have to be considered: first, the necessary short power up time of a Eurobalise, and second, the wide dynamic range of the Tele-powering signal.

Fig. 5 shows the number of transmitted bits (solid line, left axis) and the according number of whole telegrams (dashed line, right axis) against traveling speed in km/h.

Additionally, the speed of the train is also displayed in m/s (dash-dotted line, right axis). For the

calculation of the number of transmitted bits a contact length between Balise and train of 800mm was assumed what is a realistic value according to measured values of relative coupling shown in Fig. 6. The values of coupling were measured for a vertical distance of 220mm with two test antennas. The x-axes shows the horizontal shift in mm and the y-axes the coupling normalized referring to the maximum coupling at x = 0mm.
At a speed of 500km/h the contact time is about 5.7ms. If we assume that according to safety reasons at least 3 whole telegrams should be received from each Balise (this needs about 5.4ms), the Balise must be ready for transmission within 300μs. The specification (ERTMS/ETCS, 2003) goes beyond this limit: it demands that the Balise transmits the telegram 150μs after the minimum Tele-Powering flux is applied.
As defined in the Eurobalise Specification, the input signals strength assuring a proper function of the Eurobalise has a dynamic range of a little less than 30dB. This large dynamic range results on the one hand from the distance variation between train antenna and Balise (220mm to 460mm) and on the other hand from debris that can cover the Balise. For example, a Balise should work properly both if covered by 100mm of water or 10mm of iron ore. The combination of those two reasons leads to the large dynamic range of the Balise Tele-Powering.
The combination of the two aspects makes the design of the power circuit somewhat tricky: it must be fast to fulfill the timing constraints on the one hand and very efficient and able to withstand high dynamic variations on the other hand.

Additionally the Tele-Powering signal can be used to transmit downlink data from the train to the track. The modulation format of the downlink is a 10% AM of the Tele-Powering signal with a data rate of 564.48kbits/s. The telegram length for downlink is 1023bit. The processed telegram is passed on from the Balise to the interlocking. This downlink feature is mentioned just for completeness since it is not used in any state of the art system.

3.3 Transmitter

Generating the FSK uplink signal, the most common method using a switched phase locked loop cannot be applied since the characteristics of the FSK signal are not adapted for this method. According to the given FSK frequencies, the length of one bit is just seven or eight periods of the frequencies respectively. Thus, the comparable high data rate would lead to a PLL circuit that does not lock. A better method for signal generation is to generate and synchronize both frequencies permanently and switch between them according to the data bits.
Like the power circuit, the transmitter must be highly efficient to generate an output signal that is as large as possible since this will make the signal transmission and data reception more safely. Unlike to the BTM described in chapter 4, the Balise uses separate antennas for receive and transmission. This is uncritical because the design of the Balise antennas has to consider only low operating power. Using separate antennas do advantageously not require a signal combination filter.

3.4 Control Unit and Cable Interface

The control unit must process data and timing of the transmitter and receiver units. Additionally, it has to transmit data to or from memory or to the cable interface. If the Balise is programmed also the programming process has to be controlled by the control unit. The cable interface called interface "C" is used for receiving uplink data from the LEU or send downlink data to the LEU in the case of a switched or downlink Balise respectively. Additionally, it can be used to program the Balise with a telegram.

3.5 Programmer Unit

Programming an internally stored telegram into a Balise can be done in two different ways. The easier implementation is using the cable interface "C" where the telegram is directly transmitted to the Balise just applying a simple coding. The second possibility is programming by the air interface. In that case, the telegram data is transmitted with a special, not standardized downlink protocol using a 100% AM. To prevent unwanted programming, an additional 9.038MHz continuous wave signal has to be sent to the Balise that has to be evaluated and processed enabling the telegram programming sequence.

Fig. 7 – Test Setup of the BTM

4 AN OVERVIEW OF AN EXAMPLE IMPLEMENTATION OF THE TRAIN ONBOARD SYSTEM

A Balise Transmission Module (BTM) is the train onboard system used to read Balises. Fig. 7 shows a photograph of the test setup of the implemented BTM. The according block diagram is depicted in Fig. 8.
The whole system is controlled by the BTM Control Computer that acts as an interface to the Train Computer (European Vital Computer: EVC). The interface from the Control Computer to the analogue

modules is realized by the Hardware Controller. The signal processing for reading a Eurobalise works like follows. The Tele-Powering Signal is generated by the Transmitter Unit, passed to the Antenna Unit via the Combination Filter, and transmitted. The received signal from the Balise is split up with the Tele-Powering signal by the Combination Filter. Afterwards the received signal is amplified and digitized to an FSK modulated TTL signal that is processed and demodulated by the Hardware Controller. The readily processed telegram is passed to the EVC by the BTM Control Computer. Additionally there is a Test Circuit that offers possibilities to test the whole system.
For an installation on the train some special aspects have to be considered. Important aspects are: from the safety point of view, the electrical isolation between BTM and the onboard computer system, and from the installation point of view, a single cable link between the BTM and the bottom mounted Antenna Unit.

4.1 Transmitter Unit

The Transmitter Unit consists of a signal generator that generates the 27.095MHz. In the following stage the signal can be modulated with 10% AM or 100% AM to transmit either the downlink signal if applied or special Tele-Powering signals for the Balise. Those special signals are called "Toggling" or "Non Toggling" and enable special compatibility modes of the Balise that are not further important in this context. At this point it should be mentioned that the 100% AM should never be used to program a Balise by a passing train. This would be a safety risk and therefore programming is only allowed using a special programming tool.
After modulation, the output signal is amplified to a value demanded in the specification (ERTMS/ETCS, 2003). This value must be found from the input characteristic of the Eurobalise being defined in the specification. An optimal value will be reached transmitting a Tele-Powering signal being as strong that a Eurobalise gets the maximum applicable power if the distance between the Antenna Unit and the Balise has the minimum value and no debris is in between. From the values in the specification, realistic values of insertion losses of the Combination Filter and the efficiency of the Antenna Unit, an output power value of the Transmitter Unit of about 40W can be calculated.

4.2 Receiver Unit

As mentioned above, the modulation format of the Balise is a FSK with a data rate of 564.48kbit/s and a center frequency of 4.234MHz. Resulting from the specified distance range, the output signal of the Balise has a dynamic range of 20 dB. Together with the attenuation caused by debris in between Balise and BTM Antenna Unit, this leads to an overall dynamic range of 50dB of the Receiver Unit input signal.

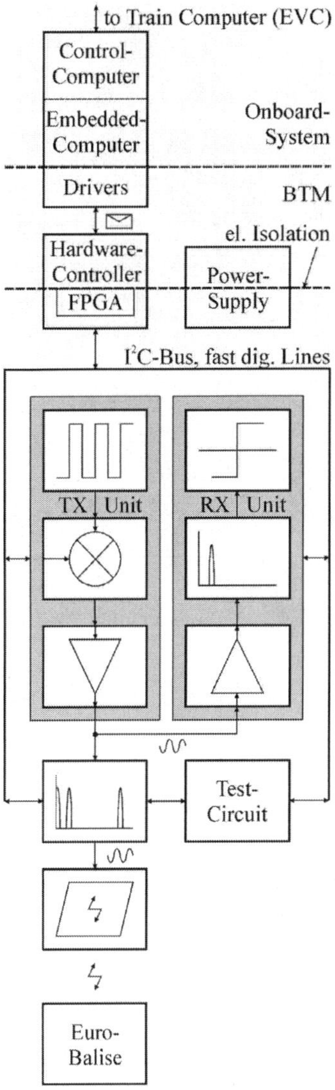

Fig. 8 - Block Diagram of a BTM

The received signal with such a large dynamic range is compressed by a logarithmic amplifier and converted to a FSK modulated digital signal. This signal is passed to the Hardware Controller where it is demodulated using a correlation demodulator. That leads to excellent and very robust demodulation results that fulfill the high safety demands of railway equipment. The classical FSK demodulation method using a phase locked loop is not sufficient because of the reasons mentioned in chapter 3.3 .

4.3 Combination Filter

For installation on the train, it is optimal using just one cable between the BTM electronic equipment located inside the train and the Antenna Unit mounted beneath the train. Therefore different signals have to be transmitted over one cable: the 27.095MHz Tele-Powering signal, the 4.234MHz FSK signal and a DC signal for Test Circuit control.
The main challenge of the Combination Filter is to decouple the Tele-Powering signal from the received FSK signal. Roughly calculated, the difference

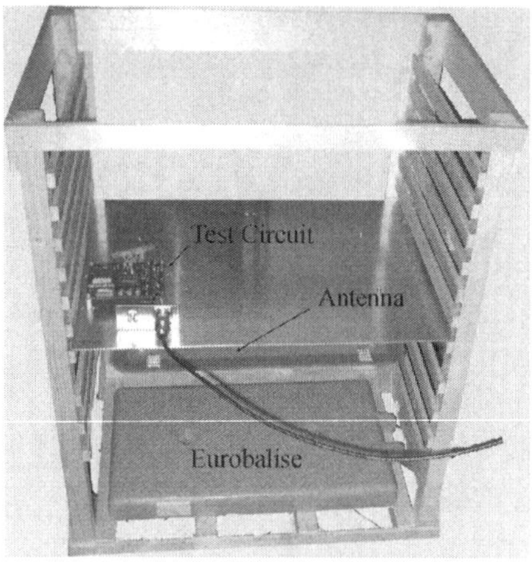

Fig. 9 – Test Setup of the Antenna Unit

between those two signals is about 80dB. To achieve a good decoupling between those two signals, a band pass filter of at least 5th order should be used. It is very important to isolate the Tele-Powering signal from the Receiver Unit. In the opposite direction, decoupling is not necessary because the small received signal has no influence on the output stage of the power amplifier.
Features of the Test Circuits are controlled by a DC signal. A 3rd order low pass filters can be used to decouple this signal from the others.

4.4 *Antenna Unit*

The Antenna Unit allows a simple construction because it is just one loop that transmits the Tele-Powering signal and receives the FSK signal. A tricky aspect of the antenna design is taking the antenna matching into consideration which depends on the distance to the Balise. The influence of the distance is caused by the magnetic coupling of the Antenna Unit loop and the Balise antenna working together like a transformer. If those loops change their distance the inductance changes too and for this reason the resonant matching changes. To decrease the influence of distance, the Antenna Unit loop is resistively damped resulting in an easier and more stable matching. However, considering resistive damping of the antenna, the power amplifier of the Transmitter Unit has to generate a higher output power compared to an only resonant matching of the antenna. Fig. 9 shows a photograph of the Antenna Unit with the Test Circuit and a Eurobalise below.

4.5 *Hardware Controller and Control Computer*

The Hardware Controller and the Control Computer work closely together managing the BTM. The Control Computer is a standard industrial PC that runs a freely available operating system. It communicates with the Hardware Controller using special drivers.
The Hardware controller is built up with a FPGA. It has the main tasks to demodulate a Balise telegram delivered from the Receiver Unit by a fast digital link with an implemented correlation receiver and to communicate with the analogue modules by I²C busses. The second main task of the Hardware Controller is the electrical isolation between the BTM and the onboard computer system which has to protect the onboard computer system from malfunctions of the analogue components.
The communication between the Hardware Controller and the Control Computer is realized as a PC/104 bus utilizing memory mapped I/O to a FPGA-realized interface.

4.6 *Test Circuit*

The Test Circuit is divided into two parts: one on the antenna unit and one separate module in the BTM rack.
The Antenna Unit additionally holds a second, much smaller antenna and some circuitry which are used to test the BTM. If the test circuitry is powered by applying a DC voltage, the small antenna transmits a test FSK signal according to a pre-defined test telegram which can be received by the main antenna, demodulated and compared with the expected telegram. If no DC voltage is applied, a passive part of the test circuitry measures the peak value of the output power and passes it to the BTM Test Circuit. That way the Control Computer can check whether the Antenna Unit works properly and get the value of

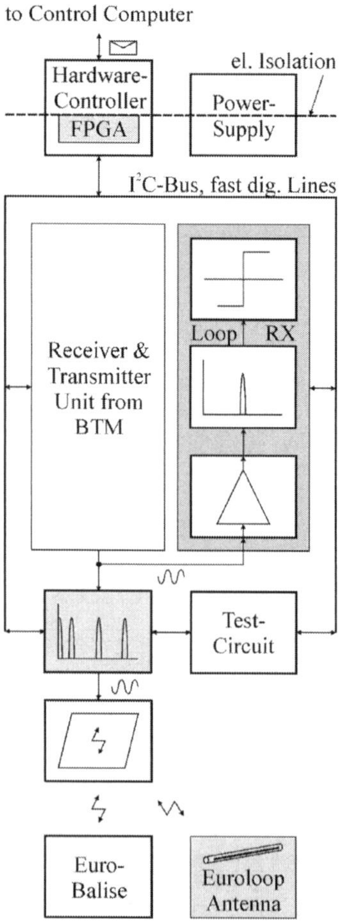

Fig. 10 - Block Diagram of an LTM

the actual transmitted power.

If initiated by the Control Computer, the Test Circuit in the BTM rack on the one hand has to generate the necessary DC voltage to power the test circuitry of the antenna unit and transmit it via the Combination Filter. On the other hand it has to measure the peak output power delivered from the Antenna Unit and pass it via I^2C bus to the Control Unit.

4.7 Power Supply

The Power Supply is also divided into two parts: one contains the Control Computer and the electrically coupled parts of the Hardware Controller and the other the electrically isolated parts of the BTM. The two power supplies have to be electrically isolated.

5 EUROLOOP – A DISTRIBUTED COMMUNICATION METHOD

The Euroloop subsystem is an extension to the Eurobalise subsystem that is used in areas where a continuous data transmission over a limited track range can improve the system performance. This chapter will focus on the additional components used to equip a BTM with Loop reading functionality. The downlink function is not explained since it is still in specification phase.

5.1 Changes to the BTM

Marked as grey shaded blocks, Fig. 10 shows the additionally used components added to an existing BTM. The loop filter must have an additional path to separate the 13.548MHz receive signal of the Loop from the other system frequencies and pass the correct signal to the Loop receiver. Furthermore a Loop receiver has to be added and the demodulation firmware of the FPGA has to be changed, what will be the subject of the next chapter.

5.2 Loop Receiver Unit

Basically the Loop Receiver Unit works similarly as the Receiver Unit of the BTM. First the signal is amplified and filtered according to the differing signal specifications explained in chapter 2.2. Then it is converted to a TTL signal that is passed to the Hardware Controller. The Hardware Controller has to demodulate the BPSK and afterwards to correlate with the spread spectrum sequences to get the data bit values. The further processing of the telegram is identical to the Eurobalise data telegram.

6 DATA PROCESSING AND TIME SYNCHRONIZATION

This chapter will sum up the kind of transmitted data and the coding of the raw information to fulfill the safety demands. Additionally, some information on necessary time synchronization between the system components is outlined.

6.1 Transmitted Data and Telegram Coding

Concerning fixed Balise telegram, data transmitted from track to train mostly contains information on track conditions like maximum line speed, absolute location of the Eurobalise or the distance to the next Balise. Data being signal dependent like a track free signal or a stop signal have to be transmitted by switched Balise or Loop telegrams. If a downlink is available, the train can use it to transmit its actual states to the interlocking.

Generally, a standard telegram consists of user data of 830 bits and a short telegram of 210 bits. Applying different standardized coding and scrambling algorithms, the overall standard telegram comprises 1023 bits and the short telegram 341 bits respectively. These telegrams are transmitted cyclically.

Outputs of the BTM are received and synchronized telegrams transmitted by the track side equipment. The final decoding and check of telegrams is done by the EVC which is considered to be a safe computer system.

The high coding and processing overhead is used because the air gap interface between train and track is error prone and on the other side safety demands for railway applications are very stringent.

6.2 Time Synchronization between the System Components

One of the most important data the Eurobalise transmits to the train is its position on the line. It is used to retrigger the onboard odometric system which accuracy should be less than 1m. For that reason the point of time when a Balise was read has to be stored together with the telegram information to allow the calculation of an actual train position by the elapsed time since reaching the Balise position.

Processing and storing the telegram in the Hardware Controller, but evaluating the telegram data in the EVC requires to synchronize all clocks accurately. Different clocks are used at the EVC and at the Control Computer. Additionally, a free running counter at the Hardware Controller delivers the time value stored together with the processed telegram. This time counter has also to be synchronized with the overall clock system.

7 SUMMARY

The Eurobalise and Euroloop subsystems are used to improve safety and efficiency of European railways. The main onboard and trackside components were presented and their interaction was explained.

Based on the experience resulting from experimental implementations of the Eurobalise communication concept, critical parts of the standard were discussed and some solutions were presented for the onboard system as well as for the trackside equipment.

Finally a short outlook to data processing, telegram coding and time synchronization demands was given. This work should help to find out details about the critical aspects of the standard and the rough conditions of air gap transmission. Finally this experience should lead to a system design that fulfills the high safety demands of railway applications on the one hand and which is easily applicable on the train on the other hand.

ACKNOWLEDGEMENTS

The project being the basis of this work is performed in co-operation between the Institute of Electrical Measurements and Circuit Design of the Technical University of Vienna and Alcatel TSD Austria. The authors would like to thank all colleagues both at the institute and at Alcatel for their assistance with the project development as well as with this article.

REFERENCES

AEIF (2005). Documents on AEIF web page, FRS 4.29 (http://www.aeif.org/ccm/doclist.asp)

ERTMS/ETCS (2003). Class 1: FFFIS for Eurobalise, SUBSET 036, 2003-09-12

ERTMS/ETCS (2005). FFFIS for Euroloop, SUBSET 044, 2005-03-11

ERTMS/ETCS (2000). Description of the Euroloop Subsystem, SUBSET 050, 2000-03-30

Finkenzeller, K. (2003). *RFID Handbook: Fundamentals and Application in Contactless Smart Cards and Identification, Second Edition,* Chapter 13.5.1. John Wiley & Sons Ltd., West Sussex

ELSEVIER
IFAC
PUBLICATIONS

PROBABILISTIC TIMING ANALYSIS OF THE h-BEB COLLISION RESOLUTION ALGORITHM[1]

Ricardo Moraes, Francisco Vasques

Department of Mechanical Engeneering, University of Porto
Rua Dr. Roberto Frias, 4200-465 Porto, Portugal,
e-mail: {rmoraes, vasques}@fe.up.pt

Abstract: Ethernet networks are becoming increasingly popular in industrial computer-controlled systems, as they allow for a single network protocol at both the higher and the lower levels of an industrial communication infrastructure. Despite the introduction in the early 90s of a full-duplex operating mode, numerous industrial Ethernet networks still operate in heterogeneous environments, with Ethernet Switching Hubs interconnecting both independent node stations and industrial Ethernet Repeater Hubs. Among node stations interconnected by a Repeater Hub, the network still operates in the traditional shared Ethernet mode; that is, collisions are solved by means of a probabilistic contention resolution algorithm *i.e.*, the medium access is inherently non-deterministic.

In this paper, it is analyzed an enhanced collision resolution algorithm for shared Ethernet networks, referred as high priority Binary Exponential Backoff (h-BEB). Such algorithm allows the coexistence of Ethernet standard devices together with modified (real-time) devices in the same network segment. Both the analytical and the simulation timing analysis show that the h-BEB algorithm guarantees a maximum access delay that is significantly smaller than for the case of standard Ethernet stations. Such enhanced collision resolution algorithm enables the traffic separation between standard and modified (real-time) stations, and is therefore able to guarantee a real-time communication behavior in unconstrained traffic environments.

Keywords: Ethernet communication; Real-time communication.

1. INTRODUCTION

Multiple fieldbus network technologies have been proposed and developed to interconnect sensor and actuators to controllers in the industrial environment, as a consequence of the need for specific communication networks in the plant floor. In spite of the adequacy of some of these fieldbus technologies for many type of applications, the use of different/ multiple technologies has obvious disadvantages: high cost, difficult integration and even the incompatibility between standard devices from different producers [1]. At the upper industrial communication levels (office domain), Ethernet has established itself as the most used communication technology, resulting in low component prices caused by the mass production of these components [2].

When Ethernet networks started to be used also in the plant floor, higher speed and low cost for the communication controllers were the major motivation. However, traditional shared Ethernet systems, with its simple CSMA/CD medium access protocol, do not allowed real-time capability. Consequently, diverse commercial companies have developed extensions to the legacy Ethernet standards and now multiple systems have the potential to fulfill real-time Ethernet specifications. A brief analysis of the state-of-the-art in Industrial Ethernet solutions is given in Section 6.

Despite the introduction in the early 90s of a full-duplex operating mode for Ethernet networks, numerous industrial Ethernet networks still operate in heterogeneous environments, with Ethernet Switching Hubs interconnecting both independent node stations and industrial Ethernet Repeater Hubs. Consequently, among node stations interconnected by Repeater Hubs, the network still operates in the traditional shared Ethernet mode; that is, collisions are solved by means of a probabilistic contention resolution algorithm. This means that heterogeneous

[1] This work has been partially supported by IDMEC and by FCT (project ADVANSYS and BD 13203/2003).

networks are not able to provide a real-time communication service.

1.1. *Rationale for the h-BEB algorithm*

Multiple techniques have been developed to provide real-time communication services in shared Ethernet networks. Such techniques are typically based on either: *avoiding collisions*, by controlling the medium access rights of each station (TDMA scheme, token passing, etc), or *ensuring a deterministic collision resolution*, by modifying the collision resolution algorithm. A third approach (that is not deterministic) is to reduce the number of occurring collisions, enhancing the network responsiveness to real-time message requests.

The drawback of such traditional approaches is that they rule out the coexistence of Ethernet standard stations together with modified (real-time) stations in the same network segment. This means that legacy shared Ethernet systems would not be able to support real-time communications without extensive modifications.

To address this problem, it has been proposed in a previous paper [3] the use of a modified collision resolution algorithm, referred as the "high priority Binary Exponential Backoff (h-BEB)"algorithm. This algorithm allows Ethernet standard stations to coexist with at most one modified (real-time) station in the same network segment, imposing a higher priority to the privileged traffic. This mechanism has been extended in a subsequent paper [4], where it has been proposed the use of a virtual token passing procedure, allowing multiple h-BEB (real-time) stations to coexist with multiple standard Ethernet stations in the same network segment, and still imposing a higher priority for the transfer of privileged traffic.

1.2. *Paper structure*

In this paper, we address the timing analysis of the h-BEB collision resolution algorithm. Section 2 reviews the BEB collision resolution algorithm used in standard Ethernet and describes the h-BEB algorithm. Section 3 addresses the timing analysis of this new algorithm. In Section 4, it is summarized the exact performance analysis in heavily loaded network scenarios. Afterwards, the comparative analysis is done in Section 5; it considers a shared Ethernet environment, where multiple stations are interconnected with a special station; the latter implementing either the h-BEB algorithm (enhanced Ethernet mode) or the BEB algorithm (traditional Ethernet mode). the maximum access delay time is then evaluated, demonstrating that the h-BEB collision resolution algorithm is adequate to support soft real-time applications. Section 6 presents a brief overview of real-time industrial Ethernet solutions. Finally, the paper is concluded in Section 7.

2. THE HIGH PRIORITY BINARY EXPONENTIAL BACKOFF ALGORITHM

The CSMA/CD (Carrier Sense Multiple Access with Collision Detection) protocol is the protocol implemented at the MAC layer of both ANSI/IEEE 802.3 and Ethernet local area networks. For a 10/100 Mbps Ethernet implementation, the following set of parameters is used:

Table 1: Ethernet parameters.

	Values	
SlotTime	512 bit times	64 byte times
InterFrameGap	96 bit times	12 byte times
AttemptLimit	16	-
BackoffLimit	10	-
JamSize	32 bit times	4 byte times
MaxFrameSize	12144 bits	1518 bytes
MinFrameSize	512 bits	64 bytes
AddressSize	48 bits	6 bytes

Basically, the CSMA/CD protocol works as follows (Figure 1a): when a station wants to transmit, it listens to the transmission medium. If the transmission medium is busy, the station waits until it goes idle; otherwise, it transmits immediately. If two or more stations simultaneously begin to transmit, the transmitted frames will collide. Upon the collision detection, all the transmitting stations will terminate their own transmission and send a jamming sequence[2]. When the transmission is aborted due to a collision, it will be repeatedly retried after a randomly evaluated delay (backoff time), until it is either successfully transmitted, or definitely aborted (after a maximum number of 16 attempts).

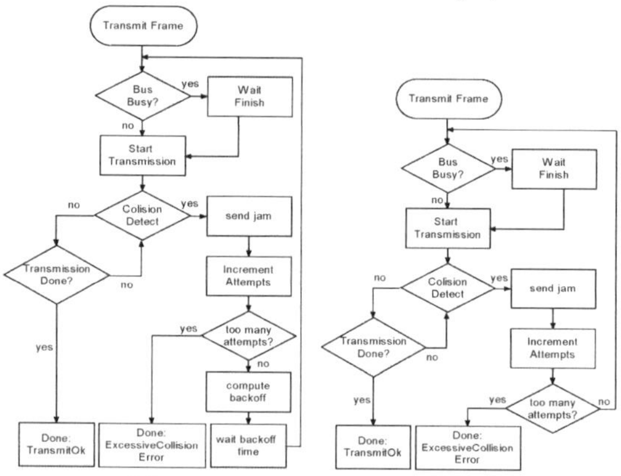

Figure 1. CSMA-CD protocol with BEB resp. h-BEB collision resolution algorithms.

The backoff delay is evaluated by locally executing the Binary Exponential Backoff (BEB) algorithm, which operates as follows: after the end of the jamming sequence, the time is divided into discrete slots, whose length is equal to the slot time. The

[2] More accurately, when detecting a collision, the station always finishes the transmission of the Preamble and the Start of Frame Delimiter (64 bits), if these have still not been completely transmitted. Afterwards, it transmits a jamming sequence (32 bits), and then stops.

backoff time is given by $t_{backoff} = r \times T$, where r is a random integer in the range $0 \leq r \leq 2^k - 1$, k is the smaller of n or 10 (n is the number of retransmission attempts) and T is the slot time in seconds. This means that the station will wait between 0 and 2^k-1 slot times. After 10 attempts, the waiting interval is fixed at 1023 slot times, and finally after 16 attempts the transmission is discarded.

On the other hand, a station implementing the h-BEB algorithm operates as follows (Figure 1b): whenever there is a collision, the station immediately starts to transmit (backoff interval equal to 0). This behavior guarantees the highest transmitting probability to the h-BEB station, as it will always try to transmit its frame in the first slot, while all the other stations will wait between 0 and 2^k-1 slot times.

The h-BEB collision resolution algorithm can be used to support real-time traffic separation, as the traffic generated by the h-BEB station will be always transferred prior to the traffic generated by the other stations. This behavior is highly adequate to, for instance, real-time video/voice transferring applications in legacy shared Ethernet networks. By simply plugging a notebook computer with the modified hardware to the network, it becomes possible to transfer traffic at a higher priority than the traffic generated by all the other stations.

3. TIMING ANALYSIS

In this section, the timing analysis of the h-BEB collision resolution algorithm is presented, for the case of a 10 Mbps shared Ethernet scenario. Such analysis can easily encompass a 100 Mbps scenario, using the timing parameters presented in Table 1.

First of all, there is the need to analyze the response time of a shared Ethernet network; that is, the time interval that it takes to transfer a message in a shared Ethernet network.

Consider a two-collision scenario (Figure 2). At instant t_0 station A has a message ready to be transferred (P_A), but at instant t_0 - ε, another station starts to transmit a 1518-byte message (P_N), which is the longest Ethernet message. Station A will wait for the completion of both the message P_N and the Inter Frame Gap (I_1: 12 byte times), before attempting to transmit again (that is, 1530 byte times). If a collision occurs during the transfer of the first 64 bytes of message P_A, a jamming sequence will be broadcasted (J_1: 4 byte times) and, according to the BEB algorithm, the stations involved in the collision will select a random backoff time (0 or 1 slot time).

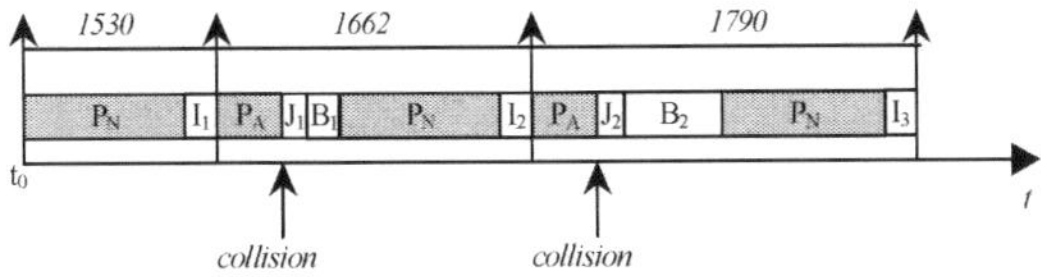

Figure 2: Worst-case 2-collision scenario solved by the BEB collision resolution algorithm.

Considering that station A selects a backoff delay of 1 slot time (64 byte times) and the other station wins the medium access, a new P_N message (1518 bytes) may be transferred. Therefore, station A will need to wait again for the completion of both the new P_N message and the Inter Frame Gap (I_2: 12 byte times); that is, it must wait (64+4+64+1518+12)=1662 byte times before attempting to transmit for the second time. If a second collision occurs, a jamming sequence will be broadcasted again and, station A may now need to wait during a backoff time of 3 slot times (192 byte times). Therefore, it may need to wait (64+4+3×64+1518+12)=1790 byte times before attempting to transmit for the third time, if a longest P_N message wins the second collision resolution round. The cumulative result (from t_0 up to the beginning of the third attempt) is then of 4982 bytes or 3,9856 ms (squared box result in Table 2).

On the other hand, a station implementing the h-BEB collision resolution algorithm is characterized by always trying to transmit its frame in the first slot (Figure 3). The worst-case scenario is when, at instant t_0, station A has a message ready to be transmitted (P_A), but at instant t_0 - ε, another station starts to transmit a 1518-byte message (P_N). In such case, the station will wait for the completion of both the message P_N and the Inter Frame Gap (I_1: 12 byte times), before attempting to transmit again (that is, 1530 byte times). If during the transfer of the first 64 bytes of message P_A a collision occurs, a jamming sequence will be broadcasted (J_1: 4 byte times) and, station A will need to wait again during an Inter Frame Gap (I_2: 12 byte times).

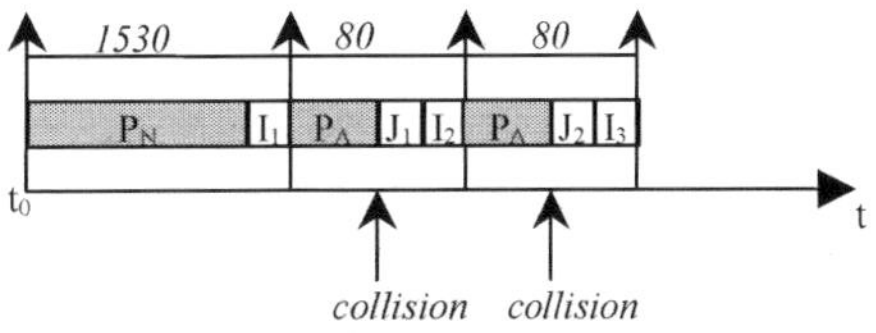

Figure 3: Worst-case 2-collision scenario solved by the h-BEB collision resolution algorithm.

Afterwards, according to the h-BEB algorithm, station A will start to transmit its message. If a second collision occurs, a new jamming sequence will be broadcasted and station A will wait during an Inter Frame Gap, before starting to transmit again. The cumulative result (from t_0 up to the beginning of the third attempt) is then 1690 bytes or 1,3520 ms (rounded box result in Table 2).

Table 2 illustrates the maximum delay to start transferring a message frame after i consecutive collisions when using, respectively, the BEB and h-BEB collision resolution algorithm.

Figure 4 illustrates the results from Table 2, in a semi logarithmic scale. For the h-BEB case, the maximum delay to start transferring a frame is significantly smaller than for the BEB case. More significantly, such maximum delay is almost constant, which is particularly adequate for the transfer of real-time messages in shared Ethernet environments.

Table 2: Maximum delay to start transferring a message frame – BEB vs. h-BEB.

Retry Number	Max delay (# slots)		Max cumulative delay (# slots)		Max delay (ms)	
	BEB	h-BEB	BEB	h-BEB	BEB	h-BEB
1	1	1	1	1	2,5536	1,2880
2	3	1	4	2	3,9856	1,3520
3	7	1	11	3	5,6224	1,4160
...						
6	63	1	120	6	15,038	1,6080
...						
10	1023	1	2036	10	118,2512	1,8640
...						
14	1023	1	6128	14	332,8752	2,1200
15	1023	1	7151	15	386,5312	2,1840
16	discard frame					

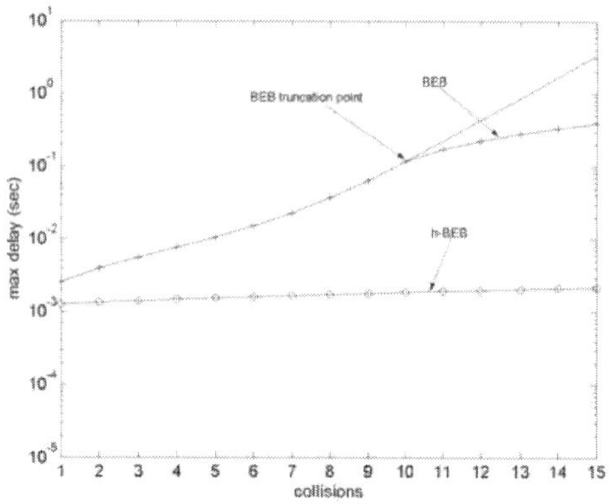

Figure 4: Maximum access delay - BEB vs. h-BEB.

However, as there is still the possibility of a message frame being discarded after 16 failed transmission attempts, there is the need to investigate the probability of such occurrence. Such probability is evaluated in Section 5, both analytically (for an heavily loaded network scenario) and by simulation.

4. EXACT PERFORMANCE ANALYSIS IN HEAVILY LOADED NETWORK SCENARIOS

One of the first Ethernet performance analysis was presented by Metcalfe and Boggs [5], where the authors presented an exact probabilistic analysis for heavily loaded network scenarios. In that analysis, a constant retransmission probability for each slot has been assumed, and the successful retransmission probability (on the next slot) has been considered to be equal to a constant: p. Such probability A is maximized when p=1/K (equal probability of successful retransmission), where K represents the number of active hosts. Such assumption is an interesting approximation for the real backoff function, as has been shown in multiple simulation studies (e.g. [6] [7]). Thus,

$$A = (1 - 1/K)^{K-1} \tag{1}$$

The probability that the contention interval will be exactly n slots is:

$$P_n = A \times (1 - A)^{n-1} \quad n \geq 1 \tag{2}$$

Obviously, the assumption that each station transmits with an equal probability is not suitable for the analysis of the h-BEB algorithm, as in the h-BEB case one of the stations (the privileged station) transmits at a higher probability.

Therefore, new and adequate formulae have been devised to perform the probabilistic analysis of the h-BEB collision resolution algorithm. In [3], it has been demonstrated that the probability of the h-BEB station sending a message up to the nth collision round (after an initial collision), is given by:

$$P(n, N) = \sum_{j=0}^{N} (-1)^j \frac{N!}{j!(N-j)!} \times 2^{-jn} \tag{3}$$

where n is the number of collision resolution rounds, and N is the number of BEB stations in the network ($N+1$ is the total number of stations).

5. COMPARATIVE ANALYSIS

A comparative timing analysis of BEB vs. h-BEB algorithms has been performed. For the case of a heavily loaded network scenario (pessimistic case), analytical results enable the evaluation of the maximum access delay vs. the transmission probability. For more realistic load scenarios (intermediate load cases), a more comprehensive analysis of the access delay is done by simulation.

Two cases are analyzed: a *small population* scenario that considers a network with 5 stations, where 4 standard Ethernet stations are interconnected with a *special station* implementing either the h-BEB (enhanced Ethernet mode) or the BEB (traditional Ethernet mode) collision resolution algorithms; a *large population* scenario extends the small population case to 65 interconnected stations.

5.1 Exact Timing Analysis for the Heavily Loaded Network Scenario

Considering the case of a heavily loaded network scenario, the transmission probability of the special station may be obtained from equations (2) and (3), for, respectively, the traditional and the enhanced Ethernet modes. Such transmission probability, P_n or $P(n,N)$, depends on the number of collision resolution rounds n. When combining the transmission probability for a given collision resolution round, with the maximum access delay (Table 2) for such collision resolution round, it becomes possible to associate a probability of occurrence to each number of maximum access delay.

Such maximum access delay vs. transmission probability is illustrated in Figures 5 and 6, for both the small and large population scenarios, comparing the traditional and the enhanced Ethernet modes.

Consider the enhanced Ethernet mode. It can be seen that, after a small number of collision resolution rounds, the transmission probability is larger than 95% for both the small and large population scenarios. From Equation (3), the transmission probability is larger than 95% when more than 6 or 10 collision resolution rounds are considered, respectively for the small and large population scenarios: $P(7,4) = 0{,}969$ and $P(11,64) = 0{,}967$. As a consequence, the maximum access delay to start transferring a message frame in 95% of the cases is 1,61ms and 1,86ms, respectively for the small and large population scenarios (Table 2). Thus, it is evident that the h-BEB algorithm is clearly adequate to support soft real-time applications.

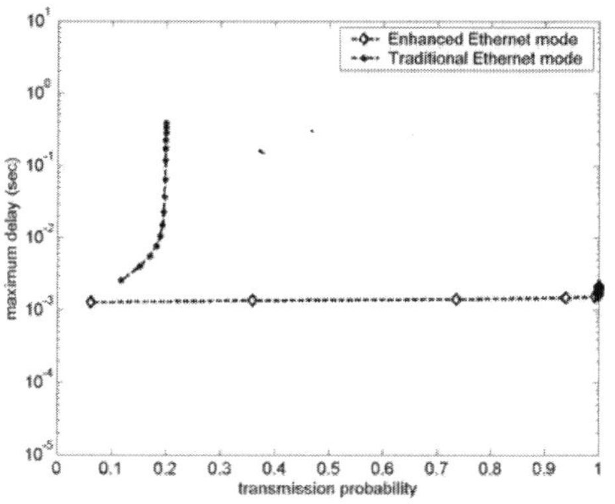

Figure 5: Maximum access delay vs. Transmission probability (small population scenario).

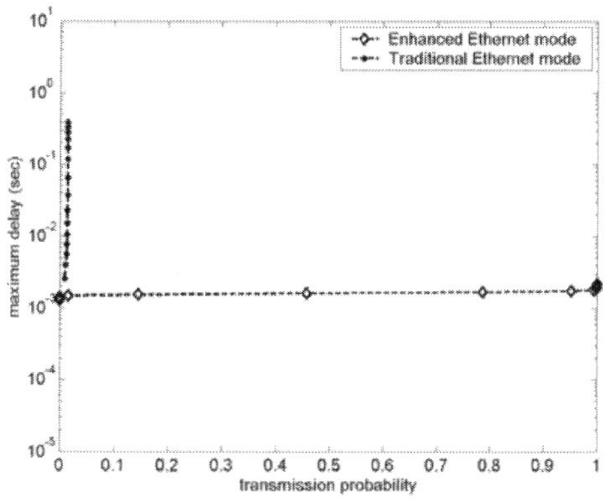

Figure 6: Maximum access delay vs. Transmission probability (large population scenario).

On the other hand, it is of utmost importance to focus on the probability of a message frame being discarded by the h-BEB algorithm, whenever the number of collision resolution rounds exceeds 15. Such probability can be easily evaluated by means of Equation (3), as the probability of a message being discarded is equal to the probability of the h-BEB station not being able to send the message up to the 15th collision round, *i.e.*, it is equal to $1\text{-}P(15,N)$. Such probability is equal to $1{,}22\times10^{-4}$ and $1{,}95\times10^{-3}$, respectively for the small and large population scenarios. Such results are consistent with the claim that the h-BEB algorithm is able to support most part of the soft real-time applications, as they confirm a rather small probability of any message being discarded.

The other set of results is related to the traditional Ethernet mode. For such case, the maximum access delay is significantly higher, even for reduced transmission probabilities. Moreover, the transmission probability is bounded to rather small numbers (0,20 and 0,015, respectively for the small and large population scenarios), as a constant retransmission probability for each slot has been assumed [5]. This means that the probability of a message frame being discarded (0,80 and 0,985) when using the BEB collision resolution algorithm impairs the support of almost any kind of application in heavily loaded network scenarios.

Nevertheless, it must be considered that such kind of exact timing analysis addresses a rarely occurring case, as it is based on the assumption that, at the start of any transmission attempt, all the network stations participate in the contention process (heavily loaded network scenario). For more realistic load scenarios (intermediate load cases), the performance analysis must be done by simulation, which enables a more comprehensive analysis of both the BEB and the h-BEB algorithms.

5.2 Timing Analysis by Simulation

A simulation model was implemented using the *Network Simulator (NS-2)* tool [8], which is a shareware discrete event simulator specially suited for the network performance analysis. For the BEB collision resolution algorithm, a station process implements directly the IEEE 802.3 standard, which is already available in the NS-2 tool. For the h-BEB collision resolution algorithm, a station process has been built according to the h-BEB specification described in Section 2. The implemented simulation model considers a 10 Mbps Ethernet network, where each station has a Poisson traffic source with a fixed packet length of 250 bytes. The total network load ranges from 40% to 110%. For each simulated load, 75×10^4 packets are successfully transmitted.

Once more, it is considered a shared Ethernet environment, where multiple stations are interconnected with a *special station* implementing either the h-BEB (enhanced Ethernet mode) or the BEB algorithms (traditional Ethernet mode). Two scenarios are assessed: the small population scenario with 5 Ethernet stations, and the large population scenario with 65 Ethernet stations.

The target of the simulations is to analyze the behavior of the h-BEB algorithm when compared to the traditional BEB collision resolution algorithm. Therefore, the *special station* is used as the test case for both scenarios. The performance measures include both the maximum access delay for 80%, 95%, 98% and 99% of messages and the standard deviation of the average access delay (transfer jitter). The maximum access delay is the maximum time required to successfully transfer a packet, measured from the first transmission attempt to the end of the packet transfer. The maximum access delay for x% of the messages is evaluated discarding the $(100\text{-}x)$% slowest messages. The standard deviation, which is related to the message transfer jitter, is given by:

$$\sigma = \sqrt{\frac{1}{N} \times \sum_{i=1}^{N} (x_i - \bar{x})^2} \qquad (4)$$

where N is the total number of simulated packets, x_i is the delay of each transferred packet and $\bar{x}$ is the evaluated average packet delay. Discarded packets are not considered for the average packet delay, as this measure deals with just the successfully transferred packets.

5.2.1 The small population case

Figures 7 and 8 illustrate the maximum access delay for 80%, 95%, 98% and 99% of the messages and its standard deviation (transfer jitter) in the small population case, for both the enhanced and the traditional Ethernet modes.

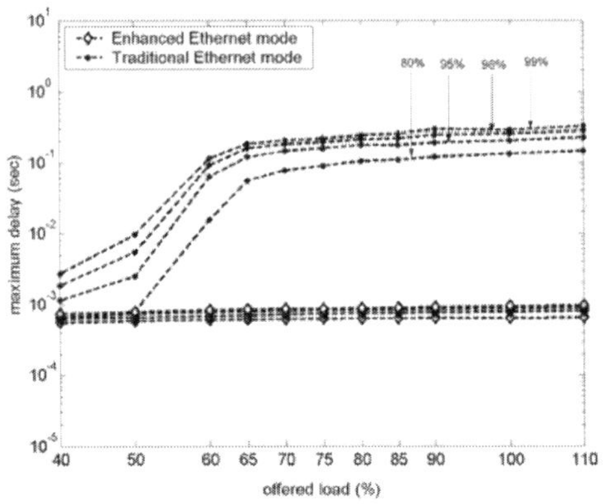

Figure 7: Maximum Access Delay for the small population case.

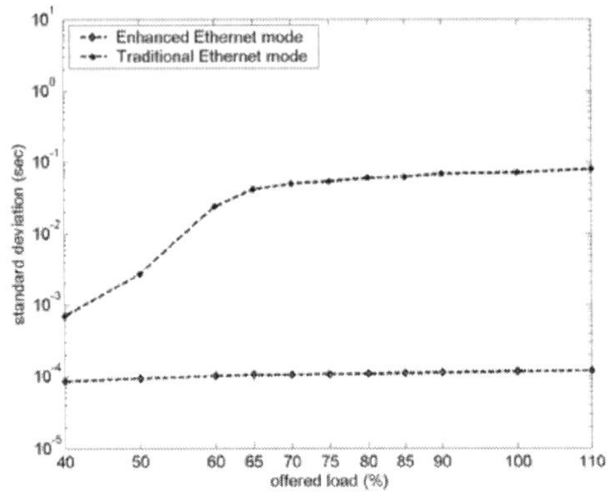

Figure 8: Standard deviation of the average delay for the small population case.

Figure 7 show that the maximum access delay for x% of the messages is nearly constant for the enhanced network case scenario, whatever the network load. More importantly, Figure 8 illustrates that the standard deviation of the average delay is one order of magnitude smaller than the maximum access delay for x% of the messages, which indicates a small dispersion of the simulated results.

Moreover, as the standard deviation of the average delay is a measure of the message transfer jitter, it becomes clear that, whatever the network load, the enhanced Ethernet mode guarantees a nearly constant message transfer jitter. This is an important result, as it forecasts a predictable communication delay when supporting real-time communications.

Finally, both Figures 7 and 8 clearly illustrate the behavior of the traditional Ethernet mode: high access delays for network loads above 60%, with a standard deviation of the average delay in the same order of magnitude of the maximum access delay; the latter indicates high message transfer jitter.

From Figures 7 and 8 it is not clear that for the traditional Ethernet mode, the packet rejection rate becomes significant for network loads above 60%, while for the enhanced Ethernet mode it was not detected any discarded packet within the 75×10^4 simulated transfers.

5.2.2 The large population case

Figures 9 and 10 illustrate the maximum access delay for 80%, 95%, 98% and 99% of the messages and its standard deviation (transfer jitter) in the large population case, for both the enhanced and the traditional Ethernet modes.

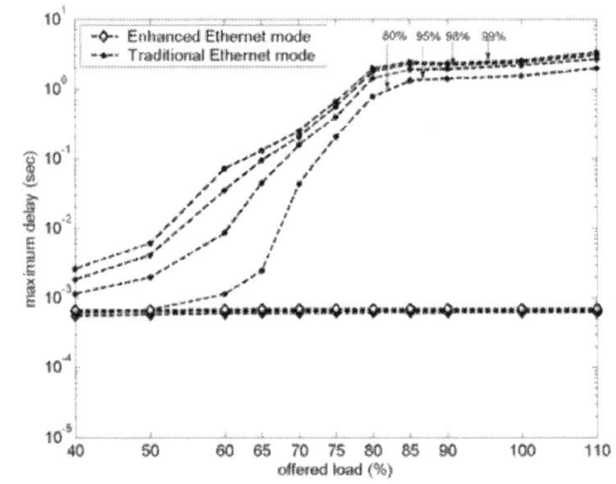

Figure 9: Maximum Access Delay for the large population case.

The presented results illustrate that, for the enhanced Ethernet mode, the results are similar for both the large and the small population cases (there is just a slight decrease in the dispersion of the results). Also, it was not detected any discarded packet within the 75×10^4 simulated transfers. These results indicate that the h-BEB algorithm behaves well, whatever the number of node stations in the network segment.

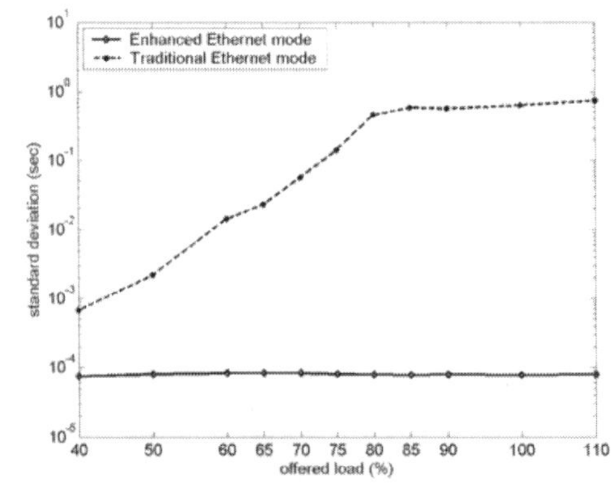

Figure 10: Standard deviation of the average delay for the large population case.

On the other hand, the results are clearly worse for the traditional Ethernet mode, when compared to those of the small population case: both the maximum access delay and the standard deviation of the average delay are one order of magnitude higher for network loads above 70%.

6. STATE-OF-THE-ART IN INDUSTRIAL ETHERNET

Basically, Ethernet networks went through a significant modification from the shared Ethernet specification [9], when the full-duplex operating mode was introduced in the early 90s (IEEE 802.1D) , specificating bridges (also referred as Ethernet Switching Hubs) to interconnect node stations. Such full-duplex operating mode enables the micro-segmentation of the network, by regenerating information only to the receiving port of the bridge, therefore avoiding collisions between messages. Additionally, when using Ethernet Switching Hubs, it is possible to manage network traffic, by means of the adequate setting of data flow permissions and priorities. The transfer of critical information was addressed both by the IEEE 802.1p and the IEEE 802.1q VLAN [10] standards; the latter extends the priority handling aspects of the 802.1p standard, by providing space in the VLAN Tag to indicate traffic priorities to support virtual local area networks (VLANs), while the former gives the ability to prioritize messages.

Nevertheless, the use of switches in an Ethernet network is not a panacea. For instance, if the traffic is sent to an output port at a higher rate than its capacity, messages must be queued. If queuing occurs in an uncontrolled way, the switch can lose messages. Another important problem concerning the use of switched Ethernet is the lack of enough priority levels to support efficient priority-based scheduling [1]. The impact of network topology and message scheduling strategies inside the switch has also been recently addressed [11].

Three approaches can be considered to support real-time communications in shared Ethernet environments [3]: either *avoiding collisions*, by controlling the medium access rights of each station (TDMA scheme, token passing, etc.), or ensuring a *deterministic collision resolution* scheme, by modifying the collision resolution algorithm. A third approach (that is not deterministic) is to reduce the number of occurring collisions, enhancing the network responsiveness to real-time message requests. Whatever the selected approach, it requires the implementation of the protocol modification *in all the interconnected node stations* (at the network adapter level or above), which makes difficult the support of real-time communications within legacy Ethernet communication systems.

The International Electrotechnical Commission (IEC) originally defined three solutions for Industrial Ethernet in the IEC standard 61158. However, there are several systems with potentials to fulfill real-time Ethernet specifications: *Profinet*, *EtherNet/IP*, *EtherCAT*, *Ethernet Powerlink* and *Modbus*, which are briefly summarized in this section.

Profinet is the Ethernet-based automation standard maintained by PROFIBUS International and more than 50 companies. In 2003 was ratified as the International Standard IEC 61158 and IEC 61784. According to Feld [12], *Profinet* provides a reaction time in the required range of 5-10 ms for factory automation and, 1 ms and below for motion control applications, which is adequate in terms of real-time responsiveness. In both *Profinet* versions (*v2* and *v3*), a middleware-scheduling layer provides the adequate priority to the real time data. *Profinet v2* can cooperate with IEEE 802.1 compatible network components. The real time channel is based on a cyclic Provider/Consumer architecture, with Ethernet layer 2 frames. *Profinet v2* can support different real-time classes for most application with cycle time in the range of 5 ms and above, using standard switch-based Ethernet technology. However, motion control applications require a cycle time in the range of 1 ms and below, with a jitter in the range of 1 μs, impairing the use of switch-based Ethernet technology, especially is standard IP traffic is scheduled in parallel to real-time data [13]. *Profinet v3* is based on TDMA scheduling that supports different real-time classes, and it is also compatible with the IEEE 802.1 standard [13]. The TDMA scheduling is based on a communication ASIC (Application specific Integrated Circuit), where a time slot is exclusively reserved for real-time communication within the communication cycle.

EtherNet/IP is an industrial communication standard originally defined by Rockwell, which is supported by ODVA and ControlNet International. It makes use of an open application layer protocol, which is based on Control Information Protocol (CIP) that is used in both *DeviceNet* and *ControlNet*. This topology implements a common set of service at all the network levels, where all the devices organize their data into a common object model. The CIP family of protocols contains a fairly large collection of commonly defined objects [14]. Ethernet/IP classifies the network nodes by device types and objects are added according to specific functionalities.

EtherCAT is an open technology for which IEC standardization is in progress. It sets new standards for real-time performance using twisted pair or fiber optic cable, and it supports line, tree or star topologies. With *EtherCAT*, the data exchange is fully based on a pure hardware machine over a logical ring structure, where a master clock determines the propagation delay. External synchronization is based on the IEEE 1588 standard. *EtherCAT* has different addressing options for different types of communication, optimized for each particular requirements [15].

Ethernet Powerlink protocol is based on the standard IEEE 802.3 layers. Deterministic time is achieved by applying a cyclic timing schedule to all the connected nodes. The schedule is divided in isochronous and asynchronous phase. During the isochronous phase, time-critical data is transferred; the asynchronous phase reserves bandwidth for non time-critical data. The node management grants the access to the physical medium via the exchange of an explicit message (token), thereby preventing collisions. The Ethernet Powerlink Standardization Group (EPSG) recommends the use of repeater hubs instead of switching hubs within the real-time domains, to minimize path delay and frame jitter.

Modbus protocol, developed by Modicon in 1979, is based on master-slave/client-server communication between devices. It is a protocol that is positioned at level 7 of the OSI model. It defines a simple protocol data unit (PDU) independent of the underlying communication layers. The Modbus messaging communication uses four type of messages: a *Modbus Request* is the message sent on the network by the client to initiate a transaction; a *Modbus Indication* is the request message received on the server side, a *Modbus Response* is the Response message sent by the Server, a *Modbus Confirmation* is the response message received on the client side.

7. CONCLUSIONS

This paper presents the timing analysis of an enhanced collision resolution algorithm for shared Ethernet networks: the high priority Binary Exponential Backoff (h-BEB) algorithm. Both the analytical and the simulation timing analysis show that the h-BEB algorithm guarantees a maximum access delay that is significantly smaller than for the standard Ethernet stations.

Two cases were analyzed. Firstly, the analytical study for a heavily loaded network scenario shows that the maximum access delay for 95% of the messages is smaller than 1,86ms. Secondly, for more realistic load scenarios (intermediate load cases), the simulation analysis shows that the maximum access delay for 98% of the messages is always smaller than 1ms. More importantly, it shows a nearly constant message transfer jitter, which is one order of magnitude smaller than the maximum access delay for 98% of the messages.

Concerning the probability of a message frame being discarded by the h-BEB algorithm, it has also been shown that, for the heavily loaded network scenario, such probability is always smaller than 2×10^{-3}. For more realistic load scenarios, the simulation analysis never detected any discarded frame.

These are important results, as they forecast a predictable communication delay when supporting real-time communications with the h-BEB collision resolution algorithm. These results are also consistent with the claim that the h-BEB algorithm is adequate to support most part of the soft real-time applications, as they confirm a rather small probability of any message being discarded.

8. REFERENCES

[1] J.-D. Decotignie, "A perspective on Ethernet-TCP/IP as a fieldbus," presented at Proceedings of LORIA. 4th International Conference on Fieldbus Systems and their Applications, 15-16 Nov. 2001, Nancy, France, 2002.

[2] P. Neumann, "Manufacturing Automation over Networks (Keynote Speech)," presented at 11th IFAC Symposium on Information Control Problems in Manufacturing, Salvador - Brazil, 2004.

[3] R. Moraes and F. Vasques, "A Probabilistic Analysis of Traffic Separation in Shared Ethernet Systems Using the h-BEB Collision Resolution Algorithm," presented at 13th International Conference on Real-Time Systems - RTS'2005, Paris - France, 2005.

[4] F. Carreiro, R. Moraes, J. A. Fonseca, and F. Vasques, "Real-Time Communication in Unconstrained Shared Ethernet Networks: The Virtual Token-Passing Approach," presented at Emerging Technologies and Factory Automation - ETFA, Catania, Italy, 2005.

[5] R. M. Metcalfe and D. R. Boggs, "Ethernet: distributed packet switching for local computer networks," *Communications of the ACM*, vol. 19, pp. 395-404, 1976.

[6] S. S. Lam and L. Kleinrock, "Packet Switching in a Multiaccess Broadcast Channel: Dynamic Control Procedures," vol. CM-23, pp. 891-904, 1975.

[7] G. T. Almes and E. D. Lazowska, "The behavior of Ethernet-like computer communications networks," presented at Proceedings of the Seventh Symposium on Operating Systems Principles, 10-12 Dec. 1979, Pacific Grove, CA, USA, 1979.

[8] "ns-2 Network Simulator," 2.27 ed, 2004.

[9] "IEEE standards for local area networks: carrier sense multiple access with collision detection (CSMA/CD) access method and physical layer specifications," in *ANSI/IEEE Std 802.3-1985*, 1985.

[10] "IEEE standards for local and metropolitan area networks: virtual bridged local area networks," in *IEEE Std 802.1Q-1998*, 1999.

[11] E. Jasperneite and P. Neumann, "Switched Ethernet for factory communication," presented at ETFA 2001. 2001 8th International Conference on Emerging Technologies and Factory Automation. Proceedings, 15-18 Oct. 2001, Antibes-Juan les Pins, France, 2001.

[12] J. Feld, "Realtime Communication in PROFINET V2 and V3 Designed for Industrial Purposes," presented at 5th IFAC International Conference on Fieldbus Systems and their applications, Aveiro, Portugal, 2003.

[13] J. Feld, "PROFINET - scalable factory communication for all applications," presented at 2004 IEEE International Workshop on Factory Communication Systems. Proceedings, 22-24 Sept. 2004, Vienna, Austria, 2004.

[14] P. Brooks, "Ethernet/IP - Industrial protocol," presented at 8th International Conference on Emerging Technologies and Factory Automation (ETFA 2001), Oct 15-18 2001, Antibes-Juan les pins, 2001.

[15] D. Jansen and H. Buttner, "Real-time ethernet the EtherCAT solution," *Computing & Control Engineering Journal*, vol. 15, pp. 16-21, 2004.

ELSEVIER
IFAC
PUBLICATIONS

IMPROVING REAL-TIME BEHAVIOR OF ETHERNET SWITCHES USING FUZZY TRAFFIC SMOOTHING

L. Lo Bello, F. Sgrò, G.A. Kaczyński, L. Di Stefano, O. Mirabella

Dipartimento di Ingegneria Informatica e delle Telecomunicazioni
University of Catania - ITALY
lucia.lobello@diit.unict.it, giordano.kaczynski@wp.pl, omirabel@diit.unict.it

Abstract: Previous work has shown that adaptive traffic smoothing can be successfully applied to a Shared Ethernet to enable it to support statistical real-time communication at the Field level. This paper addresses the case for fuzzy traffic smoothing to realize a tradeoff between bounded delivery times for soft real-time packets and throughput for non real-time traffic over non-microsegmented Switched Ethernet networks. The paper proposes an optimization technique, based on the use of multiobjective optimization, to tune the parameters of the fuzzy controller and discusses the implementation of the optimized fuzzy smoother in a real scenario, also giving comparative performance assessments. *Copyright © 2005 IFAC*

Keywords: Traffic smoother, fuzzy logic, multiobjective optimization, genetic algorithms, switched Ethernet, Real-Time communication.

1. INTRODUCTION

Recent literature on real-time communication over Ethernet networks has pointed out that simply adding a switch to an Ethernet network is not enough to make it able to provide real-time guarantees. For example, in scenarios where the producer/consumers model is adopted (e.g., at the Field level, where such a model is quite common), switches handle producer/consumer interaction as broadcast traffic, and thus one of the major benefits deriving from the use of switches, that is, the existence of multiple simultaneous transmission paths, can be affected.

In this context, (Pedreiras *et al.,* 2003) addresses the effect of what are called 'broadcast storms' in Ethernet switches. Basically, if traffic is sent to an output port at a higher rate than its capacity, long message queues can be formed inside the switch and, if queuing occurs in an unbounded way, the switch memory may be exhausted, causing message losses. A set of practical experiments carried out on an off-the-shelf Ethernet switch also showed other weaknesses affecting switched Ethernet as far as real-time behavior is concerned. These include the low number of different priority levels provided by IEEE 802.1D (ANSI/IEEE, 1998) and IEEE 802.1Q (IEEE, 1998) (up to 8 distinct traffic classes to prioritize messages inside the switches), which is not enough to support efficient priority-based scheduling in general cases (Decotignie, 2001). The experiments also showed diverse sources of blocking that higher-prioritized traffic in a switch is subject to when the switch is heavily loaded with lower-prioritized traffic. Real-time traffic may therefore be blocked by non real-time traffic. The authors of (Pedreiras *et al.,* 2003) indicate the need for some kind of traffic control, and in particular suggest traffic smoothing as a possible solution to be adopted inside switches to address this issue.

The adoption of an adaptive traffic smoother to provide soft real-time behavior over a non-microsegmented Ethernet was recently addressed in (Kweon *et al.,* 2004), where the Harmonic-Increase and Multiplicative-Decrease (HIMD) smoother originally proposed in (Kweon *et al.,* 2000) is dealt with. Here we investigate the case for exploiting the fuzzy traffic smoother proposed in (Lo Bello *et al.,* 2002) in the same scenario addressed in (Kweon *et al.*, 2004). As compared with previous work on fuzzy smoothing, there are several new insights here. First, we investigate the fuzzy smoother over non-microsegmented Switched Ethernet networks, while previous work focused on Shared Ethernet only. Second, our perspective is novel, as the explicit aim here is to achieve a tradeoff between the delivery time for soft real-time traffic and the throughput for non real-time traffic. To achieve our goal, here we adopt an optimization technique to tune the parameters of the fuzzy controller, based on the use of Multi Objective Genetic Algorithms (MOGAs) (Fonseca and Fleming, 1993; Fonseca, 1995). The paper describes the optimization procedure and addresses the implementation of the optimized fuzzy smoother in the same real scenario addressed in (Kweon *et al.*, 2004). Comparative performance assessments of the fuzzy smoother, with the HIMD smoother and without any kind of smoothing respectively, are also presented.

The paper is organized as follows. Section 2 reviews the basic concepts of traffic smoothing. Section 3 addresses tuning of the fuzzy smoother using MOGAs. Section 4 outlines the experimental environment, while Section 5 compares and discusses the performance obtained by the HIMD and the fuzzy smoother in a real scenario. Finally Section 6 gives our conclusions.

2. REMARKS ON TRAFFIC SMOOTHING

When both real-time (RT) and non-real-time (NRT) packets are transported over an Ethernet, RT packets from a node may experience a long delay due to a) contention with NRT packets in the source node and b) collision with RT and NRT packets from the other nodes. Kweon *et al.*, (1999), analytically demonstrated that, to statistically bound the medium access time for an Ethernet frame, it is sufficient to keep the total arrival rate for new packets generated by stations below a threshold called the network-wide input limit. To maintain such a global limit, in (Kweon *et al.*, 1999) each station is assigned a local threshold, called a station input limit, and a traffic smoother is implemented on each node to regulate the outgoing NRT stream, in order to maintain the traffic generation rate below the station input limit.

Traffic smoother is a software level which regulates the packet stream generated at the upper layers of the TCP/IP (or UDP/IP) stack making the packet stream as *smooth* as possible when entering the Ethernet MAC layer. Within a node, RT packets are distinguished from NRT packets using the TOS (Type Of Service) field in the IP header, and a priority queue with two priority levels, high for RT and low for NRT packets, is maintained. A RT packet is not affected by smoothing, while NRT traffic is transmitted as long as the overall station arrival rate (which comprises both RT and NRT packets) is below the station input limit; otherwise NRT packets are delayed. The traffic smoother therefore has two main effects: first it gives RT packets priority over NRT ones, in order to eliminate contention within each local node, and secondly it smoothes NRT traffic on the network so as to reduce the interference with RT packets from the other nodes.

2.1 Traffic smoothing implementation

Traffic smoothing is based on a credit bucket mechanism, which is a token bucket-based algorithm (Turner, 1986). The credit bucket has two parameters: Credit Bucket Depth (CBD), which indicates the capacity of the credit bucket, and Refresh Period (RP), which indicates the replenishment period. Up to CBD credits are added to the bucket every RP seconds. When a NRT packet arrives at the traffic smoother, if there is at least one credit in the bucket, the traffic smoother forwards it to the Ethernet NIC. Otherwise, the NRT packet is not transmitted to the Ethernet NIC until at least one credit becomes available following a replenishment.

Originally devised for Shared Ethernet in order to reduce the probability of packet collisions (Kweon *et al.*, 1999, 2000; Lo Bello *et al.*, 2002, 2005), traffic smoothing has recently been applied to non-microsegmented Switched Ethernet networks, in (Kweon *et al.*, 2004). The traffic smoother in (Kweon *et al.*, 1999, 2000; Lo Bello *et al.*, 2002, 2005) is implemented at a kernel level, as a software layer inserted between the TCP/IP (or UDP/IP) and the Data Link layer.

Implementation of the traffic smoother only requires a minimal modification of the kernel, i.e. in the device driver for Linux (or a new Network Driver Interface Specification for Windows NT), and does not entail any changes to the current standard Ethernet MAC protocol or TCP/IP (or UDP/IP) stack. What has to be modified is the Ethernet device driver to record the time when a packet in the NIC experiences a collision, so that the smoothing algorithm may read and use it.

2.2 Adaptive traffic smoothing

Adaptive traffic smoothing allows us to dynamically modify the station input limit (i.e. the CBD/RP ratio) a station is assigned every time according to the change in the network workload. In order to evaluate the current network workload, different approaches have been used, based on the measurement of either throughput (Lo Bello *et al.*, 2000) or the number of collisions (Kweon *et al.*, 2000).

The Harmonic-Increase and Multiplicative-Decrease (HIMD) approach, described in (Kweon *et al.*, 2000), applies an adaptation mechanism which reacts to the detection of a single collision over a given time α. When a collision is detected, the *RP* is increased by whichever is lower between twice its current value and a given RP_{max} value, while in the absence of collisions the *RP* is periodically decreased (with a period of τ) by a constant Δ heuristically determined down to a value of RP_{min}. The parameters α, Δ, τ, RP_{max} and RP_{min} are user-controllable, and by using different values, different delay and throughput characteristics can be obtained.

The HIMD adaptive traffic smother approach suffers from two limitations. First, it only takes collisions into account, regardless of the actual amount of network load, which could be quite low even in the presence of collisions. Secondly, the approach is not flexible, as RP regulation is based on fixed variations (doubling the RP or decreasing it by a constant Δ). These two limitations are the reason for the development of a more flexible approach, the fuzzy traffic smoother, which was proposed in (Lo Bello *et al.*, 2002).

2.3 Fuzzy traffic smoothing

The *fuzzy traffic smoother* is an adaptive traffic smoother based on a fuzzy controller. It has two inputs – the number of collisions and the overall throughput observed in a reference interval – and a single output, i.e. the quantity by which the *RP* is to be varied according to the input values, here called *VarRP*. The output *VarRP* represents the variation of the *refresh period* as compared with the current value. If RP_{old} is the current *refresh period* and RP_{new} is the new value, the formula used by the smoothing driver is:

$$RP_{new} = RP_{old} + VarRP \quad (1)$$

The fuzzy smoother improves on HIMD in two respects. First, it uses both total throughput and the number of collisions as input parameters for the smoother which together represent a more complete indicator of the actual workload than only one of them. Second, here the variation of the station input limit is not based on fixed variations as in the HIMD one, but is dynamically varied and gauged according to the actual workload by the *fuzzy controller* which, according to the total throughput and number of collisions, applies rules to choose the most appropriate RP on a case-by-case basis. Fuzzy control is particularly suitable when knowledge of the system to be controlled is insufficient or the dynamic model is too complex to model and control. This is the case of the system considered here, which, due to its non-linear and quite complex behaviour, is difficult to model and control using traditional controllers, as they rely on some knowledge of the model of the system to be controlled.

3. TUNING OF THE FUZZY SMOOTHER THROUGH GENETIC ALGORITHMS

The design of a fuzzy controller is mainly the result of subjective considerations deriving from the designer's knowledge and experience of the system to be controlled. Fuzzy control alone, however, does not provide a technique that will allow its parameters to be configured systematically. A solution which is often proposed in the literature, and has given good results, is the use of *Genetic Algorithms* (Shi *et al.*, 1999; Setnes and Roubos , 2000; Belarbi and Titel, 2000).

Genetic Algorithms (Goldberg, 1989) are search algorithms based on natural selection mechanisms. The idea is to evolve a set of individuals called *chromosomes*, which represent possible solutions to a problem, via competition and the exchange of genetic information. Each chromosome has an associated *fitness index*, which is used to activate selection, which in turn allows the creation of new individuals who will be the next generation in the evolutionary cycle. They are formed by means of operations to exchange genetic information (*crossover*) and random *mutations* of the genetic code.

In this paper Genetic Algorithms are used to set the parameters of the three membership functions for each input of the fuzzy controller, which correspond respectively to the values (*low*, *med*, *high*) of each variable. Optimization, which is performed off-line using an ad hoc simulator, is multiobjective, as the aim is to minimize the deadline miss ratio for RT traffic, while maximizing the throughput for NRT traffic.

In a problem of this kind, where the objectives are conflicting, it is not possible to define a single solution as optimal. It is rather a case of finding the set of fuzzy controller parameters that achieve a good tradeoff between the deadline miss ratio for RT traffic and the throughput obtained by NRT traffic.

3.1 Pareto Optimum and MOGAs

In a scalar problem it is simple to define the concept of maximum (or minimum) and therefore optimum. In a problem with a vectorial objective like ours, however, it is not possible to define a vector as the optimal solution to the problem. It is necessary to introduce the concept of *Pareto Optimum*, formulated by Vilfredo Pareto in 1896 (Pareto, 1896), which constitutes by itself the origin of research in multiobjective optimization.

A solution is said to be Pareto optimal if it is not possible to obtain improvements in one objective without a consequent deterioration in performance for the other objectives. To define solutions with these characteristics we have to use the concept of *dominance*.

Given a set of N objectives to be maximized, a solution x is said to be *weakly dominated* by a solution y (where x is other than y) if the following relation holds:

$$f_j(x) \le f_j(y) \quad \forall j = 1,\ldots,N \quad (2)$$

where x and y are the parameters to be optimized, whereas f_j is the objective function relating to the j-th objective. If x is weakly dominated by y, but there exists at least one index j* such that

$$f_{j*}(x) < f_{j*}(y) \quad (3)$$

x is said to be *strongly dominated* by y. It is therefore preferable to y for part of the objectives to be maximized, and is no worse for the remaining ones.

On the basis of the previous definitions, a solution is said to be a *Pareto solution* if it is not dominated (either weakly or strongly) by any other solution (Fonseca, 1995). The set of Pareto solutions to a multiobjective optimization problems define the so-called *Pareto front* (see Fig. 1); no feasible solution exists outside this front.

The Pareto front comprises an infinite number of elements. Our task is to find a subset and then choose the solution closest to the desired tradeoff between the two objectives. The technique we used to search for Pareto solutions was that of MOGAs (Multi Objective Genetic Algorithms), introduced by Fonseca and Fleming in (Fonseca and Fleming, 1993), and also used in (Lo Bello *et al.*, 2005) over a Shared Ethernet. However, unlike (Lo Bello *et al.*, 2005), where MOGAs are only used to validate results obtained with the classical *weighted sum approach*, which consists of transforming the multiobjective problem into a scalar one by means of the weighted sum of the objectives to be optimized, here MOGAs are the selection method used to form the population at each generation in the evolution process, and thus the design of the fuzzy traffic smoother.

Consider, for example, an individual x_i at generation t, which is dominated by $p_i(t)$ individuals in the current generation. Its current position in the individuals' rank can be given by:

$$\text{rank}(x_i, t) = 1 + p_i(t) \quad (4)$$

All non-dominated individuals are assigned rank 1, while dominated ones are penalized according to the population density of the corresponding region of the tradeoff surface.
Fitness assignment is performed in the following way:

1. sort population according to rank;
2. assign fitness to individuals by interpolating from the best (rank 1) to the worst;
3. average the fitnesses of individuals with the same rank, so that all of them will be sampled at the same rate.

In using MOGAs we also decided to apply *the fitness sharing* technique (Fonseca, 1995; Goldberg and Richardson, 1987). This technique penalizes the

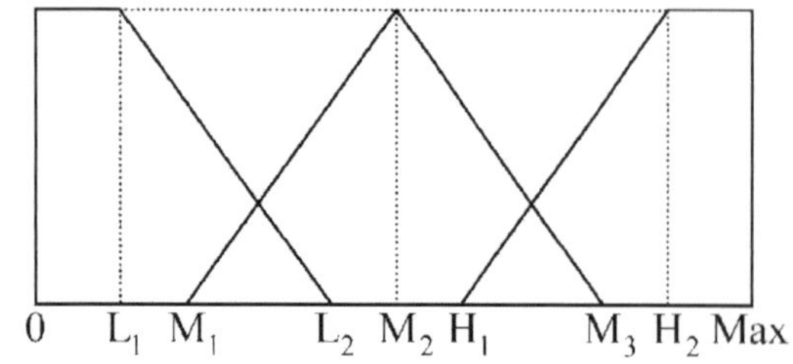

Fig. 2. Fuzzy membership parameters.

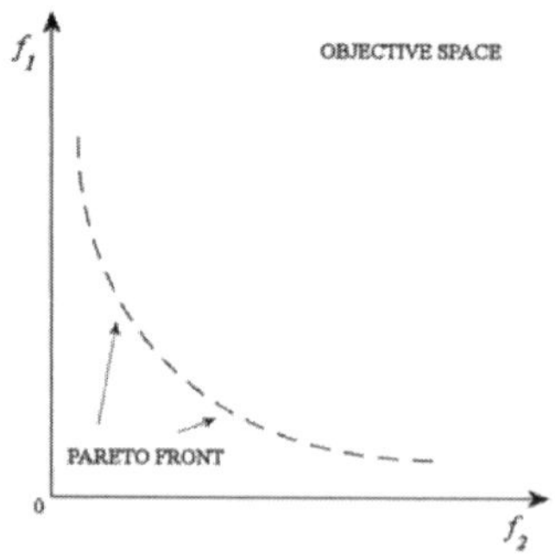

Fig. 1. Example of Pareto front in minimization of two objectives, f_1 and f_2.

fitness of individuals that are at a shorter distance from a parameter σ, so as not to concentrate the evolution in only a few areas of the search space.

3.2 Implementation details of the evolutionary process

The first step in implementing GAs is to encode the information – in our case the fuzzy membership values. With reference to Fig. 2, a generic chromosome is formed by the following vector:

$$\begin{aligned} C = (&L_{11}, L_{21}, M_{11}, M_{21}, M_{31}, H_{11}, H_{21}, \\ &L_{12}, L_{22}, M_{12}, M_{22}, M_{32}, H_{12}, H_{22}) \end{aligned} \quad (5)$$

The second index in the parameter subscript refers to the relevant input. It is 1 in the case of collisions, 2 in the case of throughput.

Using MOGAs, choice of the fitness value proved to be more simple than in a scalar problem, it is simply formed by the vector of objectives:

$$f(x) = (D_{mr}, T_h) \quad (6)$$

Each chromosome thus has an associated pair of values, D_{mr} and T_h, which respectively represent the deadline miss ratio and the total NRT throughput for the whole network.

The fundamental mechanism of the evolutionary process consists of three operations:

1. evaluation of individual fitness;
2. formation of a gene pool (intermediate population) through the selection mechanism;
3. recombination through crossover and mutation operators.

Let us describe the different steps of the algorithm used.

- *initialization*: the population is randomly generated. A number of 50 individuals was chosen, and it remains constant during the entire evolution process.
- *evaluation*: the fitness vectors of each individual are calculated. We recall that the entire evolution process is implemented in an Ethernet simulator we developed.

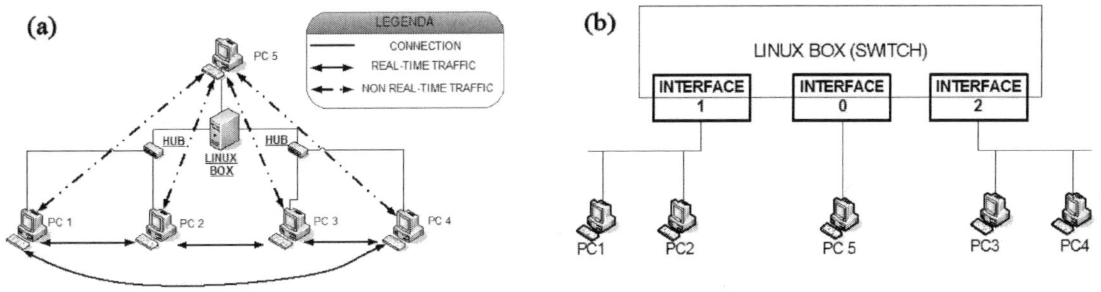

Fig. 3. Test scenario used to measure performance. (a) Communication structure. (b) Collision domains.

- *selection*: the MOGA selection technique is used in this step. The individuals are proportionally selected according to their rank. In this step *elitism* is applied: the best individual is directly copied to the new population.
- *recombination*: crossover and mutation, with their respective probabilities, are applied to generate new chromosomes that will make up the new population.
- *termination-condition*: it was decided to stop the evolution process at 100 generations, as we saw that at that point the search space was sufficiently exploited and the Pareto front was clearly defined.

The following procedure shows the structure of the GA evolution process, where p(t) denotes the population at generation t.

GA Evolution Procedure
Begin (1)
t = 0;
initialization p(t);
evaluation p(t);
While (**Not** *termination-condition*) **Do**
Begin (2)
t = t + 1;
selection p(t) *from* p(t-1);
recombination p(t);
evaluation p(t);
End (2)
End (1)

At the end of the evolutionary process using MOGAs the set of optimal Pareto solutions was formed, so the next task was to choose the best one in terms of the trade-off between the D_{mr} and T_h objectives. A set of possible solutions was chosen and then, after a series of tests, the final one was obtained.

4. EXPERIMENTAL ENVIRONMENT

To investigate the performance behavior of the optimized fuzzy smoother a testbed for a factory communication environment was set up. To make a performance comparison with the HIMD smoother, the experimental environment described in (Kweon *et al.*, 2004) was reproduced. In such an environment, a switch is used in order to separate an existing LAN into smaller collision domains. In Distributed Process Control Systems (DPCSs), the adoption of a switch as a interconnection element offers several advantages, as it makes it possible to create one separate collision domain for each cell involved in a specific phase of the control process. Cell interconnection through a switch also increases the maximum aggregated throughput obtainable. The particular scenario considered here, a non-microsegmented Switched Ethernet network, takes into account that internally cells are usually made up of sensors, actuators, PLCs interconnected by an hub.

The experimental environment, shown in Figs. 3(a) and 3(b), comprises 6 hosts running the Linux operating system (kernel 2.2.18) and interconnected via network cards with Realtek 8139 10/100 Mbps chipsets. As a network interconnection element, a PC with a minimal configuration and 3 network cards emulating a layer-3 switch, thus called a Linux Box, was used. Three separate collision domains are present, as shown in Figs.3(a) and (b).

Hosts PC_1-PC_4 exchange RT packets according to a Client/Server model, using the UDP/IP protocol. With reference to Fig. 3(a), following the real-time ring clockwise, each node (PC) represents the Server for the previous node and the Client for the next one.

The experiments were carried out with a constant RT workload of 33.2 Kbps for each RT Client (41.64 Kbps if the UDP/IP and MAC Ethernet headers are taken into account).

In addition to the RT traffic, each node generates NRT packets when a Monitor process, running on PC_5, sends them an activation message. In the tests being performed, the node labelled as PC_5, besides handling and controlling all NRT communications, is also used as the destination for the NRT traffic produced by all the other hosts. NRT packets, sent through the UDP/IP protocol, are activated in a *non-greedy* mode, i.e. the interarrival time for the bursts is exponentially distributed, with an average value of 8 ms. The total NRT workload in the network is 5.1 Mbps (including the UDP/IP and Ethernet headers).

As it will be shown in Section 4, in a configuration like the one considered here, the switch by itself is not enough to prevent NRT traffic from affecting the delay experienced by RT traffic. This is why, in scenarios like this one, an adaptive traffic smoother is advisable.

Table 1. Description of parameters of the scenario used in tests.

Name	Value
Number of hosts (PC_1-PC_5)	4
Number of switches (Linux Box)	1
Number of hubs	2
Number of RT Clients	4
Number of RT Servers	4
Number of NRT hosts	5
Collision Domain 1 (PC_1-PC_2) bit rate	10 Mbps
Collision Domain 2 (PC_3-PC_4) bit rate	10 Mbps
Collision Domain 3 (PC_5) bit rate	100 Mbps
Number of packets transmitted during the test	3100
Total RT Workload for Collision Domain 1	66.5 Kbps
Total RT Workload for Collision Domain 2	66.5 Kbps
Total NRT Workload for Collision Domain 1	5.1 Mbps
Total NRT Workload for Collision Domain 2	5.1 Mbps

For the smoother fuzzy to work properly in our scenario the switch (Linux Box) is provided with a smoother for each network card. This made it necessary to develop a *sniffer* process for each smoother, so that the data recorded refer to the network segment to which the network card is connected.

As a result, an ad hoc sniffer process, called *fsniffer*, was developed for each network interface of the Linux Box. The *fsniffer* process is a periodically activated task (the period τ_f is fixed during the setup phase - here a value of 10 ms, as in (Lo Bello *et al.*, 2002), was chosen), which records from the network collisions and throughput data to be processed to produce the *VarRP* value (Lo Bello *et al.*, 2002, 2005).

5. PERFORMANCE EVALUATION

Fig. 3 shows the experimental scenario, while Table 1 summarizes the relevant parameters. The scenario considered here is interesting because it makes it possible to analyze the network capacity to support both RT traffic exchanged in a single collision domain (*intra-domain* traffic, i.e. RT communications between PC_1 and PC_2 and between PC_3 and PC_4) and RT traffic exchanged between the two domains (*inter-domain* traffic, i.e. RT communication between PC_1 and PC_4 and between PC_2 and PC_3), in the presence of NRT traffic bursts affecting all the network segments and therefore all the switch ports.

It is also important to point out that, although small, the experimental testbed used may represent a departmental network where each subnetwork is connected to a switch that acts as the centre of a star topology.

Our measurements refer to: a) the *roundtrip delay* experienced by a complete RT transaction (Request/Reply) between the Client process on PC_3 and the Server process on PC_2 and b) the *total NRT throughput*.

A first set of experiments were performed without smoothing. Figs. 4(a) and 4(b) show the results obtained for the RT frames roundtrip delay and total NRT throughput, respectively.

The graph in Fig.4(a) shows rather high roundtrip delay values, which confirm that the use of a switch does not provide RT communications with sufficient QoS, in terms of bounded delay. On the other hand, as shown in Fig. 4(b), the switch allows to obtain a throughput value very close to the offered workload. It must be pointed out that here the total throughput is calculated as the sum of the throughput for the 3 network segments (i.e. it represents the aggregated throughput).

Other tests were performed using the HIMD smoother, which was activated in each NRT node and in the Linux Box with the same values for the smoothing parameters used in (Kweon *et al.*, 2000), and shown in Table 2. Figs. 5(a) and 5(b) show the roundtrip delay and throughput obtained.

Fig. 5(a) shows that, compared with the previous case, the average roundtrip delay is improved, as the smoother reduces the impact of NRT traffic over RT one, but the number of packets affected by a long delay is still high. On the other hand, the total throughput (shown in Fig.5(b)), is lower than the one achieved without smoother, due to quite conservative smoothing action applied by the HIMD smoother.

Finally, the behavior of the *fuzzy* smoother was investigated in the same environment and operating conditions as the previous experiments, using the smoothing parameters shown in Table 2.

Figs. 6(a) and 6(b) show the performance improvement achieved by fuzzy smoothing as compared with both the HIMD and the no smoothing approach. The average roundtrip delay is about 3 ms, with no high delay peaks, thus showing the fuzzy smoother ability to better serve RT traffic in the presence of NRT burst than the other approaches. Moreover, the average aggregated throughput is higher than the one obtained with the HIMD approach. This is because the improved fuzzy smoother, thanks to the multiobjective optimization, better adapts to workload changes.

Table 2: Smoothing parameters.

Name	Symbol	Value
Collision observation window (HIMD only)	α	10 ms
Decrease of Refresh Period (both)	Δ	100 μs
Max Refresh Period (both)	RP_{max}	100 ms
Min Refresh Period (both)	RP_{min}	3 ms
Time Invocation of refresh procedure (both)	τ	10 ms
Time Invocation of *fsniffer* (fuzzy only)	τ_f	10 ms

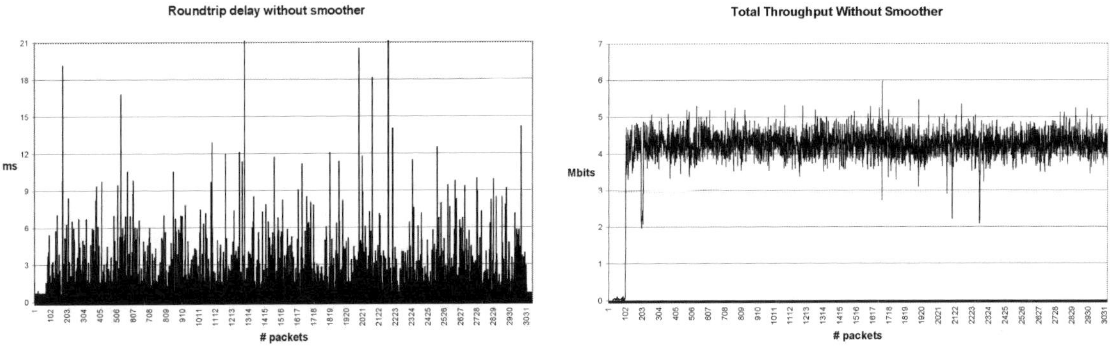

Fig. 4. Roundtrip delay and throughput without smoothing.

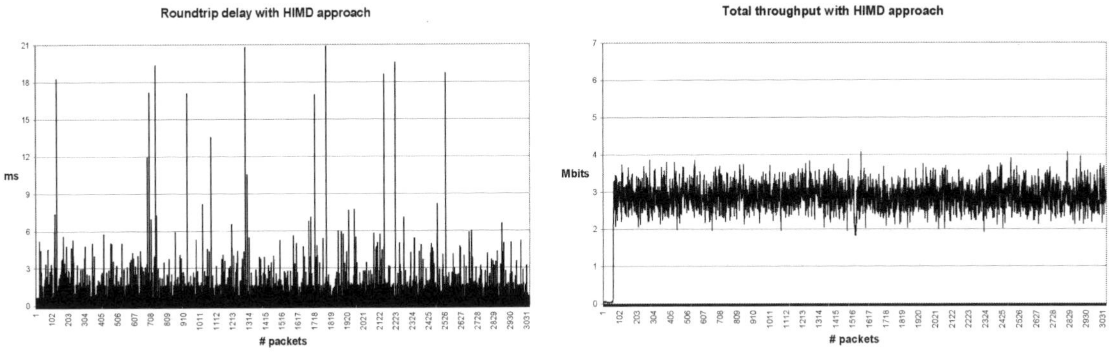

Fig. 5. Roundtrip delay and throughput with HIMD smoothing.

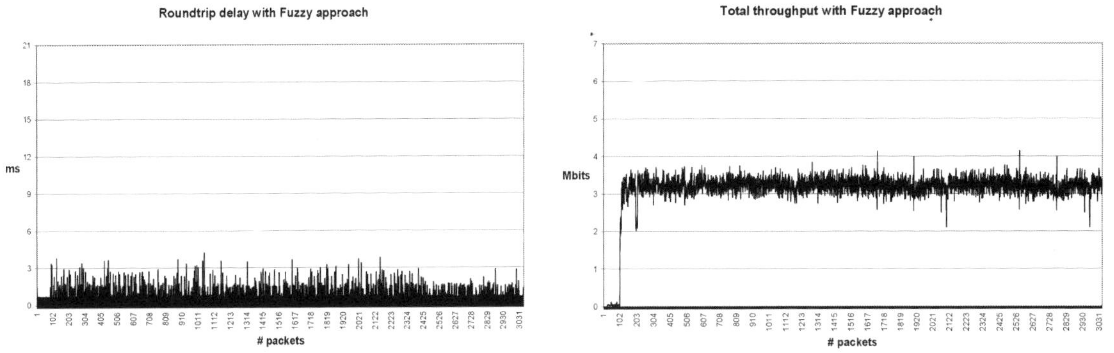

Fig. 6. Roundtrip delay and throughput with Fuzzy smoothing.

Finally, to assess the ability of supporting soft RT traffic (i.e. best effort) with very stringent timing constraints, Fig. 7 plots the deadline miss-ratio for RT messages referred to a 3 ms deadline. As the figure shows, the optimized fuzzy smoother is able to support urgent deadlines better than the other approaches, thus offering a better statistical behavior.

6. CONCLUSIONS

The results obtained not only confirm the advantages of adopting the fuzzy smoother in a Shared Ethernet (with reference to the RT traffic exchanged in each single collision domain), but also show the benefits provided by the fuzzy smoother in a non-microsegmented Switched Ethernet. It has been shown that fuzzy smoothing enhances the switch ability to support soft RT traffic (even with stringent deadlines) between the two collision domains even in the presence of a large amount of NRT traffic transiting through all the ports of the switch. Moreover, the multiobjective optimization allows us to obtain a better QoS tradeoff between roundtrip time for soft RT traffic and throughput for NRT traffic than the one obtained either in absence of smoothing or when the HIMD smoother is used. On-

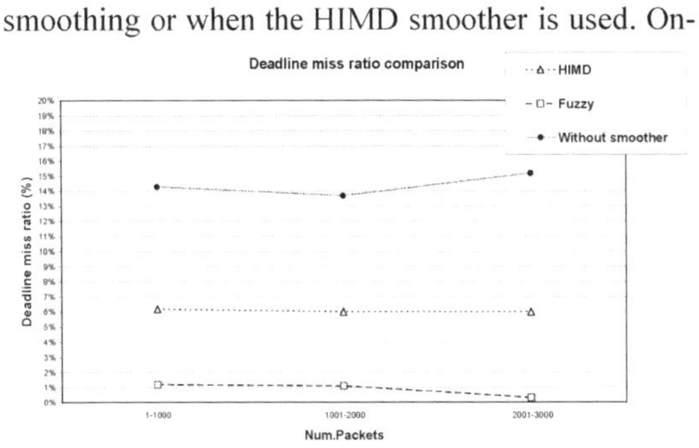

Fig. 7. Deadline miss ratio comparison

going work deals with a fuzzy smoother for a microsegmented Ethernet network. The improvement is in the controller design. In a microsegmented network no collision occurs, so the input parameters of the controller have to change, but some mechanism to enforce real-time behavior in the presence of NRT bursts is still needed. A new controller has therefore been devised, and we are currently implementing it in order to assess its performance in a real scenario.

REFERENCES

Pedreiras, P., R. Leite and L. Almeida (2003). Characterizing the Real-Time Behavior of Prioritized Switched-Ethernet. In: *Proc. of RTLIA03, the 2nd Intl. Workshop on Real-Time LANs in the Internet Age, satellite workshop of the 15th IEEE Euromicro Conference on Real-Time Systems*, pp. 59-66. Porto, Portugal.

ANSI/IEEE, 802.1D (1998). ANSI/IEEE Standard for Information Technology - Telecommunications and information exchange between systems - Local and metropolitan area networks - Common Specification - Media Access Control (MAC) bridges.

IEEE, 802.1Q (1998). IEEE Standard for Local and Metropolitan Area Networks: Virtual Bridge Local Area Networks.

Decotignie, J.D. (2001). A perspective on Ethernet-TCP/IP as a Fieldbus. In: *Proc. of the 4th FeT 2001 International Conference on Fieldbus Systems and their Applications*, pp. 138-143. Nancy, France.

Kweon, S.K., K.G. Shin and G. Workman (2000). Achieving Real-Time Communication over Ethernet with Adaptive Traffic Smoothing. In: *Proc. of Sixth IEEE Real-Time Technology and Applications Symposium, RTAS 2000*, pp. 90-100. Washington DC, USA.

Kweon, S.K., M.G. Cho and K.G. Shin (2004). Soft Real-Time communication over Ethernet with adaptive traffic smoothing. In: *IEEE Transactions on Parallel and Distributed Systems,* 15(10), pp. 946–959.

Lo Bello, L., O. Mirabella et al. (2002). Fuzzy Traffic Smoothing: an Approach for Real-time Communication over Ethernet Networks. In: *Proc. of the 4th IEEE Workshop on Factory Communication Systems*. Västeras, Sweden.

Lo Bello, L., G.A. Kaczyński and O. Mirabella (2005). Improving the Real-Time Behavior of Ethernet Networks Using Traffic Smoothing. In: *IEEE Transactions on Industrial Informatics*, 1(3), pp. 151–161.

Fonseca, C.M.M. and P.J. Fleming (1993). Genetic algorithms for multiobjective optimization: Formulation, discussion and generalization. In: *Genetic Algorithms: Proceedings of the Fifth International Conference*, pp. 416.423. Morgan Kaufmann.

Fonseca, C.M.M. (1995). Multiobjective Genetic Algorithms with Application to Control Engineering Problems, PhD Thesis, Department of Automatic Control and Systems Engineering The University of Sheffield.

Kweon, S.K., K. G. Shin and Q. Zheng (1999) Statistical Real-Time Communication over Ethernet for Manufacturing Automation Systems. In: *Proc.of the 5th IEEE Real-Time Technology and Application Symposium*. Vancouver, Canada.

Turner, J. S. (1986). New Directions in Communications (or Which Way to the Information Age?). In: *IEEE Comm. Magazine*, 24(10), pp. 8–15, Oct. 1986.

Lo Bello, L., M. Lorefice, O. Mirabella and S. Oliveri (2000). Performance Analysis of Ethernet Networks in the Process Control. In: *Proc. of the 2000 IEEE International Symposium on Industrial Electronics*. ISIE 2000, Puebla, Mexico.

Shi, Y., R. Eberhart and Y. Chen (1999). Implementation of Evolutionary Fuzzy Systems. In: *IEEE Transactions on Fuzzy Systems*, 2, pp. 109-119.

Setnes, M. and H. Roubos (2000). GA-fuzzy modeling and classification: complexity and performance. In: *IEEE Transactions on Fuzzy Systems*. 8(5), pp. 509 – 522.

Belarbi, K. and F. Titel (2000). Genetic algorithm for the design of a class of fuzzy controllers: an alternative approach. In: *IEEE Transactions on Fuzzy Systems*. 8(4), pp. 398 – 405.

Goldberg, D.E. (1989). *Genetic Algorithms in Search, Optimization and Machine Learning*. Addison-Wesley.

Pareto, V. (1896). *Cours D'Economie Politique, Vol. I and II*. F.Rouge, Lausanne.

Goldberg, D.E. and J. Richardson (1987). *Genetic algorithms with sharing for multimodal function optimization*, pp. 41-49, Grefenstette.

Copyright © Fieldbus Systems and Their Applications
Puebla, Mexico, 2005

ELSEVIER
IFAC
PUBLICATIONS

RTL-TEP: AN ETHERNET PROTOCOL BASED ON TDMA [1]

José A. Alegre, Josep V. Sala, Sergio Pérez, Joan Vila

Universitat Politècnica de València
Dept. d'Informàtica de Sistemes i Computadors
Camí de Vera s/n, 46022 València, Spain
joalsan@doctor.upv.es, josasis@doctor.upv.es, serpeal@upvnet.upv.es, jvila@disca.upv.es

Abstract: RTLinux-GPL and Ethernet networks can be effectively used for many real-time environments by using some mechanisms or protocols to provide support for real-time and QoS requirements. This paper presents RTL-TEP, a software protocol on top of Ethernet in order to make Ethernet deterministic. The main issues to face implementing TDMA in software are described, as well as the special features of RTL-TEP. Some of the questions that are discussed are clock synchronization, dynamic behaviour and fault-tolerant mechanisms. *Copyright © 2005 IFAC*

Keywords: Real-Time Communication, Distributed Systems, Ethernet, Time-slot Assignment, Clock Synchronization

1. INTRODUCTION

RTLinux-GPL (RTLinux-GPL, n.d./2005) is a real-time operating system built on top of the GNU/Linux kernel. It allows to build complex real-time systems by structuring them at two levels: hard real-time components are running on RTLinux at kernel level and soft or non real-time components run on Linux as user-level processes. There has been little network support at RTLinux level, but this support is not enough for many distributed applications with real-time or QoS requirements. Moreover, Ethernet does not provide real-time features, because it was not designed with QoS on mind, but it is certain that Ethernet provides high bandwidth at very low cost.

Although several high bandwidth networks are currently available, not all of them are suitable for accommodating real-time traffic, or traffic with QoS requirements. There exist other networks that provide QoS guarantees, like ATM, however it has proven to be too expensive for use as a LAN and is mainly used for internetwork links. On the other hand, some fieldbuses, like CAN, provide a prioritized access to the network, but they are also more expensive than Ethernet, they have lower bandwidth, and many application level protocols for distributed and multimedia computing (like CORBA, RTP, etc.) are not available for them.

The idea behind this paper is that the Ethernet network can be effectively used for many real-time environments if some mechanism or protocol is used to make it predictable. Using such a protocol usually reduces the user available bandwidth of Ethernet, but even in that case it provides a higher bandwidth than most fieldbuses and is more cost effective. There are much work dealing with Real-Time Ethernet, like adaptive traffic smoothing (Carpenzano *et al.*, 2002), RT-EP (Martínez *et al.*, 2003) or FTT-Ethernet (Pedreiras *et al.*, 2002).

The Ethernet can be wired using two main layouts: the traditional bus network configuration, using coax cabling and hubs, and the modern star-like network configuration, using universal twisted pair and switches. The reason why Ethernet is not predictable is different

[1] This work has been supported by the Spanish Government Research Office (CICYT) under grant TIC2002-04123-C03-03.

for each layout. In the first case, the source for unpredictability is in the medium access protocol: nodes access the network using the CSMA/CD (*Carrier Sense Multiple Access with Collision Detect*) technique and, although this technique does not specify a collision resolution protocol, it uses the BEB (*Binary Exponential Back-Off*) that solves it in a non-deterministic fashion. In the second case, works concentrate on bandwidth reservation schemes for switched Ethernet (Varadarajan and Chuieh, 1998). In switched Ethernet networks, collisions can no longer occur, since each node is connected with a private cable to a switch. The reason for unpredictability is in how switches manage congestion. Current switches do not provide guarantees as to which packets will be sent first and may drop messages erratically due to lack of memory capacity.

The solution we propose for solving the problem of Ethernet unpredictability is focused in the first configuration. This solution is based on using a software protocol on top of Ethernet for avoiding collisions when accessing the network. This can be done using two main strategies: a Timed-Token Protocol (Grow, 1982) and the Time Division Multiple Access (TDMA) protocol (Kopetz and Grünsteidl, 1994). We focus on making Ethernet predictable by overimposing a TDMA-based protocol on top of it. For this purpose, RTL-TEP (*RTLinux-GPL TDMA on Ethernet Protocol*) has been developed, which is a software protocol on top of Ethernet for avoiding collisions when accessing the network. This work is based in the *Time-Triggered Ethernet Protocol* (Lankes *et al.*, 2002) provided by (The ROFES Project, n.d./2005), which has been improved with dynamic bandwidth allocation and fault-tolerant mechanisms in order to allow resynchronization and node failures.

The paper is organized as follows: in section 2, we present a real-time networking architecture for RTLinux-GPL, in which RTL-TEP is integrated as a low-level layer. In section 3, some issues about design and implementation of the protocol are commented. Section 4 focuses on clock synchronization mechanisms, while dynamic configuration of the protocol is treated in section 5. Fault-tolerant mechanisms that are included in the protocol are reviewed in section 6, and section 7 presents a detailed analysis of the protocol operation. Finally, the conclusions are displayed in section 8.

2. REAL-TIME NETWORKING ON RTLINUX-GPL

A software architecture is proposed in (Pérez *et al.*, 2004), which introduces networking and distributed capabilities for real-time and QoS networking to RTLinux-GPL. This architecture (figure 1), composed of a protocol stack, provides to RTLinux-GPL a networking infrastructure based on Ethernet networks:

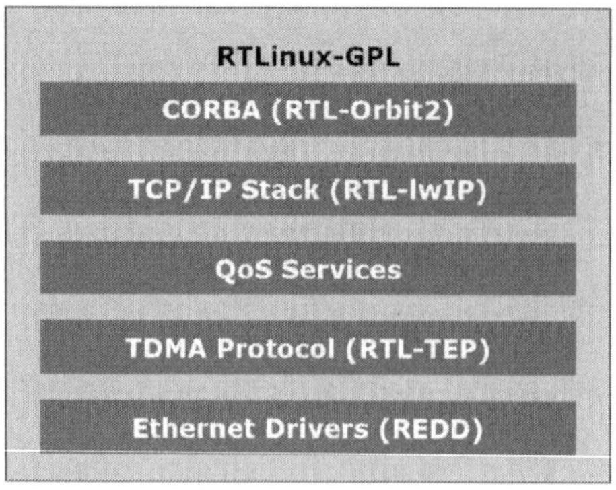

Figure 1. Protocol stack

- **Network device drivers** are the lowest layer of this architecture. Specific drivers for RTLinux are needed in order to achieve minimum latency by allowing real-time tasks direct access to the networking hardware. The main feature of these drivers is that they offer a standard POSIX interface with `open()`, `close()`, `read()`, `write()` and `ioctl()` system calls. The whole set of drivers has been gathered together in an open source project (REDD, n.d./2005).
- The **TDMA layer,** where RTL-TEP is implemented, can be put transparently on top of the Ethernet driver, providing the same POSIX interface. This way, the upper layers do not need to be aware whether they are running on Ethernet or on TDMA-Ethernet.
- The **QoS layer,** which is currently being developed, basically multiplexes different traffic types providing them with different QoS parameters.
- The **TCP/IP protocol** is a porting of the lwIP (*Light-weight IP*) project to RTLinux (RTL-lwIP, n.d./2005).
- Finally, the **upper layers** provide support for a distributed objects programming environment based on the CORBA architecture. The CORBA support is a porting of ORBit2, a GPL C-based ORB developed in the GNOME project that is CORBA 2.4 compliant (ORBit2, n.d./2005). As future work, it is expected to be also RT-CORBA compliant.

3. DESIGN AND IMPLEMENTATION OF RTL-TEP

In TDMA schemes, several nodes can share the same transmission medium by allocating each node a time slot to transmit (figure 2). A sequence of time slots forms a TDMA round, and this round is repeated periodically (cycle). In the master-slave configuration, the master inserts a special message at the beginning of the TDMA round, called the SYNC message. This way, the master node coordinates the other nodes, being a single point of failure. In a symmetric scheme

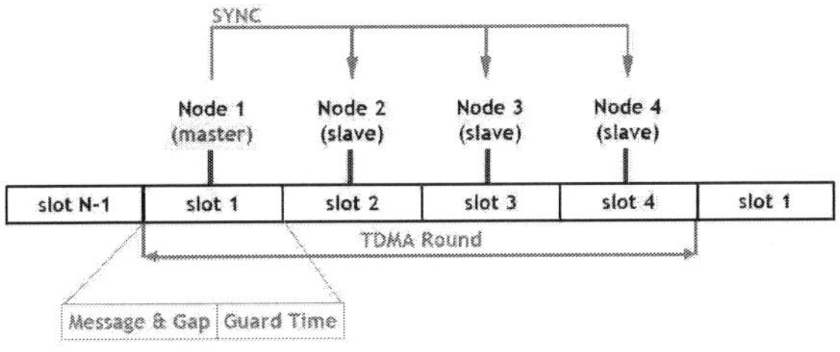

Figure 2. TDMA technology

with no mastership (Lönn, 1999), node failures do not affect the rest, but determining the beginning of the TDMA round is much more difficult, because round starts are not delineated by transmission events.

The proposed system uses a master-slave configuration because it is more efficient and less complex to implement. In this approach, one master node is responsible for all the issues that are described below.

In the proposed implementation, a TDMA round consists of $n+1$ timeslots: one slot for each node and an additional slot for the SYNC message (sent by the master node), which contains protocol state information. Each timeslot consists of a message and a guard time between time slots to prevent collisions (due to clock drift between nodes). A node can send information in its corresponding slot (sending a dummy packet if the node does not have any pending packet), while the other nodes are listening.

The main issues to face in order to implement TDMA in software over Ethernet are:

- **Clock Synchronization**: each node has to be synchronized to a global time to determine the beginning and the end of timeslots. A synchronization algorithm is needed to compensate offset and drift-rate of the node clock in respect to global-time. TDMA communication requires synchronization of both timeslot boundaries and timeslot ownership to ensure that only the node owner transmits in each slot.
- **Dynamic configuration**: in static environments, the number of nodes that can make use of the bus is fixed and known in advance. However, in the general case, a TDMA protocol should allow nodes to join or to leave the system. A special case of nodes leaving the system are node failures. In addition, RTL-TEP has another interesting feature: dynamic bandwidth allocation. Therefore, applications can change bandwidth allocation in a more flexible way, in order to adapt to workload changes or to meet QoS requirements.
- **Fault tolerant mechanisms**: It is crucial to include mechanisms that allow us to detect and to correct system errors, with the aim of maintaining the network in a stable and completely functional state. If it is necessary, the nodes can be resynchronized with the purpose of maintaining correct system operation.

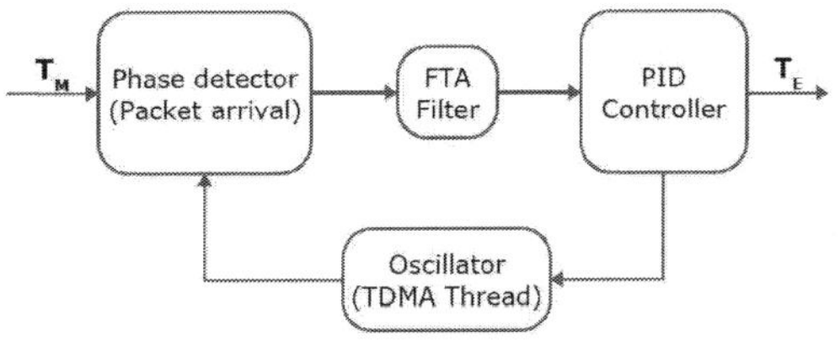

Figure 3. PLL structure

4. CLOCK SYNCHRONIZATION

Each node requires an effective synchronization to a global time-base in order to determine the beginning and the end of each slot. In our scheme, this global time-base is provided by the master node. The current algorithm is close to the Cristian algorithm, but a symmetric scheme close to NTP is being implemented as future work.

The synchronization algorithm controls the period of a TDMA thread in each node, which sends packets at the start of its allocated timeslot. The synchronization method, which is based on the synchronization algorithm used in (Lankes *et al.*, 2002), consists of performing a *Phase Locked Loop* (PLL) scheme in software (figure 3), trying to adjust the period (frequency) and phase of the TDMA thread to the timeslot boundaries of the TDMA round: every time a message is seen on the bus, each node estimates the phase difference and readjusts the thread period using a PID controller.

The synchronization algorithm is performed all the time, although during the phase of initial synchronization, the nodes only listen to the network traffic. In order to estimate the phase difference among the bus clock and the local clock (TDMA thread), the solution is to measure the propagation delay (only master node packets are considered), and substract it from a desired time difference. The resulting value is used by the PID controller, which applies the correction to the thread period.

The propagation delay can be different for each node (due to the network topology). This delay is evaluated by the master for each node joining the network. The node starts the synchronization phase considering a propagation delay of 30μs (an averaged value in Ethernet networks) and observing the delay of the node traffic. When a given number of observations is reached, the master node measures the propagation delay for the new node. This information is sent to the new node and it continues listening to the traffic and readjusting the period. The node joins to the network when it performs these observations again (now using the real propagation delay) and the error continues being small. The values obtained in these observations can be very irregular, so it is a requirement to filter them.

5. DYNAMIC CONFIGURATION

While other protocols have been developed with static characteristics, RTL-TEP has dynamic mechanisms to handle membership management and bandwidth allocation in an efficient way. In the following subsections, these features are detailed.

5.1 *Dynamic group management*

RTL-TEP allows new nodes to join and to leave the group dynamically and also to detect node failures. Membership is managed in a centralized fashion by the master, which sends a SYNC packet in the first slot of each TDMA round. This special packet contains protocol state information, including the number of network members. Each node has only an identifier when it joins the network, and this identifier is used to determine the action performed by this node.

In the first round of a TDMA cycle, there is a RESYNC slot after the SYNC slot which is used to send join requests by new nodes during the synchronization phase. Once a node has finished the synchronization phase, it sends a join request including its identifier (which is calculated using information provided by SYNC messages) to the master node, which increases the number of network members and adds a new slot for the incoming node.

When a node leaves the network, it no longer sends packets in its corresponding slot. It is considered by the master node as a node failure, and its corresponding slot is removed.

5.2 *Dynamic bandwidth allocation*

A mechanism has been introduced in the protocol for making the bandwidth allocation more flexible. This mechanism allows applications to change their allocated bandwidth, in order to adapt to workload changes or to meet QoS requirements. The idea is simple and has the advantage that does not introduce big changes in the protocol: when a node wants to get more bandwidth it requests a new slot in the TDMA round by simulating that there is a new node in the system. Using this scheme, a node may have more than one TDMA node identifier (figure 4).

Bandwidth allocation is requested by applications using the `ioctl` system call of the protocol's POSIX interface. When the protocol receives such a request, the node sends it to the master by means of a reserved field in a TDMA packet. The rest of the process is similar to the admission of a new node, and the request

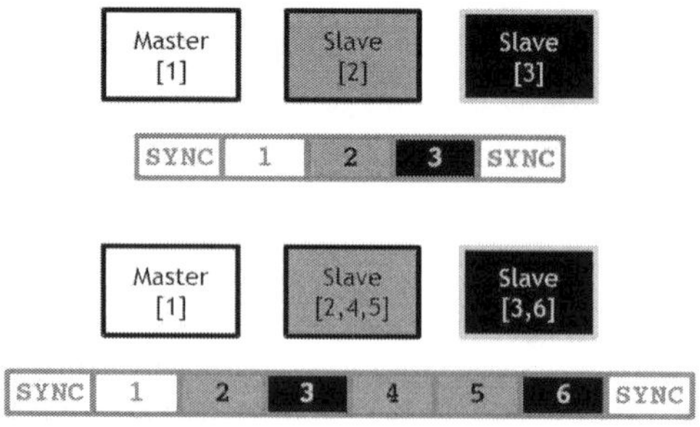

Figure 4. Dynamic bandwidth allocation

will be acknowledged by the master incrementing the number of nodes in the following SYNC packet.

Decrementing bandwidth is done by releasing timeslots or, in other words, TDMA node identifiers. When a node wants to release a timeslot, it just omits to send any packet in the corresponding slot. That will be interpreted by the master node as a node failure, and the corresponding slot will be deallocated.

In figure 4, it is shown an example of dynamic bandwidth allocation. In the beginning, each node has only one identifier, so each one has only one available slot to transmit. After varying their number of identifiers, each node has as many available slots as identifiers, so their bandwidth has been increased.

6. FAULT-TOLERANT MECHANISMS

Another special feature of RTL-TEP is the way that it handles the different errors that can occur in the system. The system errors that are considered are:

- **Loss of packets:** It is possible that one or more packets are lost. In that case, the system must maintain its stability.
- **Node failures:** If a node fails, its timeslot must be removed for an optimal use of bandwidth.
- **Collisions:** If a collision happens (due to an excessive clock drift-rate among nodes), it is desirable that the system detects and recovers from this situation.

In order to detect these errors and to make the protocol more robust, it is necessary to include some packet acknowledgement mechanism. In our case, the master node recognizes all the packets that are sent within a TDMA round. For this purpose, a vector of acknowledgment bits is included within each SYNC message. This way, each node knows if its transmitted packet has been listened by the rest in the previous round. Moreover, slave nodes recognize the SYNC message sent by the master node. It is considered that the SYNC message has been transmitted successfully if it is acknowledged at least once during the TDMA round. These fault-tolerant mechanisms have minimum spatial overhead, only a bit in each data packet header and a few bytes in the SYNC packet (these bytes have null impact in overall bandwidth, because most of the bandwidth reserved for the SYNC packet is unused).

As said before, it is mandatory for all nodes to send a packet in its corresponding slot (in spite of having pending packets or not). Thus, it is considered that a packet from a node is lost when no packet has arrived in its corresponding slot. A tolerance limit is given to all nodes (up to $n/2$ consecutive failures are allowed, where n is the number of nodes in the network), and if losses continue, it is assumed the node has failed. At this moment, the number of system's nodes decreases and the network is reconfigured (blank slots are removed and nodes fit their identifiers). If the node still remains active but its packets are not transmitted, the node leaves the system and performs the synchronization phase.

Mechanisms are more complex if the master node fails, since a new master node must be chosen (the new master node will be the one with the lowest identifier number). If two or more nodes try to transmit erroneously in the same slot, a collision will take place with an unpredictable result, although it is likely the nodes will try to relay their packets outside their corresponding slots. In order to detect collisions, the master node counts the non-received packets within a TDMA round. If the number of lost packets in the whole round is greater than $n/2$, it is considered that a collision has occurred. Then, the master node dismisses all the nodes of the network and it sends them to the synchronization phase, where they watch the network traffic without sending packets.

7. ANALYSIS

One of the most interesting metrics for evaluating the performance of the TDMA protocol is the achievable bandwidth. This can be expressed as:

$$B = \frac{n}{n + \frac{n+1}{n}} \times \frac{m \times M}{T}$$

where n is the number of nodes, M is the Ethernet packet size, m is the number of Ethernet packets that can be transmitted in a timeslot, and T is the value of the timeslot period.

The achievable bandwidth, as it is shown above, depends on several factors. The first term $\frac{n}{n+\frac{n+1}{n}}$ quantifies the effect of introducing the control timeslots for the protocol (SYNC and RESYNC). These control slots have a high influence with a small number of nodes (it can reduce the usable bandwidth to 43% with two nodes). The second term quantifies how many bits can be stuffed into one timeslot. The maximum packet size M in Ethernet is 1500 bytes. The current timeslot period is $1\,ms$. The number of packets that can be accommodated in T is currently a fix quantity, no matter how long a message is. With a 100 Mbit Ethernet,

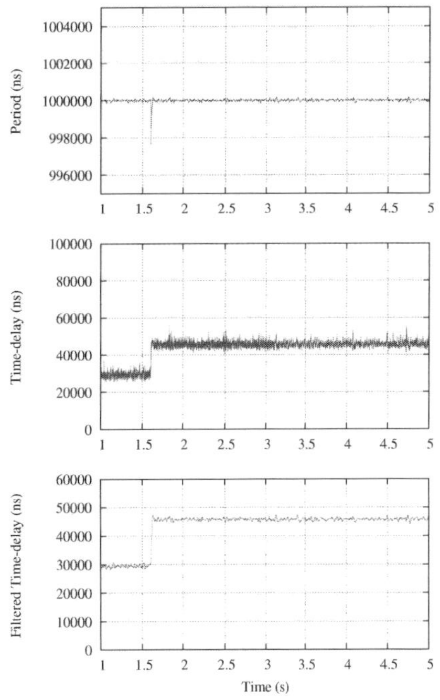

Figure 5. PID controller response

around $m = 7$ Ethernet packets of the maximum size (1500 bytes) can be fitted into a $1\,ms$ timeslot.

Another interesting feature of the protocol is how effective is the PLL and PID controller in maintaining the synchronization. Figure 5 shows the reaction of the controller after changing the value of the propagation delay from the initial estimated value of $30\mu s$ to the real measured value of $45.784\mu s$.

The jitter of the propagation delay on the network has a value of $14\mu s$, which is influenced both by the disturbance reaction on the controller and the optimization level of the network driver. In addition, the average value of the network latency (measured during 5 seconds) for a desired value of $45.784\mu s$ is $45.836\mu s$, with a standard deviation of $1.6\mu s$. During this time, the average value of the period is of $999.99155\mu s$, with a standard deviation of $0.046\mu s$. This means the controller performs an acceptable approximation to the global time, provided by the master node.

8. CONCLUSIONS AND FUTURE WORK

This paper describes RTL-TEP, a new TDMA-based protocol implemented on Ethernet. This protocol includes special and complex features, such as dynamic mechanisms to handle membership and bandwidth allocation. Using this protocol, Ethernet networks can be effectively used in real-time environments with high bandwidth and running on low-cost hardware.

As future work, new features will be added to RTL-TEP, such as an admission control for bandwidth requests, and existing functions must be improved, such as the fault-tolerant mechanisms. Finally, it is desirable to compare RTL-TEP with other protocols with similar features.

9. REFERENCES

Carpenzano, A., R. Caponetto, L. Lo Bello and O. Mirabella (2002). Fuzzy Traffic Smoothing: an Approach for Real-time Communication over Ethernet Networks. In: *Proceedings of the WFCS 2002, 4th IEEE Workshop on Factory Communication Systems*. Västeras, Sweden.

Grow, R. (1982). A Timed Token Protocol for Local Area Networks. In: *Proceedings of the Electro'82, Token Access Protocols*. p. 17/3.

Kopetz, H. and G. Grünsteidl (1994). TTP - A Protocol for Fault-Tolerant Real-Time Systems. *Computer* **27**(1), 14–23.

Lankes, S., A. Jabs and M. Reke (2002). A Time-Triggered Ethernet Protocol for Real-Time CORBA. In: *ISORC '02: Proceedings of the Fifth IEEE International Symposium on Object-Oriented Real-Time Distributed Computing*. IEEE Computer Society. Washington, DC, USA. p. 215.

Lönn, H. (1999). Initial Synchronization of TDMA Communication in Distributed Real-Time Systems. In: *ICDCS '99: Proceedings of the 19th IEEE International Conference on Distributed Computing Systems*. IEEE Computer Society. Washington, DC, USA. p. 370.

Martínez, J., M. Harbour and J. Gutiérrez (2003). RT-EP: Real-Time Ethernet Protocol for Analyzable Distributed Applications on a Minimum Real-Time POSIX Kernel. In: *Proceedings of the 2nd Int. Workshop on Real-Time LANs in the Internet Age, RTLIA'03*. Porto, Portugal.

ORBit2 (n.d./2005). (GNOME Project) Object Request Broker (ORB) compliant with CORBA 2.4. http://www.gnome.org/projects/ORBit2/.

Pedreiras, P., P. Gai and L. Almeida (2002). The FTT-Ethernet Protocol: Merging Flexibility, Timeliness and Efficiency. In: *Proceedings of the 14th Euromicro Conference on Real-Time Systems*. Vienna, Austria.

Pérez, S., J. Vila, J.A. Alegre and J.V. Sala (2004). A CORBA Based Architecture for Distributed Embedded Systems Using the RTLinux-GPL Platform. In: *Proceedings of the 7th IEEE International Symposium on Object-Oriented Real-Time Distributed Computing*. Vienna, Austria.

REDD (n.d./2005). RTLinux-GPL Ethernet Device Drivers. http://redd.sourceforge.net.

RTL-lwIP (n.d./2005). RTLinux-GPL Light-weight TCP/IP Stack. http://rtl-lwip.sourceforge.net.

RTLinux-GPL (n.d./2005). Real-Time extension to the Linux operating system. http://www.rtlinux-gpl.org/.

The ROFES Project (n.d./2005). Real-Time CORBA for embedded systems. http://www.lfbs.rwth-aachen.de/content/20.

Varadarajan, S. and T. Chuieh (1998). EtheReal: A Host-Transparent Fast Ethernet Switch. In: *Proceedings of the IEEE International Conference on Network Protocols (ICNP)*. pp. 12–21.

ELSEVIER
IFAC
PUBLICATIONS

PAWiS: TOWARDS A POWER AWARE SYSTEM ARCHITECTURE FOR A SoC/SiP WIRELESS SENSOR AND ACTOR NODE IMPLEMENTATION

Stefan Mahlknecht, Johann Glaser, Thomas Herndl†

Institute of Computer Technology, Vienna University of Technology, Austria
Gusshausstrasse 27-29, A-1040 Vienna
†Infineon Technologies Austria AG
Operngasse 20b/32, A-1010 Vienna
{mahlknecht,glaser}@ict.tuwien.ac.at; thomas.herndl@infineon.com

Abstract: The goal of the PAWiS project is to develop both, efficient system architectures and the related design methodology for power aware wireless sensor and actor network nodes that allow for capturing inefficiencies in every aspect of the system. These aspects include all layers of the communication system, the targeted class of the application itself, the power supply and energy management, the digital processing unit and the sensor-actor interface. The proof of concept will be based on a prototype system that allows a future integration in a single SiP/SoC. The project is supported by Infineon Austria and started only recently, therefore the main focus of this paper is on the design approach. *Copyright © 2002 IFAC*

Keywords: System Architectures, Design Methodology, Sensor, Integration

1. INTRODUCTION

Sensor Network Nodes are made up of microwatt radio and digital baseband transceivers that features low-duty-cycle (<1 %) low-throughput (1 bps to 10 kbps) unifying nearly all design disciplines in one package: MEMS-based sensing technology, signal conditioning, A/D and D/A conversion, digital signal processing, protocol layers such as a power-aware Media Access Controller (MAC) and routing layer, antenna design, energy management and energy scavenging.

The topic of Wireless Sensor and Actor Networks (WSANs) has been mainly researched in academia so far, however interest from industry has grown in the recent past. Most applications that make use of these networks require energy autonomy for the complete lifetime of the network which can be many years or even decades, hence the minimization of power consumption down to a few tens of μAmperes on average for a single sensor node is compulsory. Designing such a heterogeneous, extremely efficient system is a highly challenging task that requires new approaches in many different aspects of the whole system design and even the design methodology itself. A μWatt node would enable the deployment of large maintenance-free networks with numerous nodes which do not require the replacement of batteries during the lifetime. Alternatively these nodes could run from low cost energy scavenging systems extracting energy from different environmental sources (e.g. light, vibrations).

2. STATE OF THE ART SENSOR NODES

Numerous research groups and companies design, publish papers and offer wireless sensor node solutions with emphasis on one or more aspects of optimization. A very low power System-on-Chip (SoC) sensor node has been built in the course of the WiseNet research project (Enz et. al., 2004) where the receiver's power consumption is only 2 mA (permanent on) at an operating voltage of 0.9-1.5 V. The technology used is a standard low cost 0.18 μm digital CMOS process. In Berkeley Wireless Research Lab different nodes have been built, from the MICA family now commercially available to the smart dust nodes (Warneke et. al., 2001) showing some future concepts far beyond state of the art. The sensor node built at our department uses only commercially available components but relies on high bit rate transceivers with a short turnaround time and a very efficient CSMA protocol for low throughput applications (Mahlknecht and Rötzer, 2004; Mahlknecht and Böck, 2004). Within the EYES project, Infineon has developed highly efficient Wireless Sensor Network node hardware in collaboration with the project partners, particularly with TU Berlin and Universities of Ferrara and Rome (Eyes, 2005). This hardware is based on the Infineon TDA525x radio transceiver family together with a TI MSP430 microcontroller used in most sensor node implementations.

Comparing state of the art wireless sensor nodes offered by other companies (Moteiv, 2005; Sensicast, 2004; Crossbow, 2005) or nodes for research purposes in academia (e.g. Berkeley motes) evidently commercially offered nodes are neither low cost (in the order of 100 U$) nor as low power as required to run for the whole lifetime. There are also some single chip solutions on the market including a microcontroller with analog interfaces as well as a radio transceiver, (Chipcon CC1010, Nordic nRF24E1, Chipcon CC2430, CC2530). Also these implementations are not as energy efficient as desirable (< 50 μW, routing delay 10-100 ms required in many applications). This conclusion is based on datasheet information and real world experiments. Most of these nodes combine a standard 8051 CPU core as well as a radio transceiver not taking into consideration the optimization of the overall system based on a targeted class of applications. Another weakness is the missing true wakeup receiver architecture that allows a node to remain in an ultra low power listening mode. Even though proposals have been made (Gu and Stankovic, 2004; Rabaey, 2001), still no efficient implementation is available. A periodic wakeup is supported by on-chip hardware on the novel transceiver CC1100 from Chipcon; based on datasheet values, the receiver consumes only 15 μA with a periodic wakeup of 1s, however one second may be to long for short latency multi-hop applications. Shortening the wakeup period significantly increases the power consumption. The Tinymote sensor node developed at ICT has a similar wakeup scheme as described above. Experimental results showed that the average power consumption of a sensor node can be as low as 95 μW by forwarding packets at a rate of 10 packets per minute with a guaranteed hop-to-hop delay of 1 s. This includes frequent sensor measurement (light, temperature), processing and communication. Comparing available data from other research projects such as those from Berkeley (TELOS) or the TU-Berlin (Eyes 2005), the < 100 μW of power consumption based on the above scenario is excellent. However, we identified several weaknesses that we think to be able to improve significantly only with a custom SiP/SoC design.

The following technical problems of state of the art designs have been identified which have to be solved:

- Wakeup Problem: How to contact a sleeping node? Hidden terminal problem
- Voltage matching problem: How to efficiently match different voltage requirements for different modules like CPU, radio and sensors?
- Voltage converter/regulator quiescent current: How to minimize it to a sub μAmpere level for a large dynamic range?
- Oscillator start-up time: How to minimize settling time of an oscillator to reduce the turn-on time
- Find a common denominator for a generic sensor-actor interface that is flexible enough to support a number of different sensors and energy efficient enough not to compromise the overall nodes efficiency
- Process technology - Leakage current: Decreasing the feature size in a semiconductor process yields higher integration densities but unfortunately also increases leakage currents for technological reasons.

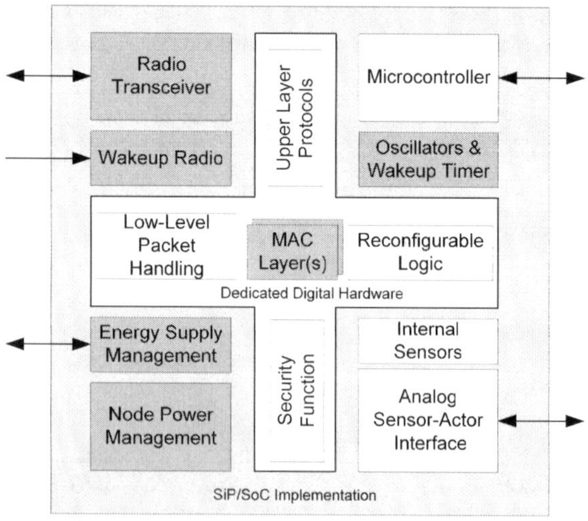

Fig. 1: Main SiP/SoC building blocks

Fig. 1 shows the main building blocks of a generic SiP/SoC sensor node implementation. Grey shaded rectangles indicate those blocks where we expect to achieve the largest efficiency gains in the overall system design. The SiP/SoC architecture will be designed to support different classes of applications where ad-hoc multi-hop communication is required as well as applications where a short latency (< 10-100 ms) between hops or real-time communication is demanded. The main challenge is to find a very efficient overall system architecture that is able to map application requirements to protocol requirements and down to the hardware with as little energy consumption as possible in a final design based on a large variety of design options. However, the question is how to find the optimum system architecture? The optimum system architecture can only be found by applying the proper design methodology. This includes identification of strategies for energy reduction at the application layer and maintaining low power constraints across all layers down to the hardware implementation. For this goal, it is mandatory to investigate the inter-dependencies between all functional units as well as between all design hierarchies.

3. DESIGN METHODOLOGY

In engineering and development often questions upon design decisions arise. The decisions are mostly driven by experience and according to instinct of technicians. Although each decision is made in an optimal manner, they only concentrate on details of the total problem. Thus the sum of all decisions leads to a local optimum for the total system but most probably overlooks other local optima, which would result in even better performance. To find the best local optimum within technological and/or physical restrictions (subsequently called "global optimum") another design approach is necessary. We apply a methodology to find this global optimum for a particular system. Therefore the total system is modeled in a so called virtual prototype at a very abstract level. The virtual prototype is a software simulation framework to model the system at a certain level of detailedness. It allows to simulate certain system properties depending on adjustable parameters. It is assembled of abstract and/or functional modules where each of these implements a model to provide results for their properties as accurate as possible.

Similar approaches have been introduced in (Silva et. al., 2001), (Tuan, 2001) and (Lizhi et. al., 2004). Where (Silva et. al., 2001) uses UML for abstract modeling and numerous universal tools for code and net list generation, we assume that the automatically generated results have to contain wrapper structures and other overhead which is not suitable for tiny embedded systems. (Tuan et. al., 2001) describes the so called "platform based design" which enables heavy module reuse by a full top-down methodology. Since they don't report about restrictions and performance properties mirrored from the bottom layer back up to the top-level optimization model, this approach seems not to be able to fully utilize capabilities of the semiconductor process. (Lizhi et. al., 2004) uses an analytical model to describe the data-link-layer (DLL) of the network protocol stack. Where this approach allows deep insight into the functionality and behavior, it only gives low accuracy.

In our design methodology all kinds of compositions and features can be simulated and compared easily parameterization. Different implementations of certain blocks (e.g. SAR- or dual-slope-ADC, different network protocols) as well as very flexible adjustment of the partitioning of a functional block (e.g. implementing parts of an algorithm in hardware or software, analog/digital partitioning of the transceiver) guarantees to avoid tabooing unusual or disliked solutions. This parameterization constitutes of system-, architecture-, cicuit-design and technology-specific parameters (e.g. ADC resolution, partitioning bounds, connectivity, bus communications protocol, leakage power, switching power). The proposed methodology is a true top-down approach. All design decisions are taken at the system level. This enables to change combinations of implementations as well as cross-layer optimization instead of just optimizing every module's implementation by its own. Nevertheless it is necessary to carefully treat the bottom layer (implementation) to find accurate simulation models. As shown in Fig. 2, the possible implementations pose constraints for the architecture which have to be considered within the virtual prototype.

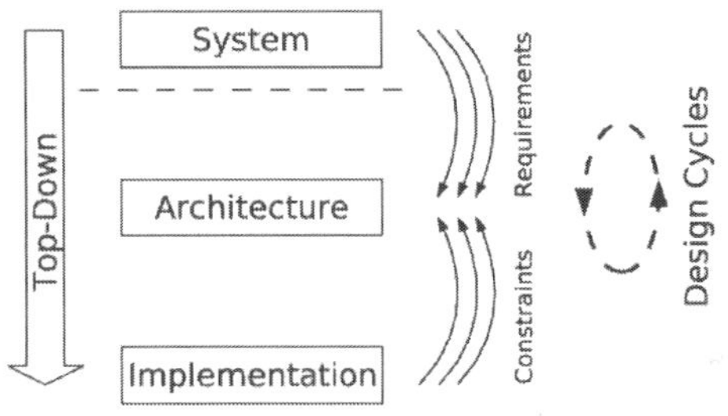

Fig. 2: Design Methodology

In the first phase one particular (sub-optimal) system architecture is selected and simulated. The models are built to estimate the power consumption and timing. The virtual prototype is extended to simulate the fully functional system. This includes an executing CPU, the radio transceiver, memories and sleep/wakeup modes. Several virtual prototypes are then instantiated and connected with a network simulator to simulate the complete network. To find an optimum system only relative accuracy throughout design changes is required to compare alternatives whereas absolute accuracy is secondary. Most model parameters will be taken from experience and raw (guided) estimations.

In the second phase the modules with most potential for energy saving are determined. Multiple types of every module, various combinations and alternatives of modules and architectures are simulated. The simulation models are refined and extended to simulate more functionality and behavior. Therefore we will have to "dive" deeply into some module's implementation details (e.g. analog leakage current of CMOS circuit prototypes, wakeup receiver implementation, e.g. (Gu and Stankovic, 2004), network protocols, e.g. (El-Hoiydi, 2003; Safwat, 2003). In this phase the system architecture is optimized even further.

In the third phase a real prototype is built. Due to financial and timing limitations, we will implement only several parts of the total system on a test chip. This chip is then mounted on a PCB which holds the residual (commercially available) parts combined with an FPGA realizing custom logic forming the total system.

The presented methodology applies at the system level as suggested by (Chou, 2005). This enables structural changes at the topmost layer and yields higher potential for improvements as opposed to optimization of the individual and predefined modules. Third party modeling and simulation frameworks will be utilized and combined to implement the virtual prototype. By forcing the design engineers to concentrate on the system level and motivating them to leave beaten tracks by introducing novel structures and architectures, we ensure a streamlined and systematic approach to achieve the overall design goal of optimization of the power consumption.

4. DETAILED FIGURES OF THE PROPOSED APPROACH

We propose to explore the following approach based on preliminary research results:

- Design Methodology: Optimization at system level before going into implementation details.
- Consider all components of the system. Understand their dependencies in terms of functionality and power consumption. For this task we first plan to develop simple energy models. Where necessary (based on the relevance of the power consumption of the functional blocks) we refine the models of the single components and subcomponents to understand how these affect the overall power consumption. Our approach is then to focus on these blocks where most of the inefficiencies can be captured.
- Explore efficient partitioning between tasks (applications, sensor reading, middleware, low level protocols) and find an adequate platform for each task.. For instance a reconfigurable platform optimized for protocol processing such as the one proposed in (Tuan et. al., 2001) will be considered for the lower level protocols that uses a combination of PAL (programmable array logic) and LUTs (look-up table) blocks. This represents hybrid cells each consisting of a small PAL block for control and an array of LUTs and flip-flops for data processing.
- Keep the design as simple as possible. We think that simplicity is one key strategy to reach the desired goal. The reason is that more complex systems tend to consume more power due to the number of transistors that will be switched and the increasing leakage current as chip area increases. This does not exclude the use of parallel very specialized hardware structures that can run at low clock speed and be turned completely powerless when not used.
- Switch off the main transceiver as much as possible. This should be possible with the help of a second receiver (wakeup receiver) that features less performance than the main receiver but only at a fraction of the energy of the latter. This radio may make extensive use of passive structures like MEMS. The wakeup receiver should only be capable of decoding incoming low bit rate wakeup preambles in order to decide whether to wakeup the main receiver or not. In scientific literature proposals for wakeup receivers have been made (Gu and Stankovic, 2004; Rabaey, 2001). One proposal from the Berkeley Wireless Research Center utilizes on-chip-BAW (Thin-Film Bulk Acoustic Wave Resonator) based super-regenerative receivers. This proposal is very simple and straight forward but lacks reliability probably generating many false wakeup alarms. An enhanced version of this architecture utilizing BAW's manufactured by Infineon Technologies could be a viable alternative, however.
- Use a simple modulation technique where a very energy efficient amplifier can be used. Very simple modulation schemes such as On Off Keying (OOK) and Amplitude Shift Keying (ASK) have the disadvantage of being not very robust in a harsh environment. Frequency Shift Keying (FSK) is more suitable in case of frequent interference. An alternative might be to use the simple modulation scheme only for the wakeup radio and make use of a very agile and more complex high bit-rate main radio transceiver that handles the actual packet transmissions and receptions. High bit-rate transceivers need to be turned on only for a very short amount of time hence higher power consumption in the active mode can be tolerated, however the turn-on time becomes a very critical parameter.
- Make use of available IP cores implementing highly energy optimized CPU cores used for application processing and integrate a very low power mode where the CPU can wakeup quickly (Wakeup based on RTC trigger, on-time: $< 10\mu s$ with DCO oscillator). We will evaluate different architectures of power aware IP cores that are commercially available such as the CoolRisc from Xemics, the eCog1 from Cyan, or the 8051 and choose an appropriate core for simulation.
- Exploit parallelism at lowest-possible clock speed. A sensor node runs different concurrent tasks with widely different requirements (sensing

-> low duty cycle, MAC -> real time, application -> dependent on the task). Hence, a bus-based heterogeneous architecture exploiting task-level parallelism is a natural choice. The components can either be a processor or configurable hardware blocks tuned to the respective application. Each processor/hardware block must be tuned to the application, with only the flexibility needed by the application.

- Dynamic voltage and frequency scaling. Based on performance requirements (real time) and operating temperature the voltage can be reduced to a minimum level in each operating stage. In order to control this a dedicated power management engine is proposed. In standby or for RAM retention the voltage of single functional blocks can be reduced below 1 V for 0.18 μm and 0.13 μm CMOS processes. Ultra low voltage CMOS technology based on SOI (Silicon on Insulator) has proven its feasibility for voltages down to 0.5 V (By Emmicroelectronics) but might not be applicable due to the lack of an available process.
- Make use of power domains: Detach any unused blocks from power supply.
- Use specialized small low power SRAM blocks. SRAM is very energy and space consuming. We expect to need no more than 512 bytes of RAM. However this would probably not allow porting the widely used operating system TinyOS developed for wireless sensor network nodes. The question of how much RAM and FLASH memory is required will be determined by the class of applications and the final sensor node system architecture.
- Minimize access to the global SRAM. Use small register banks for context switch and for keeping state information. Avoid copying data packets from the communication interface at all (zero-copy architecture).
- Minimize current consumption by aggressive use of passives. This is investigated mainly for the wakeup transceiver which will have probably the highest duty cycle of all on-chip components (highest on/off ratio). For these modules, every μAmpere of additional current significantly affects the overall energy consumption. In (Ruby, 2001) passive structures based on BAW/FBAR and RF-MEMS (Clark et. al., 2000) are proposed.
- Enable the main system blocks to trigger themselves for task execution i.e. sensor-actor interface wakes up the measurement module and only notifies the application CPU when changes based on a programmable threshold have been observed.
- Use different internal voltage levels by means of on-chip DC/DC voltage down converters or low quiescent current LDOs. Different voltages are used for different power down stages and system blocks to minimize overall power consumption.
- Use a standard digital CMOS process (most probably 0.13 μm) for most or evenall of the SiP/SoC to achieve the low cost target over a long term.
- Analog components such as wakeup radio, voltage converters or mixed signal sensor-actor interface may be integrated in a second chip based on BiCMOS depending on the simulation results comparing CMOS implementations. Investigations have to find out whether this partitioning pays off in terms of price/performance/power ratio.
- Find the right trade off between analog and digital. Especially the analog/digital partitioning of the radio transceiver is no straight forward decision. Comparison between different implementation types based on high level models will help to choose the most efficient design.
- Find a power-adaptive system-architecture with respective protocol, that is able to adapt its processing performance to the (instantaneous) available power.

5. OUTLOOK

In previous applied research at the Institute of Computer Technology of the Vienna University of Technology (Mahlknecht 2004, Rötzer, 2005) as well as in the course of the now completed EU-funded project EYES (IST 2001 34734) with Infineon as industry partner, significant experience has been gathered in the area of WSANs. Together with the many valuable solutions published in papers by the very active research community, the project consortium is very confident to develop the most efficient overall system architecture and dedicated hardware solutions for μWatt sensor and actor nodes.

REFERENCES

Chou, S. (2005). Integration and Innovation in the Nanoelectronics Era, *Session, ISSCC05.*

Clark, J. R., Hsu W.-T., and Nguyen C. T.-C. (2000). High-Q VHF micromechanical contourmode disk resonators, Tech, *Digest IEEE Int. Electron Devices Meeting, San Francisco.*

Crossbow Technologies Inc. (2005). MICA Mote Family of Wireless Sensor nodes, *MICA2 Series Online Datasheet.*

El-Hoiydi, A., Decotignie, J.-D., Enz, C. and Le Roux, E. (2003). Poster Abstract: WiseMAC, An Ultra Low Power MAC Protocol for the WiseNET Wireless Sensor Network, *in Proc. ACM SenSys, Los Angeles.*

Enz, C.C., El-Hoiydi A., Decotignie J., Peiris A. (2004). WiseNet: an ultra low power sensor network solution, *IEEE Journal Magazine.*

Eyes, (2005). The EU Project EYES (IST 2001 34734) can be found under http://Eyes.eu.org

Gu, L. and Stankovic, J. (2004). Radio Triggered Wake-Up Capability for Sensor Networks, *in proceedings of the IEEE conference RTAS-2004, pp. 27-36, Toronto.*

Lizhi, C. Z., Rabaey J. and Wolisz, A. (2004). An integrated data-link energy model for wireless sensor networks, *in Proceedings of the ICC, Paris, France.*

Mahlknecht, S. (2004). Energy-Self-Sufficient Wireless Sensor Networks for Home and Building Environment, *Dissertation Thesis at the Institute of Computer Technology, Vienna University of Technology.*

Mahlknecht, S. and Böck, M. (2004). On the use of High Bit Rate Transceivers for Low Duty Cycle Wireless Sensor Networks, *IEEE 7th Africon, Gaborone, Botswana.*

Mahlknecht, S. and Rötzer, M. (2004). CSMA-MPS: A Minimum Preamble Sampling MAC Protocol for Low Power Wireless Sensor Networks, *IEEE WFCS, Vienna, Austria.*

Moteiv Inc. (2005). Tiny Sky: Ultra low power IEEE802.15.4 compliant wireless sensor module, *Preliminary Datasheet 3/5/2005*

Rabaey, J. M. (2001). Wireless Beyond the Third Generation - Facing the Energy Challenge, *ISLPED, Aug. 01*

Rötzer, M. (2005). Routing in energieautarken Funksensornetzwerken, *Diploma Thesis at the Institute of Computer Technology, Vienna University of Technology.*

Ruby, R. (2001). Ultra-miniature, high-Q filters and duplexers using FBAR technology, *IEEE Solid-StateCircuits Conf., San Francisco, USA.*

Safwat, A., Hassanein, H. and Mouftah, H. (2003). Optimal Cross-Layer Design for Energy-Efficient Wireless Ad hoc and Sensor Networks, *The 22nd IEEE International Performance, Computing, and Communications Conference (IPCCC 2003), Phoenix, Arizona, USA.*

Sensicast Systems Inc. (2004). Datasheet EMS100 – Environmental Management Node.

Silva, J.L., Shamberger, J., Ammer, M. J., Guo C., Li, S., Shah, R., Tuan, T., Sheets, M., Rabaey, J. M., Nikolic B., Sangiovanni-Vincentelli, A. L., and Wright, P. (2001). Design methodology for PicoRadio networks, *in Proceedings of the Design Automation and Test in Europe, Munich, Germany.*

Tuan, T., Li, S., and Rabaey, J. (2001). Reconfigurable Platform Design for Wireless Protocol Processors; *ICASSP.*

Warneke, B., Atwood, B. and Pister, K.S.J. (2001). Smart Dust Mote Forerunners, *Proceedings of the 14th Annual International Conference on Microelectromechanical Systems (MEMS 2001), Interlaken, Switzerland.*

ELSEVIER
IFAC
PUBLICATIONS

THE INFLUENCE OF INTER-DOMAIN MOBILITY ON MESSAGE STREAM RESPONSE TIME IN WIRED/WIRELESS PROFIBUS-BASED NETWORKS

Luís Ferreira, Eduardo Tovar

Polytechnic Institute of Porto (ISEP-IPP)
Rua Dr. António Bernardino de Almeida, 431
4200-072 Porto, Portugal
E-mail: {llf, emt}@dei.isep.ipp.pt

Abstract: In previous works we have proposed a hybrid wired/wireless PROFIBUS solution where the interconnection between the heterogeneous media was accomplished through bridge-like devices with wireless stations being able to move between different wireless cells. Additionally, we had also proposed a worst-case timing analysis assuming that stations were stationary. In this paper we advance these previous works by proposing a worst-case timing analysis for the system's message streams considering the effect of inter-cell mobility.

Keywords: Fieldbus, Wireless, Real-time, Industrial Automation

1 INTRODUCTION

In the past years several solutions have been proposed for extending the capabilities of fieldbus networks to encompass wireless support [6-7, 9]. PROFIBUS (acronym for PROcess FIeld BUS) is a natural candidate to support such an ensemble, especially due to its market penetration and range of covered applications.

The Multiple Logical Ring (MLR) concept was introduced and discussed in [2], and further detailed in [3-4], where a bridge-based approach (thus, layer 2 interoperability) was outlined. In such an approach, each logical ring is comprised of stations that communicate via a unique medium – a domain, which can be wired or wireless. The Inter-Domain Protocol (IDP) supports the communication between stations in different domains, and the mobility of wireless stations between different wireless domains is based on the Inter-Domain Mobility Procedure (IDMP). These protocol extensions provide essential compatibility with legacy PROFIBUS technologies.

In [8], we proposed a worst-case timing analysis for transactions supported by the IDP, considering that wireless stations were stationary. In [10], that work has been applied to calculate the latencies associated with the IDMP evolution.

In this paper, we advance that previous work by analysing the impact of the IDMP on the worst-case response time (WCRT) of message streams, considering that wireless stations can move between different wireless domains.

The rest of this paper is organized as follows. In Section 2, the main concepts related to bridge-based hybrid wired/wireless PROFIBUS architectures, including the ones related to the MLR approach, are briefly presented. Then, in Section 3, we briefly present the timing analysis of the latencies associated to the mobility procedure (IDMP), which is then used in Section 4 to derive analytical formulations for the WCRT of message streams in a system allowing inter-cell (domain) mobility. Finally, in Section 5, we draw some conclusions.

2 SYSTEM ARCHITECTURE AND PREVIOUS RELEVANT WORK

2.1 *Basics of the PROFIBUS protocol*

The PROFIBUS Medium Access Control (MAC) protocol uses a token passing procedure to grant bus access to masters. After receiving the token, a PROFIBUS master is capable of processing transactions during its token holding time (T_{TH}), which, for each token visit, is the value corresponding to the difference, if positive, between the target token rotation time (T_{TR}) parameter and the real token rotation time (T_{RR}). For further details, the reader is referred to [5].

A transaction (or message cycle) consists on the request or send/request frame from a master (the initiator) and of the associated acknowledgement or response frame from a master/slave station (the responder). The response must arrive to the master before the expiration of the `Slot Time` (T_{SL}), a master parameter.

In order to maintain the logical ring, PROFIBUS provides a decentralized ring maintenance mechanism. Each PROFIBUS master maintains two tables – the *Gap List* (GAPL) and the *List of Active Stations* (LAS), and may optionally maintain a *Live List* (LL).

The GAPL consists of the address range from 'This Station' address until 'Next Station' address, i.e., the next master in the logical token ring. Every time the

Gap Update Timer (T_{GUD}) expires in a master, it starts checking the addresses in its GAPL. This is accomplished by inquiring (at most) one master on the GAPL per token visit. If a new master replies, then the requesting master passes the token to this new master and updates its 'Next Station' address. Otherwise, the requesting master continues its operation. In the MLR approach, this mechanism is used for enabling the mobility of wireless master stations, as detailed later.

The LAS is a list of all the masters in the logical ring, and the LL contains all active stations (both masters and slaves).

2.2 Basics of the MLR approach

Our hybrid wired/wireless fieldbus network is composed of wired and wireless stations. Communication is based on the PROFIBUS protocol, and the communication between different domains is supported by special-purpose bridges supporting the Inter-Domain Protocol (IDP) [4]. Fig. 1 illustrates an example network.

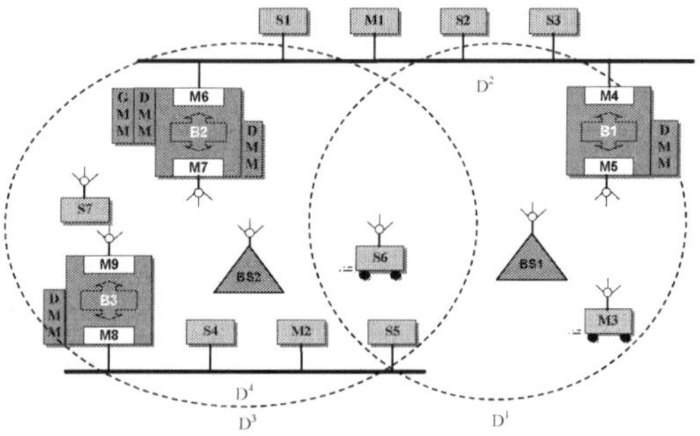

Fig. 1 – Hybrid wired/wireless PROFIBUS network

In this example, the following set of wired PROFIBUS masters (M) and slaves (S) are considered: M1, S1, S2, S3, S4 and S5. Additionally, the following set of wireless stations is considered: M3, S6 and S7. From this last set, only M3 and S6 are referred as Mobile Wireless Master/Slave station, therefore being capable of moving inside a wireless domain and between them (using the IDMP). Station S7 is referred as Domain Resident Wireless Master/Slave Station since it is stationary in a single domain. These wireless stations are standard PROFIBUS stations equipped with a radio front-end containing specific wireless extensions (as defined in RFieldbus [1]). Three bridge devices are considered: B1, B2 and B3. Each includes two modified PROFIBUS masters (denoted as Bridge Masters (BM)) implementing the required protocol extensions. In our system, the network has a tree-like topology, and bridges perform routing based on MAC addresses.

All wireless communications are relayed through base stations (BS), operating in cut-through mode. Each BS uses two channels to communicate with the wireless stations, one to receive data from the wireless stations (the uplink channel) and another to transmit data to the wireless stations (the downlink channel). Each adjacent BS (e.g. BS1 and BS2) must use a different set of radio channels. In the example each wired/wireless domain has its own logical ring, four different logical rings exist: {(M5 → M3), (M1 → M4 → M6), (M7 → M9), (M8→M2)}.

2.3 The Inter-Domain Protocol (IDP)

A consequence of the MLR approach is that when a master makes a PROFIBUS standard request addressed to a station in another domain (an Inter-Domain Request), it will not receive an "immediate" response from the responder. The IDP [4] proposes some protocol extensions suitable for handling such kind of transactions – Inter-Domain Transactions (IDT).

The IDP protocol specifies that when an initiator makes an Inter-Domain Request, only one of the BMs belonging to the initiator's domain – denoted as BM BM_i, codes the frame using the IDP, and relays it. The decision, either to receive or discard the frame, is based on a routing table contained in the BMs. Then, this Inter-Domain Request frame is relayed by the bridges until reaching bridge master BM_r (the last bridge master in the path). This bridge decodes the original request frame and transmits it to the responder, which can be a standard PROFIBUS-DP station. The response (referred as IDT Response frame) is again coded using the IDP and routed back until reaching BM BM_i, where it will be decoded and stored. The IDP assumes that the initiator Application Layer (AL) periodically repeats the same request until receiving the related response. During this period we refer to the state of the IDT in BM_i as a pending or open IDT. In Fig. 2, we illustrate this behaviour for a transaction between M3 and S7 in the example illustrated in Fig. 1.

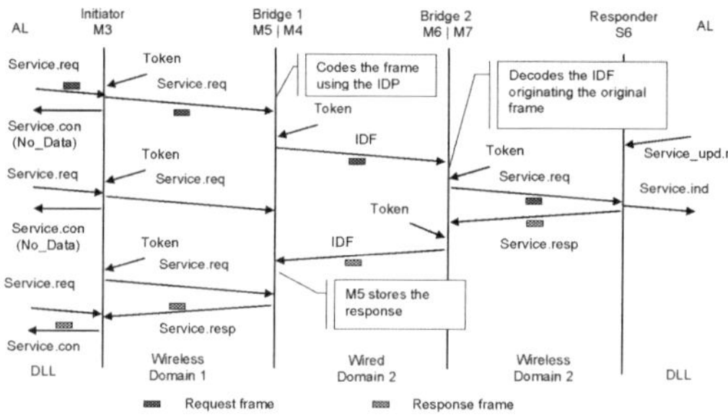

Fig. 2 – Inter-Domain Transaction (IDT) example

Note in Fig. 2 the several AL repetitions made by M3. Additionally, it is assumed that slaves read their inputs periodically, updating data structures in their DLLs, using the PROFIBUS *Service_upd.req* primitive.

2.4 Inter-Domain Mobility Procedure (IDMP)

The main objective of the inter-domain mobility procedure (IDMP) is to ensure that a wireless mobile station is able to change from one wireless domain to another, whenever it detects an adjacent wireless domain with a better signal quality. The IDMP is a hierarchically managed procedure, where one master in the system (the *Global Mobility Manager* (GMM)) is responsible for periodically starting the IDMP and controlling some of its phases. In each domain, one master controls the mobility of stations belonging to that domain – the *Domain Mobility Manager* (DMM). Finally, the bridge stations implement specific mobility services. For the network example depicted in Fig.1, M6 can assume both the role of GMM and DMM for its domain. M5, M7 and M8 can assume the role of DMM for domain D^1, D^3 and D^4, respectively.

The mobility procedure evolves through 4 phases, as illustrated in Fig. 3.

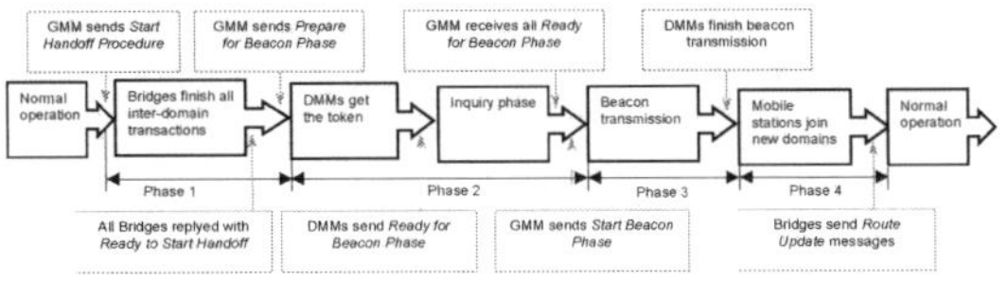

Fig. 3 – Phases of the IDMP

The GMM initiates the IDMP by sending the `Start_Mobility_Procedure (SMP)` message, which commands the system BMs to finish all pending IDTs (for which they are responsible). After receiving the confirmation that all BMs had finished their IDTs (by the `Ready_to_Start_Mobility_Procedure` (`RSMP`) message) the GMM starts phase 2. During this phase, all DMMs are commanded, using the `Prepare_for_Beacon_Transmission (PBT)` message sent by the GMM, to enter into the inquiry mode (a sort of polling mode commanded by the domain DMM), during which only mobility related messages are exchanged. This type of operation allows a minimal latency for the communication between the GMM and the DMMs, thus allowing a closer synchronization of the start of the beacon emission during Phase 3. When a DMM enters into the inquiry mode it transmits a `Ready_for_Beacon_Transmission (RBT)` message. The beacon transmission, by the DMMs, is triggered by the `Start_Beacon_Transmission (SBT)` message, sent by the GMM. This sub-phase is used by the wireless mobile stations to evaluate the quality of adjacent wireless channels and handoff if required. During phase 4, the DMMs of wireless domains try to detect which mobile stations are present on their domains. If stations are found, the DMMs inform the system BMs about the location of the mobile stations, using `Route_Update (RU)` messages.

2.5 *Previous work on timing analysis*

Related to the timing analysis approach presented in [5], the WCRT for a high priority message stream i from a master k, in a SLR network ($Rslr_i^k$), or in the case of the bridge-based approach referred as an Intra-Domain Transaction (IADT), can be computed by:

$$Rslr_i^k = nh^k \times T_{cycle}^k + Ch_i^k \quad (1)$$

where nh^k is the number of synchronous high-priority message streams generated in master k and Ch_i^k is the worst-case duration of a synchronous message cycle i issued by master k. T^k_{cycle}, the worst-case token rotation time can be computed as presented in [5].

In [8], we proposed a worst-case timing analysis of the IDP. Relevant to that analysis is the fact that the initiator of the IDT needs to periodically repeat the request until getting the actual response from the BM_i (Fig. 2). Consequently, the WCRT for a message stream i from master k on a MLR network ($Rmlr_i^k$), can be formulated as follows:

$$Rmlr_i^k = A_i^k \times T_i^k + Rslr_i^k \quad (2)$$

A_i^k is the maximum number of attempts required to obtain the actual response, which depends on the delay experienced by the IDT, from the reception of the request at BM_i, until the arrival of the respective response to BM_i ($Rbmi_i^k$). Therefore, A_i^k can be obtained by computing $\lceil (Rslr_i^k + Rbmi_i^k - C_i^k) / T_i^k \rceil$. $Rbmi_i^k$ can be obtained as follows:

$$Rbmi_i^k = \sum_{f=2}^{b+1} \left(Rslr_i^k\right)^{\Omega_{req}[f]} + \sum_{f=2}^{b} \left(Rslr_i^k\right)^{\Omega_{res}[f]} + 2 \times b \times \phi \quad (3)$$

b is the number of bridges between the initiator and the responder. Ω_{req} represents the set of BMs which relay the IDT request frame in the path from the initiator to the responder. Ω_{res} represents the set of BMs which relay the IDT response frame in the path from the initiator to the responder. The network domains are numbered from 1 to $b + 1$.

The mobility-related messages are transmitted using the PROFIBUS DLL Send Data without Acknowledge (SDN) service, which only involves the transmission of a request message. The worst-case time required by a request from a message stream i, to go from a master k to another station w ($Ru_i^{k \to w}$), can be obtained by:

$$Ru_i^{k \to w} = Rslr_i^{k'} + \sum_{f=1}^{b} Rslr_i^{\Omega_{req}[f]} + (b + di + df) \times \phi \quad (4)$$

k' is the first BM to transmit the request, which can be master k itself, when it is directly connected to the first domain in the path (in this case di is equal to 0), or can be the BM on the other side of the bridge if master k is a BM not directly connected to the first domain in the path (in this case di is equal to 1). df is equal to 0 if the destination station is a master, a slave or a BM directly connected to the last domain in the IDT Communication Path. df is equal to 1 if the destination station is a BM not directly connected to the last domain where the message is transmitted. ϕ is the internal relaying delay of the bridges in the path.

3 INTER-DOMAIN MOBILITY PROCEDURE TIMINGS

3.1 *Phase 1*

The IDMP starts with the transmission, by the GMM, of the `SMP` message, which must be received by all BMs in the system. Fig. 4, illustrates phases 1 and 2 events, assuming the network scenario in Fig. 1.

The worst-case time span for the `SMP` message to reach a BM bm (t_{SMP}^{bm}), can be calculated considering an unicast IDT (Eq. (4)): $t_{SMP}^{bm} = Ru_{SMP}^{GMM \to bm}$.

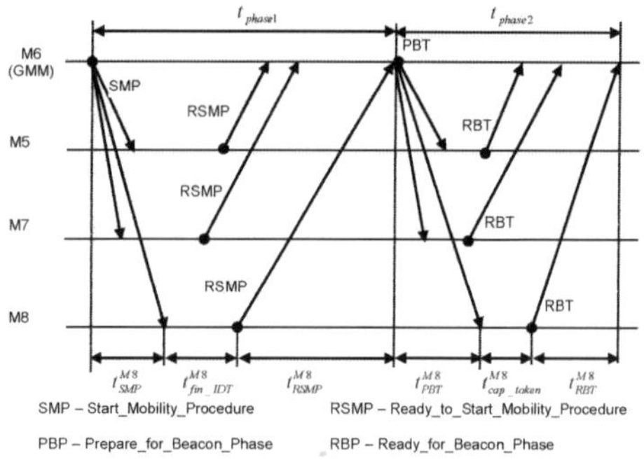

Fig. 4 – Phase 1 and Phase 2 main events timings

After receiving the SMP message, the BMs stop accepting new IDTs from masters belonging to their domains. Nonetheless, they keep handling pending IDTs and, importantly, they keep handling IDTs originated in the other domains. The following equation gives the worst-case time until all IDTs are completed for a particular BM *bm*.

$$t_{fin\_IDT}^{bm} = \max_{S_i^k \in \Psi_{IDT}} \{Rbmi_i^k + T_i^k + Rslr_i^k\} \qquad (5)$$

Ψ_{IDT} refers to the set of message streams which are also IDTs served by BM *bm*. *k* represents a master which belongs to the domain where the BM *bm* is connected, and uses BM *bm* as the first BM in the path (BM_{ini}). $Rbmi_i^k$ can be calculated using Eq. (3).

After completing all pending IDTs, the bridges signal their new state by sending to the GMM a RSMP message. The worst-case time needed by the RSMP message to go from a BM *bm* to the GMM ($t_{RSMP}{}^{bm}$) can also be calculated using Eq. (4): $Ru_{RSMP}{}^{bm\to GMM}$.

Phase 1 only stops when all RSMP messages (coming from all DMMs) have been received by the GMM. Therefore, the following equation allows computing the worst-case time span of Phase 1.

$$t_{phase1} = \max_{\forall bm}\{t_{SMP}^{bm} + t_{fin\_IDT}^{bm} + t_{RSMP}^{bm}\} \qquad (6)$$

Only at this point in time the GMM can proceed to Phase 2.

3.2 Phase 2

Phase 2 starts when the the GMM sends the PBT message. The worst-case time required for the PBT message (time span denoted as $t_{PBT}{}^{dmm}$) to reach DMM *dmm* is given by $Ru_{PBT}{}^{GMM\to dmm}$. Additionally, the DMM clears all its routing table entries related to mobile wireless stations.

The reception of the PBT message commands the system DMMs to enter into inquiry sub-phase, after which the DMMs will retain the token. The worst-case time required until capturing the token (denoted as $t_{cap\_token}{}^{dmm}$) is equal to the worst-case token rotation time of the domain where the DMM *dmm* is located, $T_{cycle}{}^{dmm}$. Following that, the DMMs send a RBT message to the GMM and enter into the inquiry sub-phase. In this sub-phase, the domain DMMs inquire, in sequence, their domain BMs, whether they have any RBT message available.

With the network operating in inquiry mode, the worst-case time required for the RBT message to go from the DMM *dmm* to the GMM can be computed as follows:

$$t_{RBT}^{dmm} = \sum_{x=0}^{b-1}(Ringq_{RBT}^{2x\to(2x+1)} + \phi) \qquad (7)$$

where $Ringq_{RBT}^{x\to(x+1)}$ is the worst-case delay experienced by the RBT message when being transmitted from a BM *x* to another BM *x+1*, in the path to the GMM. For this formulation we assume that the BMs in the path, between DMM *dmm* and the GMM are numbered as: {0, 1, 2, … ,2×*b*-1), where 0 refers to DMM *dmm* and 2×*b*-1 to the GMM. *b* is the number of bridges in the path. For further details on the reasoning on the timing analysis when the network is inquiry mode, the reader is referred to [10].

To obtain the worst-case time span for Phase 2, the following analytical formulation may then be applied:

$$t_{phase2} = \max\{t_{PBT}^{dmm} + t_{cap\_token}^{dmm} + t_{RBT}^{dmm}\}, \forall \text{ dmm} \qquad (8)$$

3.3 Phase 3

After collecting all RBT messages from all the DMMs, the GMM starts the Beacon transmission sub-phase, by broadcasting the SBT message. Fig. 5, depicts a timeline for the sequence of events during Phases 3 and 4. The worst-case time required by the SBT message to reach a DMM *dmm* is given by:

$$t_{SBT}^{dmm} = \sum_{x=0}^{b} Ringq_{SBT}^{2x\to 2x+1} + \phi \qquad (9)$$

where *x* represents the list of BMs in the IDT Communication Path, from the GMM to a DMM *dmm*, which relay the SBT message, similar to the formulation in Eq. (7). *b* is the number of bridges between the GMM and DMM *dmm*. Upon receiving this message, the DMMs start emitting Beacons. In wired domains no Beacons are transmitted, and therefore stations in these domains may resume IADTs.

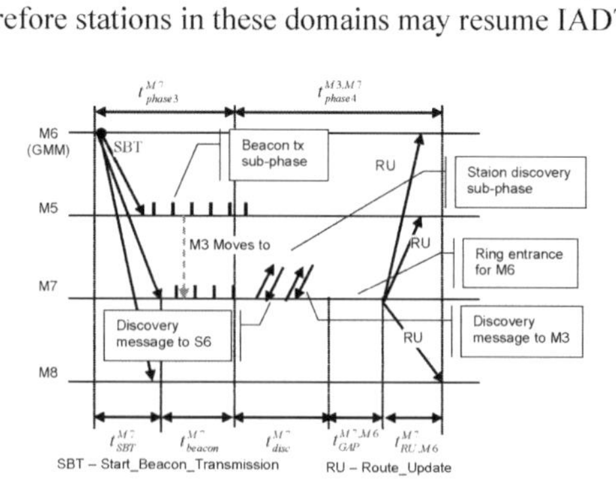

Fig. 5 – Timeline for Phases 3 and 4

The duration of the Beacon transmission sub-phase ($t_{beacon}{}^{dmm}$) is a parameter that is setup individually on every domain, according to the condition proposed in [1]. Thus, the worst-case duration of Phase 3, calculated for every wireless domain d (represented in the equation by its DMM), is given by:

$$t_{phase3}^{dmm} = t_{SBT}^{dmm} + C_{beacon} \times n_{beacon}^{dmm} \quad \textbf{(10)}$$

where, C_{beacon} is the worst-case latency associated with the transmission of a `Beacon` frame and $n_{beacon}{}^{dmm}$ is the number of `Beacon` frames to be transmitted by DMM *dmm*.

3.4 Phase 4

After the end of the `Beacon` transmission sub-phase, every wireless DMM (still holding the token) inquires all mobile wireless stations, using the `Discovery` message, in order to detect if they still belong to its domain or to detect new "entries" on its domain.

The worst-case duration of the station discovery sub-phase can be computed by:

$$t_{disc}^{dmm} = n_{mob\_stations} \times C_{dic}^{dmm} \quad \textbf{(11)}$$

where $n_{mob\_stations}$ is the number of mobile wireless stations (including masters and slaves), and $C_{disc}{}^{dmm}$ is the worst-case latency associated with the `Discovery` message on the domain represented by *dmm*, including the response from the addressed station. After this, mobile wireless slaves are capable of answering requests, but new mobile wireless masters must still enter the logical ring using the standard PROFIBUS Gap Update mechanisms (briefly described in Section 2.1).

The worst-case time for a master station j entering the ring, after master k ($t_{master\_entry}{}^{k,j}$) can be computed as described in [10].

Once the discovery of stations is complete, or a new master has entered into a different domain, the domain DMM sends a `RU` message, which will be used by the bridges to update their routing tables. The worst-case time span that the `RU` message, relative to station s, needs to go from DMM *dmm* to a BM *bm* (this time span is denoted as $t_{RU,s}{}^{bm}$) can be calculated by $Ru_{RU,s}{}^{dmm \to bm}$ (using Eq. (4)).

To summarize, the time required before a BM *bm* knows that a station s is again operational in a wireless domain, the duration of Phase 4, is given by:

$$t_{phase4}^{s,bm} = \begin{cases} t_{disc}^{dmm} + t_{RU,s}^{bm},\ s \in \Pi_{slave} \\ t_{dis}^{dmm} + 3 \times t_{cycle}^{dmm} + t_{m\_master\_entry}^{k,j} + t_{RU,s}^{bm},\ s \in \Pi_{master} \end{cases} \quad \textbf{(12)}$$

where, *dmm* represents DMM of the domain in which station s is, or to where it has entered. Π_{slave} and Π_{master} are the set of mobile wireless slaves and mobile wireless masters in the system, respectively.

4 INCORPORATING THE LATENCIES OF THE IDMP INTO MESSAGE TRANSACTIONS

When the IDMP mechanism is active, it is responsible for additional delays on the response time of the system message streams.

4.1 Intra-Domain Transactions

The period of time in which the transactions between stations belonging to the same domain are not possible, comprises the inquiry sub-phase, the beacon emission sub-phase and the identification sub-phase. Therefore, Eq. (1) must be updated by considering the period of time during which IADTs are disabled in a domain.

A worst-case condition occurs when master k queues a request related to message stream S_i^k, just before the start of the period of time during which IADTs are disabled. The following equation updates the WCRT for IADTs under these circumstances.

$$Rslr\_m_i^k = \begin{cases} Rslr_i^k + t_{IADT\_dis\_wr}^{dmm}, \text{ in wired domains} \\ Rslr_i^k + t_{IADT\_dis\_wl}^{dmm}, \text{ in wireless domains} \end{cases} \quad \textbf{(13)}$$

where $t_{IADT\_dis\_wr}{}^{dmm}$ and $t_{IADT\_dis\_wl}{}^{dmm}$ are the time during which IADTs are disabled in a wired and in a wireless domain (defined by its DMM), respectively. These time spans be calculated by:

$$\begin{aligned} t_{IADT\_dis\_wr}^{dmm} &= t_{phase2} - t_{PBP}^{dmm} - t_{cap\_token}^{dmm} + t_{SB}^{dmm} \\ t_{IADT\_dis\_wl}^{dmm} &= t_{phase2} - t_{PBP}^{dmm} - t_{cap\_token}^{dmm} + t_{phase3}^{dmm} + t_{disc} \end{aligned} \quad \textbf{(14)}$$

4.2 Inter-Domain Transactions (involving Domain Resident Wireless Stations or Wired Stations)

The IDMP requires disabling IDTs during its evolution. IDTs are disabled, on BMs, from the reception of the `SMP` message until the end of the identification sub-phase or the reception of the `SBT` message, for wireless and wired domains, respectively. IDTs involving mobile wireless stations are only enabled again at the reception of a `RU` message regarding the responder station. Therefore, in relation to transactions involving Domain Resident Wireless Stations or Wired Stations, the time during which IDTs are disabled in a BM belonging to a wired ($t_{IDT\_dis\_wr}{}^{dmm}$) or wireless ($t_{IDT\_dis\_wl}{}^{dmm}$) domain can be obtained by:

$$\begin{aligned} t_{IDT\_dis\_wl}^{bm} &= t_{phase1} - t_{SMP}^{bm} + t_{phase2} + t_{SB}^{dmm} \\ t_{IDT\_dis\_wr}^{bm} &= t_{phase1} - t_{SMP}^{bm} + t_{phase2} + t_{phase3}^{dmm} + t_{disc}^{dmm} \end{aligned} \quad \textbf{(15)}$$

To obtain the WCRT of an IDT related to message stream S_i^k, involving domain resident wireless stations or wired stations, the following situations must be analysed separately:

1. $T_i^k > t_{IADT\_dis}{}^{bm} > t_{IDT\_dis}{}^{bm}$
2. $t_{IADT\_dis}{}^{bm} < T_i^k \leq t_{IDT\_dis}{}^{bm}$
3. $T_i^k \leq t_{IADT\_dis}{}^{bm} < t_{IDT\_dis}{}^{bm}$

where $t_{IADT\_dis}{}^{bm}$ represents either $t_{IADT\_dis\_wr}{}^{dmm}$ or $t_{IADT\_dis\_wl}{}^{dmm}$ on the initiator domain, in the case when the domain is wired or wireless, respectively. *bm* represents the BM on master k domain, which is used as BM_{ini} by the message stream S_i^k, and *dmm* is the DMM in the domain to which master k belongs.

Case One. In the first case, it is obvious that at most one request related to an IDT may be lost due to the IDMP. The following equation incorporates these conditions by adding another retry on the WCRT calculation for IDTs:

$$Rmlr\_m_i^k = \left(A_i^k + 1\right) \times T_i^k + Rslr_i^k \tag{16}$$

Case Two. In the second case, several requests related to message stream S_i^k can be lost during the evolution of the IDMP. Also, since $T_i^k \geq t_{IADT\_dis}^{bm}$, it is not possible to guarantee that a request is queued on the initiator transmission queue at some point in the period of inaccessibility during which IADT are disabled. Fig. 6 depicts such kind of scenario, assuming the network depicted in Fig. 1, and an IDT between master M2 and slave S7.

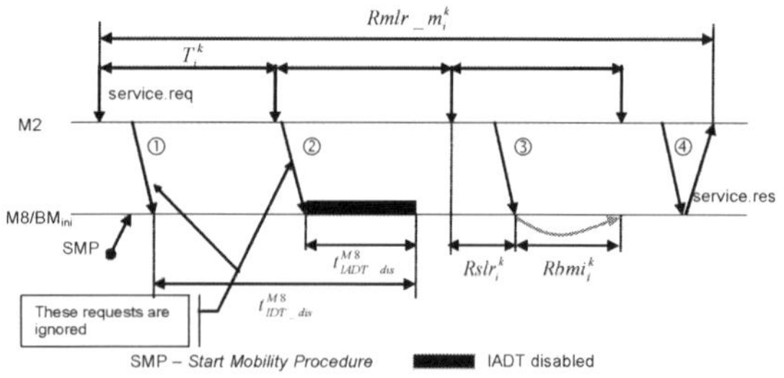

Fig. 6 – Case when $t_{IADT\_dis}^{bm} < T_i^k \leq t_{IDT\_dis}^{bm}$

In the depicted scenario, the first and the second requests arriving at M8/BM_{ini} are ignored, since M8/BM_{ini} had previously received a SMP message, and stopped accepting new IDTs. M8/BM_{ini} only opens an IDT on the third request.

To obtain the effect of IDMP on the IDT response time, when $t_{IADT\_dis}^{bm} < T_i^k \leq t_{IDT\_dis}^{bm}$, the following worst-case assumptions are made: the first request issued by master k, related to message stream S_i^k, arrives at BM_{ini} just after it had received the SMP message; another request, which initialises an IDT on the BM_{ini}, arrives after the end of the period of time during which IADTs are disabled in master k domain.

Under these conditions, the following equation accounts for the effect of the IDMP on IDTs if $t_{IADT\_dis}^{bm} < T_i^k \leq t_{IDT\_dis}^{bm}$:

$$Rmlr\_m_i^k = Rslr_i^k + t_{IDT\_dis}^{bm} + T_i^k + Rmlr_i^k \tag{17}$$

Case Three. In this case, since $T_i^k < t_{IADT\_dis}^{bm}$, master k is able to queue at least one retry related to message stream S_i^k, at some point in the period of inaccessibility during which IADTs are disabled. Fig. 7 depicts such kind of scenario.

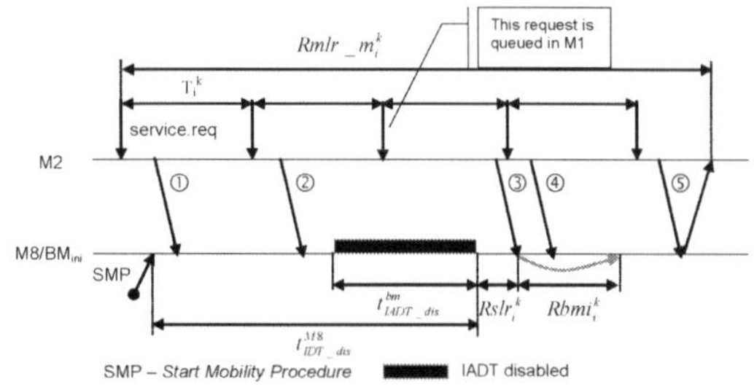

Fig. 7 – Case when $T_i^k \leq t_{IADT\_dis}^{bm} < t_{IDT\_dis}^{bm}$

In this example, the first two requests are ignored since M8/BM_{ini} previously has received the SMP message. The third request is queued on the M2 output queue at some stage in the period of time during which IADTs are disabled. As soon as this period ends, and M2 is able to contend for the medium, the request is transmitted, initialising an IDT in M8/BM_{ini}. The fourth request is ignored by M8/BM_{ini}, since it does not have any response available. Finally, only on the fifth request a response is transmitted back to M2.

To obtain the effect of the IDMP on the IDT response time, when $T_i^k \leq t_{IADT\_dis}^{bm} < t_{IDT\_dis}^{bm}$, the following assumptions are made: the first request issued by master k related to message stream S_i^k, arrives at BM_{ini} just after it had received the SMP message; another request related to the same message stream is received by BM_{ini} at some stages in the period of time during which IADTs are disabled. Therefore, the following equation holds:

$$Rmlr\_m_i^k = \begin{cases} Rslr_i^k + t_{IDT\_wr\_dis}^{bm} + Rmlr_i^k & \text{, wired domains} \\ Rslr_i^k + t_{IDT\_wl\_dis}^{bm} + Rmlr_i^k & \text{, wireless domains} \end{cases} \tag{18}$$

Accounting for the effects of the IDMP on the calculation of $Rbmi_i^k$

The analysis presented above only takes into consideration the state in the initiator domain. Nonetheless, there are no guarantees if the remaining BMs which belong to the IDT Communication Path have its IDTs enabled. In such cases, when an Inter-Domain Request frame arrives at a bridge, having one of its BMs with IDTs disabled, the IDF must wait on the BM output queue until being transmitted by the BM.

Fig. 8 depicts an example regarding an IDT between master M2 and slave S1.

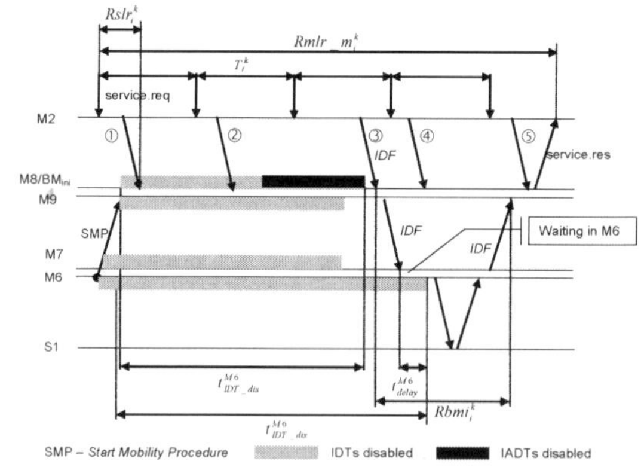

Fig. 8 – Delays due to the BMs in the communication path set not having its IDTs enabled

In this example, the two first requests transmitted by M2 are ignored, since M8/BM_{ini} has its IDTs disabled. The third request is accepted by M8/BM_{ini} opening an IDT at M8/BM_{ini}. M8/BM_{ini} receives the request and transforms it into an IDF using the rules defined by the IDP, then this frame is relayed until M6, which does not have its IDTs enabled. Consequently, the IDF has to wait (t_{delay}^{M6}) until the IDTs are again enabled in M6.

The following equation provides a new formulation to $Rbmi_i^k$, including the effects of the IDMP:

$$Rbmi\_m_i^k = \max\left\{Rbmi_i^k + pos(t_{IDT\_dis\_GMM}^{bm} - t_{IDT\_dis\_GMM}^{BM_{ini}})\right\},\quad \forall bm \in \Omega_{req}(S_i^k) \tag{19}$$

where, $t_{IDT\_dis\_GMM}^{BM_{ini}}$ is equal to the time during which IDTs are disabled on the BM which is the BM_{ini} for message stream S_i^k, plus the time required for the SMP message to reach that BM, $t_{IDT\_dis\_GMM}^{BM_{ini}} = t_{IDT\_dis}^{BM_{ini}} + t_{SMP}^{BM_{ini}}$. Similarly, $t_{IDT\_dis\_GMM}^{bm}$ refers to the time during which IDTs are disabled on a BM bm which belongs to the BMs which relay the IDT request frame in the path from the initiator to the responder related to message stream S_i^k, plus the time required for the SMP message to reach BM bm, $t_{IDT\_BM}^{bm} = t_{IDT\_dis}^{bm} + t_{SMP}^{bm}$. In the equations, the inclusion of the time spans $t_{SMP}^{BM_{ini}}$ and t_{SMP}^{bm} is required in order for $t_{IDT\_dis\_GMM}^{BM_{ini}}$ and $t_{IDT\_dis\_BM}^{bm}$ to have the same starting reference – the start of the IDMP by the system GMM. $pos(a)$ is a function that returns a when $a \geq 0$ and 0 otherwise.

4.3 Transactions Involving Mobile Wireless Stations

The problem of providing a worst-case bound for the response time of IDTs related to a message stream S_i^k involving mobile wireless stations is, in practice, similar to the scenario described in the Sections 4.2 and 4.3.

Three main sub-cases must be considered: 1 – IDTs between a wired or domain resident wireless master and a mobile wireless slave/master; 2 – IDTs between a wireless mobile master and a wired or domain resident wireless slave/master; 3 – IDTs involving two mobile wireless stations.

Case One. IDTs involving a mobile wireless station are disabled, on the BM_{ini}, from the reception of the SMP message until the reception of a RU message regarding the responder station s – $t_{IDT\_mob\_dis}^{BM_{ini},s}$. For the case of a wireless mobile master this time span can be calculated by:

$$t_{IDT\_MM\_dis}^{bm,s} = t_{phase1} + t_{phase2} + \max(t_{phase3}^{wldmm} + t_{dics}^{wldmm} + t_{m\_master\_entry}^{s,j} + t_{RU,s}^{bm}) - t_{SMP}^{bm}, \forall wldmm \tag{20}$$

where j is the address of the station that will be the predecessor of station s, when s moves into a new domain. For the case of wireless mobile slave, this time span can be calculated by:

$$t_{IDT\_MS\_dis}^{bm,s} = t_{phase1} + t_{phase2} + \max(t_{phase3}^{wldmm} + t_{disc}^{wldmm} + t_{RU,s}^{bm}) - t_{SMP}^{bm}, \forall wldmm \tag{21}$$

In both cases, $wldmm$ represents the set of DMMs which belong to all possible domains where station s is capable of belonging.

Fig. 9, depicts an example which illustrates an IDT between a master M2 and S6. In this example, the requests transmitted by M2 and received by M8/BM_{ini} are ignored from the reception of the SMP message until the reception of a RU message regarding slave S6. The third retry is successful, and opens an IDT on M8/BM_{ini}.

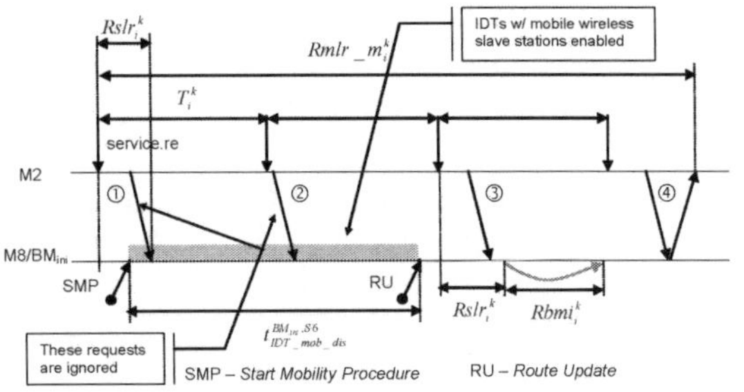

Fig. 9 – IDT example with a mobile wireless station

In this situation the WCRT for an IDT regarding a message stream S_i^k can be calculated assuming the following worst-case conditions: the first request is received by BM_{ini} just after the reception of the SMP message; no IDT request is accepted by BM_{ini} during $t_{IDT\_mob\_dis}^{BM_{ini},s}$, i.e. from the reception of the SMP message until the reception of a RU message regarding station s; a request arrives at BM_{ini} just before it receives the RU message concerning the responder station.

The following formulation gives the WCRT for IDTs between a wired or domain resident wireless master k and a mobile wireless station s:

$$Rmlr\_m_i^k = Rslr_i^k + t_{IDT\_mob\_dis}^{BM_{ini},s} + T_i^k + Rmlr_i^k \tag{22}$$

$t_{IDT\_mob\_dis}^{BM_{ini},s}$ can be calculated by Eq. (20) or (21), in the case when the mobile wireless station is a master or a slave, respectively.

Case Two. The IDT initiator is a mobile wireless master and the responder is a wired or domain resident wireless station.

Fig. 10 shows a timeline depicting an IDT between master M3 and slave S1, where master M3 moves from the original domain (D^1) to a destination domain (D^3), during the execution of an IDT with S1. After the reception of the SMP message by the IDT $(BM_{ini})^{orig}$, on the original domain (the domain D^1, to which M5/$(BM_{ini})^{D1}$ belongs to), M3 is no longer able to complete IDTs with S1. Only after entering the logical ring on the destination domain, M3 is capable of completing the transaction, using BM M7/$(BM_{ini})^{D3}$ on the destination domain.

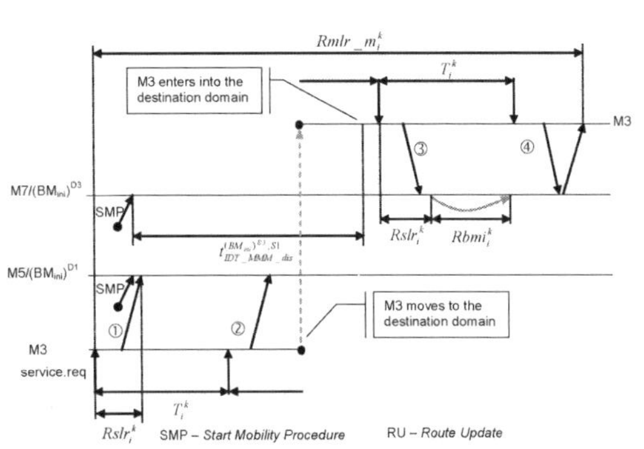

Fig. 10 – IDT example between a mobile wireless master and slave S1

To obtain the WCRT for an IDT related to message stream S_i^k, the following conditions are assumed: the first request is received by $(BM_{ini})^{orig}$ just after the reception of the `SMP` message; the first request made on the destination domain is delayed by T_i^k after master k has entered the domain.

Under these conditions, the following equation updates Eq. (13) for the case when the transaction is made between a wireless mobile master and a wired or domain resident wireless station:

$$Rmlr\_m_i^k = (Rslr_i^k)^{orig} + (t_{IDT\_MMM\_dis}^k)^{orig \to dest} + T_i^k + (Rmlr_i^k)^{dest} \tag{23}$$

$(y)^d$ represents the value for time span y in domain d. $(t_{IDT\_MMM\_dis}{}^k)^{orig \to dest}$ represents the time during which master k has its IDTs disabled when moving between the original domain and the destination domain, which can be calculated as follows:

$$(t_{IDT\_MMM\_dis}^k)^{orig \to dest} = t_{phase1} + t_{phase2} + t_{phase3}^{dest} + (t_{m\_master\_entry}^{k,j})^{dest} - t_{SMP}^{(BM_{ini})^{orig}} \tag{24}$$

Case Three. Finally, the third case occurs when the two stations move during the execution of an IDT. In this case, it is also necessary to consider two sub-cases: the `RU` message, regarding the responder station, arrives at the destination domain **before** the initiator has entered into the domain; the `RU` message, regarding the responder station, arrives at the destination domain **after** the initiator has entered into the domain.

To distinguish between the two sub-cases, the following equations allow determining the time when the initiator is operational (i.e. when it is capable of making transactions) on its destination domain ($dest_i$):

$$\left(t_{ini\_op}\right)^{dest_i} = t_{phase1} + t_{phase2} + t_{phase3}^{dest_i} + (t_{disc})^{dest_i} + (t_{m\_master\_entry}^{k,j})^{dest_i} \tag{25}$$

and when the `RU` message regarding the entry of responder (r) in its destination domain ($dest_r$) has reached the BM_{ini} on the destination domain of the initiator:

$$\left(t_{resp\_op}\right)^{dest_r} = t_{phase1} + t_{phase2} + t_{phase3}^{dest_r} + t_{phase4}^{r,BM_{ini}} \tag{26}$$

In the first sub-case, the RU message arrives at the destination domain BM_{ini} before master k is ready to make the request. Therefore, the conditions are similar to case two (i.e. when the IDT responder is a domain resident master/slave) and the WCRT can be given by Eq. (23).

In the second sub-case, the initiator has to wait until the reception of a `RU` message before being capable of completing a transaction with the responder. Therefore the WCRT can be calculated based partially on Eq. (22), in order to account for the mobility of the initiator station as follows:

$$Rmlr\_m_i^k = (Rslr_i^k)^{orig} + t_{phase1} + t_{phase2} + t_{phase3}^{dest_i} + (t_{disc})^{dest_i} + (t_{m\_master\_entry}^{k.j})^{dest_i} - t_{SMP}^{orig} + T_i^k + (Rmlr_i^k)^{dest} \tag{27}$$

5 CONCLUSION

In this paper we presented a worst-case timing analysis for transactions occurring between stations in a hybrid wired/wireless PROFIBUS network, where the interconnection between different wireless media is supported by intermediate systems operating as bridges. The presented analysis evolved from previous works which had proposed a timing analysis for the IDP [8] and for the latencies involved in the IDMP [10], to provide a complete analysis considering inter-domain mobility.

REFERENCES

1. Alves, M., Tovar, E., Vasques, F., Roether, K., Hammer, G., "Real-Time Communications over Hybrid Wired/Wireless PROFIBUS-based Networks", Proceedings of the 14th Euromicro Conference on Real-Time Systems, Vienna, Austria, pp. 142-151 June 2002.
2. Ferreira, L., Alves, M., Tovar, E., "Hybrid Wired/Wireless PROFIBUS Networks Supported by Bridges/Routers", Proceedings of the 2002 IEEE International Workshop on Factory Communication Systems, Vasteras, Sweden, pp. 193-202August 2002.
3. Ferreira, L., Tovar, E., Alves, M., "PROFIBUS Protocol Extensions for Enabling Inter-Cell Mobility in Bridge-based Hybrid Wired/Wireless Networks", Proceedings of the 5th IFAC International Conference on Fieldbus Systems and their Applications, Aveiro, Portugal, pp. 283-290, July 2003.
4. Ferreira, L., Tovar, E., Alves, M., "Enabling Inter-Domain Transactions in Bridge-Based Hybrid Wired/Wireless PROFIBUS Networks", Proceedings of the 9th IEEE International Conference on Emerging Technologies and Factory Automation - ETFA2003, Lisbon, Portugal, pp. 15-22, July 2002.
5. Tovar, E. and Vasques, F., "Cycle Time Properties of the PROFIBUS Timed-Token Protocol", Computer Communications 22 (1999), pp. 1206-1216, Elsevier.
6. Lee K., Lee S., "Integrated Network of PROFIBUS-DP and IEEE 802.11 Wireless LAN with Hard Real-Time Requirement", Proceedings of the IEEE International Symposium on Industrial Electronics (ISIE'01), Pusan, Korea, pp. 1484-1489, 2001.
7. Willig, A., "Investigations on MAC and Link Layer for a Wireless PROFIBUS over 802.11", PhD Dissertation, University of Berlin, 2002.
8. Ferreira, L., Tovar, L., "Timing Analysis of a Multiple Logical Ring Wired/Wireless PROFIBUS Network", Proceeding of the 5th IEEE International Workshop on Factory Communication Systems – WFCS'2004, pp. 81 – 90, September 2004.
9. Aslanis, S., Koulamas, C., Koubias, S., Papadopoulos, G., "Architectures for an Integrated Hybrid (Wired/Wireless) Fieldbus", EFT Journal of Electrical Engineering, October 2001.
10. Ferreira, L., Tovar, E. "Timing Analysis of an Inter-Cell Mobility Procedure for a Wired/Wireless PROFIBUS Network" Proceedings of the 10th International Conference on Real-Time and Embedded Computing Systems and Applications - RTCSA04, Gothenburg, Sweden, pp. 591 – 609, 2004

ELSEVIER
IFAC
PUBLICATIONS

A 3D-LOCATION SYSTEM OPTIMIZED FOR SIMPLE STATIC AND MOBILE DEVICES

Herbert Schweinzer, Peter Krammer

Technical University of Vienna, Institute of Electrical Measurements and Circuit Design, Gusshausstrasse 25/354, 1040 Vienna
e-mail: herbert.schweinzer@tuwien.ac.at

Abstract: This article focuses on 3D-location systems based on ultrasound time-of-flight (TOF) measurements. Commonly, location systems localize mobile devices with base stations. However, our concept is primarily oriented to simple static devices, especially sensors measuring local physical quantities. Normally the sensor-identification, established at system setup, allows to link sensor data and sensor location. Using locally measured TOFs of ultrasonic signals instead of an explicit identification, the device directly transmits measurement data together with information about its own location. This enables a receiving base station to identify the device by calculating its location. Moreover, this principle can also be applied for mobile devices. *Copyright © 2005 IFAC*

Keywords: 3D-location system, ultrasound distance measurement, trilateration, device identification, transmitter coding

1 INTRODUCTION

Location information is not only essential for mobile devices but also gets a growing importance when the number of static devices, especially sensors for miscellaneous physical quantities, increases (refer to Schweinzer and Kastner (2003a)). Applications including numerous sensors will have to associate their measurement values to the locations where they are measured. Particularly, applications in building automation systems require an appropriate resolution of the location system also in cases where the number of sensors is not high. For example, regulating a number of light sources for the illumination of a working place in principle requires only one light sensor. To achieve a sufficient regulation of illumination, however, the sensor location has to be known quite accurately (e.g. 0.1 m in an office environment).

Another reason to use a location system can be found in simplifying the setup of the sensor network: locally measured device positions are suitable to identify devices by their locations. Usually, communication requires the installation of device identifications to enable access to the devices. Performed manually by human action, this installation takes a lot of time.

However, device identification is only one method of linking sensor data with locations where they are gathered. In this case, identifications are combined with location information by a cross-reference table which has to be acquired initially. Replacing any device identification by pure location data enables a computation of the device location by a base station and allows associating this information directly with the sensor measurement data.

In case of malfunction of a sensor device, its location data deliver the necessary information to replace the sensor. Only in case of erroneous location information, an additional sensor identification will be necessary to identify the erroneous sensor and to find its location.

In the followings, the paper will discuss different variants of location systems, known from literature. Further, the concept of a location system for simple devices is presented which offers an optimization between high accuracy and error-reduced operation on the one hand, and relatively low effort on the other hand. Details of the ultrasonic measurement system are shown. System setup and outlook conclude the article.

2 LOCATION SYSTEMS USING ULTRASONIC SIGNALS

A lot of research was done in the last years, using ultrasonic distance measurement as basis of 3D-location systems. Most of these systems use time-of-flight (TOF) measurements of bursts or coded signals, utilizing transmitters for the base station and receivers for the device, or vice versa. A number of different aspects strongly influence the system design, which is briefly described in the followings.

2.1 Principle of basic operation

The conventional way to realize an ultrasonic location system, using trilateration, is as follows: An active tag transmits an ultrasonic pulse, which is received at least from three receivers of a base station. The transmission of the ultrasonic signal is marked with an additional fast signal, mostly a RF-signal, and is performed periodically or triggered by the base station. This allows the receivers to measure the TOF between the start marker and the reception of the ultrasonic signal according to the typical A-B measurement topology.

The distances between the transmitter and the receivers are directly proportional to the TOFs. Considering the speed of sound c of about 331.3 m/s (at 273 K), the TOF is about 3 ms/m which can easily be measured with simple equipment. However, a temperature-compensation is obligatory because of the high temperature dependency of c:

$$\frac{\frac{dc}{dT}}{c} = \frac{1}{2 \cdot T} \approx \frac{0.17\%}{K} \tag{1}$$

Measuring at least three distances between the transmitter and the receivers with known locations allows evaluating the 3D-location of the transmitter. Multilateration, using more than three distances for evaluation, leads to improved accuracy of results.

An example of a location system, applying this principle, is the "Bat system" (refer to Ward, et al. (1997) and Addlesee, et al. (2001)). The active tag called "Bat" is attached to a mobile device which location has to be estimated. The Bat transmits uncoded small-band ultrasonic signals, RF-triggered by the base station, which are received by receivers mounted on the ceiling. A rather similar configuration, equipped with broadband transducers, can be found at the "Dolphin system" (refer to Hazas and Ward (2002)).

2.2 Reverse signal flow systems

In location systems consisting of an unknown number of simple sensors devices, acting as transmitters, the transmission of ultrasonic signals by the devices can easily lead to signal overlapping. In this case, a reverse signal flow with broadcast activation of devices, equipped with ultrasonic receivers, is the better solution.

In mobile applications, a quite different reason is responsible for the preference of reversed operation. Transmitting signals by mobile tags has a disadvantage: The base station measures the tag locations and therefore has central knowledge about mobile activities. In applications where these tags are attached to persons, such a system will not fulfill security aspects of privacy-conscious users (refer to Langheinrich (2001) and Hazas and Ward (2003)). Therefore, mobile units of privacy-oriented systems should not emit signals, which allow detecting their locations by any observer. Thus, the direction of ultrasound signal flow should be reversed in such applications. Several transmitters with fixed locations are acting as base station while the tags receive the different ultrasound signals.

An advantage of such a configuration is the parallel access to all reachable devices. Measuring the TOFs from the different transmitters and possessing knowledge about their locations, the devices are able to compute their own positions locally.

However, the method has also the disadvantage to require different signals to characterize the different transmitters. Two alternatives can be chosen:

- Transmitters send their signals at defined points in time with a time protocol, which guarantees non-overlapping reception of the ultrasonic signals from different transmitters. Uncoded small-band signal transmission is suitable for this method (refer to Randell and Muller (2001)). However, reflections from walls or echoes, oscillating between objects, need sufficient damping. Therefore, the pause between signal transmissions has to be sufficiently large, reducing the localization speed significantly. This problem especially concerns moving devices.
- Transmitters send their signals synchronously. To allow separation, the signals have to be coded. Different methods of coding are used, shortly discussed in chapter 2.6. In any case, broadband signals have to be applied, claiming for appropriate transducers.

2.3 Realization of the base station

Because ultrasonic distance measurement requires a signal path with line-of-sight (LOS) between transmitter and receiver, devices cannot be localized in any arbitrary position. If the number of received signals is not enough for trilateration, localization fails.

In addition, multi-path propagation can be a problem. If a signal, running different paths reflected by different objects, is received multiple, there is no problem as long as the first signal is received from LOS. Thus, only the first received signal has to be considered. However, if also the first signal is not received from LOS, the localization fails and delivers wrong results. Different detection methods, especially based on multilateration, allow the recognition of this situation.

Practical realizations try to ensure LOS signal propagation by an effective placement of the base station transducers. A good solution is transducer mounting on the ceiling as proposed by Addlesee, et al. (2001) and Hazas and Ward (2003). They are networked, utilizing a common time-base for TOF-measurement. For private applications, a single compact base station as proposed by Dijk, et al. (2003, 2004) is advantageous, housing three transducers and enabling measuring locations in a restricted area.

2.4 *Trilateration*

Using trilateration for locating a device with the unknown spatial position (x,y,z), the distances d_i to well-known locations (a_i,b_i,c_i) are given with:

$$d_i = \sqrt{(x-a_i)^2 + (y-b_i)^2 + (z-c_i)^2} \qquad (2)$$

Presupposing a minimum of three given locations, x, y, and z can be computed based on the equation above. Multilateration uses more than three reference locations, which can be used to improve the accuracy. Distances d_i usually are measured by TOF. Another possibility is *pseudo-ranging* where time-intervals between received signals are measured instead of TOFs. Analogue to GPS systems, pseudo-ranging uses an asynchronous receiver without knowledge about the start of signal transmissions. Assuming a synchronous start of all transmissions, distances can be measured with a common unknown offset d_0

$$d_i = \sqrt{(x-a_i)^2 + (y-b_i)^2 + (z-c_i)^2} - d_0 \qquad (3)$$

Using at least four transmitters on fixed locations, d_0 can be eliminated and x, y, z of the receiving device can be computed.

In Hazas and Ward (2003) a comparison of the two methods is presented where the accuracy of pseudo-ranging is shown to be worse by a factor of about 5. Moreover, the vertical and horizontal accuracy depend on the geometric arrangement of receiver and ceiling transmitters.

2.5 *Bandwidth of ultrasonic signals*

To achieve a sufficient location accuracy, the accuracy of TOF-measurement is crucial, especially for pseudo-ranging. Standard signals for ultrasonic ranging are mono-frequent bursts. TOF-measurement is performed by means of a constant threshold, leading to a systematic distance measurement error, depending on the amplitude of the echo signal. Mono-frequent bursts enable the use of low-cost resonant piezoelectric transducers, but several further arguments request for broadband transducers:

- distance measurement accuracy can significantly be increased by a short signal rise time
- duration of ultrasonic signals can be shortened, increasing accuracy of locating moving devices,
- coded signals can be used

Adequate broadband transducers are typically electrostatic devices (e.g. Polaroid), but there are also other types, e.g. based on piezoelectric PVDF (Polyvinylideneflouride) foils available.

2.6 *Signal coding*

Simultaneous transmission of all transmitters is necessary to achieve short location times. To allow independent TOF-measurements and to avoid problems with parallel operation, coded signals have to be used. A coding method, known from GPS-applications with direct sequence spread spectrum / code division multiple access (DS/CDMA), can also be applied with ultrasonic ranging (refer to Hazas and Ward (2002, 2003)). In addition, frequency-hopping signals can be used for coding.

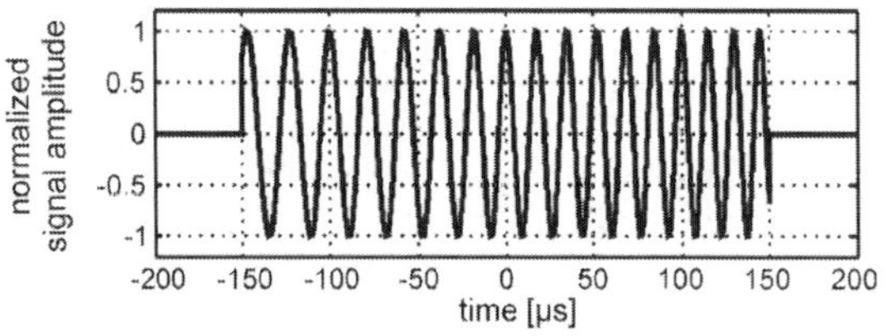

Fig. 1. Example of a chirp signal

A continuously frequency-modulated signal, especially designed for pulse-compression, is the chirp (Fig. 1)[1]. Firstly used in radar technique, chirps can be used effectively for ultrasonic ranging to achieve high resolution and accuracy. Moreover, a set of chirps with different frequency modulations can be defined to achieve coding (refer to Elmer and Schweinzer (2002) and Schweinzer and Elmer (2003b)).

2.7 *Signal processing within the receiver*

As mentioned in 2.5, threshold based burst detection provides bad accuracy because of its amplitude dependency. Using chirps or coded signals enables the application of correlation-based receivers (Fig. 2), delivering significantly improved distance measurement accuracy.

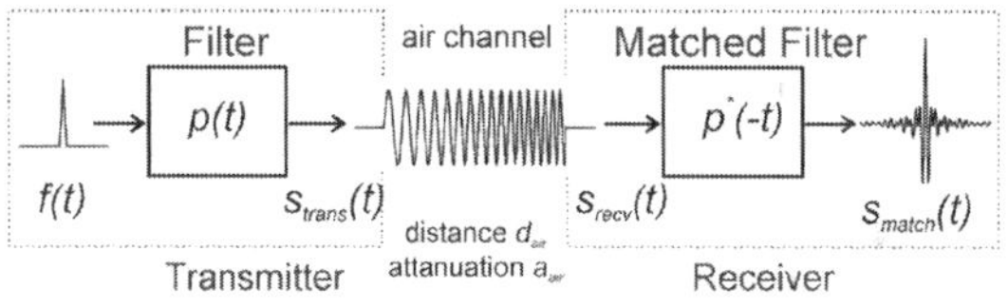

Fig. 2. Pulse compression to achieve high accuracy of TOF measurement

An important aspect, considering the expense, is whether one or more reference signals have to be correlated with the received signal to identify the signal transmitter. Processing correlations by software requires a local storage of the received signal and typically much processing time. Hardware embedded correlation can be performed without signal storage but increases the hardware expense.

Real-time hardware-correlation with limited expense is possible by "one-bit correlation", based on a one-bit-digitalization of the received signals (refer to Elmer and Schweinzer (2002)). Delivering a high ranging accuracy, the method also shows a good signal-separation capability for partial signal overlapping. Full overlapping of multiple signals, however, provides different amplitudes and delivers problems of echo separation. The use of chirps therefore can be very effective for pseudo-ranging,

---

[1]In Randell and Muller (2001), the term "chirp" is used erroneously for a burst signal with constant frequency of 40 kHz.

especially when several chirps are transmitted from different fixed locations with a pre-defined time sequence, avoiding signal overlapping as far as possible.

3 CONCEPT OF A 3D-LOCATION SYSTEM FOR SIMPLE SENSORS AND MOBILE DEVICES

3.1 *Motivation*

Based on the design criteria summarized in chapter 2, a system concept was developed, especially suitable for simple, low-cost sensor devices. Only one single ultrasonic receiver within the sensor device has to process signals of multiple transmitters at fixed locations. The activation of the receiver unit has to be performed only by ultrasonic signals without external synchronization with the transmitters.

For simple sensor devices, the measured time-intervals are passed to the base station without local computation of the device location. Together with this information, also the device data will be transmitted. Thus, the data receiving base station will get consistent data sets of device data together with locally measured TOFs of ultrasonic signals, which allow computing the device location. Location information can immediately be used as device identification. Information about the device type can be transmitted together with all the other device data, enabling a proper insertion of the sensor in the network.

In addition, mobile tags with identification and medium motion speed can participate of this localization method. Of course, they are not restricted to deliver unprocessed location information. They can behalf according privacy demands. Thus, this system combines advantages of 3D-localization of simple sensor devices by a base station and a privacy-oriented system for mobile devices of medium speed.

3.2 *Design characteristics*

The demand of a low-cost receiver claims for easy signal processing. However, pseudo-ranging can significantly be improved by the good distance measurement accuracy of correlation based signal processing, using chirp signals. Moreover, this method is characterized by very good noise suppression and effective separation of partially overlapped signals.

Although the expense in signal processing is not minimized, there is no need for further local computing of location on the other hand. The measured time-intervals just have to be transmitted together with the device data and device type.

Embedded correlative signal processing within the receiver adds some effort, but allows reducing additional location information from the same transmitter. This information may follow the LOS-signal and may be an oscillating echo or received from multi-path propagation. A minimum-cost receiver may trigger received signal frames and transmit the one-bit digitized data stream together with the other devise data. , Any received ultrasonic signal is passed to the base station without any preprocessing.

Transmitting device and location data primarily requires the implementation of a RF-transmitter. To handle collision problems of RF-communication of different devices, a simple RF-receiver can support collision detecting or collision avoiding strategies.

3.3 *Device activation*

Transmitters of the base station will be mounted near the ceiling to ensure signal propagation in LOS for the most receivers.

The devices are active, powered by battery supply. They will be activated by the ultrasonic signals. Doing so, the initiation of location measurements defines a polling rate for all sensors. Nevertheless, sensor devices can also use their own rate of activity. In that case, they have to transmit the current time-intervals of location measurement together with actual device data.

4 ULTRASONIC MEASUREMENT

4.1 *Transmitter coding*

Ultrasonic signals from the transmitters consist of a unique chirp to perform an accurate TOF-measurement, followed by one or several sequences of individual frequencies, identifying the transmitter. Signal overlapping may result from both, overlapping of signals of different transmitters, and overlapping by multi-path propagation and oscillating echoes. To minimize signal overlapping, transmitters send their signals in a predefined sequence.

4.2 *Measurement sequence*

The location measurement cycle (refer to Fig. 3) is started by a wake-up burst signal with constant frequency, performed by several or all transmitters in parallel. This frequency is determined by a passive band-pass filter. After an intermission (30-50 ms), the measurement cycle continuous with the transmitter identifying frames.

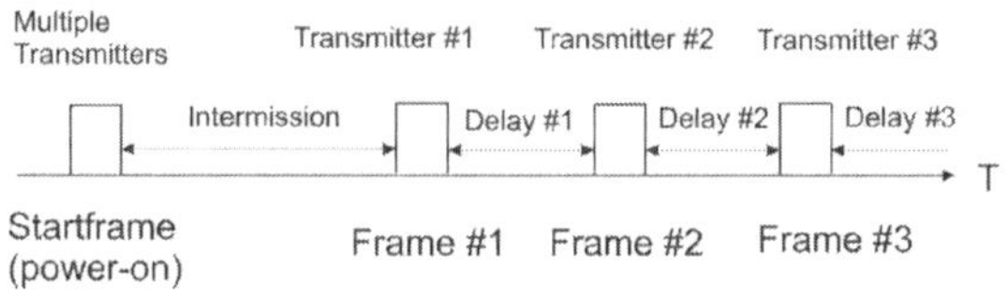

Fig. 3. Sequence of transmission to ensure device power-on and non-overlapping reception of transmitter identifying frames

At least four ultrasonic transmitters at different ceiling locations have to participate with the measurement sequence. They transmit frames, mainly consisting of a constant chirp signal with defined length and its transmitter coding, and a delay to the next frame transmission afterwards. The delay is defined according to the distance of successively activated transmitters: the minimum is given from the ultrasound propagation time between these transmitters at lowest temperature (typically 10-30 ms). Thus, this delay has to prevent "race hazards" or signals of a succeeding transmitter to be passed by signals from the previous transmitter for any location of the receiver.

4.3 *Setup of transmitted frames*

Transmitted frames start with a lead-in part of constant frequency which allows the decay of transient effects of the transducer (refer to Fig. 4). Three periods of 30 kHz are used for this task.

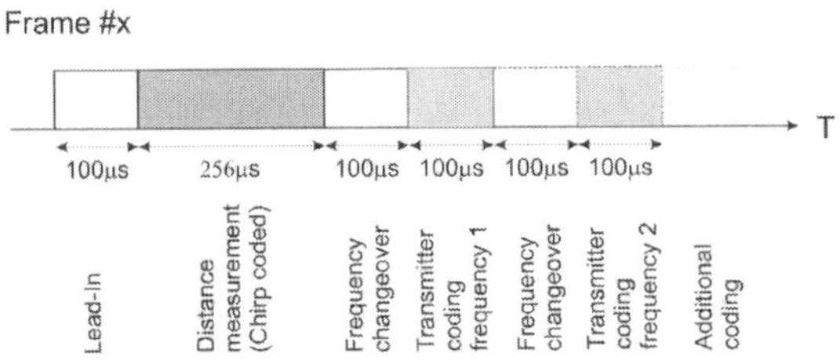

Fig. 4. Setup of a transmitted frame consisting of lead in, well defined chirp and one or more transmitter coding time slots with certain constant frequencies

The following part is a chirp signal enabling high resolution and precision distance measurement. The chirp of 256 µs duration is identical for each transmitter, which allows calculating the TOFs from all transmitters with the same signal processing hardware or software. The frequency change of the chirp is linear and running from about 30 kHz to 60 kHz.

To be able to include the fixed transmitter position in the trilateration algorithm, the chirp part is followed by one or several parts of transmitter identification. To enable continuous frequency changes, the chirp and the constant coding frequency parts of defined length (100 µs) are connected with changeover phases of 100 µs as illustrated in Fig. 4. The main reason of frequency changeover phases is to enable the coding frequencies to start the coding time window with zero phase shift.

Table 1 Frequencies used for transmitter coding and number of periods within 100 µs

Frequency	Number of periods
f(1) = 29.999 kHz	P(1) = 3
f(2) = 33.852 kHz	P(2) = 3.385
f(3) = 38.146 kHz	P(3) = 3.815
f(4) = 42.808 kHz	P(4) = 4.281
f(5) = 47.676 kHz	P(5) = 4.768
f(6) = 52.466 kHz	P(6) = 5.247
f(7) = 56.770 kHz	P(7) = 5.677
f(8) = 60.096 kHz	P(8) = 6.01

Transmitter identification is performed by frequency coding with one or several frequencies. Eight frequencies can be used (see Table 1), thus allowing the identification of eight different transmitters with one identification part, or up to 64 by use of two identification parts (in practice there are about 50 combinations usable) and so on. The number of coding frequencies is chosen in order to enable a good, noise suppressing frequency detection in the 100 µs time window. Performing cross-correlation of the eight signals of Table 1, the correlation result is about 60% maximum relating to 100% auto-correlation.

The number of periods of the used frequency normally is not an integer number. This means the frequency evaluation within the time-interval of 100 µs has to be started at a given point of time with respect to the accurate position of the chirp. After the 100 µs time-interval, the remaining part of the last period can already be changed to perform an appropriate frequency changeover to the next coding frequency.

4.4 *Tasks of signal processing*

Each activated device gets an individual sequence of received signals. It has to perform the task of measuring time-intervals between signals of different transmitters with high resolution and accuracy.

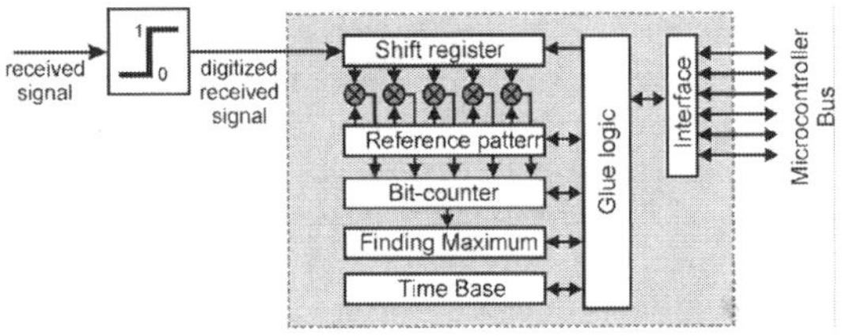

Fig. 5. Principle of FPGA realization of a one-bit correlator

The chosen method is a one-bit correlation of the chirp part of incoming signals. Fig. 5 shows a hardware realization of the correlation utilizing a FPGA (in Elmer and Schweinzer (2002), a ALTERA EPF10K30). However, also a software solution of the correlation algorithm may be implemented.

Reception-times of signals are measured in relation to a local time-counter with sufficient accuracy. However, received signals have to be treated with respect to their transmitter code. Only the first incoming signal of an individual transmitter has to be used for device localization.

Three types of receivers can be realized in principle, featuring decreasing effort in hardware:

Performing the correlation in hardware enables an effective preprocessing with fast reaction time. Because only one type of chirp is used for accurate time measurement, a minimal digital hardware-correlator with 64, 128, or 256 stages, according to the desired time resolution of 4, 2, or 1 µs, can be implemented. Performing correlation in real-time, the receiver is able to sample the transmitter-code on-line, immediately after the chirp as a digital sequence.

Performing the correlation locally in software requires storing the received bit-stream before processing. Sufficient RAM-space and relatively long processing time is the main disadvantage of this solution, although being able to deliver the same results as the hardware solution at the end.

Triggering received frames by use of a defined threshold has the advantage that the frames can be completely stored in maximum 1024 bits (at a sample rate of 1 μs). Since triggering takes already place within the lead-in phase, the relevant information will be stored and can be processed concentrated to the presumed frame. Stored information can be processed locally or transmitted directly to the base station. The disadvantage of the last method is the amount of data of all received signals which has to be transmitted. Furthermore, the trigger level has to be set to a constant value and cannot be adapted to the signal amplitude.

Transmitter codes are stored in combination with the signal time-counter value. Thus, post-processing allows to find and to eliminate received signals with identical transmitter codes after the first one.

Receiving at least four different signals delivers enough information for 3D-localization. However, processing of all received time-intervals is performed, both in case of more then four signal receptions, but also if a fewer signal number than four was received. For that purpose, a timeout of at least two further signal transmission periods is initiated after each signal reception. A timeout overflow finishes the signal reception and initiates the preparation of the information for RF-transmission. Time intervals between received signals are computed relatively to the first received signal and are combined with actual sensor measurement values, sensor type information, and eventually supplementary sensor identification.

5 SYSTEM SETUP

5.1 *System configuration*

The experimental system setup contains five or six transmitters, mounted on the ceiling of a big laboratory (see Fig. 6). Starting the signal transmission and controlling the sequence of transmitters will be performed by a base station as master-device. The transmitters are triggered with RF-communication or by wire.

The transmitters are approximately cylindrically shaped (or quarters of a cylinder). They are made of PVDF-foils and show a radiation pattern with an opening angle of more than 50° perpendicular to the direction of the cylinder axis.

The sensor devices are equipped with small electret microphones as receivers. Location information, sensor data, and sensor type are transmitted to the base station by means of RF-communication.

The received information will be analyzed in the base station. Frames are checked with respect to the transmitter code. Finally the device location is processed by use of pseudo-trilateration.

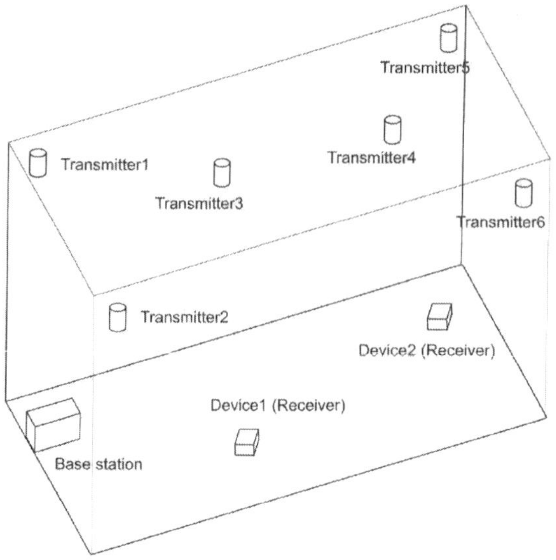

Fig. 6. Arrangement of six cylindrical transmitters at the ceiling of a laboratory to get sufficient LOS-transmissions to receiving devices

The base station as master-device is able to control several rooms in parallel by means of RF-communication. This meets the possibility to code a lot of transmitters, thus allowing to control and evaluate location measurements in several rooms. The co-operation of transducers within a room has to be considered only at the activation of measurement. The trilateration algorithm automatically considers the predefined positions of co-operating transmitters by the received transmitter codes.

5.2 *Device prototypes*

Standard test equipment with Digital-Storage-Oscilloscope and PC is used to test the characteristics of ultrasonic signals. For device prototypes, a FPGA development board is used. To simplify the RF-communication in the experimental setup, data transmission will be realized by a standard transmit / receive link.

6 CONCLUSION AND OUTLOOK

The presented system was designed for locating simple static and for mobile sensor devices. 3D-localization is realized by multilateration based on pseudo-ranging. Processing received chirp signals by one-bit correlation allows reducing the efforts for the local devices and simplifies them. However, the precision of device localization is significantly improved.

A laboratory setup of an experimental system will be started in the next months. Primarily it will allow studying problems of proper receiving ultrasound signals by the device receivers. Another focus of research will be the recognition of missing line-of-sight cases and the treatment of oscillating echoes.

REFERENCES

Addlesee, M., R. Curwen, S. Hodges, J. Newman, P. Steggles, A. Ward and A. Hopper (2001). Implementing a sentient computing system. *IEEE Computer*, 34(8):50–56, Aug. 2001.

Dijk , E., K. van Berkel, R. Aarts and E. van Loenen (2003). Ultrasonic 3D Position Estimation Using a Single Base Station. EUSAI 2003, pp. 133–148, 2003.

Dijk , E., K. van Berkel, R. Aarts and E. van Loenen (2004). A 3D Indoor Positioning Method using a Single Compact Base Station. In *Proceedings of PERCOM 04: Pervasive Computing and Communications*, pp. 101–110, March 2004.

Elmer, H. and H. Schweinzer (2002). High resolution ultrasonic distance measurement in air using coded signals. In *Proceedings of IMTC/2002: 19th IEEE Instrumentation and Measurement Technology Conference*, Anchorage, USA, May 2002, pp.1565-1570.

Hazas, M. and A. Ward (2002). A novel broadband ultrasonic location system. In *Proceedings of UbiComp 2002: Ubiquitous Computing*, pages 264–280, Göteborg, Sweden, Sept. 2002.

Hazas, M. and A. Ward (2003). A high performance privacy-oriented location system. In *Proceedings of PERCOM 04: Pervasive Computing and Communications*, pages 216–223, March 2003.

Langheinrich, M. (2001). Privacy by design—principles of privacyaware ubiquitous systems. In *Proceedings of UbiComp2001: Ubiquitous Computing*, pages 273–291, Atlanta, Georgia, USA, Sept. 2001.

Randell, C. and H. Muller (2001). Low cost indoor positioning system. In *Proceedings of UbiComp 2001: Ubiquitous Computing*, pages 42–48, Atlanta, Georgia, USA, Sept. 2001.

Schweinzer, H. and W. Kastner (2003a). Systems with numerous low-cost sensors – new tasks and demands for fieldbusses. In *Proceedings of FeT'03: 5th IFAC Int. Conference on Fieldbus Systems and their Applications,* pp. 73-78, Aveiro, Portugal, July 2003.

Schweinzer, H. and H. Elmer (2003b). High resolution ultrasonic distance measurement systems using pulse compression and their application. In *Proceedings of IMTC/2002:ISMTII 03: Measurement Technology and Intelligent Instruments VI,* Hongkong, Nov. 2003, pp. 301-306.

Ward, A., A. Jones and A. Hopper (1997). A New Location Technique for the Active Office. IEEE Personal Communications Magazine, Vol. 4, No. 5, October 1997. pp. 42–47

ELSEVIER
IFAC
PUBLICATIONS

A MODEL BASED ON A STOCHASTIC PETRI NET APPROACH FOR DEPENDABILITY EVALUATION OF CONTROLLER AREA NETWORKS

Paulo Portugal, Adriano Carvalho, Francisco Vasques

University of Porto, FEUP, Rua Dr. Roberto Frias s/n, 4200-465 Porto, Portugal
Tel.: +351.22.5081815, Fax: +351.22.5081443, E-mail: {pportugal,asc,vasques}@fe.up.pt

Abstract: The paper proposes a dependability model to evaluate the behavior of a CAN network in scenarios of transient faults which affect data communications. Fault occurrence is modeled by a Markov Modulated Poisson Process (MMPP) which is capable to describe the typical behavior of electromagnetic interferences (EMI) that occur in industrial environments. An accurate and efficient representation of the network behavior is achieved by adopting a set of assumptions that reduce the pessimism level and which are closer to the real operating conditions. The model is based on Stochastic Petri Nets, which are a high-level modeling formalism able to produce very compact and efficient models, supporting both analytical and simulation solutions. Dependability measures are established from the fulfillment of the real-time constraints (deadlines) defined on messages exchanged between network nodes. Analytical and simulation solutions are both investigated. A case study is proposed to asses both model performance and network dependability. Copyright © 2005 IFAC.

Keywords: Petri-nets; Reliability; Real-time communication; Safety; Stochastic modelling.

1. INTRODUCTION

Dependability attributes, like safety, reliability and availability, have become essential parameters on industrial automation systems design. Nowadays fieldbuses have a central role in these systems, with large application domains which extend to almost any area in manufacturing and process industries. They are presently the backbone of distributed industrial control architectures, providing a communication infrastructure which supports control, monitoring and supervision applications (Thomesse, 2005).

Industrial environments are characterized by the existence of a high diversity of equipments which are source of large patterns of electromagnetic interferences (EMI). These interferences induce faults in electronics circuits that disturb their normal operation. In communication systems these types of faults usually affect the transmission medium and related circuits, since, in most situations, these are the system components most exposed to them. EMI faults are generally characterized by occurring in bursts, a long latent period followed by relatively short period of presence, and by having a short duration (transient faults) (Kim *et al*., 2000).

In this context faults produce errors on transmitted messages by corrupting their contents. To recover from these situations fieldbus networks implement several fault-tolerant mechanisms. However, this creates a communication overhead by introducing delivery delays in messages which could imply performance degradation in the control system. When messages have real-time requirements, which is common in control systems, these problems can seriously disturb the system operation and can even lead to its failure (Shin and Kim, 1992; Kim and Shin, 1994).

The importance assumed presently by these control systems compels to evaluate their dependability (Navet, *et al.*, 2005). In a distributed system, determining the dependability of the communication channel is of particular importance, especially when this component is susceptible to EMI problems. Therefore in these systems it is vital to evaluate how the system dependability is affected by faults on communication (Broster, *et al.*, 2002).

This paper deals with a specific fieldbus: CAN (Controller Area Network) (Bosh, 1991). It proposes a model that enables to evaluate the network dependability with respect to deadline failures in the presence of transient faults induced by external sources (EMI). Shortly, the effect of faults on the real-time properties of the network is investigated.

In contrast to most of the previous works (Tindell, *et al.*, 1995; Punnekkat, *et al.*, 2000; Hansson *et al.*, 2002) a stochastic model is used to describe the fault occurrence and its duration, which guarantees a better representation of the phenomena involved. Although some recent works have already included this aspect (Navet, *et al.*, 2000; Broster, *et al.*, 2002; Broster, *et al*,. 2004), this paper provides a more realistic fault model, a representation of the network behavior much closer to the real operation conditions and the use of less pessimistic assumptions. The combination of all these aspects will provide more accurate dependability results.

2. ANALYSIS OF THE PREVIOUS WORK

The growing diffusion of CAN in safety- or mission-critical applications (Navet, *et al.*, 2005) asks for suitable techniques to assess the dependability of CAN networks. In the last decade, these aspects have motivated the academic community to study these problems and to propose answers to them. The following subsections present the most important works concerning the CAN behavior in fault scenarios.

2.1 *General Behavior in Fault Scenarios*

An analysis of the efficiency of the error detection mechanisms was performed by (Charzinski, 1994). By assuming a two-state channel model for the physical transmission medium, a set of expressions are derived which permits to quantify the probability for errors to be undetectable at receivers (residual error probability). Although (Bosh, 1991) defines this value as 4.7×10^{-11}, due to the combination of several aspects: error bursts, bit-stuffing, frame structure, spatial node distribution and efficiency of error detection mechanisms, it is shown that in scenarios with a high *Bit Error Rate* (BER) this value is overestimated (takes lower values).

The previous work is extended by (Tran, 1999), where the effects of multi-bit errors on those mechanisms are studied. By adopting a simulation model many of the restrictions imposed in (Charzinski, 1994) are removed, which permits to cover a larger number of fault scenarios. The results obtained confirm that the value presented in (Bosh, 1991) is in fact overestimated, as well as some scenarios evaluated in (Charzinski, 1994). As an example, a double-bit error (that should be detectable by the CRC mechanism) could result in a 1.3×10^{-7} residual error probability.

In a different context (Barrenscheen, *et al.*, 1997) performs a study about behavior of the CAN physical layer in environments with electromagnetic interferences. This work focuses in the electrical aspects of the physical layer (terminating resistors, stubs, cables, signal levels, etc.) and permits to establish the best conditions, from an EMI viewpoint, for data communications.

In a near context (Rufino, *et al.*, 1999) identifies the main causes that lead to a medium failure (permanent and transient faults, e.g. one-wire interruption or short-circuits), and proposes a fault-tolerant communication architecture to cope with these problems. This solution is based in the assumption of several failure mode scenarios and it uses redundancy (dual) both at medium and transceiver level.

The execution of the CAN error recovery mechanisms results in an inaccessibility period where the network isn't able to provide its service. The consequences of this behavior are studied by (Rufino and Verissímo, 1995). Several error scenarios are defined and for each one an expression for the inaccessibility time is derived.

The problem of message inconsistencies or omissions in CAN networks is addressed by (Rufino, *et al.*, 1998), where the occurrence of these scenarios is quantified by a simple probability model based on the BER. To cope with this problem, it is proposed a communication stack, on top of CAN, which supports a set of fault-tolerant broadcast protocols (atomic, reliable, etc.) which provides different degrees of group communication services.

On the sequence of the previous work (Ferreira, *et al.*, 2004) proposes a fault-injection experiment to evaluate the BER in different scenarios: aggressive and normal environments. The results confirm that in practice the effective BER is very small: normal (10^{-11}), aggressive (10^{-7}).

The behavior of the CAN fault-confinement mechanisms (TEC and REC counters) in fault scenarios is addressed by (Guajal and Navet, 2001). By using mean value rates both for message traffic and fault occurrence, the temporal evolution of these counters is modeled by a Markov chain. Two scenarios are analyzed: Bus-Off hitting time (TEC Markov chain: time until Bus-Off state) and Error-Passive hitting time (REC Markov chain: time spent in Error-Passive state). This study shows that these values strongly depend from the BER. For low / medium BER, Bus-Off hitting times are very high and Error-Passive hitting times are very small. Meanwhile, if the BER takes high rates the previous values can suffer a dramatic reduction or increasing respectively. Some improvements to these mechanisms are also presented.

2.2 *Message Scheduling in Fault Scenarios*

A simple schedulability analysis of CAN in the presence of faults is provided by (Tindell, *et al.*, 1995). Faults are incorporated into the traditional analysis by introducing an additional term, called *error recovery overhead function*, which is the upper bound of the overhead due to error recovery that could occur in a time interval. The fault model is very simple and based on a minimum inter-arrival time between faults. The main drawback of the analysis is the use of a deterministic model to describe fault occurrence. A model with these characteristics is not suitable for two reasons: **(i)** In a realistic scenario faults have a random nature (Kim, *et al.*, 2000); **(ii)** It assumes that the number of faults that can occur in a time interval is bounded, which is in contradiction with the previous statement.

The previous work is extended by (Pinho, *et al.*, 2000) by introducing the inaccessibility periods obtained in (Rufino and Verissímo, 1995). The fault model is slightly changed by introducing new terms (e.g. an erratic transmitter), but maintains its determinist nature.

Another extension to (Tindell, *et al.*, 1995) is presented by (Punnekkat, *et al.*, 2000), by providing a more general fault model which can deal with interferences (faults) caused by several sources. This model assumes that every source of interference has a typical pattern, consisting of an initial group of bursts with a fixed period and a distribution of single interferences, with a know minimum inter-arrival time. Since it uses a deterministic fault model, it suffers from the same problems as discussed previously.

Unlike the previous works, (Navet, *et al.*, 2000) proposes a stochastic fault model which is closer to the typical EMI behavior. This model considers both the frequency of the faults, modeled as a homogeneous Poisson Process, and its gravity (burst or single errors) modeled by a distribution function. This work doesn't try to determine whether a systems is schedulable (as the previous works), but it calculates the probability (WCDFP – *Worst Case Deadline Failure Probability*) that a message doesn't meet its deadline. Although this work is an important improvement face the previous ones, it includes two inaccuracies that increase the pessimism in the estimation of the WCDFP. First, the WCDFP definition doesn't reflect properly the conditions in which a message can miss its deadline (see (Broster, *et al.*, 2002) for details). Second, a burst of *n* errors is treated as a sequence of *n* single errors, each causing a maximum error overhead. This causes pessimism of several orders of magnitude.

An extension to (Punnekkat, *et al.*, 2000) is presented by (Hansson, *et al.*, 2002), by using a fault model which reduces the pessimism of the analysis (a lower error overhead during bus idle time). To guarantee this aspect, faults are modeled as fixed patterns of interferences (with different phases) which simplifies the model. The analysis is carried out only during the least common multiple (LCM) of the messages periods, being the remaining behavior extrapolated from that. Simulation is used to obtain the probability of having a message that misses its deadline. This approach has important limitations. First, it is necessary to determine the fault pattern for each source of interference. Second, the combination of several fault sources increases the complexity of the analysis to such an extent that it becomes infeasible, so random sampling is used. Third, the fault model is deterministic.

A stochastic fault model is again proposed by (Broster, *et al.*, 2002) by modeling their occurrence through a Poisson Process (single bit errors). The analysis is similar to (Tindell, *et al.*, 1995) but assumes that each fault occurs only in the last bit of the longest frame, imposing therefore the highest overhead. This analysis is naturally pessimistic. A probability tree is used to represent the occurrence of faults during non-overlapping intervals where the error overhead function is evaluated. By exploring the tree, an analytical model is obtained which permits to obtain the distribution function of the message response time. Although the tree has the potential to grow exponentially, measures are taken to avoid it by ignoring low probability branches. In (Broster, *et al.*, 2004) this work is improved by adopting a more efficient computation algorithm.

3. MODELLING FRAMEWORK

From the discussion presented in the previous section it becomes clear that some works (Navet, *et al.*, 2000; Broster, *et al.*, 2002; Broster, *et al.*, 2004) have already focused the same objectives which are intended for this paper. Therefore it is necessary to discuss the need for a new methodology to evaluate the behavior of a CAN network in the presence of faults. This analysis can be performed from two viewpoints:

- **Pessimism Level**. As stated previously the analysis presented by (Navet, *et al.*, 2000) is in fact very pessimistic. As an example, (Broster, *et al.*, 2002) evaluates the message set proposed by (Navet, *et al.*, 2000) using different models (analytical and simulation) and concludes that the probability that any message misses its deadline is insignificant (≈ 0), while in (Navet, *et al.*, 2000) the results have significant values ($\approx 10^{-2}$). In (Broster, *et al.*, 2002; Broster, *et al.*, 2004) the pessimism is reduced when it compared to (Navet, *et al.*, 2000), but the analysis is still pessimist because it considers that faults occur in a worst scenario (last bit of the Data frame) and causes the maximum possible overhead. Although authors say that the analysis in only slightly pessimistic (they use a simulation model to compare results), it is clear from the results that the pessimism level strongly depends from the message set and the fault parameters. Nevertheless, it is necessary to recognize that this work has very important merits;
- **Fault Model**. Since faults typically occur in random bursts, (Navet, *et al.*, 2000) presents a fault model that is closer to the real environmental conditions and therefore will permit to obtain more accurate results. In order to obtain an analytical model (Broster, *et al.*, 2002; Broster, *et al.*, 2004) adopts a simplified model where it is assumed a constant fault rate and that fault effects result only in single errors. It is clear that this model isn't appropriate to evaluate burst scenarios since it will produce very pessimistic (or even inaccurate) results.

To cope with these problems, this paper proposes a model that represents more accurately the fault occurrence and makes use of less pessimistic assumptions by considering a closer representation of the real network behavior. The combination of all these aspects it will provide more realistic results.

3.1 Stochastic Petri Nets

Dependability evaluation is necessarily associated with the development of a model that describes the system behavior in the presence of faults and which enables to retrieve dependability measures from it. Petri Nets (PN) are a graphical and mathematical modeling tool which enables the description and analysis of the system dynamics, where concurrent, asynchronous, distributed, parallel and stochastic behaviors are present (Murata, 1989). Over the last two decades Stochastic Petri Nets (SPN) has become a widely used framework for the performance and dependability evaluation of various kinds of systems by several reasons (Marsan, *et al.*, 1996; German, 2000): **(i)** An intuitive description of the system behavior; **(ii)** Representation of complex systems by very compact models; **(iii)** A formal basis; **(iv)** Full representation of the stochastic processes; **(v)** Large number of evaluation tools are available.

The use of SPNs to derive dependability models can be performed from two viewpoints, according to the type of solution required:

- **Analytical**. When the SPNs transitions obey to certain conditions an analytical stochastic process can be automatically generated and solved. If time is continuous, among these conditions the most important one imposes that the number of timed transitions (not exponential ones) concurrently enabled at each marking is limited to one (Bobbio, *et al.*, 1998) (there are some exceptions, but impose very strict structural conditions to the models). When these conditions aren't fulfilled it is possible to obtain approximate analytical solutions by using exponential transitions arrangements (phase-type distributions) to approximate the behavior of (non exponential) timed transitions. Another alternative is to use a discrete time approach where all the previous restrictions can theoretically be relaxed;
- **Simulation.** In this case neither of the previous limitations are present and a significant number of SPNs modeling extensions are available to reduce the model complexity. Since the SPNs semantics are formally well-established, models are easily constructed and less error-prone than simulation programs.

Due to their flexibility and advanced characteristics SPNs have become a popular formalism to evaluate communication systems. They have been widely used in performance and (Marsan, *et al.*, 1996; Lindemann, 1998; Billington, *et al.*, 1999) and dependability studies (German, 2000; Trivedi, 2002).

3.2 Model Assumptions

Dependability models were developed accordingly the following assumptions:

- There is a single transmission medium without any type of redundancy;
- Faults occur according to a Markov Modulated Poisson Process (MMPP) (Trivedi, 2002) and their effects are equally observed by all the network nodes. Faults occur in bursts of random length and have a random duration (Fig. 1). This model compels with the analysis performed by (Kim, *et al.*, 2000);

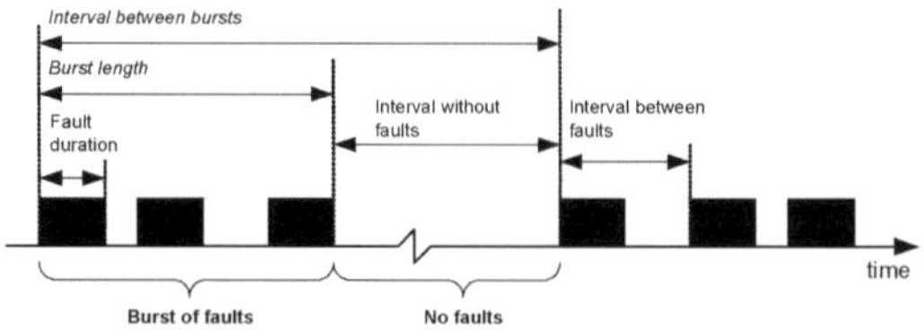

Fig. 1 - Fault occurrence.

- Nodes only transmit Data (messages) and Error frames. Overload frames are also considered but only in an indirect way. Remote frames are not explicit included because from a modeling viewpoint they can be represented as Data frames;
- Messages (Data frames) have real-time requirements which are represented by mean of deadlines. It was assumed a classical scenario similar to (Navet, *et al.*, 2000): messages have their deadlines equal to their production period and there is no jitter. It is important to notice that other assumptions can also be considered (deadline > period or deadline < period). However, the inclusion of all these variants in the same model would be cumbersome. In fact it would be necessary to develop a specific model for each situation. Therefore and in order to simplify the presentation it was chosen the most common scenario. It is also assumed that messages have a fixed length;
- If a fault occurs it is assumed that it corrupts the bus contents by introducing errors (frame transmissions, bus idle or intermission times). Those will trigger the CAN error recovery mechanisms (Error frame transmission). It is also assumed that errors are always detected by the nodes. This assumption is supported by (Charzinski, 1994; Tran 1999) where it is shown that the probability of undetected errors is very small. Although it could be important to model this behavior this is not addressed by this model. This approach was also assumed by (Navet, *et al.*, 2000; Broster, *et al.*, 2002);
- A network failure occurs if a message misses its deadline, and its probability is the most important dependability measure to obtain.

4. DEPENDABILITY MODELS

In order to improve the presentation, the complete dependability model is split into 3 small models, each one presented in different subsections: fault occurrence, message transmission and error signaling.

4.1 Fault Occurrence Model

The use of a Markov Modulated Poisson Process (MMPP) enables to define a Poisson Process whose arrival rate is "modulated" by a Markov chain (German, 2000; Trivedi, 2002). With this process is possible to model a time variable fault rate, which makes it ideal to represent burst scenarios.

Although it would be simple to model multiple fault rates, it was assumed in our model (Fig. 2) to represent only two levels: with faults (fixed rate) and without faults (null rate). This option was motivated by two reasons. First, several levels it would introduce to much complexity in the model, which in many cases is unnecessary. The use of multiple rates also implies that their values need to be known, which in practical terms is difficult to achieve. Second, the model can be easily adapted to include multiple levels if necessary. This can be performed by a simple model replication. Therefore this model should be seen as a building block for more sophisticated fault models.

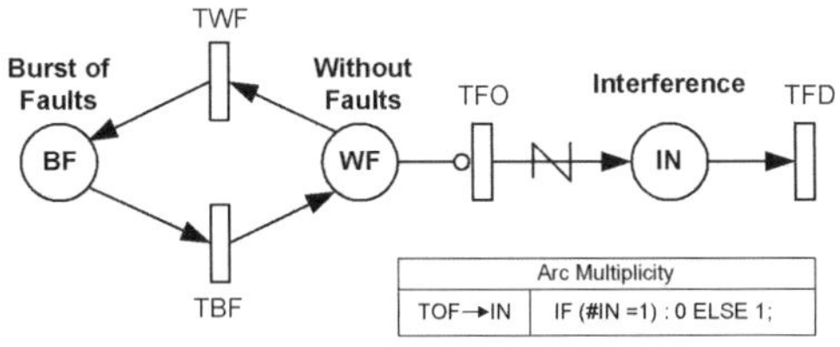

Fig. 2 - Fault occurrence model.

The places BF and WF represent respectively the occurrence and absence of bursts of faults. The alternation between these two states is ruled by two exponential transitions TBF and TWF, which represent the burst length (interval) and the interval without faults respectively (these places and transitions implement the Markov Chain). The expect time of those intervals correspond to the transition duration's. The place WF is initially marked with one token.

During the burst interval (BF is marked) faults occur according a Poisson Process, which is the common modeling assumption (Kim, *et al.*, 2000; Navet, *et al.*, 2000; Broster, *et al.*, 2002). This is represented by the exponential transition TFO whose rate is the same of the Poisson Process. This rate can be adjusted to represent the effects of several fault sources. Transition TFO is only enabled if BF is marked, which is assured by the inhibitor arc that connects WF to TFO. When this transition fires the place IN is marked. This place indicates that a fault (interference) had occurred. To guarantee that the number of tokens in place IN never exceeds 1, a variable multiplicity arc is used (a zig-zag line). Therefore if a fault occurs during the occurrence of another fault it is considered as a single fault.

Fault duration is modeled by the timed transition TFD. Since there is little information about this aspect, it was assumed that it could be modeled by an exponential distribution (other assumptions could also be adopted). This transition fires only if IN is marked (a fault exists). After its firing, the marking of IN is removed which indicates that the fault (interference) has ended.

4.2 Message Transmission Model

The model represented in Fig. 3 describes the network behavior during the transmission of one message. Subsection (§4.4) presents how this model can be extended to include all transmitted messages.

Message production is represented by the deterministic transition TMP, whose duration is equal to the message production period. When this transition fires, a token is put in MP to represent a message production. This transition is always enabled to indicate a continuous message production. In this marking one of following scenarios can happen:

- **Place TB is marked**. Since TB represents the transmission buffer, its marking indicates that the previous message (its latest instance) wasn't yet transmitted and therefore the message deadline was missed. In this situation the immediate transition TDM fires, removes the token from place MP and puts a token in place DM. Place DM is used to indicate that the message has missed its deadline. A variable multiplicity arc is used to assure that the number of tokens in DM never exceeds 1. It is assumed that a message production always overwrites the its previous instance in the buffer;
- **Place TB isn't marked**. In this case the previous message was already transmitted. In this situation the immediate transition TTB fires and removes the token from place MP and puts it in place TB, indicating that there is a new message in the buffer ready to be transmitted. By using guard functions the conflict between TDM and TTB transitions is avoided. Notice that after TMP firing, TB or DM places are marked is zero time.

The transmission medium is represented by place TM. When this place is marked the medium is free (idle bus). If there is a message to transmit (TB is marked) two conditions must be fulfilled to initiate its transmission (notice that in this model it is assumed that there is only one message): **(i)** The medium must be free (TM is marked); **(ii)** There aren't interferences (IN isn't marked). The latter condition is only necessary to prevent an eventual conflict with the TID transition in (§4.3) model. When the previous conditions are gathered, the immediate transition TPR fires and removes the token from TM (the medium is being used) and puts a token in MT. This transition has a guard that depends from TB and IN places, and also has a priority. This priority is the same of the message. This behavior will be detailed further (§4.4).

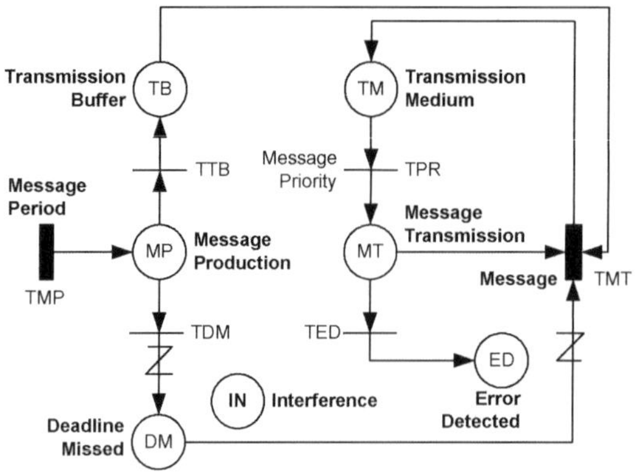

Arc Multiplicity	
TDM→DM	IF (#DM =0) : 1 ELSE 0;
DM→TMT	#DM

Guards	
TDM	#TB = 1
TTB	#TB = 0
TPR	(#TB = 1) AND (#IN=0)
TED	#IN = 1

Fig. 3 - Message transmission model.

When place MT is marked the deterministic transition TMT is enabled (TB and DM also enable this transition). This transition represents the time necessary to transmit the message (Data frame) plus the *intermission time* (3 bits). This last term models the minimum *interframe space* which is necessary to maintain between consecutive frames.

If there weren't faults during the message transmission, then when TMT fires the following sequence happens: **(1)** a token is removed from MT; **(2)** a token is removed from TB, indicating that the message was successfully transmitted; **(3)** If there are any tokens in DM they are removed. This indicates that the message deadline was fulfilled. A variable multiplicity arc is used to implement this aspect; **(4)** Place TM is marked, indicating that the transmission medium is free again.

If faults (interferences) occur during message transmission the place IN is marked (§4.1). In this case it was assumed that the error is immediately detected by the nodes which initiate (immediately) the error recovery mechanisms. This behavior is represented by the firing of the immediate transition TED, which has a guard function that depends from the IN marking. Since immediate transitions have always a higher priority that timed ones, when IN and MT are both marked the transition TED fires immediately. In this case, it removes a token from MT, which disables TMT and interrupts the message transmission, and puts a token in ED initiating the error recovery mechanisms (notice that TB and DM markings are unchanged).

There are some final aspects that should be discussed about this model: **(i)** Since faults can occur anywhere during the message transmission this results in a non pessimistic approach; **(ii)** According to (Bosh, 1991) it is not possible to guarantee that errors (due to faults) will be immediately detected by the nodes. In fact if a fault occurs in certain fields (e.g. *Identifier*) it is possible that the error only be detected a few bits latter. To model accurately this behavior it would be necessary to take into account those fields and the fault characteristics (instant of occurrence and duration). Although this could be implemented, it would introduce to much complexity into the model, which in most cases is unnecessary. Besides, in most of the fields error signaling begins in the next bit and these fields represent, typically, most of the situations; **(iii)** If faults have a very small duration it is possible that in some situations errors aren't detected and signaled (e.g. *Identifier*). However, these are marginal cases which have only a minor contribution to the results.

Therefore, it is possible to conclude that inaccuracies that result from our assumption (immediate detection and signaling) will be very small and can be ignored. Notice that this assumption leads to a slightly pessimistic approach, which guarantees that results aren't overestimated (optimistic).

4.3 Error Signaling Model

The marking of place ED (Fig. 4) indicates that an error was detected and its signaling begins. The behavior in this situation was defined based on the following assumptions:

- The assumption that a fault has duration (e.g. how many Error frames are transmitted?) with the possibility of faults occurring during an Error frame transmission can lead to an unpredictable behavior, which is difficult to foreseen and to model. To cope with this problem the following behavior was assumed: **(i)** During interference period (faults) the bus is considered inaccessible; **(ii)** After the end of the interference an Error frame is transmitted. This behavior is equivalent to observe a sequence of corrupted Error frames followed by a last, and not corrupted, Error frame;
- Network nodes are always in the *Active-Error* state. This assumption is supported by the results presented by (Gaujal and Navet, 2001). The behavior of the TEC and REC counters need only to be included if fault sources have high rates and are continually disturbing the transmissions, which is uncommon. Therefore Error frames are always Active ones;
- The problem of the incertitude related how signaling happens is solved by assuming that Error frames have always the maximum possible length (20 bits). Therefore all possible situations are covered;
- In (§4.2) the TMT transition duration includes the intermission time (3 bits). If a fault occurs during this period an Overload frame is transmitted. Since Overload frames have the same structure of the Active Error frames the model can incorporate both behaviors without the need of any modifications.

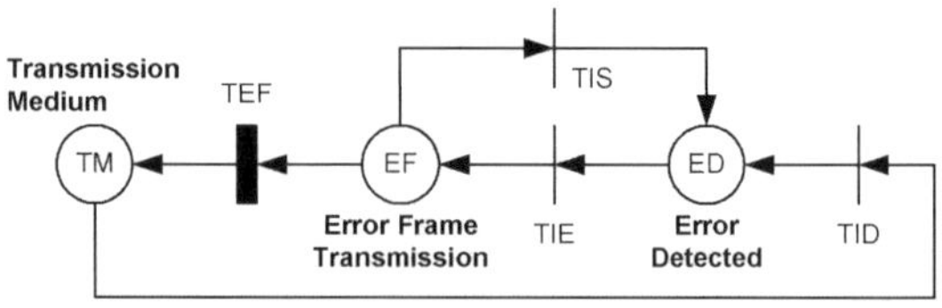

Guards	
TID	#IN = 1
TIE	#IN = 0
TIS	#IN = 1

Fig. 4 - Error signaling model.

When place ED is marked, the immediate transition TIE is used to "wait" for the end of the interference (IN isn't marked). When this transition fires it moves the token from ED to EF. This marking enables the TEF deterministic transition, which represents the transmission of an Active Error frame (or an Overload frame) plus the intermission time. When this transition fires it moves the token from EF to TM, indicating that the transmission medium is free again. After this, a new transmission can initiated according to (§4.2).

While an Error frame is being transmitted (EF is marked) if a fault occurs (IN is marked) the transition TIS fires and moves the token from EF to ED. Consequently, the TEF transition is disabled and the Error frame transmission is interrupted.

If during the bus idle time (TM is marked) a fault occurs (IN is marked), then an Error frame is transmitted. This is due to the fact that the fault will cause an erroneous *start of frame* signal, which will lead to an erroneous (virtual) Data frame. In this case, the immediate transition TID fires removing a token from TM (bus isn't free) to ED. In the last two scenarios, after ED marking the net evolves according to the previous discussion.

4.4 Extension to Several Messages

To extend the model to *n* messages it is only necessary to replicate the message transmission model *n* times. During the replication process places IN, TM and ED should be shared between models. The use of a priority in the TPR transition automatically implements the CAN arbitration mechanism. When there are several messages ready to be transmitted (several TB places are marked is different models) and the medium is free (TM is marked), only the highest-priority TPR transition will fire (while the others wait), representing the transmission of the highest priority message.

4.5 Dependability Measures

Dependability evaluation is performed by defining a set of measures in the model. In the context of SPNs this measures are derived from the concept of *reward* (Malhorta, 1995; Lindemann, 1998; German, 2000; Trivedi, 2002). Two types of rewards are defined: **(i)** *Rates*, associated with markings of the SPN which are collected during the time the SPN resides on the marking; **(ii)** *Impulses*, associated with transitions firings which are collected when the transition fires. From these definitions other important measures can be derived such as: marking probability, expected number of tokens in a place, expected number of firings of a transition, etc. These measures are typically obtained considering two scenarios: transient or steady-state analysis.

The probability of a message to miss its deadline can be obtained by defining a reward rate of 1 in the DM place (§4.2) and by computing the stationary expected instantaneous reward (this is equivalent to the probability of #DM=1). When it is necessary to consider all the messages, this measure it is defined as the sum of all rates. The model can be also used to obtain other types of measures, such as performability ones (Marsan, *et al.*, 1996; Lindemann, 1998; Trivedi, 2002). This flexibility results from the manner how the model was developed.

4.6 Model Solution

As discussed previously SPNs models support both analytical and simulation solutions. In the following subsections this topic is discussed briefly.

4.6.1 Analytical

An analytical solution is possible if certain structural conditions are met (Bobbio, *et al.*, 1998; Lindemann, 1998; German, 2000) (see §3.1). It is clear from the proposed model that TMP transitions (one for each message transmission model) are always concurrently enabled, which hampers any analytical solution. However if approximate methods are employed it is possible to obtain an analytical solution.

Since the complete model (using several messages) has almost deterministic transitions, Discrete Deterministic and Stochastic Petri Nets (DDSPN) (Zijal, *et al.*, 1996) can be employed to obtain an analytical solution. DDSPNs use a discrete time approach (ΔT) where discrete phase-time distributions are used for modeling transitions firing times. In this case deterministic transitions have an exact representation, while exponential ones are approximated by a geometric distribution. The principal drawback of this approach is the uncontrolled increase of the state-space when the time step (ΔT) is reduced. Several experiences (using this model) had been carried out with this formalism. The results obtained show that a practical solution exists only if high ΔT values are used. This results in a poor approximation, since deterministic transitions must have its duration "adjusted" to a ΔT multiple.

Another alternative is to use a continuous-time phase-type expansion. In this case timed transitions (non exponential ones) are approximated by a combination of exponential transitions (phases) (Malhorta and Reibman, 1993; Bobbio, *et al.*, 1998). The quality of this approximation depends of the number of phases used. The principal drawback of this approach is also the problem of state-space explosion when the number of phases increases. Several experiments were performed and the results show that is possible to obtain a practical solution if the number of phases didn't exceed 3. Since this approximation implies a very high variance, dependability measures will suffer from the same problem.

4.6.2 Simulation

The use of simulation removes all the limitations discussed previously and guarantees that model's solution always exist independently from its size or structure. However, the use of simulation for dependability evaluation arise some important problems. For an accurate estimation of dependability measures it is necessary frequent observations of the system-failure event, which by definition are rare events. This results into a substantial increase of the simulation time, which could lead to impractical values. To attack this problem there has been considerable and successful efforts to develop fast simulation techniques (Nicola, *et al.*, 2001). Among these techniques the most important are: *importance sampling* and *variance reduction* (e.g. control and antithetic variables). Their main aim is to reduce the simulation time necessary to obtain the results.

These techniques have been systematically incorporated into several SPNs tools (Haverkort and Niemegeers, 1996; German, 2000), allowing that these can be used successfully for dependability evaluation. Meanwhile, most of these tools provides also a distributed or parallel a simulation environment, which permits a further reduction of the simulation time. Besides, SPNs are also an adequate formalism to capture the behavior of *Discrete Events Systems* (Haas, 2002), which are the basis of simulation environments.

5. CASE STUDY

A case study was chosen to assess the proposed model.

5.1 Message Set

In previous works two alternatives have been used for the message set: the Peugeot-Citröen set (Navet, *et al.*, 2000) and the SAE Benchmark (Tindell, *et al.*, 1995). Since the former fits better in our the model assumptions it was the chosen one. However, since (Broster, *et al.*, 2002) shows that the probability of any message misses its deadline in insignificant, it was necessary to reduced their deadlines. It was chosen a 3.5 reduction to impose a high bus utilization (75,45%), which permits to evaluate the CAN behavior is an high-demand scenario. The message set is presented in Table 1 for a data-rate of 250Kbit/s. All values (except Priority) are expressed in bit units.

Table 1. Message set

Priority	Length DLC	Period	WCRT
1	64	714	257
2	24	1000	342
3	24	1428	427
4	16	1071	502
5	40	1428	607
6	40	2857	712
7	32	1071	942
8	40	3571	1047
9	32	1428	1397
10	56	7142	1502
11	40	3571	1987
12	8	7141	1990

5.2 Fault Parameters

The parameters for the fault model were chosen to represent real environments. Three levels of fault rates were evaluated: 50, 150 and 500 faults/s. This choice was only motivated by the fact that these values represent 50Hz (line frequency) multiples and therefore are closer to the typical EMI in industrial environments. Although fault rates are somewhat high, the intention is to represent burst situations. Fault duration was chosen to be: 5, 15 and 50 bits (mean values). Here, the intention is to assess the influence of different durations. The intervals with and without faults (§4.1) are defined as 1 min. and 5 min. respectively. Due to lack of data arbitrary values were chosen. Only by performing an environmental analysis it is possible to obtain reliable values.

5.3 Model Performance

From the previous discussion (§4.6) it becomes clear that due to model characteristics its solution (dependability measures) can only be obtained with efficiency and reliance if simulation is used.

The use of simulation poses some problems in the manner how the solution is obtained. First, simulation produces results that are just an approximation of the real ones. A confidence interval is used to characterize the accuracy of the results. Second, in many situations the results are obtained through the use of exigent computational resources (CPU time and memory). While the former problem is inherent to the simulation process, the latter can be minimized by adopting the following measures: **(i)** Developing the model in a way that maximizes its execution performance; **(ii)** Choosing an adequate modeling tool.

The execution of the model during the interval without faults degrades simulation time, since during those intervals no data is gathered (assuming that without faults message deadlines are always fulfilled). To overcome this problem a slight modification of the fault occurrence model is necessary (§4.1). The probability of a deadline to be missed can be defined as:

$$P\{\text{Deadline Missed} \mid \text{Faults Occur}\} \times P\{\text{Faults Occur}\} \qquad (1)$$

where $P\{\text{Faults Occur}\}$ can be obtained as the probability of place BF to be marked. Since BF, WF, TBF and TWF (§4.1) represent a two-level Markov chain, this probability can be easily obtained as:

$$P\{\text{Faults Occour}\} = \frac{E[\text{Interval with faults}]}{E[\text{Interval without faults}] + E[\text{Interval with faults}]} \qquad (2)$$

Therefore, if the previous places and transitions are removed from the model, the first term of expression (1) can be obtained directly from a simulation where only transitions TFO, TFD and place IN are present. In this case faults are always present when the model is simulated and the simulation time is strongly reduced. Note that the results of (1) are only valid for a steady-state analysis. All data presented in the following subsection was obtained with this modification.

The experiments were performed using the TimeNET tool (Zimmermann, *et al.*, 1999). This tool provides state-of-the-art analytical and simulation solutions, including a distributed simulation environment (not used in the experiments).

The model presents very good performance from a simulation viewpoint. As an example, the computation time necessary to obtain the results presented in this section are in 40s to 120min. range (PIII@730MHz), which are very reasonable values for a simulation.

5.4 Dependability Evaluation

The network behavior was investigated as function of the fault rate and fault duration. Due to lack of space only the results for the message of priority 9 are presented (Fig. 5). The results were obtained from a steady-state simulation using a confidence interval of 95% with a relative error of 10% (half width).

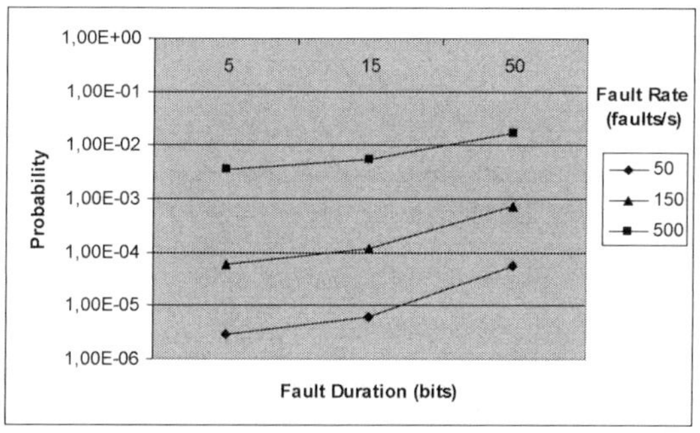

Fig. 5 - Probability of missing a deadline.

From a global perspective it is possible to conclude that the probability of missing a deadline in a typical industrial environment (50 faults/s) is very small. This conclusion is reinforced by two facts: **(i)** The message set used has high bus utilization, imposing therefore a high load **(ii)** The chosen message has the smallest *slack time* of all messages, and therefore is the closest to miss its deadline. Both aspects contribute to increase the probability of missing a deadline. Therefore the results should be interpreted as a conservative scenario.

As expected, the fault rate has a major influence in the results. It is possible to observe that for an increase from 50 to 150 the probability increases ≈10 times, while an increase from 50 to 500 the probability increases ≈1000 times. This behavior is maintained even if the fault duration changes. It is interesting to notice that is almost possible to establish a mathematical relationship between failure probability and fault rate.

Fault duration presents two distinct behaviors. For small values (5, 15 bits) the impact in the results is very small, while for higher values (50 bits) the impact cannot be ignored. This behavior can be explained as following. When a fault occurs during a frame transmission and has a long duration, their effects will cause a considerably delay in the next transmission. A similar behavior will happen when a fault occurs during the bus idle time. If a fault has a long duration the probability of delaying a future transmission will increase, while if it has a short duration it is quite probable that its effects ended before the next transmission.

6. CONCLUSIONS

A model was proposed to evaluate CAN dependability in scenarios of transient faults that occur during communications. The model represents accurately the fault occurrence and makes use of less pessimistic assumptions by considering a closer representation of the real network behavior. The combination of all these aspects will provide more realistic results. Although the results are obtained by means of simulation, the model presents a good performance which makes it useful to evaluate the CAN behavior in typical fault scenarios. A case study was performed to evaluate CAN dependability and to assess qualitatively and quantitatively the influence of fault parameters. The results show that the fault rate has a major influence and the fault duration is only relevant if it assumes high values.

REFERENCES

Barrenscheen, J. and G. Otte (1997), Analysis of the Physical CAN Bus Layer, *Proceedings of the 4th CAN Conference*.

Billington, J., M. Diaz and G. Rozenberg (Eds.) (1999), *Application of Petri Nest to Communication Networks, Lecture Notes in Computer Science Vol. 1605*, Springer.

Bobbio, A., A. Puliafito, M. Telek and K.Trivedi (1998), Recent Developments in Non-Markovian Stochastic Petri Nets, *Journal of Systems Circuits and Computers*, **Vol. 8**, No. 1, pp. 119-158.

Bosh, R. (1991), *CAN Specification – Version 2.0*, Robert Bosch GmbH.

Broster, I., A. Burns and G. Navas (2002), Probabilistic Analysis of CAN with Faults, *Proceedings of 23rd Real-Time System Symposium*.

Broster, I., A. Burns and G. Navas (2004), Comparing Real-Time Communication under Electromagnetic Interference, *Proceedings 16th Euromicro Conference on Real-Time Systems*.

Charzinski, J. (1994), Performance of the Error Detection Mechanisms in CAN, *Proceedings of the 1st International CAN Conference*.

Ferreira, J., A. Oliveira, P. Fonseca and J. Fonseca (2004), An Experiment to Assess Bit Error Rate in CAN, *Proceedings of the 3rd International Workshop on Real-Time Networks*.

Gaujal, B. and N. Navet (2001), Fault Confinement Mechanisms on CAN: Analysis and Improvements, *Proceedings of 4th IFAC Conference on Fieldbus Systems and their Applications*.

German, R. (2000), *Performance Analysis of Communication Systems – Modeling with Non-Markovian Stochastic Petri Nets*, Wiley.

Haas, P. (2002), *Stochastic Petri Nets: Modelling, Stability, Simulation*, Springer-Verlag.

Hansson, H., T. Nolte, C. Norström and S. Punnekkat (2002), Integrating Reliability and Timming Analysis of CAN-Based Systems, *IEEE Transactions on Industrial Electronics*, **Vol. 49**, No. 6, pp. 1240-1250.

Haverkort, B. and I. Niemegeers (1996), Performability Modelling Tools and Techniques, *Performance Evaluation*, **Vol. 25**, pp. 17-40.

Kim, H. and K. Shin (1994), On the Maximum Feedback Delay in a Linear/Nonlinear Control System with Input Disturbances Caused by Controller-Computer Failures, *IEEE Transactions on Control Systems Technology*, **Vol. 2**, No. 2, pp. 110-122.

Kim, H., A. White and K. Shin (2000), Effects of Electromagnetic Interference on Controller-Computer Upsets and System Stability, *IEEE Transactions on Control Systems Technology*, **Vol. 8**, pp. 351-357.

Lindemann, C. (1998), *Performance Modelling with Deterministic and Stochastic Petri Nets*, Wiley.

Malhorta, M. and A. Reibman (1993), Selecting and Implementing Phase Approximations for Semi-Markov Models, *Stochastic Models*, **Vol. 9**, No. 4, pp. 473-506.

Malhotra, M. and K. Trivedi (1995), Dependability Modeling Using Petri-Nets, *IEEE Transactions on Reliability*, **Vol. 44**, No. 3, pp. 428-440.

Marsan, M. A., A. Bobbio and S. Donattelli (1996), Petri Nets in Performance Analysis: An Introduction, *Advanced Course in Petri Nets*, Gajstuhl, Germany.

Murata, T. (1989), Petri Nets: Properties, Analysis and Applications, *Proceedings of the IEEE*, **Vol. 77**, No. 4, pp. 541-580.

Navet, N., Y. Song and F. Simonot-Lion (2000), Worst-Case Deadline Failure Probability in Real-Time Applications Distributed over Controller Area Network, *Journal of Systems Architecture*, **Vol. 46**, No. 1, pp. 607-617.

Navet, N., Y. Song, F. Simonot-Lion and C. Wilwert (2005), Trends in Automotive Communication Systems, *Proceedings of IEEE*, **Vol. 93**, No. 6, pp. 1024-1223.

Nicola, V., P. Shahabuddin and M. Nakayama (2001), Techniques for Fast Simulation of Models of Highly Dependable Systems, *IEEE Transactions on Reliability*, **Vol. 50**, No. 3, pp. 246-264.

Pinho, L., F. Vasques and E. Tovar (2000), Integrating Inaccessibility in Response Time Analysis of CAN Networks, *Proceedings of 3rd IEEE Workshop on Factory Communication Systems*.

Punnekkat, S., H. Hansson and C. Norström (2000), Response Time Analysis under Errors for CAN, *Proceedings of IEEE Real-Time Technology and Applications Symposium*.

Rufino, J. and P. Verissímo (1995), A Study on the Inaccessibility Characteristics of the Controller Area Network, *Proceedings of the 2nd International CAN Conference*.

Rufino, J., P. Verissímo, G. Arroz, C. Almeida and L. Rodrigues (1998), Fault-Tolerant Broadcasts in CAN, *Proceedings of 28th International Symposium on Fault-Tolerant Computing*.

Rufino, J., P. Veríssimo and G. Arroz (1999), A Columbus' Egg Idea for CAN Media Redundancy, *Proceedings 29th International Symposium on Fault-Tolerant Computing*.

Shin, K. and H. Kim (1992), Derivation and Application of Hard Dead-lines for Real-Time Control Systems, *IEEE Transactions on Systems, Man and Cybernetics*, **Vol. 22**, No. 6, pp. 1403-1413.

Tindell, K., A. Burns and A. Wellings (1995), Calculating Controller Area Network (CAN) Message Response Times, *Control Engineering Practice*, **Vol. 3**, No. 8, pp. 1163-1169.

Thomesse, J.-P. (2005), Fieldbus Technology in Industrial Automation, *Proceedings of IEEE*, **Vol. 93**, No. 6, pp. 1073-1101.

Tran, E. (1999), Multi-Bit Error Vulnerabilities in Controller Area Network Protocol, *Technical Report*, Carnegie Mellon University, 1999.

Trivedi, K. (2002), *Probability and Statistics with Reliability, Queuing and Computer Science Applications – 2nd Edition*, Wiley.

Zijal, R., G. Ciardo and G. Hommel (1996), Discrete Deterministic and Stochastic Petri Nets, *ICASE Technical Report 96-72*, NASA Langley Research Center.

Zimmermann, A., R. German and J. Freiheit, G. Hommel (1999), Time-NET 3.0 Tool Description, *Proceedings of International Conference on Petri Nets and Performance Models*.

ELSEVIER
IFAC
PUBLICATIONS

SECURITY CONSIDERATIONS FOR ENERGY AUTOMATION NETWORKS

Albert Treytl[1,2], Peter Palensky[1], Thilo Sauter[1,2]

[1] *Vienna University of Technology Institute of Computer Technology*
Gußhausstraße 27-29/E384, A-1040 Vienna, Austria
{treytl, palensky}@ict.tuwien.ac.at
[2] *Austrian Academy of Sciences Research Unit for Integrated Sensor Systems*
Viktor Kaplan Strasse 2, A-2700 Wiener Neustadt, Austria
thilo.sauter@oeaw.ac.at

Abstract: Modern society depends on a reliable energy distribution network. Recent incidents such as the infiltration of a U.S. nuclear power plant together with the impacts of big power outages call for security measures to guarantee supply with energy. This article deals with security goals, attacks, and protection mechanisms for energy automation systems. Nevertheless many of the discussed issues and solutions also apply to other large scale automation systems. *Copyright © 2005 IFAC*

Keywords: security, power distribution and automation, security threats, access control, security requirements.

1. INTRODUCTION

Power outages during the last years clearly show that modern society depends on a reliable electric energy distribution network. Although most of these outage disasters stem from natural sources such as ice storms or dropped-out network components such systems are also vulnerable to malicious intentional attacks.

Many research activities deal with the problem of reliably distributing energy, preventing undervoltage and overvoltage as well as failure of components. Solutions to the manifold threats result in automatic generation control (AGC), energy management systems (EMS), or special protection and remedial action systems (SPS/RAS) (Tomsovic, 2005). All these systems increasingly rely on communication networks that allow collecting data and sending appropriate commands. In general there exist two parallel trends in energy automation (Bertsch, 2005):

1. centralization by moving network control to regional or even nationwide control centers,
2. decentralization by deploying "intelligent" components and "decision authority" locally.

Both trends demand for communication networks to allow for transport of the necessary information. At the moment there is still little awareness of security in these communication networks – i.e., measures that ensure a state of inviolability from hostile acts or influences – although risks are known, e.g., (U.S. Nuclear Regulatory Commission, 2003). At the moment many systems rely on the principle of "security by obscurity" by simply keeping information about the network undisclosed.

The integration of energy automation networks into public networks such as the Internet (e.g. tunneling of SCADA (Supervisory, Control and Data Acquisition) messages) and the increasingly distributed nature of network equipment increases the importance of security. The authors are currently managing a project in this area called REMPLI (Remote Energy Management over Power Lines and Internet[1]), which connects automatic meters at the customers' premises via the medium and low voltage power lines with regional control centers (Sauter, et. al., 2005). A special issue of this project is the integration of a complete security concept covering the complete system.

This article will analyzes the situation of security in energy automation systems ranging from security on

---

[1] The REMPLI project is supported by the European Commission NNE5-2001-00825; (www.rempli.org)

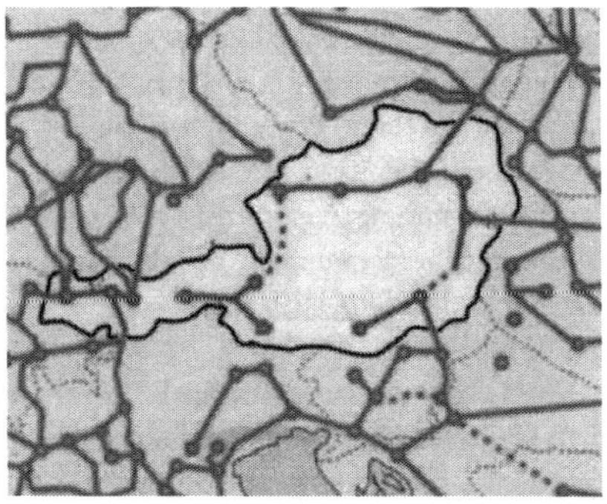

Fig. 1 The Austrian high voltage 380kV grid, (Fischer-Drapela, 2003)

low-level automation networks to the security of energy trading networks. Although the focus is on energy automation most issues also apply to classical (vertically integrated) automation systems. Section 2 describes threats and risks and the resulting security goals. Section 3 shows common attacks. Finally section 4 indicates possible solution to the mentioned problems.

2. SECURITY GOALS

Electric power systems rely on a highly distributed and complex infrastructure. This includes the distribution lines, power plants, protection systems, SCADA systems, but also financial mechanisms like trading agreements, schedules or balance groups (Werner, 2002). Traditionally, all these systems were isolated, with no or no common security concept behind. Recent developments like the liberalization of the energy market, increased competition, and the need for cutting costs lead to two trends that both increase the need for sophisticated security measures.

First, in order to stay competitive, infrastructural investments were minimized. Unlike in former times the European power grid is no longer an over-engineered "copper plate". The utility companies and transmission and distribution network providers are operating their equipment on the edge. Situations that were unthinkable before are now normal, like exceeding the thermal load limits of transmission lines multiple times per year (Fischer-Drapela, 2003). Such a system is weak and fault-prone. The increase of formerly unknown blackouts shows this very clearly.

Second, IT is changing the energy business in every aspect. Electronic bills, remote administration of equipment, automated meter reading and other IT disciplines are used for faster and more efficient data processing. The link of all these IT systems to globally available communication infrastructure like telephone networks or the Internet lead to a new situation in which intruders do not need to be physically present: they can attack remotely. These two ingredients ultimately yield a weaker system with more points of attack than before.

Each subsystem of the energy business might take measures for securing their processes, but the "big picture" is sometimes not considered. Obviously non-critical events can be composed to a catastrophe with the right coordination applied: Let us assume the following – intentionally incomplete – example: A large city – the subject of our attack - is supplied via two non-redundant lines, each with its own transformer station. The IT equipment of the local utility is infiltrated by a sleeping computer virus that can be activated remotely. Via this virus, it is possible to get information from the utility's file servers such as telephone numbers of insecure AMR and SCADA equipment, trading partners and load plans, passwords, etc. Using this information, it is possible to estimate when the system is on its limits. Combined with faked trading and scheduling requests, attacks to the SCADA equipment and a collapsing IT department (computers, telephone, etc.), the utility would neither be able to avoid a provoked overload nor to react on it in the appropriate manner. Automatic, and non-networked, protection mechanisms would deterministically switch off parts of the distribution network. If the attacker would plan and coordinate this very carefully, this could lead to an arbitrary chain reaction, as it sometimes happens coincidentally.

This example naturally requires the application of multiple types and steps of attacks, but is not unrealistic. The U.S.-Canada power system outage in August 2003, although not initiated by an active attack, is a vivid example for the consequences of multiple failures inside the power grid. Sometimes there is even no need to find out weak points of the energy grid via hacking the servers of a utility. Instead it is possible to find out potential points of attack via journals (Brauner, 2004). Fig. 1 shows for instance the well-known weak points of the Austrian high voltage grid. Due to missing major 380kV lines between the power generation and main backbones in the North and South, the 220kV connections of these two networks (not shown in the figure) could easily overload and cause multiple consecutive failures up to a breakdown of supply. The dotted lines indicate 380kV connections which are in planning to securely connect the two main networks.

It is important to note that the entire attack affecting widely distributed components is done by means of information technology. There is no need to recruit and coordinate a large amount of persons to cause all these disturbances that finally lead to the collapse.

To protect the automation systems and IT infrastructure of (energy) automation networks in general the following security goals can be identified:

- Confidentiality (privacy or secrecy) prevent unauthorized disclosure or traffic flow from analysis by unauthorized entities
- Integrity: no unauthorized entity (including accidental alterations) must be able to change data without the change being detected
- Availability: data is at hand when needed
- Authentication: origin of data is proofed

Table 1 Threats and risks to powerline based energy automation system REMPLI

Thread/Risk	Metering	SCADA
direct manipulations of input and output values	high	high
manipulation or replacement of equipment	medium	medium
manipulation and insertion of data into the Private Network	medium/ high	medium/ high
manipulation and insertion of data into the PLC network	low/ medium	low/ medium
denial of service	high	high

- Authorization and access control: determines what an entity allowed to do once you are authenticated and allowed access
- Non-repudiation: allows to legally prove that a certain event or action was done by a certain entity.

Following the first three most important goals often the abbreviation CIA is used.

It is important that all threats to a system security can use the energy network as well as the superposed communication network. Although there are a lot of publications dealing with the energy aspect, there is only little awareness for the security of the communication system. Table 1 exemplarily lists the threats and risks to the above mentioned security goals for the REMPLI project that uses a medium and low voltage power line based communication system as well as a private IP-based network to access electricity meters at the customers' premises on a broad scale. The risks analysis of the REMPLI project (Treytl and Sauter, 2005) indicates the following important issues:

1. the most relevant goals for today's communication networks are integrity and authorization preventing the active misuse of the infrastructure
2. Attackers will choose the easiest point. E.g., in REMPLI an attacker would rather disconnect the meter from the communication system than hacking the communication system.

Confidentiality and non-repudiation are not important goals since the systems are usually only used by one company, but will gain importance in the future when communication infrastructure are shared between multiple companies. Similar research projects such as (Selma consortium, 2005) obtain similar results.

Even if attacks by means of information technology are not yet as developed as they could be, they will certainly gain more and more relevance, since the costs of such attacks are very low compared to traditional physical attacks like blowing up selected transformer stations in a coordinated way. In general there is an increasing need to protect systems actively. The still very common attitude of security by obscurity – i.e., keeping essential information about a system confidential – is no longer an appropriate measure against serious attacks.

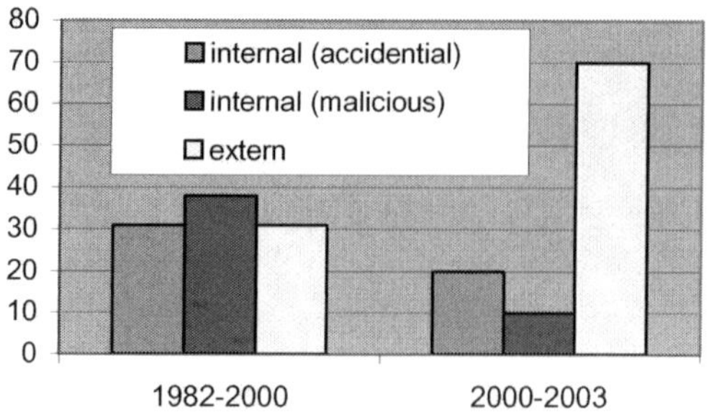

Fig. 2 Attack sources for automation systems [% of incidents] (British Columbia Institute of Technology, 2005)

3. ATTACKS

Attacks to automation systems can be classified in various ways. One classification uses the origin of attacks which can be divided into internal and external sources as well as intentional and unintentional attacks. Another possibility of classification is the kind of harm done to the system. In this rating, attacks are in general associated with one of the following classes:

1. eavesdropping of data,
2. modification of data,
3. fabrication of data, and
4. interruption of communication

Finally, attacks could also be classified by the technology used such as Viruses, Worms, Trojans etc. or by the impact such as in (Tomsovic, et. al. 2005), who classifies by attacks upon the power system, attacks by the power system or attacks through the power system. For this article the focus is on the first two classes since these show technology independently the threats to automation systems.

3.1 Source of Attack

For regular office communication systems there are various sources available concerning security, e.g., CERT Coordination Center (www.cert.org); generally information about recent attacks becomes public quite fast. For automation systems the situation is different and information is hardly made public. The industrial security incident (ISI) database maintained by the British Columbia institute of technology (British Columbia Institute of Technology, 2005) and the reports of the U.S. nuclear regulatory commission are two of the rare yet restricted sources for information.

Owing to the fact that automation systems have been located in restricted areas and that knowledge about the systems was not made public, attacks from insiders have been a major source of threat. In the last years this situation changed and attacks from the outside are becoming more frequent.

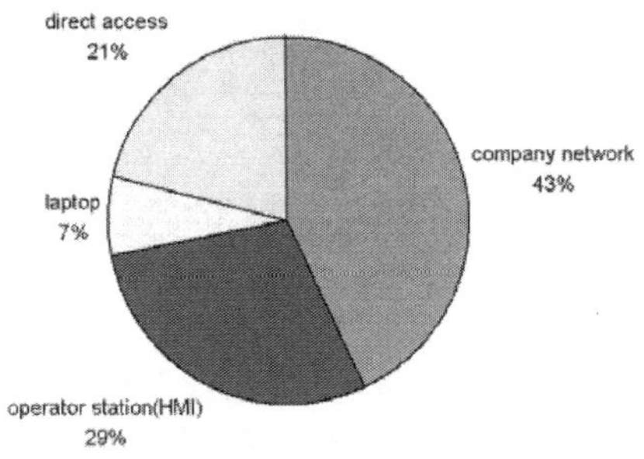

Fig. 3 Infiltration of automation networks listed by intrusion path

Two trends caused this increase of external attacks: first, automation systems became interconnected with other automation networks (horizontal integration) as well as with management and administration networks (vertical integration) – automation networks are no longer island solutions. Second, standardized components are replacing proprietary solutions. Typical examples are operator stations which are based on common operating systems such as MS Windows or Linux.

Both facts allow for electronic remote – no physical presence in a restricted area is necessary – and automated attacks such as worms or viruses. Nevertheless it must be clearly stated that standardized components are not less secure than proprietary ones. In general, standardized and wide spread components undergo a better security audit, but once a weakness is discovered it can be used more easily since it can be applied not only to one system but to multiple ones. As an example the misuse of the Maroochy Shire sewage system [14] – the attacker flooded parks and a river in an Australian town with sewage – was based on an exploit of a weakness of the WEP encryption algorithm of the IEEE 802.11b wireless LAN used in the system. For the intrusion a tool available from the Internet was used.

Another important issue is whether an attack was done intentionally or accidentally. Especially in view of remote maintenance this is a critical issue since the majority of systems, once access is granted, do not apply any further security measures such as restricting access to units that an engineer is responsible for. Fig. 2 shows clearly that the number of accidental attacks is already beyond the number of malicious insider attacks. Since the estimated number of not reported incidents is around 90% (for all kind of attacks) the share of accidental wrong operation will most likely be much higher.

3.2 Attacks on Automation Level

Attacks on the automation level are manifold and it would go far beyond the scope of this article to list them all. Beside attacks which are tailored to a particular system – the range begins at knowing the telephone numbers of SCADA stations and ends at complex tasks like protocol re-engineering - a special focus should be set to automated attacks and exploitation of standardized protocols.

Automated attacks, which at the moment are not directly aiming at automation systems, use means of Trojan horses or Internet worms to intrude utility equipment. The Internet worm Slammer (2003) or Code Red (2001) are typical representatives of such automated attacks. It infiltrated control systems of the U.S. nuclear power plant Davis Besse and other industrial automation systems. The typical points of infiltration were Internet connections (36%), dial-up IP connections (12%), wireless systems (8%), plain telecom networks (8%), trusted connections (4%) or SCADA networks (4%) (Byres and Lowe 2005). Fig. 3 shows the paths of intrusion. Interesting are the high rates of intrusion via the company network but also the direct infiltration via operator control stations. The conclusions that can be draw are that companies as well as automation networks are no longer isolated networks and no special knowledge is necessary to initiate such attacks. With the availability of tools to create viruses and worms by click and drop and irrespective whether the malicious code is introduced by an unprotected laptop or directly via the Internet, the network, or relevant network segments must be protected.

Another sensitive area is the denial of service (DOS) attack which cuts off control stations from the network. In general three kinds of DOS attacks can be distinguished:

1. DOS by overloading the device
2. DOS by overloading the connecting network
3. DOS by congesting a "parallel" network

Whereas the first type of attack blocks a device by requesting the intended service of the machine too frequently, the other types block the network by congesting or interrupting the network. Especially type three is hard to account for: A particular incident of this type has been the loss of monitoring capability due to the side effects of an overload cause by the Slammer worm: although the utility company had a frame relay connection with guaranteed bandwidth, one segment which was tunneled over an ATM line broke down by the unlimited increase of a parallel Internet connection. The same might also happen to normal GSM or POTS systems when in emergency situation or during extraordinary traffic situations no free entrance points of the telephone system are available.

Additional to the total denial of service discussed before, heavily congested networks also introduce additional delays that might severely affect SCADA operations which sometimes require soft-real-time behavior. Concerning these delays the experience gathered by the authors in the REMPLI project showed that timing requirements are seldom clearly specified for remote access in energy management applications. Often the capacity of a 9600 bit per second modem line is demanded although – seen from the application level – much higher delays would be acceptable. This fact is quite noticeable if security devices need to be integrated in low-cost devices like meters.

3.3 Attacks on Energy Management Level

Beside the on-line connectivity of distribution equipment, transformer stations and energy meters there are plenty of other processes in the energy business that more and more rely on global communication channels with questionable IT security.

Some utilities still use plain text e-mails with spreadsheet documents to exchange energy business data such as load charts or load estimations. Such e-mail based, non-secured communication can easily be intercepted, manipulated, or faked.

Business processes such as exchanging roadmaps will in future be more and more based on standards in order to achieve a higher level of interoperability and efficiency. The natural level of obscurity that today's proprietary intermediate solutions offer will then be lost. A top-level candidate for exchanging business data, especially for the energy business, is ebXML (van der Togt, 2003). ebXML uses XML as a means of transport and a means of interoperability for electronic business (Patil et al. 2003). It is clear, that there are entire teams working on making such an important business tool more secure.

(OASIS security team, 2001) names the Security Assertion Markup Language (SAML), XML encryption, WebTrust principles, XML Key Management Specification (XKMS) and various public key infrastructures (PKI) as necessary countermeasures to potential security risks of electronic business with XML-based data.

Business applications that act as an Internet service platform like the JEVis system (Palensky, 2005) face an additional problem. Applications are typically hosted on an external server providing all databases, applications, and connectivity. When multiple users host their proprietary data on this server, they want to be assured that the other users – possible competitors – get no access whatsoever to their data, under no circumstance. Therefore, beside the proper user management, access rights and database design, the usage of virtual private databases within one database, as Oracle9i offers it, or even more sophisticated measures are necessary (Dwivedi et al., 2005) to achieve the need of privacy and protection of business critical data.

4. SYSTEM PROTECTION

This section will deal with the protection of (energy) automation systems. Since security measures must be adapted to each particular system only general recommendations for the planning of security and selected problems existing in many energy automation systems will be discussed.

4.1 Security Architecture

The most important first step to introduce security is to define the security architecture and policy to allow effective security management. Security should be introduced top-down to prevent flaws due to unconsidered threats.

A security policy is a formal statement of rules through which people are given access to an organization's assets (information as well as hardware). It defines business and security goals and contains a description of the implemented security measures. The security policy is an organization's approach to risk. It is important to note that only 20 percents of security are technological aspects like username/password or cryptography. Hence most (80%) aspects of the security policy will be dealing with procedural, organizational, and cultural aspects of the system. These areas can be characterized by the "4 Ps" of security – people, policy, processes (description of the system), and procedures.

In the further steps of setting up the security architecture the requirements of the security policy will be implemented in appropriate security measures: These measures can range from explicit physical protection of equipment, to protection of information by cryptographic means, as well as the enforcement of simple procedures such as "only employees might enter the office".

Security is always a compromise between the costs caused by an attack and the costs of the countermeasures. Hence, measures have to be selected individually for each use case.

It is obvious that it will not be feasible to physically protect all components of an energy distribution center. Rather only key infrastructure will be protected in order to obtain the needed security level. The omnipresent dilemma of security is that it should make a system more secure without losing productivity. Being only economically feasible, i.e. providing reasonable protection is not enough. In order to prevent the security architecture from being circumvented by the users it must also be understandable, consistent, and most important it should not interfere with normal operation.

Special issues for security in automation networks and in energy automation networks in particular are maintenance and the long life time of components. For maintenance the wide spatial distribution and the human-less remote control are limiting constraints which, in general, require adaptation of common IT security measures that often include human participants. Concerning the life time (up to 30 years) technology migration, capital expenses and limited life time of cryptographic algorithms are boundary conditions that are not known to classical IT systems. An additional (open) challenge is to establish integrated security for all participants of a liberalized energy market.

The next subsections address some particularly important aspects in energy automation systems and list approaches for security.

4.2 Securing remote access

A common measure to secure remote access is still the use of username and passwords. Nevertheless this technology has its limitations. Besides improper transmission via plain text, passwords introduce heavy requirements on distribution and memory of the user if used for large systems. From the security point of view unique passwords with big length and a long character set are desirable. Yet such passwords cannot be remembered by the user and often will be replaced by simple mnemonics such as names or number passwords which are more vulnerable to lexical attacks.

To solve the problem of secure transmission in remote access virtual private networks (VPN), http authentication, SSL (Secure Socket Layer), and TLS (Transport Layer Security) are common measures that use cryptographic operations to secure communication and to authorize users (Sauter and Schwaiger, 2002). Also access portals such as the Multi-tier architecture or the virtual private infrastructure (VPI) (Sikora and Brügger, 2005) allow handling administration in an efficient and secure way.

A still pending issue for most remote access systems is that once access is granted no further access restrictions exist. In particular during maintenance of bigger units a defense in depth that allows only access to a certain subgroup of components of the unit will increase the resistance against accidental failures as well as intentional attacks and therefore increase the robustness of the system.

4.3 Security in automation networks

Typical protocols for energy automation do not implement any security measures. The most used standards such as IEC 60870, IEC 62056 (also known as IEC 1107), or M-BUS offer no security at all. Also industrial fieldbus systems have no serious security measures built-in (Treytl, et. al., 2004) – they are mostly limited to simple UNIX-like access control and plaintext passwords. For the building automation networks BACnet and LonWorks the situation is a little bit better due to the usage of cryptographic measures, but also these systems have their vulnerabilities (Schwaiger and Treytl, 2003).

Ethernet-based solutions on the other hand base their security on network address and port numbers. Switched networks offer additional security against eavesdropping since traffic is separated. Nonetheless, with regard to a planned attack these measures are far too week, since it must be assumed that an attacker will have the computational resources common in the IT world and not the ones of limited embedded systems.

Today, calling line identification and plain text passwords are common measures for field level automation networks but also for remote administration and monitoring. Only in the case of IP-based networks and connection over the Internet (direct telephone connections still use the above mentioned simple measures) advanced measures such as SSL/TLS are used.

A general approach to secure existing networks is tunneling. Comparable to secure web browsing, where the unprotected http protocol is encapsulated in SSL/TLS packets and therefore secured, tunneling can also be applied to automation networks. A general approach is the usage of specialized security modules as suggested in (Naedele, 2004, Palensky, et. al., 2000), which allows to efficiently handle security functions that usually consume more computational power than is available in resource-limited devices such as meters or SCADA actuators. Such an approach was selected for the PROFInet security concept.

Another approach, yet rarely applied, is to directly integrate the security measures in the communication protocol. One of the few examples for this approach is the integration of authentication and encryption using DES (data encryption standard) in the M-Bus protocol (Ziegler, 2001) but this is still a working document for standardization. On higher levels where the Internet Protocol is the base for communication secure derivates of the standard protocols are used. This is commonly done only for more powerful services such as embedded web servers or web services.

The complete integration of security for all parts of a system is almost only applied in (industrial) research systems such as SELMA or REMPLI. The SELMA system (SELMA consortium, 2005) introduced elliptic curve cryptography (ECC) to create a chain of trust between metering and billing. REMPLI (REMPLI consortium, 2004) on the other hand set up a secure communication between the metering and control equipment at the premises of the customer and the background application such as SCADA, metering or billing. REMPLI facilitates symmetric cryptography to achieve this. For SELMA even a dedicated meter was designed to guarantee non-repudiation. The REMPLI system tries to be more open by using standardized metering protocols to connect meters to the security token. This decision was taken with respect to retain investments (e.g., meters, SCADA equipment) and to ease legal cost-intensive issues (certification). The lack of security on the unprotected connection to the meter or control device is reduced by plausibility checks on the input data.

4.4 Intrusion Detection and Access Control

The advantage of automation networks for intrusion detection systems lies in the static nature of the network, well defined communication and the limited group of users and devices. Based on these (semi-) static patterns it is easy to identify malicious activities. Systems can use well-known strategies such as network-based or host-based intrusion detection systems, but already simple plausibility checks increase the overall security, e.g., metering values that result in a negative increment are suspicious, the same happens if the input in a low-voltage segment differs from the sum of consumed energy. In the REMPLI system such measures are used to detect manipulation on the connected meters since the used M-Bus and IEC 62056 protocols do not offer any security measures.

Another advantage of (energy) automation networks is that they are monitored by control centers that are manned 24 hours a day and allow for fast reaction in case an intrusion is detected. On the other hand the rising personnel costs also result in a reduction of remote staff at remote sites make equipment installed in the field more vulnerable to physical attacks. In order not to jeopardize the overall security of the system it is important to store security-relevant information in special tamper-proof devices. Since it is not feasible to physically protect all components in the field, a layered approach should be selected: A physical housing that is a first barrier against vandalism and simple attacks; a second layer that prevents the remainder of security-relevant information in the memory of the device if the device is powered off; finally, a reliable security token such as a smart card known from banking or mobile communication applications that can retain the secret information even if the entire device is stolen.

As the deregulated market forces distribution network providers to deliver energy from various producers to the connected consumers, metering and information transport also become a multi-user problem. Previously, all infrastructure – energy as well as information technology – belonged to one single company. Now, it is more and more common that infrastructure must be shared: a further challenge to IT security, which has to guarantee fairness and confidentiality of transmitted data.

4.5 Overhead of Security Measures

Security measures usually introduce an overhead compared to insecure systems; this is due to additional administration (e.g. distribution of access credentials), additional equipment (e.g. security modules), or additional transmission bandwidth (e.g. due to appended message authentication codes).
In the following three specific overheads for the REMPLI system are analyzed. Nevertheless it must be said that REMPLI is a yet industrial oriented but still a research project and tries to cover a broad range of use cases.

The additional overhead concerning equipment is the installation of a smart card as security token that performs the necessary security operations and serves as a secure storage for the keys distributed over the system. This will introduce a cost overhead of approximately 3 Euros per REMPLI boards. These costs are based on official prices for production quantities of 10000 units.

Overheads caused by management of keys cannot be quantified seriously, because this depends on the integration in the particular SCADA or metering application which is not a direct goal of REMPLI. Key management for daily operations is done automatically by the system components. Only the management keys and initial distribution must be done by a security center. In (Sauter, et. al., 2005) the details on key management are described.

Another important issue is the overhead to communication that is caused by the application of cryptographic operations. Table 2 shows the overhead of different cryptographic algorithms used in the REMPLI system. Compared to other systems the power line communication system implies very strict requirements concerning payload size and troughput. Especially for small size single metering values the overhead is of the magnitude of the payload or well beyond.

Table 2 Overhead of Security Services (Treytl, et. al., 2004)

Data	**Size [byte]**					
Data	10	5k	50k	10	5k	50k
PLC packet	32	32	32	64	64	64
PLC payload	20	20	20	51	51	51
Service	**Overhead [byte]**					
(E – Encryption, A – Authentication						
Sec. headers	5	5	5	5	5	5
3-DES A	8	8	8	8	8	8
3-DES E&A	14	8	8	14	8	8
AES A	16	16	16	16	16	16
AES E&A	22	16	16	22	16	16
RSA A	384	384	384	384	384	384
RSA E&A	512	512	512	512	512	512

Concluding, the range of traffic caused by security measures for network protection range from below 1% for monthly metering profiles up to 90% for singular small size meter readings.

5. CONCLUSION

With respect to the serious damage that misuse of energy automation systems can cause the still widely applied policy of "security by obscurity" offers not enough protection against threats of the (near) future. Also the increasing interconnection of units, usage of standard components such as operating systems and the trend towards remote control higher security measures must be integrated. Such measures usually also increase the safety of the communication system (e.g. prevents accidental misuse by unauthorized maintenance personal).

Initiatives such as SELMA or REMPLI show ways to integrate security into automation systems, although it should be pointed out that security measures will introduce additional overhead. Hence, the overall security policy must define the balance between the risk and the overhead. Special attention must be drawn to integrate the system environment into the security policy to avoid indirect attacks such as the indirect DOS attack described in section 3.2.

The lesson learnt from the Internet and also from incidents in energy automation networks is that well designed security measures must be applied for vital systems. The old scenario of an isolated and physically protected system is no longer true.

REFERENCES

Bertsch, J., et. al. (2005). Wide-Area Protection and Power System Utilization. *Proceedings of the IEEE*, **VOL. 93**, NO.5, p. 997-1003.

Brauner, G. (2004). Simulation for congestion management to avoid blackouts. Elektrotechnik und Infromationstechnik, **Vol 11**, p. 425-429.

British Columbia Institute of Technology (2005). *Industrial security incident knowledgebase* [online]. [available at] http://www.bcit.ca/appliedresearch/security/services.shtml

Byres, E., Lowe J. (2005). Real World Cyber Security Risks For Industrial Control Systems. *The Online Industrial Ethernet Book*, **Response No. i22 35** [online]. [available at] http://ethernet.industrial-networking.com/origarticles/i22cyber.asp

Dwivedi, S. Menezes, B. and. Singh (2005), A. Database Access Control for E-Business – A case study. In: *Proceedings of the International Conference on Management of Data*, Haritsa and Vijayaraman (Eds.), 6.-8.1.2005, Goa, India

Fischer-Drapela, B. (2003). Nadelöhr für Österreich und Europa? In: *VEOE Journal*, **March 2003**, p. 4-7. (in German)

Naedele, M. (2004). Innovative Lösungen für die Informationssicherheit in Automatisierungs-systemen In: *Proceedings of VDE Congress 2004*, Berlin, ISBN: 3-8007-2827-3, p. 233 – 238. (in German)

OASIS security team (2001). Technical Architecture Risk Assessment V1.0, technical report, OASIS http://www.ebxml.org/specs/secRISK_print.pdf

Palensky, P., Sauter, T. , Schwaiger, C. (2000). Security and fieldbus systems – a contradiction? it&ti, 42, 2000, 4, p. 31 – 37.

Palensky, P. (2005). The JEVis Service Platform - Distributed Energy Data Acquisition and Management. In: *The Industrial Information Technology Handbook*, Zurawski, R. (ed.), CRC Press, Boca Taton, Florida, 2005, p. 111-117.

Patil, S. and Newcomer, E. (2003). ebXML and Web Services. In: IEEE Internet Computing, **Vol. 7**, Issue 3, May-June 2003

REMPLI consortium (2004): White Paper REMPLI Security Concept [online]. [Available at] http://www.rempli.org

Sauter, T., Schwaiger, C. (2002). Achievement of secure Internet access to fieldbus systems. *Microprocessors and Microsystems,* **Vol. 26**, p. 331–339.

Sauter, T., Pratl, G. Treytl, A., Bumiller, G. (2005). Secure and Reliable Wide-Area Power-Line Communication for Soft-Real-Time Applications within REMPLI. *Proceedings of 2005 International Symposium on Power Line Communications and Its Applications*", IEEE 05EX981, p. 57 - 60.

Schwaiger, C., Treytl, A. (2003). Smart Card Based Security for Fieldbus Systems. *Proceedings of 2003 IEEE Conference on Emerging Technologies and Factory Automation*. p. 398 - 406.

SELMA consortium (2005). SELMA - Sicherer ELektronischer Messdaten-Austausch (secure and reliable exchange of metering data) [online]. [Available at] http://www.selma-project.de

Sikora, A., Brügger, P. (2005). Secure Architecture for Embedded Web Servers. *The Online Industrial Ethernet Book* [online]. **Response No. Issue 23:32**. [Available at] http://ethernet.industrialnetworking.com/ieb/articledisplay.asp?id=39

Tomsovic, K., et. al. (2005). Designing the Next Generation of Real-Time Control, Communication, and Computations for Large Power Systems. *Proceedings of the IEEE*, **VOL. 93**, NO.5, 965-979. p. 393 – 396.

Treytl, A., Sauter, T., Schwaiger, C. (2004).Security Measures for Industrial Fieldbus System - State of the Art and Solutions for IP-based Approaches. *Proceedings of the 2004 IEEE International Workshop on Factory Communication System*. p. 201 - 209.

Treytl, A., Roberts, N., Hancke, G., P. (2004). Security Architecture for Power-line Metering System. *Proceedings of the 2004 IEEE International Workshop on Factory Communication System*.p. 393 – 396.

Treytl, A., Sauter, T. (2005). Security Concept for a Wide-Area Low-Bandwidth Power-Line Communication System. Proceedings of 2005 International Symposium on Power Line Communications and Its Applications. IEEE 05EX981, p. 66 - 70.

U.S. Nuclear Regulatory Commission (2003). NRC Issues Information Notice on Potential of Nuclear Power Plant Network to Worm Infection. *NRC News*, **No. 03-108**, 2. September 2003. [available under] http://www.nrc.gov/reading-rm/doc-collections/gen-comm/info-notices/2003/

van der Togt, Ted (2003). Standardisation and security in message exchange. Metering International, Issue 3/2003, p. 44, 2003

Werner, T. G. (2002). Load profiling in Germany. 1st Telemark Discussion Forum: Technology Evolution and Future European Electricity Markets, 2-4 September 2002, London, UK, http://www.telmark.org/

Ziegler, H. (2001). Dedicated Application Layer (M-Bus). Working document of CEN TC 294 WG 4, 2001

ELSEVIER
IFAC
PUBLICATIONS

SECURITY SERVICES IN FIELDBUSES: AT WHAT COST?

Miguel León Chávez[1], and Francisco Rodríguez Henríquez[2]

[1]Benemérita Universidad Autónoma de Puebla
Facultad de Ciencias de la Computación
14 Sur y Av. San Claudio, CP 72570, Puebla, México
Tel. (52) 222 229 55 00 ext. 7213 Fax (52) 222 229 56 72
E-mail: mleon@cs.buap.mx
[2]CINVESTAV-IPN
Sección de Computación
Av. Instituto Politécnico Nacional No. 2508, Col. San Pedro Zacatenco
México, D.F. 07300
Tel: (52) 52 55 5747 3800 ext. 6570 Fax: (52) 555 747-7002
E-mail: francisco@cs.cinvestav.mx

Abstract: *This paper discusses the security in the fieldbuses. The discussion takes into account, on one hand, the security services defined by the ISO Security Architecture, and on the other hand, the security mechanisms defined by some fieldbuses. From our analysis, it is shown that there are two critical points for attacking these networks: the bus and, in the case of centralized networks, the master node. This paper presents then the possible solutions to protect the fieldbuses and discusses their associated computing cost. Copyright © 2005 IFAC*

Keywords: Fieldbus, Security.

1. INTRODUCTION

Fieldbuses are special purpose Local Area Networks (LAN) used to connect all kinds of devices in a factory, such as sensors, actuators, transmitters, programmable controllers, (C)NC machines, processors, and so on (Thomesse, 2002). These networks usually define the services of three OSI layers, namely the physical, the data link and the application layers. Although some services of the missing layers are still present in some fieldbuses.

Typically, the fieldbuses are used by the distributed manufacturing applications in order to monitor and control the processes taking place in the factory. Examples of such applications are: factory automation, automotive industry, textile machinery, electronics manufacturing, food and beverage, chemical processing, and so on.

Up to now, the security in the fieldbuses, such as the IEC 61158, has only been considered for access protection on some objects. This is not for protection against intentional misuse of the communication facilities of a field device but in order to protect a system of accidental erroneous use of the objects.

However, there exist at least two possible security attacks that fieldbuses can suffer: Non-authorized users gaining access to the communication channel and non-authorized human operators accessing the master node, if it exists.

Clearly, there exists always the possibility for non-authorized users to gain access to the communication channel. If that happens, then the intruders can launch a passive attack by eavesdropping all or part of the information exchanged among the network's entities. Even worse than that, active attacks are also possible as hackers can maliciously insert or modify the data traveling through the communication channel at will.

On the other hand, usually the master node stores the entire network configuration as well as other important system global information, such as the presence variables in WorldFIP, i.e. variables containing summarized information on the node's global operating state. Hence, mechanisms of user identification should be put in place in order to avoid leakage of valuable data to non-authorized human operators.

As fieldbuses become more and more diverse, complex and integrated into other kind of networks (Gordeev, 1999; Decotignie et al., 2001), potential attacks to the security of a fieldbus network increase in the same rate. Hence, it is of the utmost importance to incorporate security mechanisms on fieldbus communication protocols so that such kind of security attacks can be avoided, prevented and/or thwarted (Gordeev, 1999; Morris and Koopman, 2003; León and Rodríguez, 2004a)

However, most of the fieldbuses were designed to meet real-time constraints, such as bounded end-to-end delay, periodicity, jitter, and coherence (León and Thomesse, 2000). No matter what security mechanism is instrumented, it will require valuable processor time for its execution, and therefore some real-time constraints might be missed.

This paper discusses the security in the fieldbuses, according to the classes of security service defined by ISO. The discussion is focused on TS 61158, WorldFIP, and CAN because they are typical example of centralized and distributed fieldbuses. The paper analyses then the proposed security mechanisms for fieldbuses and discusses the associated computing cost of such solutions.

The remaining of this paper is organized as follows: section 2 presents the ISO security services; security mechanisms defined by TS 61158, WorldFIP and CAN are presented in section 3; section 4 presents some proposed security services; section 5 discuses their computing cost; finally some future work directions and conclusions are given in section 6.

2. ISO SECURITY SERVICES

The Security Architecture of the OSI Reference Model (ISO 7498-2) considers five main classes of security services: authentication, access control, confidentiality, integrity and non-repudiation. These services are defined as follows: The *authentication* service verifies the supposed identity of a user or a system. The *access control* service protects the system resources against non-authorized users. The *confidentiality* service protects the data against non-authorized revelations. The *integrity* service protects the data against non-authorized modifications, insertions or deletions. The *non-repudiation* service prevents an entity from denying previous commitments or actions.

All the security services defined by ISO can be achieved in a centralized fieldbus by using public key cryptography. That can be accomplished by assigning to each slave node in the network a unique private key and a master node's public key. During communication, slave and master nodes may mutually authenticate each other with these keys using well known protocols. To provide confidentiality, nodes may encrypt their contents using a random session key and a symmetric crypto-algorithm specially tailored for constrained environments. Integrity and non-repudiation can be obtained by signing/verifying all the messages transmitted between a particular slave node and the master node.

The integrity service can be achieved also by using a one-way hash function optimized for heavily constrained environments, as those typically found in fieldbuses. The hash functions accept a variable-size message as input and produce a fixed-size code, called the hash code or message digest. The verification of the hash code is designed to detect intentional and unauthorized modifications of the data, as well as accidental modifications. Whereas the verification of a checksum value or an error detecting code, as those produced by the CRC algorithms or the frame check sequence (FCS), is designed to detect only accidental modifications of the data.

However, strong public key cryptography is in general an expensive fancy solution for fieldbuses because, on one hand, most of the field devices have limited capacities, such as processor speed and memory. And on the other hand, public key cryptography requires complex algorithms, large key-sizes, and management of the public keys.

Moreover, some of the security services defined by ISO are probably not very likely to be useful on the context of some fieldbuses. Particularly, non-repudiation seems to be not suitable for the centralized fieldbuses since the master node "gives permission to speak" to each slave node.

If for a given fieldbus public key cryptography solutions are too expensive, we can still design limited security schemes for fieldbuses at a cheaper price, i.e. fast security algorithms requiring a small amount of memory. For instance, data confidentiality can be achieved by using some lightweight cryptographic stream cipher, such as RC4 or A5/1 GSM, or even a reduced version of traditional symmetric algorithms such as DES or AES, which can be obtained by reducing the size of the encryption key or by limiting the standard number of rounds used during the encryption/decryption processes (16 in the case of DES and 10 for AES).

Although the previous limited security schemes have a cheaper price, some fieldbuses may not be able to afford them.

3. FIELDBUS SECURITY MECHANISMS

This section presents the security mechanisms defined by TS 61158, WorldFIP, and CAN standards jointly with suggested security improvements.

3.1 Security in TS 61158

TS 61158 (IEC 61158) provides users with the authentication and access control services for some objects. These services are implemented at the application layer, at this level, the application process objects (APOs) are defined, which are components of the application process and are visible across the network. An APO is identified by one or more key attributes, such as access privilege. This attribute specifies the access controls defined by APO, and is composed of the following fields:

- Password, it contains the password for the access rights. Its value is null if it is not used.
- Access Groups, it identifies which of the user defined access groups are defined for the object.
- Access Rights, it defines the type of access attributes authorized for the object. Legal examples of such attributes are as follows: right to read/write for access groups, for the registered password, and for all communication partners.

The access privilege attribute is defined for the following APOs: variables, events, and load regions. Nevertheless, TS 61158 specifies that an authorized human operator must invoke the management system in order to configure the initial schedule in LAS (Link Active Scheduler). Therefore, this fieldbus must provide a user authentication service in order to put in place some mechanisms of user identification in LAS, such as logging and password, to avoid access from non-authorized human operators.

On the other hand, the TS 61158 does not provide the following ISO security services: node-authentication, integrity, confidentiality, and non-repudiation (León and Rodríguez, 2004a).

According to standard's specification, the LAS node uses the notion of Delegated Token (DeT) to transfer the right to transmit to another node for a specified duration time. Usually, the token has two fields for addressing the source and destination nodes. With the purpose of thwarting a masquerade attack, the TS 61158 should provide a node-authentication service.

The integrity service can be achieved by using the CRC field of the MAC frame to store the hash code produced by a one-way hash function.

In order to avoid passive and active attacks from intruders that have managed to gain access to the communication channel, the TS 61158 should instrument a data-confidentiality service via feasible encryption/decryption schemes.

Regarding the non-repudiation service, it seems to be not very useful for this kind of networks due to the centralized nature of the protocol where the LAS node "gives permission to speak" to each node.

3.2 Security in WorldFIP

WorldFIP provides some security mechanisms implemented in its protocol or in its components. The mechanisms considered by the protocol are (EN 50170-3): medium redundancy, errors in the physical layer, data link layer status machines, frame check sequence (FCS), bus arbitrator redundancy and variable validation.

Clearly, WorldFIP security mechanisms do not provide the ISO security services, even though FCS allows to detect accidental modifications of the data (Erdner et al., 2001), i.e. FCS is calculated when the frame is transmitted and when it is received. If the code received matches the code calculated there is a very high probability that the frame is correct.

It is worth to mention that once again some of the ISO security services are not likely to be useful on the context of WorldFIP. Particularly, in the case of non-repudiation service it seems that since there exists a Bus Arbitrator (BA) node "gives permission to speak" to each producer node, it would not be useful at all.

According to standard's specification, in WorldFIP an authorized human operator configures the BA, which stores the network configuration into a table and several queues.

Therefore, WorldFIP (León and Rodríguez, 2004) should provide some sort of authentication service in order to put in place user identification mechanisms in the BA, such as logging and password, with the purpose of avoiding access from non-authorized human operators. Access-rights can also be added to the authentication mechanism so that authorized human operators can have access rights to configure, read and modify the network configuration. Furthermore, in order to avoid passive and active attacks from intruders that have managed to gain access to the communication channel, WorldFIP should instrument a data-confidentiality service via feasible encryption/decryption schemes.

3.3 Security in CAN

Not all ISO security services are useful in the context of CAN (León and Rodríguez, 2005) due to the following considerations. Both, node and message authentication together with non-repudiation services are not needed in CAN due to the fact that CAN nodes do not make use of any information about the network configuration, e.g. node addresses. Instead, in CAN, all the messages have been assigned a unique identifier which is used as a static priority for bus access. The identifier does not indicate the destination of the message, but describes the meaning of the data, so that all the nodes in the network are able to decide by message filtering whether the data is intended for them or not and to be acted upon by them or not. As a consequence of the concept of message filtering any number of nodes can receive and simultaneously act upon the same message. Message filtering is based upon the whole identifier, although optional mask registers may be used to select groups of identifiers to be mapped into the attached receive buffers.

The access control service may be implemented at the higher layers protocols (e.g. application). At network configuration time, all the messages have been assigned a unique identifier. Therefore, the higher layer protocols based on CAN should provide this service by using some mechanisms of user identification, such as logging and password, so that access of non-authorized human operators to the CAN based system configuration would be restricted.

CAN (Bosch, 1992) provides users with a special kind of service for data transfer, namely safety service, which includes the following procedures: error detection, error signaling, and self-checking.

For error detecting the following measures have been taken into account: Monitoring (transmitters compare the bit levels to be transmitted with the bit levels detected on the bus), Cyclic Redundancy Check (CRC), Bit stuffing, and Message Frame Check.
Nevertheless, these procedures do not provide the integrity security service, which can be achieved by using cryptographic mechanisms such as one-way hash functions.
Finally, CAN does not provide the confidentiality service. All data transfers are made in plaintext. Therefore, in order to thwart possible passive and active attacks from intruders that have managed to gain access to the bus, CAN should instrument a data-confidentiality service via feasible encryption/decryption schemes.

4. SECURITY SERVICE FOR FIELDBUSES

This section presents the proposed security mechanisms and discusses their computing cost.

4.1 Security goals

As it was mentioned in section 2, all customary security services for fieldbus protocols can be achieved by using public key cryptography schemes. However, public key cryptography requires a processing power that is typically well beyond the reach of many field devices. Fortunately we can still design limited security schemes for a fieldbus at a cheaper price by using alternative cryptographic options.
Data confidentiality can be achieved by using some lightweight cryptographic stream cipher such as RC4 or A5/1 GSM, or even a reduced version of traditional symmetric algorithms such as DES or AES.
Due to the fact that password-based security mechanisms are needed only for data protection at the master node (i.e., LAS or BA), where typically processing power is not a concern, we can use any of the traditional schemes based on symmetric ciphers.
Summarizing, papers published in (León and Rodríguez, 2004, 2004a, 2005) have proposed to achieve the above stated security features by incorporating the following security mechanisms into the Fieldbus protocols:

- A lightweight stream cipher in order to guarantee data confidentiality by encrypting all the relevant data to be transferred by the network's entities.
- A password-based security mechanism to prevent non-authorized users to gain control in the master node.
- One-way hash function in order to provide the integrity security service.

4.2 Computing Cost of the Security Services

Offering security services for resource-constrained platforms does not have a long history in the cryptographic community. Most research work has targeted the design of efficient symmetric stream ciphers under the assumption that this type of ciphers is much more economical than their relatives, symmetric block ciphers (Kumar et al., 2004). Among the few examples of lightweight block ciphers available in the open literature, one can mention the Tiny Encryption Algorithm TEA (Wheeler, Needham, 1994), which is already an old proposal, and more recently the SEA algorithm proposed by (Staandert et al., 2005). Present block ciphers, such as the Advanced Encryption Standard (FIPS 197; Daemen and Rijmen, 2001) are generally considered too costly for constrained platforms.

Using well-known stream cipher algorithms, the confidentiality security service for CAN was achieved in (León and Rodríguez, 2005) by means of RC4 and A5/1 stream ciphers, with RC4 showing the best performance. In that work the platform targeted was the Intel MCS®96 microcontroller running at a clock frequency of 16 MHz. This microcontroller can be considered typical of CAN applications. However obtained timing performances were not quite promising. The encryption time, using the full RC4 algorithm (i.e., including the initialization phase), goes from 7.5 ms to 7.8 ms for 1 byte to 8 bytes in the CAN frame, respectively, and the encryption time using the A5/1 algorithm (also considering its full version) goes from 64.1 ms to 80.1 ms for 1 byte to 8 bytes, respectively.

Clearly, those times are very high regarding the CAN transmission time, which operates in the range that goes from 5 Kbits/s to 1 Mbit/s, and therefore there are some doubts about the feasibility of offering security services without affecting the real-time constraints.

It should be noticed, however, that the overwhelming majority of clock cycles for ciphering the data field of the CAN frames were spent during the initialization phase. For example, RC4 takes in this phase from 99.28% to 95.49% for 1 byte to 8 bytes, respectively. For the A5/1 algorithm, this phase takes from 99.61% for 1 byte to 97.56% for 8 bytes.

The initialization phase is required by RC4 in each encryption because it is strongly recommended that no two messages should be encrypted using the same key. Otherwise the message can usually be broken. Indeed, if the two encrypted messages are XOR-ed together, the result is the XOR of the original messages (Dawson, and Nielsen, 1996).

Due to the fact that the overhead introduced by the initialization phase is too large for both algorithms, it was proposed in (León and Rodríguez, 2005) to include the concept of a session key. That would imply to open a new session (and thus to generate a new key) each time that the CAN protocol is initialized by the application. As long as the session is still active that same session key will be used to encrypt all CAN frames. According to results presented in (León and Rodríguez, 2005), RC4 can be implemented at a cost of some 6245 clock cycles, which would be attractive compared with block cipher proposals such as SEA that needs no less than 17745 clock cycles (Standaert et al., 2005).
However, a session-key feature will require a specific protocol among the parties involved in order to

resolve issues related to key generation and management. Such issues include: key generation and renovation; opening and closing sessions, etc. It is still an open question how these amendments could be introduced in fieldbus protocols such as CAN without loosing compatibility with former versions.

5. CONCLUSION

In this contribution we have discussed the security in fieldbus protocols, such as TS 61158, WorldFIP, and CAN. Our analysis has shown that these fieldbuses provide only limited security, namely the ISO integrity security service. WorldFIP implements that service, until certain extent, by using a frame check sequence (FCS), CAN uses CRC and message frame check.

However, these fieldbuses are vulnerable to at least two possible security attacks: Non-authorized users gaining access to the communication channel and non-authorized human operators accessing the master node.

Both types of security attacks can be avoided using public key cryptography schemes. However, public key cryptography requires a processing power that is typically well beyond the reach of many field devices. To overcome this difficulty, previous works have proposed security mechanisms that provide the authentication, confidentiality, and integrity services at a cheaper price by using alternative cryptographic options.

Naturally, implementing those security services introduces an extra overhead that must be quantified and measured in order to verify that the timing constraints on the messages are met.

Experimental times of two stream ciphers algorithms, such as RC4 and A5/1, on the Intel MCS®96 microcontroller have shown that the confidentiality service is still an expensive fancy solution for fieldbuses. Therefore, new cipher algorithms are required for small data sizes and real-time constraints. We believe much work can still be done in this area.

ACKNOWLEDGMENTS

Authors would like to acknowledge support from CONACyT through project number 45306.

REFERENCES

Bosch, R. GmbH (1992). CAN Protocol Specification V2.0 (A,B).

Daemen, J., and V. Rijmen (2001). The Design of Rijndael. *Springer-Verlag.*

Dawson, E., and L. Nielsen (1996). Automated Cryptanalysis of XOR Plaintext Strings. *Criptologia*, **vol.** XX, No. 2.

Decotignie, J.D., P. Dallemagne, and A. El-Hoiydi (2001). Architectures for the Interconnection of Wireless and Wireline Fieldbusses. In *4th IFAC International Conference on Fieldbus Systems and their Applications (FeT'2001)*, Nancy, France, Nov. 15-16, pp. 285-290.

EN 50170-3 (1995). WorldFIP, General Purpose Field Communication System, CENELEC *EN 50170-3.*

Erdner, T., W.A. Halang, K.C. Chan, and J.K. Ng (2001). Secure Data Communication over Fieldbus Systems. In *4th IFAC International Conference on Fieldbus Systems and their Applications (FeT'2001)*, Nancy, France, Nov. 15-16, pp. 37-44.

FIPS 197 (2001). Advanced Encryption Standard. Federal Information Processing Standard, NIST, U.S. Dept. of Commerce.

Gordeev, M. (1999). Security Architecture for Field Area Networks Connected to Internet. In *3th International Conference on Fieldbus Systems and their Applications (FeT'99)*, Magdeburg, Germany, Sep. 23-24, pp. 69-75.

IEC 61158 (1999). International Electrotechnical Commission. Digital Data Communications for Measurement and Control – Fieldbuses for use in Industrial Control Systems. *IEC 61158 Type 1.*

ISO 7498-2 (1989). International Organization for Standardization. Information processing systems - Open Systems Interconnection - Basic Reference Model - Part 2: Security Architecture.

Kumar, S., K. Lemke, and C. Paar (2004). Some Thoughts about Implementation Properties of Stream Ciphers. In *SASC - State of the Art of Stream Ciphers Workshop*, Brugge, Belgium, October 14-15.

León, M. and J.P. Thomesse (1999). Fieldbuses and Real-Time MAC Protocols. In *4th IFAC International Symposium on Intelligent Components and Instruments (SICICA'2000)*, Buenos Aires, Argentina, pp 51-56.

León, M. and F. Rodríguez (2004). SDL Specification of a Security Architecture for WorldFIP. In *IEEE XIV International Conference on Electronics, Communications, and Computers (CONIELECOMP'2004)*, Veracruz, México, February, 2004, pp. 149-154.

León, M. and F. Rodríguez (2004a). SDL Specification of a Security Architecture for the IEC 61158. In *11th IFAC Symposium on Information Control Problems in Manufacturing (INCOM'2004)*, Salvador da Bahia, Brazil, April 5-7.

León, M. and F. Rodríguez (2005). Performance Analysis of the Confidentiality Security Service in CAN. In *XVI IFAC World Congress*, Praha, Czech Republic, July, 2005.

Morris, J. and P. Koopman (2003). Critical Message Integrity over Shared Network", in *5th IFAC Conference on Fieldbus Systems and their Applications (FeT'2003)*, Aveiro, Portugal, July 7-8, pp 145-151.

Standaert, F.-X., G. Piret, N. Gershenfeld, and J.-J. Quisquater (2005). SEA - a Scalable Encryption Algorithm for Small Embedded Applications. In *Workshop on RFID and Lightweight Crypto*, Graz, Austria.

Thomesse, J.P. (2002). A Review of the FieldBuses. *Annual Reviews in Control*, **vol.** 22, pp. 35-45.

Wheeler, D.J., R. Needham (1994). TEA, a Tiny Encryption Algorithm. In *Proceedings of FSE 1994, Lecture Notes in Computer Science*, **vol.** 1008, pp. 363-366, Leuven, Belgium, Springer-Verlag.

ELSEVIER
IFAC
PUBLICATIONS

A NOVEL REQUIREMENTS METAMODEL FOR AUTOMOTIVE ELECTRONIC NETWORK DESIGN

Dra. Liliana Díaz-Olavarrieta
Dr. David Báez-López

Fundación UDLA.Puebla, Dept. Electronics
100 Sta. Catarina Mártir,
San Andres Cholula, Puebla, Mexico,
liliana.diaz@udlap.mx, dbaeziec@mail.udlap.mx

Abstract: Automotive electronic subsystems are resource constrained, heterogeneous, distributed, real-time systems, and may implement safety-critical, X-by-wire applications, requiring fault-tolerance. Specifying requirements for an automotive application is a decision making problem, where perfect rationality does not exist, and thus need to help, within our limited rationality, the decision making process. The purpose of this article is to make a step in that direction, by proposing a novel meta-model that offers four main analysis perspectives (User, Application, Development Process and Competitive Industry Analysis) to explore the automotive requirements space in a level of detail that permits attention management, problem decomposition and decentralized decisions.

Keywords: Automotive Requirements Specification Design Space, User-centred, Application Based Requirements, Component Integration Based Development, Competitive Industry Context, Distributed, Real-time Systems, Fault Tolerance, Time-Triggered, Event-triggered protocols, Safety Critical Applications, In-Vehicle Networks.

1. INTRODUCTION

The global demand for vehicle electronics is forecast to reach nearly $75 billion by 2005, and the percentage of automotive electronics cost in 2010 will grow from 12 % to 30 % of a mid-range car's total cost (Mayer, 2005).

The automotive electronic control applications range from non-critical comfort level functions such as doors, lights, mirrors, window and seat control, to critical-safety applications (where human life is at risk if the electronic system fails) or image-critical functions, like being able to get into a locked car through the door, and not through an open window. In critical activities, generically *X-by-wire* applications, (Kopetz, 1995), taking their name from the first "Fly-by-Wire" (FBW) Aircraft systems, fault-tolerance has to be guaranteed. The first all digital FBW application without mechanical backup was the F-8 military aircraft (1972), while the first commercial aircraft with Fly-by-Wire technology was the A320 which entered service in 1988. At Boeing, research on FBW prototypes began in 1986 led by GE & Allied Signal, and the first full FBW civil commercial aircraft was the Boeing 777, which started flying in 1995, and used the fault-tolerant communications protocol SafeBus™, with 3 primary flight computers, 3 completely redundant physically and electrically separated ARINC 629 Databuses, 4 Actuator Control Electronics modules (the equivalent of 4 *ECUs* or Electronic Control Units), Pilot interfaces, Sensors, and an Airplane Information Management System (Ong, 2003).

In contrast, a high-end automobile today has more electronic functionality than a fault-tolerant aircraft had a decade ago: a BMW Mini Cooper, has between

7 and 23 *ECUs* (Electronic Control Units), a VW Bora has more than 30 Electronic Control Systems, and a BMW Series 7 car has more than 70 *ECUs* with a higher degree of integration (Mayer, 2005). This generates completeness, consistency, and dependability challenges, for *IVN* specifications.

2. SPECIFYING IVN REQUIREMENTS IS A DECISION MAKING PROBLEM

Specifying requirements for an automotive application is a decision making problem. Perfect Rationality in decision making does not exist, as Herbert Simon pointed close to 50 years ago, leading to the Limited Rationality concept in decision making (Sterman, 2000): "The capacity of the human mind to formulate and solve complex problems is indeed very small compared to the size of the problem whose solution is required to produce an objectively rational behaviour, or one which approximates in a reasonable manner this objective rationality." Our limited rational capacity comes from our limited knowledge, our cognitive inability to learn in "big chunks", and our selective perceptions that yield an incomplete view of the real world, which in the case of automotive embedded networking applications is at the intersection of many engineering domains (Pimentel, 2003). Our mental models are oversimplified in the decision making process in many cases, to yield with complexity, and our deductive powers are fallible and sometimes tainted with emotional, unconscious, and irrational factors.

Moreover, to decide is to select between multiple alternatives, which takes time and more often than not we are called to decide before the examination of all alternatives has been done. Thus, we need to use ways to help the decision making process, within our limited rationality, and be able to answer some or all of the following questions:

a) How does one ensure the requirements' completeness, consistency and correctness?
b) What are the user's expectations and service trends that one should consider to explore the requirements' design space in a holistic sense, as defined here, in addition to the engineering domains required (Pimentel, 2003) ?
c) What perspectives should be analyzed to produce a requirement specification for an electronic automotive application, which is "strategically consistent", in a competitive industry view (Porter, 1988)?
d) What is the best way to categorize or organize the multitude of details and specific requirements for *non-critical*, or *Y-critical* applications? (Y= safety, image, cost or any other requirement from the design space that has the highest priority to be implemented).
e) What automotive electronic requirements are derived from external (to the company) analysis and which are derived from internal analysis?
f) Can Resource constraints make the requirement "customizable" to different reality scenarios?
g) Can one generate a "generic" automotive IVN specification, and perhaps find "patterns" amongst the distinct applications domains being specified separately now?

The current way of specifying, designing and implementing automotive in-vehicle networks in the automotive industry is partitioned or domain specific: there are currently power train control specifications, body and chassis specifications with comfort application networks, and safety critical XBW applications such as Brake-by-Wire, Steer-by-Wire, Cruise control or collision Avoidance. However, it would be interesting if the specifications for a certain application, say Brake-by-Wire, were parameterized / customized with different "context or user parameters" (i.e., weather and road conditions, person's height, size and weight) as is done with comfort applications such as detecting the weight of the person to ensure that the safety belt force or the airbag force is not more than that required?

This paper attempts to give answers to these questions, through the proposal of a novel requirements perspective framework, which can serve as a reference meta-model for specifying requirements – both for automotive applications, but indeed in a more general sense for systems which are distributed, real-time, fault-tolerant, resource constrained, user-defined, and strategically consistent within the competitive industry context.

The paper is organized as follows: *Section 3* presents Requirements Analysis: Perspectives and Design Space Exploration; *Section 4* will present a brief overview of In-Vehicle Networks, and automotive protocols; *Section 5* presents the USER requirements perspective; *Section 6* presents the APPLICATION requirements. *Section 7* present the INDUSTRY perspective; *Section 8* the (CBD) Company Development Process perspective; *Section 9* presents Conclusions and further work, *Section 10* includes the main references consulted for this work.

3. REQUIREMENTS ANALYSIS: FOUR PERSPECTIVES FOR DESIGN EXPLORATION AND SPECIFICATION

The analysis of requirements will be done through a user-guided perspective kaleidoscope, with the high level bird's view perspective inspired from competitive business analysis (Porter, 1988), and the lower requirements perspective driven from safety critical and non-critical applications. The four main requirements perspectives to examine are:

1) USER
2) APPLICATION
3) DEVELOPMENT PROCESS
4) INDUSTRY CONTEXT

a. The User Perspective

Requirements derived from the USER, in any of the roles of a final Client, can be obtained through market surveys. Another common source of user requirements are those defined by the Automotive manufacturer, which may take into account Service Trends (multimedia, connectivity with local and global networks), Market Specific User Resource Constraints (i.e. Selling Cost, Speed Limits, Minimum Features, Financing, Re-configurability) or Context-Based constraints (rain, ice, temperature).

b. *The Application Perspective*

APPLICATION: Requirements derived from the Nature of the Application: Distributed, Real time, Safety-Critical, Resource Constraints (Standards, Regulations, Supplier Offerings) can be dealt with in a conceptually orthogonal way, keeping in mind that simultaneous optimization of requirements in all perspectives is not always possible due to conflict.

One must remember that the domain of automotive IVNs applications is at the crossroads not only of a variety of domain engineering knowledge, networking engineering, dependability engineering (Pimentel, 2003), and a variety of implementation engineering frameworks (hydraulics, mechanics, electronics, electrical, communications), but within each, a particular application (critical vs. non-critical), will influence the choice of appropriate architecture modelling frameworks and languages.

Software engineering, hardware engineering (FPGA-DSP semi-custom hardware design and custom IC for embedded engineering solutions) or hybrid solutions, require to be specified in a high-level architectural language such as VHDL (Very High Level Description Language), AADL (Avionics Architecture Description Language), MetaH (from Honeywell), or SysML (Systems Modelling Language), employed by HIRTS (Radjenovic, et al., 2004), for example, prior to the modelling, simulation and synthesis of a solution, which has be proven correct, dependable (reliable, safe, and available), testable and composable upon integration.

c. *The CBD Process Perspective*

Automotive DESIGN & DEVELOPMENT PROCESSES which yield Requirements specific to the way the application is produced, are usually derived from a *Component Based Design Integration Process or* CBD (CBDP, 2005). A *CBD Process* consists of 3 stages: *component architecting*, *provisioning* and *assembly* (taking variant handling into account).

As mentioned above, one of the most important issues in CBD Processes, is the integration and testing of the system, which has to have the property of *composability*.

This means that each individual component – previously tested and checked to specifications- will not interact in an unexpected or negative way upon integration as part of the whole system, but that rather, the integration has to yield a *no-surprises* working system, with the only requirement that the individual components are within specification.

In this context, the natural variability of the performance of different Integrated circuits from one run to another, for example, has to be managed, through sensitivity analysis, statistical modelling of the expected performance, or discarding "outliers" which may later deviate in performance.

d. *The Industry Context Perspective*

INDUSTRY: Requirements derived from the automotive industry competitive environment according to Michael Porter's Competitive Strategy model (Porter, 1988): Suppliers, Substitute Products-Technologies, Competitors and Potential Entrants, Clients (considered in User Requirements), the Company itself represent the five perspectives which have to be considered in order to assess the competitive industry context of the automotive company (both in a static and dynamic sense).

Setting the industry context perspective during the specifications or requirements design by the automotive system engineers, may help to define strategically consistent specifications, which are aligned with the resource constraints, market target of a particular automobile model, functionality and technology available.

This would ensure that the implementation is within the cost target set, and the service expectations of a client in that segment are fulfilled. Indeed, from this viewpoint, engineers could design automobiles which are reconfigurable to the segment market being targeted, without having to reengineer a new version or model, for each new market segment being considered.

4. FOUR PERSPECTIVES METAMODEL: ADVANTAGES FOR SPECIFICATION

4.1 *An organizing framework to ensure completeness*

As mentioned in the abstract, providing an "organizing framework", or meta-model as the one proposed here, offers –on the one hand- a set of perspectives which are coherently united by common attributes in a large domain (user, product- the automotive application, process, and context), gives added conceptual value to the completeness of the specification design process, and provides help in this value-analysis decision making process that constitutes the specification or requirements design process. On the other hand, this meta-model helps in the decomposition of the decision problem and decentralized requirements decision making on the other. This multi-perspective meta-model serves as a thinking guide within the multiplicity of "multi-valued, context-defined, temporally defined information objects", and their inter-relationships.

4.2 Exploration of the Requirements Design Space

The organized exploration of the requirements design space thanks to the guide provided by each of the four perspectives in the requirements meta-model is the first step to design user oriented electronic automotive control applications. The design of a specification requirement for a particular application is the second step, and once the requirement specification has been decided upon, the third step in the process is that a system engineer must match the application requirements to a small subset of communications protocols, to implement the *IVN*.

The exploration of the requirements design space is implicitly tied to the exploration of reality or rather, to its modelling and interpretation. It has been said (Sterman, 2000) that decision should not be based on the assumptions we may have of reality, but only on certainties about reality itself.

However, we know that often assumptions are made with ambiguous or incomplete information, and thus, we should make these assumptions explicit and relate them to the degree of information certainty (or ambiguity) to avoid making unconscious decisions that may change when better information becomes available about reality. Another property of a requirement set, is that they are context dependent, which makes them difficult to reuse. However, if one views the set of requirements as a set of individual requirements with certain relative weight or priority, when this weight is made explicit, the requirement set may be reused in other reality contexts, with a change in assigned relative priorities and certainty attributes.

Thus, we can view requirements as a constraint to be honoured by a component in the system, with honours a set of constraints that form a contract to be respected by the component at its defined interfaces following the HIRTS constraint model (Radjenovic et al., 2004):

The components will have certain properties, which honour assertions or constraints at specified moments in time, among which we could define a certainty attribute and a priority. These certainty attributes could multi-valued to account for the context-dependence of the value, which depend on the temporal and space characteristics of a certain property. Thus, we can expect to produce through this requirements meta-model, a "context dependent, probabilistic, priority directed specification", to explore the requirements design space.

In the section that follows, we briefly introduce the concept of In-Vehicle Networks and give standard protocol examples used in this application domain.

5. IN-VEHICLE NETWORKS (IVN)

There is a large variety of in-vehicle automotive protocols and structural topologies (nominally called "busses") allowed for in-vehicle communication, which depend on the final application such as Emissions/Diagnostics, Mobile Media and "X-By-Wire" protocols, which are used for different applications within the automobile sector (*Automotive Buses*, 2005) for both non-critical and critical applications.

Protocols may be categorized by the data bit rate they handle as SAE Class A, B, C or D. Class A if for low speed applications, (bit rate < 10 Kb/s), Class B (medium speed, between 10kb/s and 125 kb/s for general information transfer), Class C (high speed, bit rates higher than 125 Kb/s), and Class D protocols are used for speeds > 1 Mb/s -though there are no SAE implemented Class D protocols (Bell, 2002) yet, though in practice operation of a protocol may exceed the published requirements.

There exist currently in 2005, more than 42 automotive/industrial communication application protocols in use, but only a few are being considered as de-facto or by-design standard protocols: CAN, LIN, MOST, TTP and FlexRay (with some variations) , for the reasons we briefly detail below.

5.1 Standard Automotive Protocols

There seems is a growing consensus within the automotive industry that the communications protocols that will prevail are (Alford, 2003): LIN (LIN, 2005), for low cost applications which are event-triggered, CAN, for event triggered applications, TT-CAN (TTTech, 2004) for time-triggered CAN compatible applications, TTP/C (Kopetz, 1995), FlexRay (FlexRay, 2004) for fault tolerant, safety-critical applications, and MOST (MOST, 1999) for multimedia applications.

CAN is an event-triggered protocol Class A protocol with a very widespread use in the automotive industry. LIN is a recent low cost Class A alternative to CAN for comfort applications. An adaptation of the CAN standard protocol with a superimposed time/triggered layer to accommodate synchronous applications has produced the Time-triggered TT-CAN protocol, now standardized as ISO/DIS 11898-4.

TTP (TTP/C and its low cost alternative TTP/A) are time-triggered protocols designed specifically for safety-critical applications and have been used in the avionics and automotive industry extensively. TTP has received a DOD safety certifications for avionics applications, which makes it suitable for safety-critical applications, as it is the only protocolo which has been proven to be correct with formal verification methods, and can be considered to be reliable, safe and fault tolerant "by design".

TTP is based on the notion of a static frame, with a global time base, which determines a fixed frame length. Access fairness is ensured through a daisy chain (round robin) access to all nodes, and handles bus, star and star couplers to extend the network. A star coupled topology, with only two central guardians is capable of managing the communication, scheduling and synchronization of all nodes in a TTP network. TTP was originally specified for copper interconnect, but fibre interconnect is also allowed for greater bandwidth.

FlexRay was designed by a large consortium of automotive industry manufacturers, the FlexRay Consortium, as a more flexible approach to networking applications than TTP. From a behavioural viewpoint, FlexRay allows both time triggered as well as event-triggered applications, and from a structural viewpoint it allows more varied topologies than TTP, and specifies both copper or fibre. Fibre interconnects are considered more reliable with respect to EMI susceptibility, which leads to certain types of faults in IVN networks.

FlexRay has a static frame of constant length across the network nodes for time-triggered applications, a dynamic frame, of variable length, for event-triggered applications, and an idle frame, to allow for timing synchronization "realignment" among nodes. FlexRay allows a wide range of network topologies, such as bus, stars, and combinations of buses and stars, united through bus couplers.

Another fibre based interconnect protocol is MOST, which was specifically designed for broadband multimedia applications (MOST, 1999).

5.2 Safety-Critical Protocols

An important consideration to match a protocol to the application is if the protocol is apt to implement a safety-critical fault tolerant application. There is a general opinion that time-triggered protocols are better suited than event-triggered protocols for safety-critical applications, given the deterministic nature of synchronous protocols needed to guarantee fault tolerance in safety-critical applications (Kopetz, 2003). In this respect, the only protocols which are currently accepted as fault-tolerant for safety-critical applications in 2005 are: TTP and SAFEBus™, (used in the avionics and automotive industries), SPIDER (non-commercial), and FlexRay (Rushby, 2001).

In what follows, we present in more detail the properties of the requirements for IVN networks in each of the four meta-model perspectives: user, application, CBD process, and industry context.

6 USER REQUIREMENTS PERSPECTIVE

The user which drives the design and implementation of automotive products and services. User requirements are usually derived from use-case scenarios, which when written in UML modelling language, can then be used to develop the software directly form this UML specification. The use-case scenario process has to be as general as possible in terms of imagining scenarios (both normal and failure scenarios) in order to approach reality. They also have to represent a "generic user", defined by the strategic direction of the company, within the "market segment target". Use case scenarios can be either goal-driven or context-based driven, and are designed based on the service being designed.

Goal-Driven Use Cases: Defined by hierarchical successive refinement of the goals and sub-goals.

Context-driven Use cases: Sub-goals are reviewed in the light of differing environment or context scenarios such as Weather, Traffic Situation, Control Lever, Brake Position, Accelerator position, to refine use cases for the distinct context scenarios.

Reconfigurable (Both in Goal and in Context) Use Cases. As mentioned above, would use certainty attributes and priority weights to adapt a requirement to the context and the information available, improving the specification in time, as more information about reality becomes available.

Use cases have to be centred in the ways in which a potential user might utilize an automotive feature, or enjoy an IVN "service", in a very general sense. In-Vehicle-Networks services, can draw inspiration from the so-called "*5Ms* for Service Extension": Movement, Moment, Me, Money and Machine, originally used within the 3G cellular *UMTS* standard, and known as the "UMTS´s 5M´s.

6.1. The "5M's" for Service Extension

User expectation trends in terms of service for multimedia wireless communications –voice, data, video- have been named by the 3G UMTS Forum as the "*UMTS*" "*5Ms* for Service Extension": Movement, Moment, Me, Money and Machine:

Movement: To escape a Fixed place, a memory, virtually and literally in a car, while keeping connected. A recurrent user requirement is to be always connected to the large variety of LANs, WANs, MANs, and global external networks to enable personal mobile communications, such as Bluetooth,WiFi (802.11), GSM/GPRS/EDGE (for 2 and 2.5G) and UMTS / IMT-2000 3G standards.

Moment: Comfort Function Control to improve the experience of present and Moment. Also, it means to expand the concept of time, from Discrete to Continuous / Past, present, future / Scenarios / Experiences into the Memory. Memory is enabled with emotion, and emotion with 5-sense involvement (eyes, ears, taste, smell, touch) to create a better "moment" or "infotainment" experience. Mostly Eyes and Ears have been catered to in automotive applications such as Digital TV, DAB Digital Audio Broadcast / Music download capability, and the use of CD/DVD players.

However, *Touch, Smell* and *Taste* have been forgotten (we could not identify one application – except for the rear-seat videogame capability for children which involves Touch as well. Thus, Touch/Taste/smell is still open to new "better or expanded moment" creation, within New Magic Worlds applications. ("Mobile Virtual" Eating/ Drinking/ Smelling experiences, such as "perfume catering" while on the road?)

Me: The person and its expansion to a Community, extension of home or office. Shared Access (Business Broadband Internet capability for Videoconferencing and mobile Multi-site meetings).

Interactivity, Gaming or Collaboration.. Branding and Self-configurability, as expression of oneself, not only of "settings", but with a personal "look" and functionality, perhaps reconfigurable, based on electronic added-value.

Money: Financial Services, E-mobile commerce, and Banking applications, which also have to be made fault-tolerant and safe. Money means also Cost to the User, in life-cycle terms (acquisition cost, operation, maintainability, insurance - prices should be lower for certified fault tolerant cars-, disposal/ recycling cost).

Machine: "*More Car*". Empowering Gadgets & Devices. Added Processing Intelligence, with Power and a "universal dock connection capability", to connect to *PDAs,* to Tablets, to *iPods*, Cell phones.

7 NATURE OF THE APPLICATION REQUIREMENTS PERSPECTIVE

The four characteristics that emerge as defining the "nature of the application" are: 1) the *distributed* nature of the network, 2) The *real time* application requirements for some of the subsystems, 3) The *safety-critical* requirements for X-by-wire applications, for example, and 4) The *Resource Constraints*, which is derived from the implementation of the application.

7.1 Distributed Networks

A distributed network (Kopetz, 1997), (Tanenbaum, 2002), (Coulouris, 2001) is usually recognized because there are *concurrent processes* running in parallel on various processors, there is a *distributed memory or shared state*, and *data is communicated* through an *interconnection medium* that links the *multiple processors and the storage recipients*, be they volatile storage, non-volatile storage, or stable storage (a mixture of both).

The medium for the interconnection network can be either copper wire, fibre or an RF link, and the topology is constrained by the communication protocol used.

Concurrency of processes over a distributed network implies that communication and access to the controllers has to be arbitrated. Concurrent access is made through the shared medium –copper, fibre, wireless- and can be either *programmed* and *without contention*, such as in TDMA, FTDMA, FDMA, CDMA, DAMA access schemes, or random assigned schemes with resource contention and possible collisions, such as in CSMA/CD/CA/CR access schemes.

These access schemes motivate a triggering classification for protocols, related to the synchronicity or *periodicity of events*: time-triggered or synchronous, or periodic protocols, vs. event-triggered, or asynchronous protocols.

A distributed network implementation should also be "invisible" to the user, i.e., *transparent* in the way processes *communicate*, are *scheduled* and *synchronized* over a shared medium, independently of what functionality the interconnected *ECUs* have.

7.1.1 Transparency

The transparency requirement means that the user should not be able to distinguish between the performance of a uniprocessor central controller architecture, and a multiprocessor distributed architecture, except perhaps for increased efficiency. Various types of transparency follow:

Access: Local and Remote Resources are accessed using identical operations

Location: Users cannot tell where HW and SW resources are located

Migration - Mobility: Resources should be able to move without having their "names" changed.

Replication: System replicates critical data, without the user noticing it, for increased performance and reliability.

Concurrency: Users- processes will not notice the entry of other users in the system, even if they share the same resources.

Failure: Failure transparency implies fault independence, fail-silence, fail-operational, and fail-safe modes, in increasing order of fault-tolerance.

Performance: Load variation should not lead to performance degradation. Automatic Reconfiguration of BW allocated to multimedia processes (DAMA).

7.1.2 Inter-Process Communication

The separation of concerns between the Functionality of Processes vs. their Communication (which require Scheduling and Synchronization among them) is an important requirement for later reusability of the designs.

All four types of behavior: *function execution in controllers, synchronization, scheduling and finally the communication itself, take time*. That is, one should consider for the requirement specification that communication (sending messages across an interconnect network) takes time due to signal propagation delays, processor "interpretation" and execution of processes, and synchronization and scheduling delays.

The transmission of the data or messages per se between transmitter and receiver are what constitutes the communication within a protocol. It is affected both by the size of the messages, and the number of simultaneous messages that the network can handle. The topology of the network is also involved (there may be bus, stars, networks, star-couplers, and combinations of these in different protocols). The communication management processes (communication, synchronization and scheduling), have to be separated from the exchange of messages related to functionality per-se in order to improve reusability, predictability and growth of systems, in particular for real time systems.

7.1.3 Inter-Process Synchronization

There are two types of process synchronization: synchronous or periodic, also called Time-triggered (where a clock transition is the trigger to move information across a communications network) and asynchronous or aperiodic also called Event-triggered, (where specific Event signals act as the triggers to change state).

Time-triggered systems have an internal locus of control through the global clock, while Event-triggered protocols have an external locus of control. Thus, time-triggered protocols are more predictable (in value, time and space) than event-triggered protocols, and are preferred for critical applications.

Inter-process synchronization is obtained by a global clock for time-triggered protocols, and by an arbiter, a "bus guardian" or "central guardian" (using the FlexRay or TTP terminology) which controls the handshaking communication for an event-triggered protocol. In both cases, there is a structural entity, the clock generator, in the synchronous case, or the arbiter, in the asynchronous case.

For process synchronization, it is important to know the various types of delays: real message propagation time, scheduling time and synchronization time, will make up a realistic "communication delay".

7.1.4 Inter-Process Scheduling

Scheduling amongst processes refers to the way tasks or processes are prioritized to give a fair share of access to all the processes from *ECU* nodes to the shared distributed, interconnection network. There exist centralized (i.e. Daisy-chain) and distributed scheduling algorithms (i.e. Token passing methods).

Concurrency of processes over a distributed network implies that communication and access to the controllers has to be arbitrated. Concurrent access is made through the shared medium –copper, fibre, wireless- and can be either *programmed* and *without contention*, such as in TDMA (TTP), FTDMA (FlexRay), FDMA (Bluetooth), CDMA (WCDMA - 3G cellular), or *random* assigned schemes *with resource contention* and possible collisions (CSMA/CD/CA/CR) or DAMA.

7.2 Real-Time Requirements

Real-time requirements are often related to Safety Critical applications (Kopetz, 1995), (Dilger, 1997), Timeliness of Response, Protectiveness, Real-time Scheduling and Communication, Clock Synchronization, Membership Services within fault containment regions (FCR/FCU), Composability, Error-Detection, Robustness, & Fault Independence.

8 CBD: DEVELOPMENT PERSPECTIVE

A component based design process methodology is used in the automotive industry for the construction of communications and electronic control functions. The CBD can be a mixture of component *architecting*, component *assembly* provided by first-tier suppliers and component.

A *CBD Process* consists of 3 stages (CBDP, 2005): *component architecting* (specifying to second-tier development firms in the semiconductor business and constructing their own design*)*, component *provisioning* (often by third parties) to form subsystems with integrated circuits from second tier suppliers such as Motorola, Philips, Texas Instruments, Hitachi & ST Microelectronics and component *assembly* (parts provided by first-tier suppliers such as Bosch, BMW, Siemens and Magneti-Marelli).

Not only does the partitioning of the system into components, and the process used by each industry to assemble, and test these components have to impact on the system requirement specifications, but also have to take into account the variant handling and variability points inherent in CBD design.

9 INDUSTRY PERSPECTIVE

The "Nature of the Industry" can be static or dynamic (trends and direction). The static perspective from a strategic point of view considered was developed by Michael Porter to analyze the competitive environment of a company in a given industry. It is formed by five elements that interact with the company: Suppliers, Clients, Competitors and Potential Entrants, and Substitute product or services (Porter, 1985).

Furthermore, the "Nature of the Industry" can be static or dynamic (trends and direction). The static perspective from a strategic point of view considered was developed by Michael Porter (Porter, 1988) to analyze the competitive environment of a company in a given industry. The dynamic model considers each of these views and their evolution in time, following market trends, to produce strategically relevant, scalable and updateable automotive *IVN*s.

10 CONCLUSIONS

This paper has introduced a novel, "automotive requirements multi-perspective meta-model" to analyze requirements for the design of a distributed, real-time, heterogeneous system, for safety-critical and non-critical applications to create complete, strategically consistent requirements. The perspectives used are the User perspective, The Nature of the Application perspective, the Nature of the Process Development Perspective, and finally, the Competitive Industry Context perspective. The conceptually orthogonal perspectives help in improving the requirements specification decision.

11 REFERENCES

Alford, C., Paskvan, J., 2003. *"Local Interconnect Network: Hands-on LIN Training"*, retrieved March 2005, from Volcano Automotive Group Website.

Automotive Buses, 2005. "Logic Design Information: Automotive Buses", retrieved Feb 4, 2005 from http://www.interfacebus.com/Design_Connector_Automotive.html/

Bell, J., 2002. "*Network Protocols used in the automotive industry*", Doc. Ref SD/TR/PRO/01, July 24. 2002.

Byteflight, 2005. Website available at http://www.byteflight.com/

CBDP, 2005. "*Chapter 30 - Component-Based Development*", SEPA 6/e, available from http://www.rspa.com/reflib/CBSE.html, April 2005.

Chisalita, I., Shahmehri, N., 2004. "*A Context-Based Vehicular Communication Protocol*", IEEE Conf. Proc., ISBN 0-7803-8523-3/04, pp. 2820 – 2824.

Coulouris, G., Dollimore, J., Kindberg, T., 2001. "Distributed Systems – Concepts and Design", Addison Wesley Publ. Comp., 3ed., 2001.

Demmeler, T., Giusto, P., 2001. "*A Universal Communication Model for an Automotive System Integration Platform*", IEEE Proc. 1530-1591/01, pp. 47-54, 2001.

Dilger, E., et al., 1997. "Towards an Architecture for Safety Related Fault Tolerant Systems in Vehicles", On-line document available at http://www.vmars.tuwien.ac.at/projects/xbywire/projects/new-esrel97.html

ETAS GmbH, 1998. "Whitepaper ASCET-SD, ETAS GmbH, 1998", mentioned (Demmeler, 2001) above.

FlexRay, 2004., "*FlexRay Communications System, Protocol Specification, Version 2.0*", June 2004, retrieved on February 1, 2005 from FlexRay site at http://www.flexray.com/specification_request.php.

Kopetz, H., 1995. "*A Comparison of CAN and TTP*", available from TTTech website: http://www.tttech.com/technology/articles.htm

Kopetz, H., 1997, "*Design Principles for Distributed Embedded Applications*", Kluwer Academic Publishers, 1997.

Kopetz, H., 1998, "*TTP – A New Approach to Solving the Interoperability Problem of Independently Developed ECUs*", On-line Document Available from http://www.vmar.tuwien.ac.at/projects/xbywire/projects/new-ecu.html, on February 1st, 2005.

Kopetz, H., 2003. "*Fault Containment and Error Detection in the Time-Triggered Architecture*", Proceedings of the Sixth Internation Symp. On Autonomous Decentralized Systems (ISADS´03). IEEE Computer Society, 2003.

LIN, 2005. Local Interconnect Network protocol website available at http://www.lin-subbus.org/.

Mayer, A., 2005, "*Innovation is accelerating in the Automotive Electronics*", Compiler, retrieved from http://www.synopsys.com/news/pubs/compiler/art1lead_bmw-1101.html

Ong, E., 2003, "*From Anonymity to Ubiquity: A Study of Our Increasing Reliance on Fault-TolerantComputing"*, MIT SERL, NASA Goddard OLD, Dec. 9, 2003.

Porter, M., 1988. *"Competitive Strategy"*, Free Press Editorial, 1988.

Rushby, J., 2001. "*Bus Architectures for Safety-Critical Embedded Systems*", Proceedings of EMSOFT 2001: First Workshop on Embedded Software, 8-10, October 2001, Lake Tahoe, CA. Springer-Verlag LN in C.S.

Sterman, John, D., 2000. "Business Dynamics: Systems Thinking and Modeling for a Complex World", Irwin-McGraw-Hill Ed., 2000.

TTTech 2004, "*CAN/TTCAN – Byteflight-FlexRay-TTP: Technical Comparison of protocol properties with a focus on safety-related applications*", PPT presentation available at TTTech Website on http://www.tttech.com

X-by-Wire Consortium, 1998. "*X-by-Wire: Safety Related Fault Tolerant Systems in Vehicles*", Final Report, XbyWire-DB-6/6-24, Nov26, 1998, 2.0, http://www.vmars.tuwien.ac.at/projects/xbywire

ISO/DIS 11898-4; " Road Vehicles – Controller Area Network (CAN) - Part 4: Time Triggered Communication.

ELSEVIER
IFAC
PUBLICATIONS

DESIGN-PATTERNS BASED DEVELOPMENT OF AN AUTOMOTIVE MIDDLEWARE

Ricardo Santos Marques - Françoise Simonot-Lion

LORIA - INPL
Campus Scientifique, BP 239
54506 Vandoeuvre-lès-Nancy - France
{santos,simonot}@loria.fr
tel: +33 3 83 58 17 28, fax: +33 3 83 58 17 01

Abstract: An automotive middleware layer masks the heterogeneity of platforms, and provides high level communication services to applicative tasks. In addition, this layer is a software architecture, shared between car makers and third-part suppliers, ensuring the portability and interoperability of the applicative tasks. In this study, a method aiming at developing the middleware's software architecture, and obtaining a set of tasks well characterized representing the middleware's implementation, is presented. This architecture is built with a set of design patterns, and identifies a set of middleware tasks whose characteristics allow the execution of an algorithm trying to determine a feasible priority allocation for the set of applicative and middleware tasks.

Keywords: Design Patterns, Scaling, Embedded Systems, Real-Time Middleware.

1. INTRODUCTION

Context of the study. On each node of an in-vehicle network, a set of applicative tasks execute control algorithms. Automotive functions may be performed by several distributed applicative tasks, and thus, these tasks communicate by producing and consuming signals (e.g. the number of RPM of the engine) that are sent over the network. On each node, the goal of a middleware layer is, on the one hand, to mask the heterogeneity of communication platforms. On the other hand, to offer communication services independent of applicative tasks location, and other more specialized such as diagnostic modules or I/O abstraction. In this study, the emphasis is given on the following set of communication services: sending of produced signals, and reception of signals to be consumed.

Since car makers purchase components developed by third-part suppliers, this middleware layer becomes a software architecture, shared between these actors, which ensures the portability and the interoperability of the applicative level code. Moreover, the execution of the communication services provided by the middleware interferes with the tasks running in a node, and hence, increases the probability of the timing constraint, named relative deadline, associated to the execution of applicative tasks not being met.

Problem definition. The problem faced by car makers and third-part suppliers is, on the one hand, the development of a middleware's software architecture that improves the maintenance and the reusability of the software components, and can be easily documented. Note that if these characteristics are achieved, the middleware's software is easily exchanged between car makers and third-part suppliers, and can be adapted to different car makers needs. On the other hand, there is a problem of starting from this software architecture, and obtaining a middleware's implementation that allows to verify that the relative deadline imposed on the execution of tasks is respected.

Goal of the study. The objective of this paper is illustrated in figure 1. Precisely, it presents a method aiming at developing the middleware's software architecture, and obtaining a set of characterized tasks representing the middleware's implementation. The software architecture is composed of:

- a class diagram built from a set of design patterns, which specifies the code sequences executed to accomplish the middleware's communication services, and
- a set of tasks capable of executing on the OSEK/VDX Operating System (OSEK/VDX OS (OSEK Consortium 2005)), which is becoming the standard operating system for event-triggered automotive applications. These tasks are identified using a strategy whose criterion is adapted to the properties of OSEK/VDX OS, and implement the sequences of code identified in the class diagram.

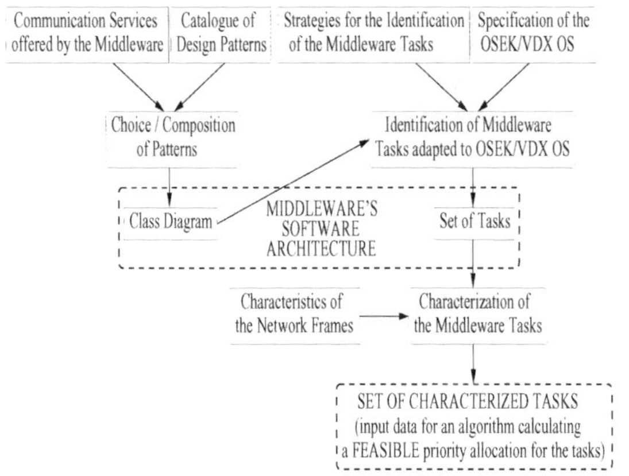

Figure 1. Objectives (represented by dashed boxes) of this study. The steps and the intermediary results are illustrated by rectangles with and without rounded corners respectively.

To obtain a set of middleware characterized tasks (characteristics like the execution time and the activation period), we use the parameters of the frames transmitted over the network. From these parameters (signals composing each frame and their emission period) we derive the work and the activation rates of the middleware tasks. From this point, one is able to quantify the interference of middleware tasks, and the entire set of tasks (applicative and middleware) form the input data for an algorithm that tries to calculate a priority allocation allowing the respect of the tasks relative deadline.

Previous work. To our best knowledge, design patterns (Gamma *et al.* 1995, Buschmann *et al.* 1996, Schmidt *et al.* 2000) have not been yet applied in the automotive systems development, but some work exists concerning their application to the design of a real-time middleware, TAO (**T**he **ACE ORB** (Schmidt and Cleeland 1999)). This middleware, specified using patterns, offers services for applications with real-time QoS requirements like video-on-demand or teleconferencing. However, it is designed to be dynamically configurable, and due to its resources consumption, is not a feasible solution in the automotive systems context, where the costs pressure is very strong. Moreover, there is no identification of the tasks that will actually implement TAO in the system.

The construction of a configuration of in-vehicle network frames has been studied in (Marques *et al.* 2003, Saket and Navet 2003). These proposed algorithms construct a set of frames, such that, the timing constraints associated to the signals are met, and the bandwidth consumption is minimized. However, these studies do not deal with the development of a middleware capable of performing the transmission and reception of these frames.

Organization of the study. The reminder of this paper is organized as follows: section 2 explains how the class diagram of the middleware's software architecture is obtained, and particularly, lists the set of used design patterns. Section 3 introduces the strategy allowing the identification of a set of middleware tasks able to execute on top of the OSEK/VDX OS, and whose task model permits the calculation of their interference on applicative tasks.

2. CLASS DIAGRAM OF THE SOFTWARE ARCHITECTURE OF THE MIDDLEWARE

A usual method for the design of software architectures is based on UML (OMG 2004). In particular, the identification of the structural components of the architecture can be achieved through the use of class diagrams (see figure 2 for an example). For this purpose, we propose a method based on design patterns, whose structural representation is done using this kind of diagram. We therefore present, on the one hand, the benefits of using design patterns for the development of the middleware's software architecture, and, on the other hand, the class diagram identifying the components of the architecture, as well as, the design patterns used to achieve it.

2.1 *Benefits of using design patterns*

A design pattern (Gamma *et al.* 1995, Buschmann *et al.* 1996, Schmidt *et al.* 2000) identifies the main aspects of a given object-oriented design structure: the participating classes and objects, their roles, and relations. The goal is to solve design problems arising in a certain context, to make these designs more flexible and reusable, and to improve the documentation and maintenance of existing systems by creating a pattern language. Numerous problems are addressed by design patterns: structural (architecture and organization of classes), behavioral (event-handling, synchronization, concurrency), etc.

In-vehicle embedded software, and particularly the middleware, should take advantage of the use of patterns: increased reusability and improved maintenance of software efficient solutions in order to better react to the demands of new automotive functions. Moreover, design patterns are a good solution to provide portability and interoperability between separately developed software components, which are faced with crucial issues typical of a multi-task context: concurrency and synchronization.

2.2 *Design patterns for the software architecture*

The class diagram representing the software components of the middleware is shown in figure 2. It is composed of the set of classes that participate in the design patterns used to build the software architecture of the middleware. In order to obtain this class diagram from the set of design patterns, a "composition" activity is needed. Note that there is, for the present, no formal technique allowing to accomplish this activity. Following an intuitive rule, we selected in the structural description of each design pattern, the class that represents the core functionality of the middleware. Such a class is present in each used design pattern, and hence, we "merged" them in a unique class termed *Core* in figure 2. Obviously, the role of this class is different in each design pattern. In the following, the used patterns, as well as their application to the middleware's context, are introduced:

- *Adapter* (Gamma *et al.* 1995): this pattern allows classes to cooperate together when their interfaces are incompatible. It is composed of an abstract class defining a standard interface to be used by client classes, and of an adapter class that makes the translation between the standard interface and the incompatible one. In figure 2, this pattern is illustrated by a set of adapter classes (*AdMOST* and *AdCAN*), which adjust the interface of in-vehicle networks (MOST (MOST Cooperation 2004) and CAN (ISO 1994) in this case) to a standard set of network services defined in the abstract class *Comm*. This pattern helps the middleware to handle the heterogeneity of communication platforms, and allows the middleware's main class, named *Core*, to be developed and modified independently of the underlying communication network.
- *Observer* (Gamma *et al.* 1995): it should be used when an object must notify other objects without making assumptions about which these objects are. This pattern creates a loose dependency between objects, such that, when the state of an object changes, all its dependents (or observers) are immediately notified. It is represented in figure 2, firstly, by classes *Core* and *Comm* that must be immediately notified when a new frame arrives (class *Comm* must notify class *Core*) or is ready to be sent (class *Core* must notify class *Comm*). Secondly, by the abstract class *SubjObs* defining the interface that each observer and observed class must implement (both classes *Core* and *Comm* are "observer" and "observed"). This pattern permits classes *Core* and *Comm* to evolve independently without hindering the possibility of passing data between them.
- *Asynchronous Completion Token* (Schmidt *et al.* 2000): the purpose of this pattern is to allow an object to efficiently demultiplex the responses of asynchronous services invoked on other objects. For that, when an asynchronous service is invoked, the invoker passes a token (under the form of an object) containing information that identifies the function responsible for processing the service's response. When the service terminates, the response contains the token and thus, the invoker object can identify the function that will process the response. In the middleware's context, this pattern lets class *Core* (see figure 2) efficiently manage the frame transmission completion events dispatched by the network adapter (class *Comm* in figure 2). If the used communication platform does not provide this type of event, or the service cannot be implemented as asynchronous, the pattern can still be used with the purpose of encapsulating the information exchanged between these two actors. Hence, this pattern contributes to the creation of a loose coupling between middleware classes and still allowing an efficient exchange of data.
- *Integrated Scheduler, variant of the Active Object* (Schmidt *et al.* 2000): this pattern

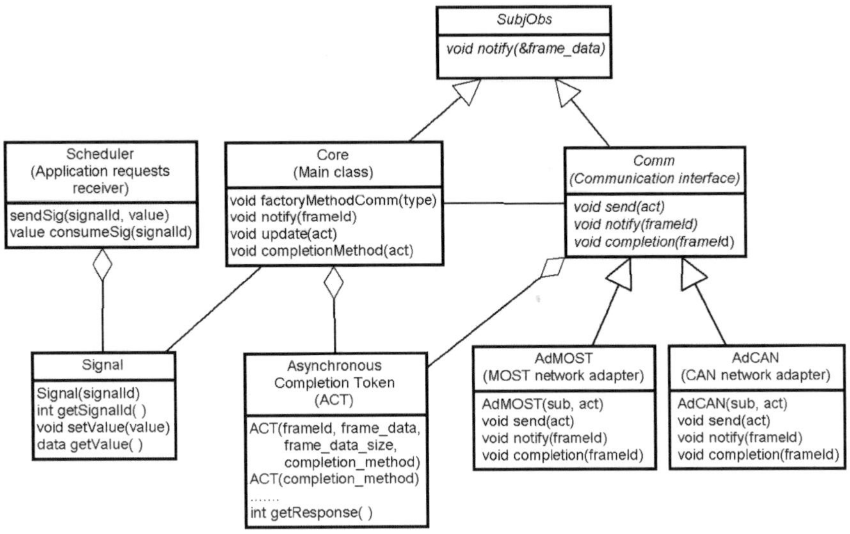

Figure 2. UML class diagram representing the software specification of the middleware. The classes are the actors of the used design patterns

addresses a concurrency aspect by decoupling the service invocation (occurring in the client's task) from the service execution (happening in a separate task). In the middleware's class diagram of figure 2 the pattern is composed of:

- a service provider represented by class *Core*,
- a service requests receiver specified by class *Scheduler* that defines the communication interface provided by the middleware, and
- a service requests repository depicted by class *Signal*, where applicative tasks store the produced signals and retrieve the signals to be consumed.

While class *Scheduler* is executed in applicative tasks, class *Core* is ran in a separate set of tasks, and class *Signal* represents a shared memory area. Therefore, the communication services provided by the middleware are executed asynchronously from applicative tasks. The main consequence is that the functionalities executed by the tasks running class *Core* simply become to, one the one hand, construct and send frames containing the produced signals, and, on the other hand, receive and handle the frames carrying the signals to be locally consumed.

Moreover, the fact that the production of signals (performed by applicative tasks) and their transmission is carried out by different tasks, has the advantage of allowing the middleware to send several signals in each frame. For this purpose, a frame packing algorithm can be used to determine a configuration containing information like the distribution of the signals among the frames, and the instants when these frames must be transmitted. Some frame packing algorithms exist applying optimization strategies aiming at, for example, minimizing the bandwidth consumption (Marques *et al.* 2003, Saket and Navet 2003).

3. IDENTIFICATION AND CHARACTERIZATION OF THE MIDDLEWARE TASKS

From the software architecture presented in section 2, one can determine the sequences of code that implement the services of the middleware, and can conclude that the functionalities allowing to accomplish these services are executed by a set of tasks. The next logical step is to identify this set, as well as to specify the sequence of code that will be executed by each task. Moreover, the identified tasks must be able to run on the OSEK/VDX OS, and must be characterized (activation period, execution time, ...) in order to allow the execution of an algorithm for the calculation of a feasible priority allocation for the entire set of tasks (applicative and middleware).

This section begins with the presentation of the activation mechanisms that trigger the execution of the middleware functionalities, followed by a set of strategies applicable in this context and based on the activation events (instances of the activation mechanisms) handled by the middleware. Next, the chosen strategy is introduced, the technique used to retrieve the code sequence executed by each task is given, and finally, a discussion of the chosen strategy is shown.

3.1 Activation mechanisms triggering the functionalities of the middleware

The two functionalities that the set of middleware tasks must perform are the construction and sending, and the reception and handling of frames. Since these functionalities are executed asynchronously from applicative tasks (see section 2.2), the tasks supporting their execution need activation mechanisms that can be provided by the OS. OSEK/VDX OS offers different means, and among them, we keep the following: hardware interrupts, and timing alarms.

The functionality responsible for the construction and sending of frames is executed periodically according to the frame packing configuration. This functionality can be efficiently activated through a cyclic timing alarm. The execution of the functionality receiving and handling frames can be either triggered by a cyclic timing alarm (polling period), or by a network controller interrupt. The former activation mechanism degrades the middleware's performance by increasing the time delay between the arrival of the frame and its handling. Thus, in this study, we consider the following types of activation events:

- *time-triggered*: OSEK/VDX OS cyclic timing alarms for the periodic activation of the functionality responsible for the construction and transmission of frames, and
- *event-triggered*: network controller interrupts indicating the sporadic arrival of frames, and triggering the functionality in charge of receiving and handling those frames.

3.2 Different strategies for the identification of middleware tasks

We have specified above the activation mechanisms for each functionality. The problem now is to determine how many tasks have to be identified according to the set of activation events. From the work of (Douglass 1999) and (Saksena *et al.* 2000), one can construct a list of strategies based on the set of activation events and applicable in the middleware's context:

(1) *one task for each event*: this strategy assigns one task for each frame that is received (if for each different frame there is a different interrupt), and for each cyclic timing alarm (assuming an alarm for each different frame emission period). The number of middleware tasks depends on the number of different frames that are received, and on the number of distinct transmission periods.

(2) *one task for each type of event*: this strategy identifies one task to handle all cyclic timing alarms, and one task to manage all network controller interrupts. The amount of middleware tasks is dependent on the different types of activation events. In this case, there are only two types and thus, two tasks.

(3) *one task for each purpose*: one example of purpose in the middleware's context is the set of signals that the nodes exchange for operation mode management (e.g. Pre-Run-Mode for node testing and network initialization, Run-Mode for full functionality of the in-vehicle system, ...). For instance, one task can be periodically activated by a cyclic timing alarm in order to send a frame containing the signal indicating the current mode, and can be activated by an interrupt caused by the arrival of the frame carrying the signal informing on the new mode. The number of middleware tasks identified by this strategy depends then on the purpose of the signals.

3.3 Chosen strategy

The specification of the OSEK/VDX OS advises, according to the used conformance class, to limit to 8 or 16 the number of priorities and the number of tasks (the one executing plus those in the ready queue). Otherwise, the portability of the software components is not assured. From this limitation, one must minimize the amount of middleware tasks, allowing the execution of a maximum number of applicative tasks. One can therefore exclude the utilization of the strategies 1 and 3. The chosen strategy is the one that assigns one task to each different type of activation events. There is then one task responsible for the construction and sending of frames, activated by cyclic timing alarms, and another in charge of handling the newly arrived frames, triggered by network controller interrupts.

Note that from this point, a feasible priority allocation for the entire set of tasks (applicative and middleware) has to be determined. This can be achieved with the optimal Audsley algorithm (Audsley 1991): we recall that if a solution exists then it will necessarily be found. The feasibility test must however calculate the worst-case response time of the middleware tasks. To perform this calculation, the characteristics of the tasks are needed.

3.3.1. Characteristics of the task handling frames
In OSEK/ VDX OS, this task would be most efficiently implemented as an interrupt service routine (ISR) activated by network controller interrupts, decreasing even more the number of tasks necessary to implement the middleware. In this OS, the ISRs have a higher priority than any other regular task, hence, it is not necessary to determine its priority. To quantify its interference

on other tasks, one needs to calculate its execution time and activation period. These values depend on the type of the underlying network.

On both event-triggered and time-triggered types of networks, the time interval between the arrival of any two frames is not constant. The ISR is thus considered as sporadic (Liu and Layland 1973, Mok 1983), and its activation period is set to the smallest value possible in its context. In a event-triggered network, the activation period of the ISR is equal to the time needed to transmit the smallest frame that is received. Note that in this case, this estimation is very pessimistic. In a time-triggered context, the activation period is equal to smallest time interval between the emission of two frames received by the ISR. In both cases, the execution time is assigned to the time necessary to handle the largest frame received. The worst-case response time calculation for this type of task model is detailed in (Tindell 1992).

3.3.2. Characteristics of the task sending frames
Being responsible for the transmission of frames, the characteristics of this task depend on the frame packing configuration. The frames to transmit can however be assigned a different emission period, and therefore, the activation rate of the task is obliged to respect all those periods. Consequently, we cannot use the usual task model where tasks have a unique activation period and execution time (Liu and Layland 1973). We have to study extended models as multiframe (Mok and Chen 1996) and generalized multiframe (GMF) (Baruah *et al.* 1999):

- if one assumes that the first emission request of all frames is issued by the first instance of the task, the multiframe task model can be used. A multiframe task ϕ_i is characterized by a set C_{ϕ_i} composed of N execution times such that $C_{\phi_i} = (c_{\phi_i}^{(0)}, c_{\phi_i}^{(1)}, ..., c_{\phi_i}^{(N-1)})$, and by a unique activation period T_{ϕ_i} and relative deadline $\bar{D}_{\phi_i}$. The worst-case response time calculation method for this type of task model was introduced in (Takada and Sakamura 1997).

  From the frame packing configuration, one derives the characteristics of a multiframe task ϕ_i as follows. Let the set $Q_i = \{(Q_{f_{i,1}}, T_{f_{i,1}}), ..., (Q_{f_{i,k}}, T_{f_{i,k}})\}$ where $Q_{f_{i,k}}$ is the time needed to construct and request transmission of frame $f_{i,k}$, and $T_{f_{i,k}}$ is the transmission period of the frame. The activation period and relative deadline, T_{ϕ_i} and $\bar{D}_{\phi_i}$, are simply $\gcd(T_{f_{i,1}}, ..., T_{f_{i,k}})$. For each activation during the first hyperperiod, $\mathrm{lcm}(T_{f_{i,1}}, ..., T_{f_{i,k}})$, one determines the frames that are to be sent and, thus, the set of execution times for ϕ_i:

  $$\forall\, 0 \le a \le \mathrm{lcm}(T_{f_{i,1}}, ..., T_{f_{i,k}})/T_{\phi_i} - 1,$$

  $$c_{\phi_i}^{(a)} = \sum_{\{k \,|\, a \cdot T_{\phi} \bmod T_{f_{i,k}} = 0\}} Q_{f_{i,k}}$$

- if one of the several execution times is greater than the relative deadline, then one can try to overcome this problem by implementing this task as generalized multiframe (GMF) (Baruah *et al.* 1999). Again, we assume that the first emission request of all frames is issued by the first instance of the task. The main difference from the multiframe task model is that the activation period and relative deadline also become a vector composed of N elements. We have then $T_{\phi_i} = (t_{\phi_i}^{(0)}, t_{\phi_i}^{(1)}, ..., t_{\phi_i}^{(N-1)})$ and $\bar{D}_{\phi_i} = (\bar{d}_{\phi_i}^{(0)}, \bar{d}_{\phi_i}^{(1)}, ..., \bar{d}_{\phi_i}^{(N-1)})$. To determine the worst-case response time of GMF tasks one can use the algorithm presented in (Takada and Sakamura 1997).

  From the same set $Q_i = \{(Q_{f_{i,1}}, T_{f_{i,1}}), ..., (Q_{f_{i,k}}, T_{f_{i,k}})\}$ (see the configuration of a multiframe task), one constructs the vector of activation periods (and relative deadlines) of a GMF task ϕ_i as follows:

  $$t_{\phi_i}^{(0)} = \min(T_{f_{i,1}}, ..., T_{f_{i,k}}),$$

  $$t_{\phi_i}^{(j+1)} = \min_{\forall k}(a \cdot T_{f_{i,k}} - \sum_{l=0}^{j} t_{\phi_i}^{(l)}),$$

  $$a = \min\{i \in \mathbb{N}^+ \,|\, i \cdot T_{f_{i,k}} > \sum_{l=0}^{j} t_{\phi_i}^{(l)}\}$$

  This vector is built until the following expression becomes true:

  $$\sum_j t_{\phi_i}^{(j)} = \mathrm{lcm}(T_{f_{i,1}}, ..., T_{f_{i,k}})$$

  For the vector of execution times one acts in the following way:

  $$c_{\phi_i}^{(0)} = \sum_k Q_{f_{i,k}},$$

  $$c_{\phi_i}^{(j+1)} = \sum_{\{k \,|\, \sum_{l=0}^{j} t_{\phi_i}^{(l)} \bmod T_{f_{i,k}} = 0\}} Q_{f_{i,k}}$$

  Furthermore, since GMF tasks do not have a unique activation period, their implementation on top of OSEK/VDX OS is not trivial. Appendix A illustrates the problem and proposes a solution.

If however one of the execution times is still greater than its corresponding relative deadline, the solution is to split the work in two tasks (either multiframe or GMF). One task, having a higher priority, would be responsible for the transmission of the frames with smaller deadline, while the other task would be in charge of sending the frames with larger deadline. This solution splits the work among two tasks, and increases the probability of respect of frames deadline, by delegating the transmission of those with a stricter timing constraint to the higher priority task.

3.4 Generation of the code executed by the tasks

The sequence of code that each task must execute contributes for the task's execution time. Recall that one must well characterize each task, in order to allow the Audsley algorithm to determine the worst-case response time of each task when trying to calculate a feasible priority allocation. The code executed by each task is then retrieved by simulating the triggering of a network controller interrupt and a timing alarm, and performing a run-to-completion through the set of classes. This procedure identifies the set of objects necessary to instantiate in each task.

3.5 Discussion of the proposed strategy

This strategy based on the type of the activation events, is, in our opinion, the more suited to identify a set of tasks adapted to the middleware's context. The reasons justifying this choice are:

- the minimization of the number of middleware tasks. This feature is important because OSEK/VDX OS specifies a maximum number of tasks in order to guarantee the portability of the software components;
- if new frames with different transmission periods must be sent, the characteristics of the middleware tasks change but not their task model. Nevertheless, the amount of tasks may increase if one of the execution times of the multiframe or GMF task is greater than the relative deadline (see section 3.3.2);
- if a new service is included in the middleware, these two tasks are capable of executing it if its activation events depend on interrupts or timing alarms.

Besides the maximum bound on the amount of tasks, other problems of the OSEK/VDX OS are the limit of one alarm, and the restriction of one task activated per alarm. If at least one applicative task uses the alarm, no other task, applicative or middleware, is able to employ this mechanism for its activation. An inconvenient of our strategy relies then on the fact that at least one alarm is needed (activation of the middleware task sending frames). Without any other mechanism that might be used for periodic task activation, we are obliged to take the risk of non-portability of the middleware's software.

CONCLUSION

This study proposes a method for the development of the software architecture of an in-vehicle communication middleware. It presents a class diagram built from a set of design patterns, and introduces a strategy for the identification of a set of middleware tasks adapted to the characteristics of the OSEK/VDX Operating System.

The proposed architecture implements communication services provided to applicative tasks, masks the heterogeneity of in-vehicle networks, and specially, benefits from the advantages of using design patterns: increased reusability, improved maintenance and evolution, and easier documentation. The identified tasks accomplish the middleware's communication services, and are characterized in order to allow one to run the Audsley algorithm, aiming at determining a set of priorities that permits the respect of the relative deadline of applicative and middleware tasks.

Future work consists of the definition of an improved frame packing algorithm that tries to build the set of network frames and the priority allocation for the tasks of each node, such that, the timing constraints of applicative and middleware tasks, and signals are met. The goal is to implement the middleware's software architecture presented in this study, and to be able to generate its configuration, and the one of OSEK/VDX OS, in conformity with a given set of characterized applicative tasks and signals.

REFERENCES

Audsley, N. (1991). Optimal priority assignment and feasibility of static priority tasks with arbitrary start times. Technical Report YCS164. University of York.

Baruah, S., D. Chen, S. Gorinsky and A. Mok (1999). Generalized multiframe tasks. *Real-Time Systems* **17**(1), 5–22.

Buschmann, F., R. Meunier, H. Rohnert, P. Sommerlad and M. Stal (1996). *Pattern-Oriented Software Architecture - A System of Patterns.* John-Wiley and Sons.

Douglass, B. (1999). *Real-time UML Second Edition - Developing efficient objects for embedded systems.* Addison-Wesley Longman Publishing Co., Inc.

Gamma, E., R. Helm, R. Johnson and J. Vlissides (1995). *Design Patterns: Elements of Reusable Object-Oriented Software.* Addison-Wesley.

ISO (1994). *ISO 11898 - Road vehicles - Interchange of digital information - Controller Area Network for high-speed Communication.* International Standard Organization. ISO 11898.

Liu, C. L. and James W. Layland (1973). Scheduling algorithms for multiprogramming in a hard-real-time environment. *J. ACM* **20**(1), 46–61.

Marques, R. Santos, N. Navet and F. Simonot-Lion (2003). Frame packing under real-time constraints. In: *5th IFAC International Conference on Fieldbus Systems and their Applications - FeT'2003, Aveiro, Portugal.* pp. 185–192.

Mok, A. (1983). Fundamental Design Problems for the Hard Real-Time Environments. PhD thesis. Massachusetts Institute of Technology (MIT).

Mok, A. and D. Chen (1996). A multiframe model for real-time tasks. In: *RTSS '96: Proceedings of the 17th IEEE Real-Time Systems Symposium (RTSS '96)*. IEEE Computer Society. p. 22.

MOST Cooperation (2004). MOST Media Oriented Systems Transport specification, version 2.3. Available at http://www.mostcooperation.com.

OMG (2004). *OMG Unified Modelling Language: Superstructure.* version 2.0 ed.

OSEK Consortium (2005). OSEK/VDX operating system, version 2.2.3. Available at http://www.osek-vdx.org.

Saket, R. and N. Navet (2003). Frame packing algorithms for automotive applications. Technical Report INRIA RR-4998.

Saksena, M., P. Karvelas and Y. Wang (2000). Automatic synthesis of multi-tasking implementations from real-time object-oriented models. In: *Proceedings of the Third IEEE International Symposium on Object-Oriented Real-Time Distributed Computing.* IEEE Computer Society. p. 360.

Schmidt, D. and C. Cleeland (1999). Applying patterns to develop extensible ORB middleware. *IEEE Communications Magazine.*

Schmidt, D., M. Stal, H. Rohnert and F. Buschmann (2000). *Pattern-Oriented Software Architecture.* Vol. 2: Patterns for Concurrent and Networked Objects. John-Wiley & Sons.

Takada, H. and K. Sakamura (1997). Schedulability of generalized multiframe task sets under static priority assignment. In: *RTCSA '97: Proceedings of the 4th International Workshop on Real-Time Computing Systems and Applications (RTCSA '97).* IEEE Computer Society.

Tindell, K. (1992). An extendible approach for analyzing fixed priority hard real-time tasks. Technical Report YCS189. University of York.

Appendix A. IMPLEMENTATION OF GENERALIZED MULTIFRAME TASKS ON OSEK/VDX OPERATING SYSTEM

Since generalized multiframe (GMF) tasks do not have a unique activation period, in OSEK/VDX OS this value can only be assigned dynamically. Each instance of a GMF task, just after its beginning of execution, cancels the previous alarm, and sets a new one equivalent to the next activation period. This procedure however, does not guarantee the respect of the set of activation periods. Figure A.1 describes this problem.

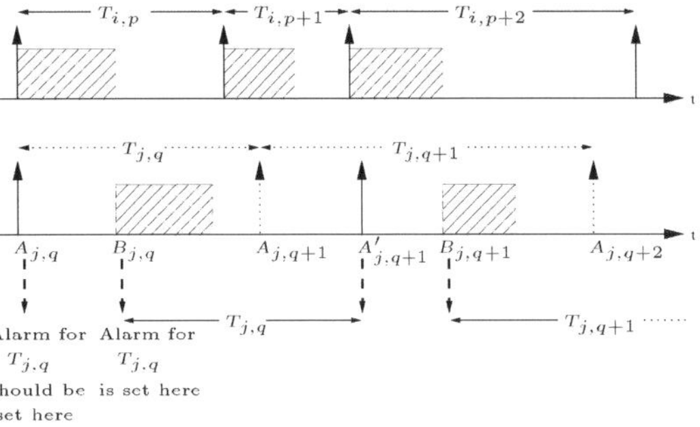

Figure A.1. This figure demonstrates the difficulty in the setting of an OSEK/VDX Operating System alarm that respects all activation periods of a GMF task. For task τ_j, one could set an alarm at the activation instant of its q-th instance - $A_{j,q}$ - in order to activate the $(q + 1)$-th instance $T_{j,q}$ units of time later. Since the q-th instance cannot start its execution when activated (higher priority task τ_i is running), the alarm will be set too late, at the beginning of the execution of the instance - $B_{j,q}$. The $(q+1)$-th instance is therefore activated $(B_{j,q} - A_{j,q})$ units of time too late. In the figure, instant $A_{j,q+1}$ is the activation instant that would allow the respect of $T_{j,q}$, while instant $A'_{j,q+1}$ is the activation instant that effectively occurs

Task τ_j, whose q-th activation takes place at instant $A_{j,q}$, cannot begin its execution since an instance of an higher priority task τ_i is executing. The setting of the new alarm that should take place as sooner as possible after instant $A_{j,q}$, is effectively set at instant $B_{j,q}$ the beginning of execution of the q-th instance of τ_j. Future activations of τ_j are now delayed of $B_{j,q} - A_{j,q}$ units of time. Note that this delay is increased each time an instance of τ_j cannot begin its execution at its activation instant. To overcome this problem, when the q-th instance of task τ_j begins its execution, it must calculate the value of $B_{j,q} - A_{j,q}$. Instead of setting the alarm with $T_{j,q}$, it sets with $T_{j,q} - (B_{j,q} - A_{j,q})$.

ELSEVIER
IFAC
PUBLICATIONS

LARGE-SCALE DATA ACQUISITION AND SCADA APPLICATIONS OVER POWER LINE NETWORKS

Maxim Lobashov[†], **Gerhard Pratl**[†], **Thilo Sauter**[‡]

[†] *Vienna University of Technolgy*
Gusshausstr. 27-29/384
A-1040 Vienna, Austria
{lobashov,pratl}@ict.tuwien.ac.at

[‡] *Austrian Academy of Sciences*
Viktor Kaplan Str. 2
A-2700 Wiener Neustadt, Austria
thilo.sauter@oeaw.ac.at

Abstract: A system that enables narrow band power line communication to be used for data acquisition applications over large areas, such as a district or a town, is presented in its design and architecture. We present the capabilities of the system, which include the possibility to use existing applications and field-level equipment and extend their communication ability to the wide-area power line network. The core of the communication system is designed to employ not only power line communication, but also other communication channels. *Copyright © 2005 IFAC*

Keywords: telecontrol, communication networks, protocols, communication systems, automation, data acquisition

1. INTRODUCTION

The communication systems used for many purposes in contemporary energy management are rather conventional with respect to networking capabilities and complexity. Applications like SCADA (Supervisory Control And Data Acquisition) or Automated Meter Reading (AMR) typically require a designated communication channel that they may use exclusively). Standards like EN 62056 (also known as IEC 1107), IEC 60870, or M-Bus are widely used, but are based on the availability of point-to-point connections and allow – if at all – only a limited number of participants on a bus. Addressing and managing of comprehensive networks comprising an adequate number of devices embedded in a hierarchical structure is not foreseen in the protocols. The simplicity of the communication system is also reflected on the physical level. In practice, this is typically a serial line (RS 232 or RS 485). Hence, the implementation of larger systems requires two essential ingredients: a communication channel to bridge the distance between the utility company (the respective control room, to be precise) and the devices or device clusters; and the massive use of parallel channels to overcome networking limitations of the protocols.

A typical way to connect utility servers and field equipment is to use existing telecommunication channels – either dial-up or mobile – or wireless networks. In most cases, however, such infrastructures are provider-based, which reduces in particular the economical control the utility company can gain over a communication system vital for their business. It is therefore obvious that means are sought to use a different infrastructure, which is under the control of the company: the power distribution grid. Contrary to the highly disputed and not overly successful attempts to provide Internet over the power line, SCADA and AMR systems have lower data rate requirements and can be operated using narrowband power line communication, which is much less problematic in terms of electromagnetic interference. Nevertheless, the power line is a rather hostile communication channel, and if it is to be used with state-of-the-art SCADA and AMR protocols and field devices, special precautions have to be taken.

This paper presents the approach to using power line communication (PLC) for low-level data transmission together with standard protocols developed in the EU project REMPLI (Real-time Energy Management with Power-Lines and Internet, project identifier NNE5-2001-00825). The technology of power line communication is covered in (Dostert, 2001). The goal of REMPLI is to use existing applications or equipment (utility meters, switchgear equipment and the like) as much as possible. Therefore we have to provide a channel that is transparent to the application while at the same time fulfilling all its communication needs. A key issue in the communication system is flexibility: due to its modular design it is possible to not only use power line communication as the data transport channel, but also other kinds of communication networks such as the GSM network, wireless networks or dial-up connections. Therefore we have introduced a communication interface that shields the properties of the underlying medium from higher layer communication services. Another aspect is modularity with respect to the higher-level protocols used on top of PLC. The primary goals are SCADA and AMR, but extensions to other applications like energy management or domotic applications are possible and can easily be adopted.

This paper focuses on the power line related parts of the communication system. The Internet Protocol (IP) network, which is also part of the communication system, is not considered here. This is also true for the security related issues that have to be taken into account when using a public network like the power grid for exchanging sensitive data. The security concept of the REMPLI system is described in (Treytl and Sauter, 2005).

The rest of the paper is organized as follows: in section 2 the applications that benefit from the communications system are presented, section 3 explains the details of power line communication layout, while section 4 takes a look at protocol stack. Section 5 describes how to implement different fieldbus protocols seamlessly into the communication system and section 6 shows the communication services that are available for the applications. In section 7 we examine the issue of a dynamically changing power grid and the implications for PLC and section 8 sums up the paper and gives conclusions.

2. REMPLI APPLICATION AND BOUNDARY CONDITIONS

The main goal of the REMPLI project is to design and implement a remote metering and SCADA control network, using power line communication (PLC) as a medium. The use of power line is necessitated by the main target application of the system: it is designed for utilities that automate meter reading in private households and service provider companies, operating in the areas of domotics, in-house security and the like. The only economically-feasible way of providing these applications, especially in those parts of the world, where broadband Internet connectivity in private households is uncommon, is to use the power line, so that any additional wiring is avoided.

Additional to this main goal, the REMPLI system is designed such that it is suitable not only for the aforementioned metering/SCADA tasks, but can also be used for virtually any control and data acquisition application, where power line communication is deemed necessary.

A significant economy-related aspect is that utility companies, service providers and many other control network operators already have some sort of metering and actuator equipment installed. Usually such solutions allow for direct connectivity between the equipment and software (metering or SCADA) using a limited-length physical network, such as RS 232 or RS 485, running some standard or proprietary protocol. Even if this is not the case, the market currently offers a large variety of such field devices and software packages. Thus it makes no sense to develop custom hardware or software for operating over PLC. Instead, REMPLI is thought as a snap-in replacement for the existing physical wire between field devices and control software, as it is shown in Figure 1.

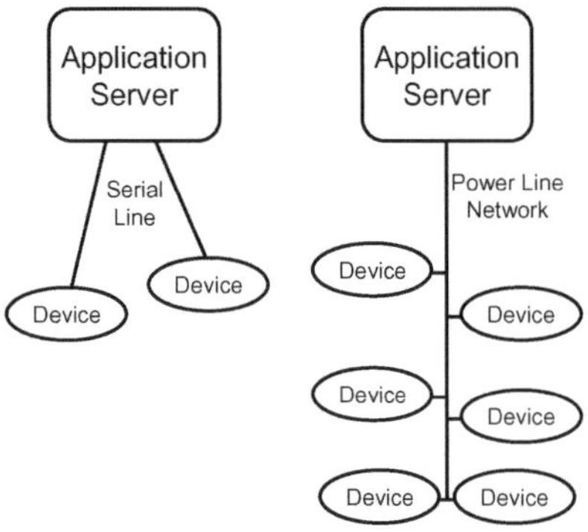

Figure 1: Using PLC network to replace direct serial line communication.

With REMPLI the control application can be substantially increased in scale (meaning the number of attached devices) and distance between them and the control center.

A main boundary condition in energy management applications is to maintain investments in application server software and field devices; the application server is in most cases already installed at the utility side and must be re-used, introducing new software would not be cost effective. Even more, an adaptation of the software to a particular communication system is not feasible either. At best it is possible to do small adaptations to the drivers that transport the metering or SCADA protocol between application server and access point.

Although the protocols used in SCADA and AMR are standardized, there is still a large variety within these protocols, because many of them are specified too loosely (especially at the application layer). For example, IEC 1107 do not ensure interoperability between solutions from different vendors (equipment as well as metering software). If the power line communication system was to be widely adopted without additional adaptations it must exhibit utmost flexibility to serve as a drop-in replacement for the currently dominating direct cabling between application servers and field equipment.

This effectively prohibits the implementation of a gateway solution: if a protocol can be used differently in different implementations (using other semantics at the application layer, as is the case in IEC 1107), it would be necessary to implement a dedicated translator module. Depending on the varieties in semantics this would be necessary for each combination of protocol and vendor. This solution is not reasonable for complexity and software maintenance reasons.

Finally, simple protocol and data translation is the preferred solution, because of national regulations, which impose stringent rules on the communication system: if, for example, metering data are used for automatic billing, the data must not be altered on their way from source (i.e., the meter) to sink (the billing application) or, if there is any equipment in the data path which touches the data as such, it must be certified separately, which is a costly procedure. Therefore it is advisable not to convert the data. Instead we have decided to tunnel all data between the end points of the communication.

3. POWER LINE COMMUNICATION

The layout of the REMPLI power line communication network is based on the requirements set by a number of utility companies – participants of the project. These requirements can be considered as "worst-case": most other foreseen control networks will utilize PLC to a lesser extent or provide a more "friendly" environment to data communication.

The power line network is comprised of two distinct segments. The first one spans from a private household or an apartment, over a low-voltage (LV) power line, to the medium-to-low voltage (MV/LV) transformer station, also called *secondary transformer*. The second segment continues PLC communication further, over medium-voltage lines, to the *primary transformer* (high-to-medium voltage). At this point utilities tend to have some sort of data communication network already installed and running (typically it is based on Internet Protocol (IP) – either via dedicated cables or wireless, such as GSM or GPRS). Thus the last communication segment, between the primary transformer and the utility or service-provider control center, is not PLC-, but rather IP-based. A single installation can encompass several MV segments, each, in turn, attached to several LV grids (the control network can cover a whole district, or even a small town).

The length of both segments, depending on the deployment area, can be rather substantial – up to several kilometers. The low-voltage part can also span across several dozen consumers (in Western Europe) or even up to 500 (in Eastern Europe) consumers (Figure 2). Both factors essentially prohibit use of existing broadband power line communication systems, such as HomePlug (Homeplug, 2005). Instead, the REMPLI project has developed its own narrow-band PLC system, suitable for the described harsh environments and distances of several kilometers (Bumiller, 2005). Conventional power line communication systems have data rates between 300 and 9600 bit/s and are designed for reading actual meter values at most once a day. These data rates are too small for transporting load profiles (where data rates between 30 and 100 kbit/s are required) and much too small for real-time energy consumption monitoring or online trading of energy in an open market via power line (which requires calculated data rates between 100 and 300 kbit/s). The REMPLI system aims at providing 500 kbit/s (100-200 kbit/s on currently regulated CENELC bands) raw bandwidth. While this raw bit rate is of course not available as payload bandwidth for the end user of the system (due to protocol overhead, repetitions and the like), it is still an impressive progress compared to existing systems.

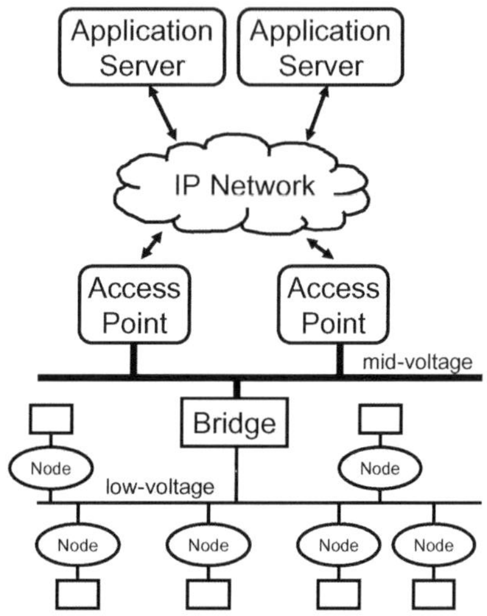

Figure 2. REMPLI communication architecture.

According to the described layout, the PLC system consists of two independent master-slave networks, operating in LV and MV segments respectively. The secondary transformer prohibits transparent data communication between the segments; thus REMPLI introduces a PLC *bridge*, located between the LV and MV parts. The bridge behaves as a slave on the MV side and as a master on the LV side.

For the reasons of reliability and security, a single low-voltage network is usually supplied from two, or even more, secondary transformers. Normally, at a given time only one of these transformers is active (in operation); other are in cold-standby. Since inactive transformers are physically disconnected from the grid, the only way to provide uninterruptible data communication is to install bridges at every secondary transformer, supplying the given LV net. From the upper (MV) side this poses no problem: every bridge acts as a regular slave. However, on the LV side it means that multiple masters are communicating on the same network (under normal conditions it is only one master, installed at the currently active transformer, but under special conditions more than one master can be active). Different masters transmit at different frequencies.

Slaves, upon loosing connection to a current master, perform a frequency scan of the medium and "reattach" themselves to the other master.

Slave devices in the LV network are PLC *nodes*. Every node incorporates a number of interfaces: RS 232/485, pulse inputs, digital outputs, etc., which enable it to communicate with locally attached field devices (such as energy meters and SCADA equipment). The node also has a certain amount of computing power, sufficient to implement communication protocols of the attached devices and installation-specific local control algorithms.

The MV-side master is an *access point*, which routes traffic between the PLC and Internet Protocol (IP) networks. The latter connect access points with application servers (control software, metering systems, SCADA, etc.).

The situation in mid-voltage network is similar to that on the low-voltage side: for redundancy, the same grid is connected to multiple primary transformers, each equipped with an access point.

The dual power supply redundancy, described above, results in a rather awkward situation for data communication:

- in the LV network nodes are uncontrollably roaming between the bridges;
- in turn, in the MV network bridges uncontrollably roam from one access point to another.

Further, due to switching processes in the power line grid, it is also possible, that during some periods of time the same LV node is reachable via multiple different paths (via different bridges and/or access points).

Neither application software, nor field devices can, or should, implement special handling for the switched and meshed PLC medium. This is done internally by the PLC communication stack, as described below. Each field device is connected to only one node; every application server also communicates to only one of the access points (any of them).

In order to cover a wider range of potential applications, the REMPLI system also supports installations with only one power line segment (either LV, or MV). In this case PLC bridges are unnecessary, since direct communication between access points and nodes is possible.

4. PLC COMMUNICATION STACK

The communication protocol stack, implemented by all devices in the REMPLI network (access points, bridges and nodes) is comprised of the following layers (Figure 3).

- **PLC physical and link layers**, which implement the actual power line coupling. An in-depth description of these two layers is beyond the scope of this article; for more details refer to (Bumiller, 2002) and (Bumiller, 2005).

- **PLC network layer**, responsible for providing datagram-oriented communication in a master/slave fashion with reliable delivery. The network layer communication runs between devices in the same PLC segment; that is, PLC bridges terminate network layer from both sides. At the masters this layer also handles logon/logoff procedures for the slaves (i.e. slaves that "attach" to a master), directly attached to the underlying PLC segment. Upper layers can query the list of slave addresses, which are currently in operation, at any time.

- **PLC transport layer** adds the end-to-end communication capabilities: from an access point, through bridges, to the target nodes and backwards. In a switched/meshed PLC this is the layer that makes a decision, which of the available communication paths should be used to deliver a packet. Information, required for such decisions, is available from the network layer (a special status-of-link table, where all available nodes, along with "link quality" characteristic values, are listed).
  The transport layer also extends network discovery mechanisms such that access points in the MV segment become aware of attached node addresses in the LV segment (information from all bridges is collected together).
  In order to perform packet forwarding between the LV and MV networks, a special transport layer architecture is required at the bridge: both master and slaves sides of the bridge share the same transport layer (unlike the network layer, which exists in two instances: one at the LV side, one at the MV side).
  Similar to the network layer the transport layer communication is reliable and datagram-oriented (connectionless).

- **Communication multiplexer/demultiplexer**. On top of the transport layer, the de/multiplexer merges (and separates on the other side) streams of packets, related to different fieldbus protocols. Another function, performed by this layer only at the access points, is re-routing of packets to nodes, currently unreachable via any of the connected slave-bridges due to PLC switching, through other access points.

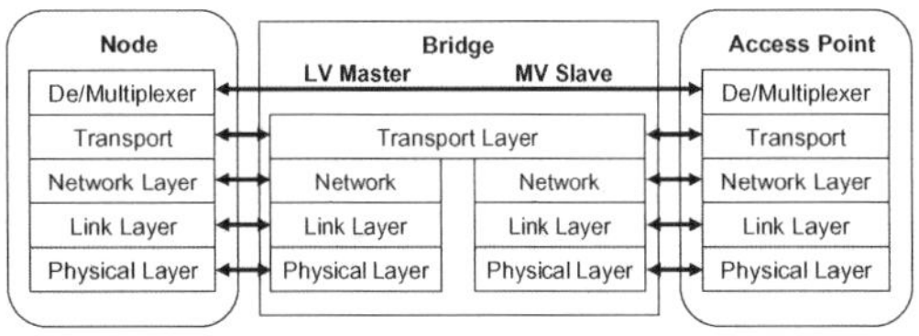

Figure 3. Protocol stack and layer-to-layer communication between different devices in the network.

In some applications the communication de/multiplexer will not be present at the bridge, since this device doesn't need to specifically process application traffic and only forwards packets up and down. However, many utility companies find it useful to install field equipment at the secondary transformers; in this case the de/multiplexer is required at the bridge as well (its architecture and functionalities are then exactly the same as at the nodes).

5. FIELDBUS PROTOCOL HANDLING

A single REMPLI installation can contain different types of field devices at the same time. For instance, one node can be connected to several IEC 1107 (IEC 62056) energy meters, while a second node provides access to M-Bus meters and an IEC 870-5-101 SCADA control box. On the other side of the network every type (group) of devices is typically controlled by a dedicated software application: e.g., all IEC 1107 meters in the network will be polled by a respective metering system.

As described above one of the design criteria for the REMPLI network was to enable transparent transmission of multiple different fieldbus protocols (Lobashov 2003). The number and types of these protocols are not limited to those integrated during system design (currently these are mainly SCADA and metering protocols). Rather, it should be possible that any third party user of the system transmits their application-specific protocols over PLC, using REMPLI as a communication platform.

Due to this requirement, the application layer of REMPLI devices is designed in a modular way. Every fieldbus protocol is handled by a "triple" of communication protocol drivers:

- At the node side the protocol driver interfaces on the one side with the field equipment (via RS 232/485, digital and analog I/Os, etc.) and on the other side exchanges application Protocol Data Units (PDUs) of the fieldbus protocol over PLC with the access points.

- The access point side driver interfaces with the control software (application server) over the IP network and also communicates with all of its "siblings" (drivers for the same protocol type) at the nodes.

- The application server uses a specific communication driver to exchange data with the respective driver at one of the Access Points over IP.

Depending on the protocols that the connected field equipment support the node can run one or more protocol drivers – each for a single fieldbus protocol (one driver can be, of course, responsible for more than one device of the same type). Every access point always runs a superset of drivers – for all fieldbus protocols, transferred over this PLC network. The application server software typically integrates only one driver, which enables it to transmit application PDUs over IP to the access points. The logical structure of the resulting communication network is shown in Figure 4.

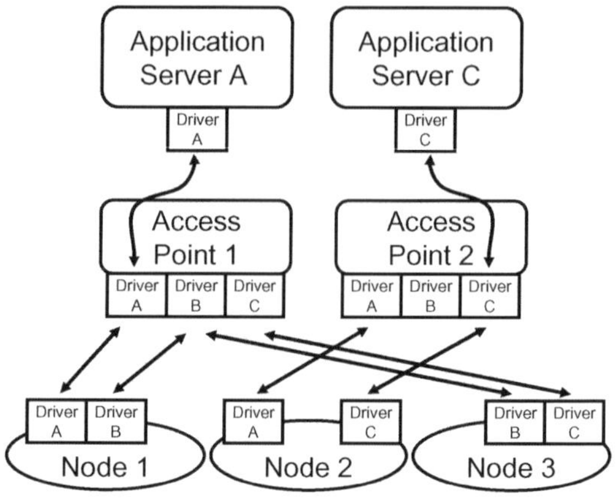

Figure 4. Communication between drivers shown as logical connections between drivers on access points and nodes.

On the PLC side the underlying protocol stack (de/multiplexer and layers below it) fully isolates communication between different types of drivers. Every driver group communicates as if it were the only user of the PLC data transmission services. On the IP side communication is also "isolated": every application server establishes its own TCP connection to one of the access points.

Functionality of the access point and node protocol drivers is not limited to simply tunneling fieldbus protocol PDUs over the PLC. Depending on the protocol, the driver design can include more or less sophisticated parsing and processing of PDUs. The main goal is to reduce bandwidth utilization in the PLC network to a minimum. For instance, an IEC 1107 driver couple will, likely, compress load profile data from meters, while it is transmitted over the power line (decompression will be performed at the access point side, so that the application server receives the original PDU). IEC 870-5-101 drivers, similarly, will suppress transmission of keep-alive polling from SCADA server.

6. COMMUNICATION SERVICES

The modular and open architecture of the application layer (communication protocol drivers) requires "standardizing" the interface, provided to it by the communication system. This interface is located above (or below, in case of a node) the de/multiplexer layer, as shown in Figure 5.

To allow for fieldbus protocol tunneling, the following data transmission services are considered necessary at the access point side, and provided to drivers by the communication system.

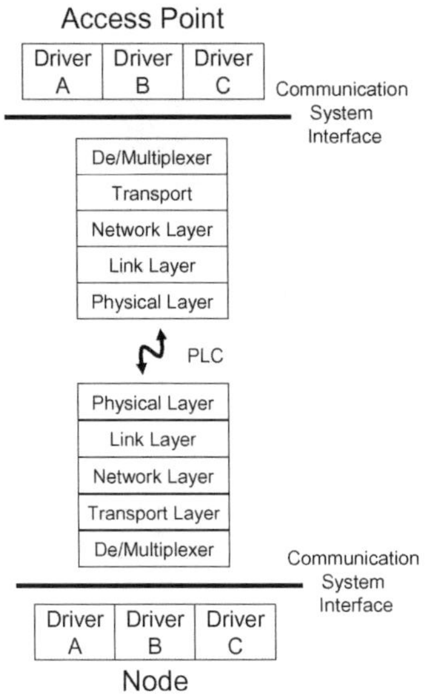

Figure 5. The communication system interface allows using different communication media without changing the system core.

- **Request/response**. Driver can generate a request, which is sent to a certain node. The driver receives a unique transaction ID for the request, after it is submitted, from the de/multiplexer layer. The communication system then waits for a response to be generated by the node. After the response is delivered back to the access point, it is returned to the driver along with the same transaction ID, which was returned upon sending the request (in this way driver can match requests against responses, in case multiple parallel request/response transactions are opened at the same time). If a node does not respond to the request within a certain amount of time, the communication system cancels the request/response transaction and informs the access point side driver about this fact.
  The request/response service is typically used by drivers for handling their respective application protocol transactions. Most metering and SCADA protocols are request/response in their nature, thus they perfectly fit to the master/slave concept of the PLC communication.

- **Request/no-response.** This service is similar to the request/response service, but a node driver does not generate any response to requests when using this service. After transmission of the request is completed, the transaction is considered to be finished, and the requester (driver) is informed about this fact.
  It's worth mentioning that, while no response is generated by the node driver, for reliability reasons the request/no-response service is still internally acknowledged within the communication system. The de/multiplexer layer reports "success" to the driver only after a request message has been successfully delivered to the destination, and this fact has been acknowledged by the node. The request/no-response service can be used by access point drivers to send additional control information to their siblings at the nodes, or to handle some specific application protocol transactions.

- **Multicast and broadcast.** This is an unacknowledged service, used to send messages to a group (or all) nodes, connected to the REMPLI network. The communication system reports that a transaction is completed after the message is transmitted to all specified nodes, but without waiting for any acknowledgement from them. If transmission did not succeed for one (or all nodes) in the multicast group, the driver has no way to learn about this fact. This service can be used, for example, to upgrade software at all nodes in the network: first, access point broadcasts new software images to all nodes and then verifies upload status with each node separately using the request/response service, completing any missing parts of the image.

- **Network discovery.** Depending on the fieldbus protocol and driver functionality, the driver at the access point may need to query information about the list of nodes, currently attached to the network. In particular, this is the case for add-on services, implemented as "virtual protocol drivers" (e.g., system configuration and monitoring "driver"). The communication system offers two such lists: one containing all nodes that are currently reachable directly from this access point and another one, where all nodes in the PLC network (including those that are only reachable via other access points) are included.

Due to master/slave communication in the PLC network, the communication system interface at the node side has to offer a different set of data transmission services:

- **Response.** A counterpart of the request/response service at the access point. When a node driver generates response to a previously received request from the access point, it sends it via the response service. Response messages are always delivered to the same access point where the request was received from (this fact is important in meshed PLC environments, where a single node is "visible" from more than one access point).

- **Alarm.** A node driver can generate an asynchronous alarm message, and send it to the access point driver (without the access point driver issuing a request first). Such messages are delivered to all access points that can currently reach this node. The communication system returns a successful completion result after the

alarm message has been transmitted; i.e., this service is unacknowledged. In a master/slave PLC system implementing acknowledged alarms would require an additional request/response transaction, which is, in most cases, unacceptable.
Since the node has no way of determining, which access point actually receives the alarm message (i.e., where the target application server is connected to), and the message is delivered to all access point protocol drivers within a group, drivers have to coordinate their "efforts" and suppress duplicate alarms. This can require additional driver-to-driver communication over the IP network.

- **Fast status transmission.** Every node driver in the PLC network has a possibility to transmit a few bits of status information to its access point siblings as out-of-band data, included into unused space of the PLC control packets (periodic slave polling, as a part of the internal network management). The transmission is asynchronous: first, the node driver sets the value of its bit field; then, as soon as the polling cycle reaches this node, the bit field is actually transferred to the access point driver. Clearly, if a node driver changes its status bit field faster than the polling cycle period, intermediate bit field values will not reach access points.

The described interface is designed such that it can accommodate different types of protocol drivers (and, respectively, fieldbus protocols) – as long as they fit the master/slave communication principle and do not require too much asynchronous traffic from nodes to access points. Introducing the "standardized" communication interface above and below the de/multiplexer also allows, in principle, to replace PLC network with other communication media (wireless networks, analogue telephone systems, etc.). The latter will not affect communication protocol drivers, field equipment and application software.

7. COMMUNICATION IN SWITCHED PLC NETWORKS

The network discovery service at the access point does not provide any information regarding meshed or switched communication paths. The only information that is available to drivers is the list of nodes that can currently be accessed via this access point directly. In case of PLC network, this list is comprised of nodes that have logged in and can be physically reached at the moment. As the network switches and changes its characteristics, this list will change accordingly: new nodes appear, others disappear.

In case of GSM, ISDN or analogue telephone networks this list is more static. It can include, for example, those nodes, for which telephone numbers have been configured into the access point.

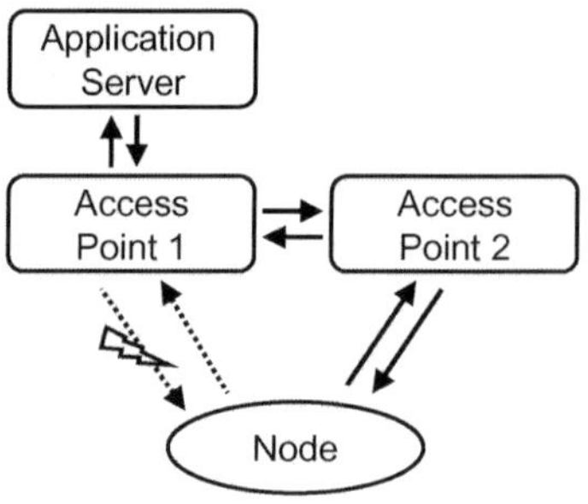

Figure 6. Re-routing of application PDUs through other access points.

Apart from addresses of reachable nodes, the "live-list" also includes connection qualities for each of them (e.g., a relative parameter, expressed in percents). Using this information the de/multiplexer can make decisions on whether a packet, sent by an access point driver has to be re-routed to another access point or not.

As shown in Figure 6 re-routing of messages between access points occurs at the de/multiplexer level and only at the access point side. In fact, each de/multiplexer layer "talks" to de/multiplexers in all other access points over the IP network (communication there is TCP-based and uses point-to-point links).

Horizontal communication between de/multiplexer layers serves three main purposes.

- Every access point periodically informs its siblings about any changes to its live list. Therefore, each de/multiplexer layer knows about current live lists of all other access points in the REMPLI network. Transferred live lists include not only addresses of nodes, but also the connection qualities.

- In case this access point cannot deliver message to a target node (node is currently not on its live list), the de/multiplexer layer can optionally forward the message to a sibling that currently has a link with the node. If none of the access points has the destination node address in their live lists, an error is returned to the driver. Re-routing can also occur in case another access point has a better connection quality to the destination node (more precisely, if connection quality to the node from the other access point is higher by a certain threshold: re-routing of a PDU over the IP network introduces a latency, which can be in some cases higher than that due to retransmissions in the PLC network).

- Responses from nodes, received by a certain access point, are re-routed back to the originating access point, so that its de/multiplexer layer can forward the response to the appropriate driver and close the transaction.

Delivery of multicast/broadcast messages over a switched/meshed network has to be handled in a special way. For instance, in order to cover all nodes in the network, every broadcast message has to be transmitted from, potentially, all access points. This, naturally, causes duplicates appearing at the nodes. Thus every broadcast message needs to be assigned with a system-wide unique ID, transmitted along with it over the PLC. Using these IDs de/multiplexers at nodes maintain a backlog of the recently received broadcasts and, thus, filter out duplicates.

Re-routing of PDUs is optional and always controlled by a communication protocol driver at the access point. Drivers that implement application protocols with integrated link control can prohibit re-routing of messages at the de/multiplexer level, since link management is done by the upper-level software (such an application will then connect to all access points in the IP network).

In order to simplify the design of communication protocol drivers, the communication system interface attempts to "hide" the switched/meshed nature of the PLC network. A typical driver will simply utilize the "automatic" re-routing capabilities, in which case it doesn't need to handle PLC switching at all.

8. CONCLUSIONS

We have presented the design and architecture of a communication system that is on the one hand able to employ narrow band power line communication for SCADA and similar applications, and is on the other hand flexible enough to also support other communication media such as GSM or dial-up lines. The key to this ability is the driver concept, which requires a driver for each application layer protocol to be available at node, access point and application server.

The design of the system reflects requirements by industrial partners in the REMPLI project, namely the possibility to reuse existing investments such as switchgear equipment, utility meters and application software. First tests have shown good results, two upcoming field tests scheduled for the end of this year in Portugal and Bulgaria will show first real-world results of the capabilities of the system.

REFERENCES

Bumiller, G. *Influence of Single Frequency Network Transmission on the Physical Layer of a Multi Carrier Modulation System.* In: Proceedings of the 9th International Symposium on Power-Line Communications and Its Applications (ISPLC 2005), Vancouver, pp. 80

Bumiller, G., Sauter, T, Pratl, G, and Treytl, A (2005). *Secure and Reliable Wide-Area Power-Line Communication for Soft-Real-Time Applications within REMPLI.* In: Proceedings of the 9th International Symposium on Power-Line Communications and Its Applications (ISPLC 2005), Vancouver, pp. 57

Bumiller, G. *Single Frequency Network Technology for Medium Access and Network Management,* In: Proceedings of the 6th International Symposium on Power-Line Communications and Its Applications (ISPLC 2002), Athens

HomePlug Powerline Alliance, http://www.homeplug.org, visited on May 12th, 2005

Lobashov, M., Pratl, G., Sauter, T. (2003). *Implications of Power-line Communication on Distributed Data Acquisition and Control Systems*, In: Proceedings of the 9th IEEE International Conference on Emerging Technologies and Factory Automation (ETFA2003), Lisbon.

Dostert, Klaus (2001). *Powerline Communications.* Prentice Hall PTR

Treytl, A., Sauter, T. (2005). *Security Concept for a Wide-Area Low-Bandwidth Power-Line Communication System*, In: Proceedings of the 9th International Symposium on Power-Line Communications and Its Applications (ISPLC 2005), Vancouver, pp. 66

The Development of Smart Differential Pressure Transmitter Based on WorldFIP

Bai Yan, Zhai Wei Xiang, Han Yu

Dept. of Automation, North China Electrical Power University
Beijing 102206, China. E-mail: baiyan@email.com.cn

Abstract: The developing method of smart differential pressure transmitter based on WorldFIP protocol is proposed in this paper. First, the features of WorldFIP are introduced. Second, a novel measuring circuit of capacitive cell differential pressure sensor is presented. Finally, the overall structure is discussed in detail. It can be proven from the experimental results that this new pressure transmitter can work as subscriber of WorldFIP and the measuring accuracy meets the designing requirement. Copyright © 2005 IFAC

Keywords: Fieldbus, Smart differential pressure transmitter, WorldFIP protocol, Capacitive cell measuring

1. INTRODUCTION

Fieldbus technology has become the development trend of control system, which changed the structure of the traditional control system and formed a new-type network-integrated distributed control system. With the development and popularization of control system based on Fieldbus, the new requirement on measuring and transmitting instrument appears. Besides basic measurement function, the new instrument should have the ability of control, bi-directional digital communication, remote configuration and diagnostics. Differential pressure transmitter plays an important role in industrial process automation, which can be used to measure differential, absolute and gauge pressure, flow and level. A smart differential pressure transmitter based on WorldFIP is proposed in this paper, which adopt a novel measuring circuit of capacitive cell. Compared with the traditional pressure transmitter, this new one has several features such as flexibility, high accuracy and reliability.

2. FEATURES OF WORLDFIP

WorldFIP is one of the eight Fieldbus protocols of IEC. Besides the common features of Fieldbus, it has the following unique features:

WorldFIP adopts the international IEC standard for the physical layer, supports physical layer redundancy and is compatible with the EMC standard of IEC.

No matter what the WorldFIP is used for, there is only one protocol. So the connection between the low-speed and high-speed network can be done through software instead of the interface of gateway or gatebridge.

The communicating mode of Producer/Consumer and Bus Arbiter is suitable for the process control, which is adopted by the international IEC standard and FF standard with different name.

3. THE PERFORMANCE OF TRANSMITTER

This transmitter has the following performances:

Sample Mounting and economical wiring cost. As WorldFIP instrument, several devices can be connected on a single pair of wires.

Digital measurement provides high accuracy. Linear output accuracy: 0.1%; range radio: 10:1; Non-linear compensation function and temperature compensation function improve the measurement precision.

Multi-variable access. It can measure temperature, pressure, flow and level. All the measured values can be transmitted through bus at the same time.

Configurable instrument. The parameters of function block configuration, range, alarm and input selection can be set up by the engineer workstation and be loaded into instrument through bus.

4. THE MEASURING PRINCIPLE OF TRANSMITTER

The differential pressure transmitter uses filed-proven capacitive sensor (capacitive cell) as pressure sensing element, as shown in Fig1.

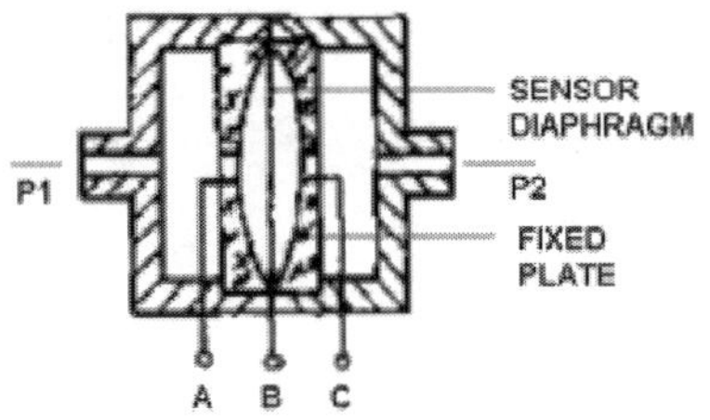

Fig. 1.The structure of capacitive cell

There is an elastic diaphragm in the middle of sensor. On the both sides of sensor, two glass plates with gold plating work as fixed plates. Wires are respectively guided from the diaphragm and the two plates. Thus two capacitors are formed. Silicone oil is filled between the diaphragm and the fixed plates to increase the capacitance value. When the two sides of sensor undergo different fluid pressure, the fluid pressure is transmitted to elastic diaphragm through ripple effect and then the diaphragm bends to the low pressure side. The capacitance values of two sides vary with their distance. On the side with high pressure, the distance increases and capacitance reduces. While on the side with low pressure the distance reduces and capacitance increases. The relationship between the capacitance of the two sides and the differential pressure can be expressed as follows:

$$\Delta P = K\frac{C_1 - C_2}{C_1 + C_2} \quad (1)$$

Where, K is a constant relevant to original capacitance dielectric constant of the medium inside of capacitive cell.
C_1= Capacitance between the plate and the elastic diaphragm with low pressure.
C_2= Capacitance between the plate and the elastic diaphragm with high pressure.

Thus the pressure difference can be figured out after the capacitance of high side and low side are measured by using of the expression (1).

The routine measuring circuit of this capacitive sensor often adopt analog amplifier. The useful signal can be obtained after A/D converting. This kind of circuit is complicated and is easily affected by temperature. A new measuring method is presented in this paper, that is, the capacitance can be obtained through time measurement. This measuring method bases on an assumption that the period of oscillating circuit has linear relationship with capacitance when it oscillates under the action of signal. The effect of nonlinear element and distributed capacitance can be corrected by calculation or software adjustment.

The measuring circuit consists of capacitance-time converting, control logic and counting sections. The capacitance relevant to differential pressure can be obtained from measuring the charging and discharging time. The measuring circuit is shown in Fig.2.

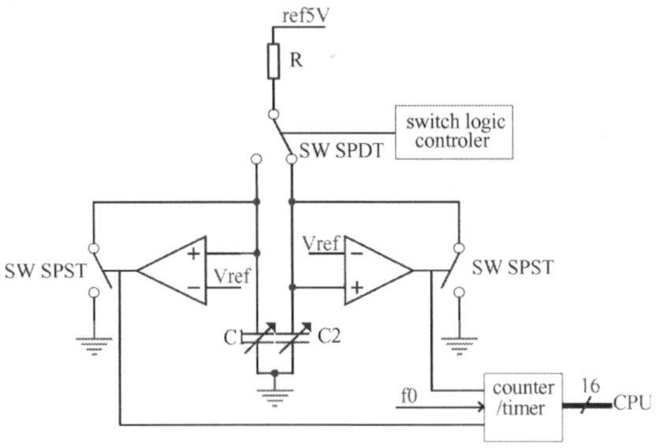

Fig.2. Capacitive cell measuring circuit

The two capacitors of capacitive sensor are charged and discharged by using single resistance. The voltage comparator controls the upper limit voltage. When the voltage of capacitor reaches the ref5V, the voltage comparator outputs high level, which causes the analog switch closed. Then the capacitor is discharged quickly. After a delay time, voltage comparator opens the analog switch and capacitor resumes charging state. The interval of level pulse outputted by the voltage comparator can reflect the charging and discharging time of capacitive sensor. The single pole double throw switch is used for selecting which capacitor is charged or discharged. The charging and discharging time can be gained through a counter. Regarding 5 times as a circulation in actual measurement, the counter takes count of a circulation. Thus the time quantity (digital quantity) that is proportional to the capacitor can be obtained. The differential pressure can be figured out according to expression (2) (3).

$$\frac{T_1 - T_2}{T_1 + T_2 - 2T_s} = \frac{C_1 - C_2}{C_1 + C_2 - 2C_s} \quad (2)$$

$$\Delta P = K\frac{C_1 - C_2}{C_1 + C_2 - 2C_s} \quad (3)$$

Where, K is the scale coefficient. T_1 = Time that is proportional to the C_1 . T_2 = Time that is proportional to the C_2 . T_S = Time that is proportional to the parasitic capacitance C_S .

In Capture CIS environment of ORCAD, the measuring circuit that consists of voltage comparator max907, high-speed analog switch TCL4066 and voltage reference REF02 is simulated. The voltage wave of capacitor is shown in Fig.3. To verify the relationship between capacitance variation and cycle of pulse output, let capacitance change from 128pF to 144pF with the increment of 0.02pF. Voltage comparator will have a pulse output corresponding to every capacitance variation. After amplifying the pulse output of voltage comparator, it can be seen from the simulation wave as shown in Fig.4 that the pulse is almost equally spaced. This means that the period of pulse output is proportional to capacitance. The simulation result proves that the circuit design is feasible.

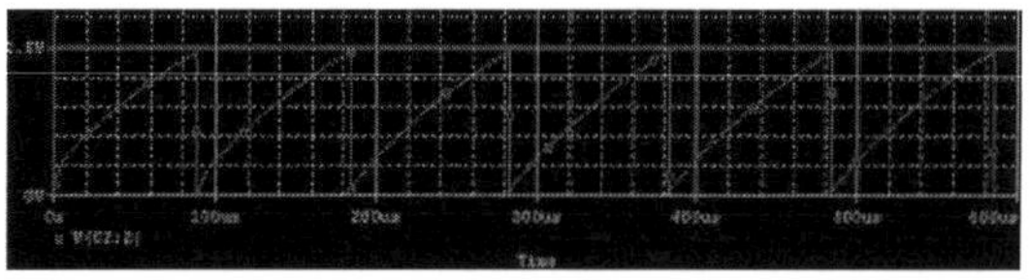

Fig.3. Voltage wave of capacitor

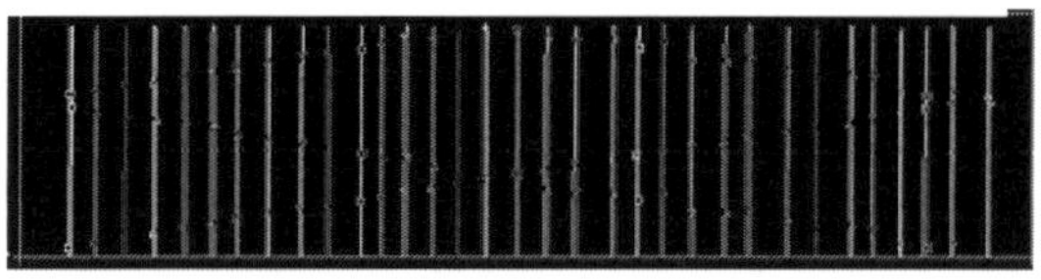

Fig.4. Amplified output wave of voltage comparator

The programmable logic device EPM7128, which can provide higher resolution than the interior timer of MCU, is responsible for the control logic and counting function. To minimize the influence of quantization uncertainty and guarantee the measuring accuracy, an internal counter of 32 bits is designed for measuring the intervals of five continuous pulses. The quantization uncertainty is limited within 1/5 cycle.

5. SYSTEM HARDWARE STRUCTURE

Hardware structure is made up of four modules as shown in Fig.5, including sensor assembly, central processing unit, communication round card and display controller. The sensor assembly includes capacitive sensor transducer and temperature measuring circuit with thermal resistance.

MCU is the intelligent portion of differential pressure transmitter. It is responsible for the management and operation of measurement, self-diagnostic and communication. The AT91M40800 chip is used in MCU, which has integrated ARM7TDMI core and effective RISC and is suitable to real time control. The operation system and application program are stored in a Flash memory and run in a SRAM memory. The parameters used for calibrating, configuring and identifying are stored in a nonvolatile E^2PROM memory.

The MICROFIP chip designed by ALSTOM is used in the round card. The FIELDRIVE chip and FIELDTR chip are adopted in the communication interface. MICROFIP is an ASIC solution implementing the WorldFIP protocol provides data link layer, application layer and network management services. MICROFIP is designed to support and manage medium redundancy without the need for another chip. The FIELDRIVE chip is a high-immunity line driver dedicated to interface a protocol component to a copper twisted pair through an insulating transformer. The FIELDTR components supply a galvanic isolation between the FIELDRIVE line driver and the Fieldbus physical medium. [3] [4]

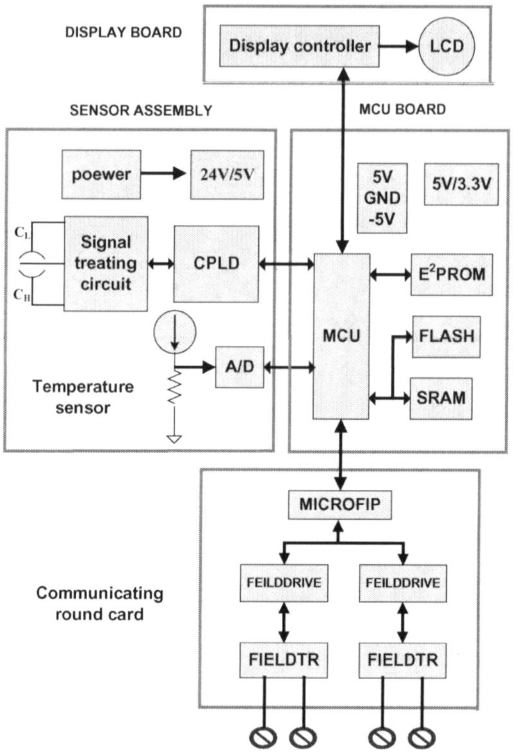

Fig. 5.Hardware structure

6. SOFTWARE

The embedded operation system that is implanted into the transmitter is in charge of resource dispatching, time synchronizing and function block scheduling. In configuration mode, the transmitter receives device configuration and makes the priority table and operation schedule. In normal mode, the function blocks work in the prescribed order. The differential pressure transmitter has several main function blocks as follows:

Analog Input: receives the digital of differential pressure and temperature and transforms the scale.

Signal Characteristic Description: the actual value of differential pressure is figured out through temperature compensation based on the curved surface fitting theory.

Common Calculation: provides many kinds of arithmetic, such as the flow calculation.

Integrator: accumulates the flow and calculates the total volume and mass passing through pipeline in specific time.

PID Controller: carries out PID according to the difference between PV and SP. It also has functions of rate limiting, output tracking and anti reset wind-up.

Self-verification: the curved surface fitting coefficients $\alpha_0 \sim \alpha_5$, which are used for temperature compensation, are determined for each sensor before the pressure transmitter is finished.

$$P = \alpha_1 M + \alpha_2 U_t + \alpha_3 M^2$$
$$\alpha_4 U_t M + \alpha_5 U_t^2 \quad (4)$$

Where, M is the output of pressure sensor; U_t is the output of temperature sensor; ε is infinitely small quantity.

Factory calibration: every pressure transmitter needs be calibrated during the test in factory. The measured signals are stored in E^2PROM memory, which are used as the calculation standard.

Display: converts the format of variable and sends the data to the display controller.

Communication: initializes the communication and network and exchanges the variable and message periodically or aperiodically.

7. TESTING

In order to test the performance of this pressure transmitter, a little WorldFIP networks that includes single level measurement is constructed, as shown in Fig.6. Taking the minimum level of container as testing point, the pressure is guided to high-pressure cell of pressure transmitter, whose low-pressure cell is exposed to atmosphere. So this pressure can be measured. Because the liquid density ρ and acceleration of gravity g are definite value, the height of the liquid level can be calculated according to the measured pressure.

In this Fieldbus networks, a PC with Windows NT4.0 works as bus arbitrator, which is connected with bus through CC121-ISA card. As a subscriber, the pressure transmitter is connected to trunk bus through the branch device named of FIELDTAP. The line termination (LT) is mounted at each end of trunk cable. The network communication speed is 1Mbit/s.

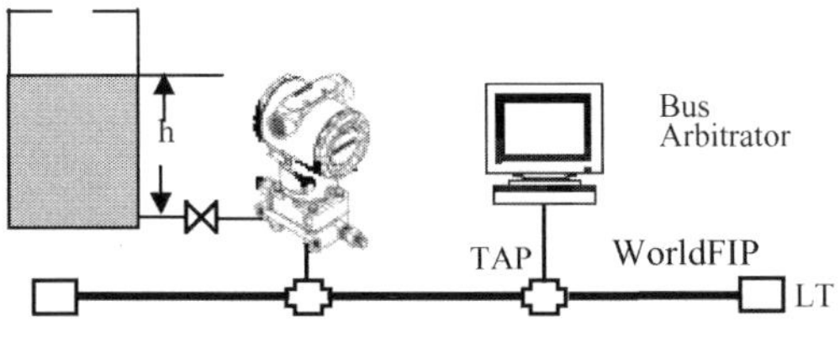

Fig.6.Testing chart of the transmitter

8. SUMMARY

The design of smart differential pressure transmitter based on WorldFIP is discussed in detail in this paper. It can be seen from the experimental results that the expected propose has been arrived. The advantages of Fieldbus control system are obviously and its application relays on the development of Fieldbus intelligent instrument. The success of this kind of differential pressure transmitter will greatly improve the spread of WorldFIP in China.

REFERENCES

ALSTOM (France) (1999). *MICROFIP User Reference Manual ALS 50280 b–en [EB]*. ALSTOM (France).

ALSTOM (France) (2000). *FIELDRIVE User Reference* Manual *ALS 50261 d-en [EB]*. ALSTOM (France).

Bai Yan and Wu Hong (2001) *Distributed Control System and and Fieldbus Control System –Foundation, Design and Application*, pp. 211–218. China Power Publishing House.

SMAR Company (1996). *Smar Fieldbus 302 series Instruction Manual*. SMAR Company.

S.M.Huang,C.G.Xie,R.Thorn,D.Snowden and M.S.Beck (1992): Design of sensor electronics for electrical capacitance tomography, *IEE PROCEEDINGS-G*,**Vol.139,** pp.83-88.

Yang Xianhui (1999) *The Basic and Application of Fieldbus Technology*. Tsinghua university publishing House.

ELSEVIER
IFAC
PUBLICATIONS

RECONFIGURABLE DISTRIBUTED SYSTEM BASED ON SOM NETWORK APPROACH

Benítez-Pérez H.*, Ceballos Miguel. **

(*) Departamento de Ingeniería de Sistemas Computacionales y Automatización, IIMAS, UNAM, Apdo. Postal 20-726. Admón. No. 20. Del. A. Obregón, México D.F., CP. 01000, México.
Fax: ++52 55 5616 01 76, Tel: (*) ++52 55 5622 36 39
(**)Universidad Autónoma de Querétaro, Posgrado en Ciencias de la Computación, Facultad de Informática, Centro Universitario, Querétaro, Querétaro, México.
Email: (*) hector@uxdea4.iimas.unam.mx (contact author)

Abstract: Online reconfiguration is a multidisciplinary approximation where areas such as computing and control theory need to be combined from the point of view of their respective effects. There are several kind of algorithms from computing areas in order to bound time effects over control systems. There are several types of schedulers like dynamic and static where characteristics as feasibility and safety play an important role. This paper proposes a quasi-dynamic scheduling algorithm based on self-organizing maps (SOM) in order to avoid misclassification of plan selection and to provide a fast and valid response during plan evaluation. *Copyright © 2005 IFAC.*

1. INTRODUCTION

Several strategies for managing time delay within control laws have been studied for different research groups. For instance (Nilsson, 1998) proposes the use of a time delay scheme integrated to a reconfigurable control strategy based upon a stochastic methodology. On the other hand, (Wu, 1997) proposes a reconfiguration strategy based upon a performance measure from a parameter estimation fault diagnosis procedure. Another strategy has been proposed by (Jiang, et al., 1999) where time delays are used as uncertainties, which modify pole placement of a robust control law. (Izadi, et al., 1999) present an interesting view of fault tolerant control approach related to time delay coupling. Reconfigurable control has been studied from the point of view of structural modification since fault appearance as presented by (Blanke, et al., 2003). From the point of view, reconfigurable control performs a combined modification of system structure as studied by (Benítez-Pérez, et al., 2005) and (Thompson, 2004).

Some considerations need to be stated in order to define this approach. Firstly, faults are strictly local in peripheral elements and these are tackled by just eliminating the faulty element. In fact, faults are catastrophic and local. Time delays are bounded and restrictive to scheduling algorithms. Global stability can be reached by using classical control strategy for online time delays.

The objective of this paper is to allow safe online reconfiguration based on previous knowledge of valid plans, evaluated by planning scheduler algorithm, by using a trained SOM network.

This paper is divided in six sections. First section is current introduction. Second section presents a general background of the techniques used in this paper. Third section presents a review of the proposed algorithm. Fourth section shows current case study. Fifth section depicts some relevant results. Finally sixth section presents current conclusions.

2. BACKGROUND

Two strategies need to be reviewed in order to pursue this approximation self organizing maps and planning scheduler.

The purpose of Kohonen self-organizing feature map is to capture the topology and probability distribution of input data (Kohonen, 1989 and Hassoum, 1995) (Fig. 2.1). Firstly, a topology of self-organizing map is defined as a rectangular grid (Nelles, 2000) (Fig. 2.2). Different types of grid may be used, although Fig. 2.2 presents an homogenous response suitable for noise cancellation. The neighborhood function with respect to a rectangular grid is based upon bi-dimensional Gaussian functions shown in eqn. 2.1.

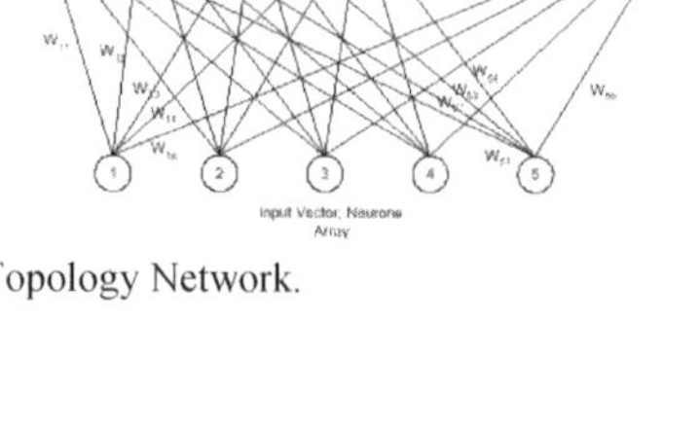

Fig. 2.1 Topology Network.

Eqn. 2.2 is the basis of the neural network, in this equation the weight matrix is updated based upon bi-dimensional indexing named $h(i_1,i_2)$. This equation is used during training (off-line) stage.

$$h(i_1,i_2)=\exp\left(-0.5*\frac{\left(i_1^{win}-i_1\right)^2+\left(i_2^{win}-i_2\right)^2}{\sigma^2}\right) \tag{2.1}$$

where i_1 and i_2 are the index of each neuron. σ is the standard deviation from each Gaussian distribution. This distribution determines how the neurons next to winner neuron are modified. Each neuron has a weight vector (w_i^j) that represents how this is modified by an input updating. $h(i_1,i_2)$ is the Gaussian representation that permits the modification of neighbor neurons.

This bi-dimensional function allows the weight matrix to be updated in a global way rather than just to update the weight vector associated to the winner neuron.

An inner product is performed between weight matrix W and input vector (I) in order to define the winner neuron. Having calculated this product, the maximum value is determined by the comparison between each scalar from resultant vector. This value is declared as winner as in the technique named the winner take all. The related bi-dimensional index (Fig. 2.2) is calculated in order to determine how the weight matrix is modified.

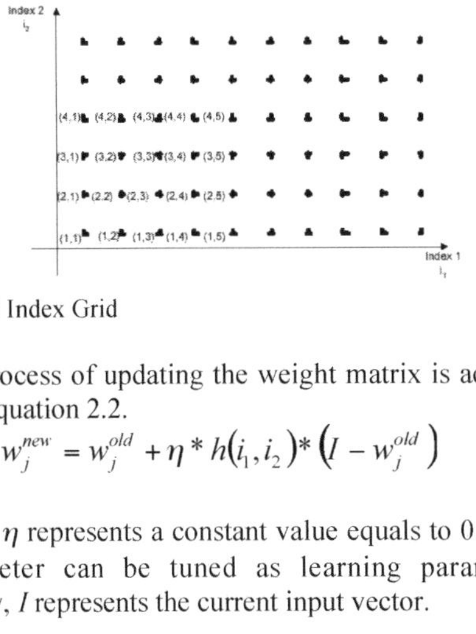

Fig. 2.2 Index Grid

The process of updating the weight matrix is achieved as in equation 2.2.

$$w_j^{new}=w_j^{old}+\eta*h(i_1,i_2)*\left(I-w_j^{old}\right) \tag{2.2}$$

where η represents a constant value equals to 0.7. This parameter can be tuned as learning parameters. Finally, I represents the current input vector.

Furthermore scheduling algorithms play an important role in order to pursue predictability during computer network behaviour. Several algorithms can be used, such as rate monotonic (RT), earliest deadline first (EDF), flexible time triggered (FTT) (Almeida *et al.*, 2002), and least slack time (LST) (Cheng, 2002). The difference between them is marked by the way tasks are ordered. It depends on the application which method for ordering tasks is the most suitable for a particular example. Those algorithms already mentioned are divided into two categories as static and dynamic schedulers. The main difference is that the static scheduler defines during the off-line process the allocation of task, whereas the dynamic scheduler allocates tasks based on current conditions considering a time slot.

For real-time purposes, it is best to use static schedulers because of its deterministic behavior. Recently, quasi-dynamic scheduling algorithms have been defined to give certain flexibility to the static communication approach. An example of this sort of algorithm is the planning scheduler (Almeida *et al.*, 1999). The planning scheduler is a pseudo-dynamic scheduler, in the sense that it presents some dynamic properties but is not fully dynamic. The underlying idea is to use the present knowledge about the system (in particular, the variable set) to plan the system activity for a certain time window in the future. Such a time window is fixed, and independent of the periods of the variables, and it is called a plan.

The scheduler must, then, be invoked once in each plan to build a static schedule that will describe the bus allocation for the next plan. The potential benefit of the planning scheduler in terms of run-time overhead is revealed by the following reasoning: Within a fixed time window of duration P_i, such as the period of variable i among a set of N variables, there are at most S transactions

$$S=\sum_{i=1}^{N}\left(\left\lceil\frac{W}{P_i}\right\rceil+1\right) \tag{2.3}$$

When idle time is manipulated to give an opportunity to sporadic tasks, preemptable tasks are to be expected. To perform task re-allocation, the macro-cycle of N tasks is divided into smaller windows called elementary cycles (ECs) that are divided into basic units that are multiples of consumption times of every task. The only condition for an elementary cycle is that it has the same period as the fastest task. As this partition is proposed, the group of tasks conformed by N elements is re-organized according to these time restrictions, taking into account periodic time sizes to define priorities of execution.

If there is one who cannot fit in any EC, it is said that this group of tasks cannot be scheduled. The main plan consumption time (W) is divided into several elementary cycles (EC) where each EC is divided into local time windows named as esp_i. These last divisions result into a more efficient time managing based upon a preemptive strategy. Fig. 2.3 shows how this proposal (planning scheduler) divides a time window into a more complex time division to that presented by rate monotonic.

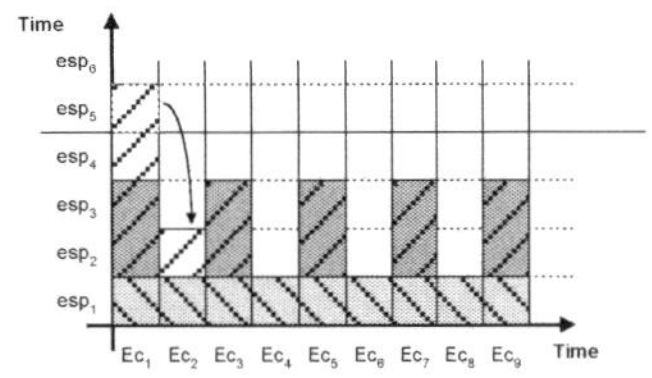

Fig 2.3 Time Division proposed in Planning Scheduler

This planning scheduler is based upon eqns 2.4, 2.5 and 2.6 where U is the total consumption time with respect to related periods. N is the total number of tasks, X is the maximum wasted time between time windows EC's.

$$U = \sum_{i=1}^{N} \frac{c_i}{T_i} \tag{2.4}$$

$$U = \sum_{i=1}^{N} \frac{c_i}{T_i} \left(N\left(2^{1/N} - 1\right)\right) * \frac{E - X}{E} \tag{2.5}$$

$$X = \max_{i=1...N}(X_i) \le \max_{i=1...N}(C_i) \tag{2.6}$$

In this case, time performance is increased in comparison to Rate Monotonic due to re-order of useless time spaces. It presents the advantage of a possible dynamical modification every time window W who is defined as the time window where a very task is executed at least one time. This characteristic makes the system pseudo dynamic in terms of reconfiguration. This algorithm (planning scheduler) is enhanced in order to incorporate new measures such as system performance. As explained in first section, these measurements are taking into account during off-line performance in order to define a suitable control law for those valid plans. This implementation is further reviewed in next section.

Fig. 2.5 shows a schematic structure of planning scheduler. Where a descriptive table is proposed who is checked by planning algorithm. If this table is validated, a new plan is created (plan i, plan i+1, and so on). This new plan is dispatched afterwards.

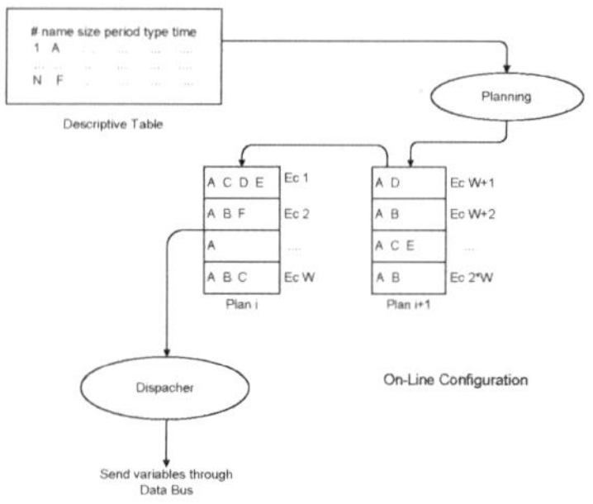

Fig. 2.5 Schematic Structure of Planning Scheduler

3. PROPOSED ALGORITHM

The core of this algorithm is to perform on-line reconfiguration based upon a review of the proposed plan. The review uses a SOM network in order to classified valid and non-valid plans. First, the SOM network is trained offline using valid and non-valid plans from planning scheduler evaluation and case study response. Based on this training procedure two main regions are determined, one related to suitable reconfigurations and other that holds non-trustable reconfigurations. During online stage SOM network allows classification from new plans if the response of this network is within the region of valid plans the computer network will be reconfigured otherwise the proposed plan will be rejected. Current approach is based on two stages as shown in Fig. 3.1.

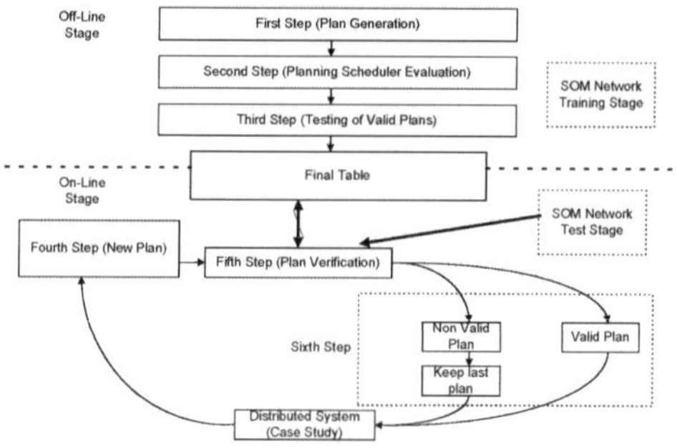

Fig. 3.1. Algorithm Proposal

Fist stage is an offline procedure to train SOM network by selection of valid plans according to planning scheduler. Second stage is an online procedure to test system response where a reconfiguration plan is proposed and evaluated to planning scheduler. Fig 3.2 shows time window selection on order to implement valid reconfiguration performance. It is important to mention that SOM network cannot learn new plans during online stage as safety precaution.

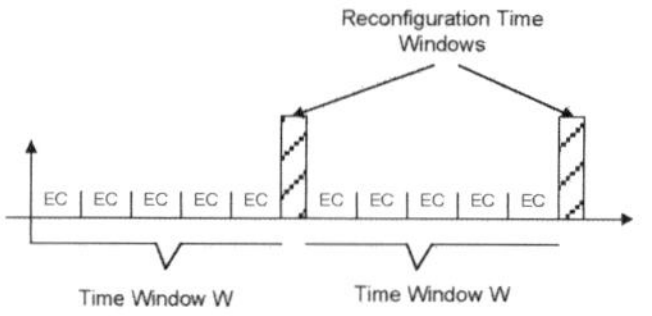

Fig. 3.2 Time Window Managing

4. CASE STUDY

Case study is based on three conveyor belts (Gudmundsson, 1999) integrated as follows, it comprises 3 conveyors belts, 4 actuators, and 12 sensors. It has 16 computing elements that considers the controller and the bus controller. Figure 4.1 shows a diagram of this implementation. The procedure of the example is next; conveyor belt 1 detects a box on it as soon as this is presented, and it modifies its speed up to a certain level to transport it in a faster way than the box.

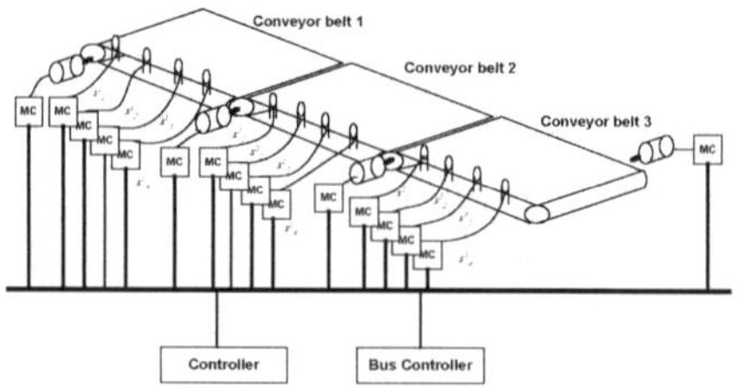

Figure 4.1. Conveyor belt example

When this box arrives to conveyor belt 2, its speed is modified up to another level to transport this box in a faster manner. Similar behavior is presented at conveyor belt 3. The sensor vector is used to detect the current position of box in any of these conveyor belts. Furthermore, actuator 4 has the task of pushing the current box when it arrives at this position.

For the case of multiple boxes in this example, the aim is that the boxes never crashed between each other. From this explanation, Table 4.1 shows the modification of speeds.

This case study is peculiar in that each conveyor belt has two different speeds as shown in Table 4.1.

Table 4.1. Speed selection

	Conveyor belt 1	Conveyor belt 2	Conveyor belt 3
Sensors S_*^1 = Low	Low Speed	Low Speed	Low Speed
Sensors S_*^1 = High	High Speed HS^1	Low Speed	Low Speed
Sensors S_*^2 = Low	Low Speed	Low Speed	Low Speed
Sensors S_*^2 = High	Low Speed	High Speed HS^2	Low Speed
Sensors S_*^3 = Low	Low Speed	Low Speed	Low Speed
Sensors S_*^3 = High	Low Speed	Low Speed	High Speed HS^3

These speeds are dependent on the sensor situation. This sensor situation is depicted as low and high, which is a semaphore for determining the presence of an object.

The second peculiarity is related to the difference between *HS* as follows:

$$HS^1 < HS^3 < HS^2$$

where the middle conveyor belt is the fastest, then third conveyor belt, and so on. As the reader may realize, there are four motors, three for the conveyor belts and the fourth is to pull any object presented at its region.

Based on this case study, the dynamic response of the three actuators is shown in Figure 4.2. Different speed-ups are shown assuming that a box is presented during a certain time. For instance, the first conveyor belt presents a faster speed-up during the first 3000 seconds, in comparison with the low speed-up during 3000 to 6000 seconds. This speed-up is shown as a change of slope of the current graphic. Similar behavior is presented for both conveyor belts as 2 and 3 are modified, because Hs^2 is bigger than Hs^3.

Dynamic system implementation is based on following equations considering system response and control implementation.

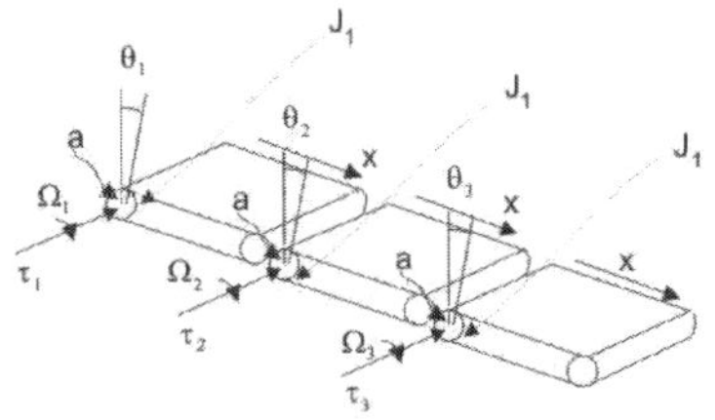

Fig. 4.2. Dynamic System Implementation

In Figure 4.2,

is the linear displacement

is the angular displacement

*

is the angular velocity

*

is the lineal force

is the lineal inercy

*

is the parameter

is the torque

*

is the radius

is the lineal velocity

In this case, the plant presents two cases with or without a box per belt. As the second case is trivial, the first case is expressed per belt considering the mass of the box (referred to as *m*). The first conveyor belt is expressed as

$$\begin{bmatrix} \ddot{x}_1 \\ \dot{\theta}_1 \end{bmatrix} = \begin{bmatrix} J/m & 0 \\ & J_1 \end{bmatrix} \begin{bmatrix} \theta_1 \\ \dot{x}_1 \end{bmatrix} - \begin{bmatrix} 1/m \\ 0 \end{bmatrix} \tau_1 \qquad (4.1)$$
$$y = \dot{x}_1$$

The second conveyor belt is expressed as

$$\begin{bmatrix} \ddot{x}_2 \\ \dot{\theta}_{21} \end{bmatrix} = \begin{bmatrix} J/m & 0 \\ & J_2 \end{bmatrix} \begin{bmatrix} \theta_2 \\ \dot{x}_2 \end{bmatrix} - \begin{bmatrix} 1/m \\ 0 \end{bmatrix} \tau_2$$
$$y_2 = \dot{x}_2$$

(4.2)

and the third conveyor belt is

$$\begin{bmatrix} \ddot{x}_3 \\ \dot{\theta}_3 \end{bmatrix} = \begin{bmatrix} J/m & 0 \\ & J_3 \end{bmatrix} \begin{bmatrix} \theta_3 \\ \dot{x}_3 \end{bmatrix} - \begin{bmatrix} 1/m \\ 0 \end{bmatrix} \tau_3 \qquad (4.3)$$
$$y_3 = \dot{x}_3$$

From these considerations, discrete plants are defined next by considering the presence of the box

$$x(k+1) = Ax(k) + \sum_{i=0}^{l} B_i^k u(k-i)$$

$$B_i^k = \int_{t_i^k}^{t_{i-1}^k} exp(A(T-\hat{o}))Bd\hat{o} \quad (4.4)$$

where $l=1$ because the maximum number of sensors with delays is just one. Therefore, the A matrix is expressed as

$$A_* = \begin{bmatrix} \exp(J/m) & 0 \\ 0 & \exp(J_*) \end{bmatrix} \quad (4.5)$$

where T is the inherent sampling period, and t_0^k, t_1^k, and t_2^k are the related delays of the plant. For the case of local control laws, these are expressed next as

$$\begin{aligned} x_c(k+1) &= A_c x_c(k) + B_c u_c(k) \\ y_c(k) &= C_c x_c(k-\tau_c) - D_c u_c(k-\tau_c) \end{aligned} \quad (4.6)$$

giving the delays as a result of decomposition from sensor and actuators, which are expressed as τ_{sc} and τ_{ca}, respectively. The augmenting representation is given next:

$$\begin{aligned} u_c(k) &= y_p(k-\tau_{sc}) \\ u_p(k) &= y_c(k-\tau_{ca}) \end{aligned} \quad (4.7)$$

where states are augmented as

$$z = \begin{bmatrix} x_p(k) \\ x_c(k) \end{bmatrix} \quad (4.8)$$

and expressed as

$$z(k+1) = \begin{bmatrix} A_p & 0 \\ 0 & A_c \end{bmatrix} z(k) + \begin{bmatrix} 0 & 0 \\ B_c C_p & 0 \end{bmatrix} z(k-\hat{o}_{sc})$$

$$+ \begin{bmatrix} B_p D_c C_p & 0 \\ 0 & 0 \end{bmatrix} z(k-\hat{o}_{sc}-\hat{o}_{ca}-\hat{o}_c)$$

$$+ \begin{bmatrix} 0 & B_p C_c \\ 0 & 0 \end{bmatrix} z(k-\hat{o}_{ca}-\hat{o}_c) \quad (4.9)$$

After the stability expression expression in Equation 4.10, the local control laws on a numerical basis are given next:

$$\begin{aligned} A_c &= \begin{bmatrix} 1 & 0 \\ 0 & 1 \end{bmatrix} \\ B_c &= \begin{bmatrix} 2 & 0 \end{bmatrix} \\ D_c &= 1 \\ C_c &= \begin{bmatrix} 1 & 0 \end{bmatrix} \end{aligned} \quad (4.10)$$

The only considered delays are τ_{sc} equal to 110 ms. Having shown local control laws structures, the global control law, taking into account the first and second cases on a fault-free basis, is shown in Figure 4.3, where reconfiguration is expressed for the formal event manager. In this case, two states are possible with several events, which are managed by the sensor vector for each belt (first, second, and third belts) and expressed as $S^1_{1\le i\le N}$, $S^2_{1\le i\le N}$, and $S^3_{1\le i\le N}$, respectively.

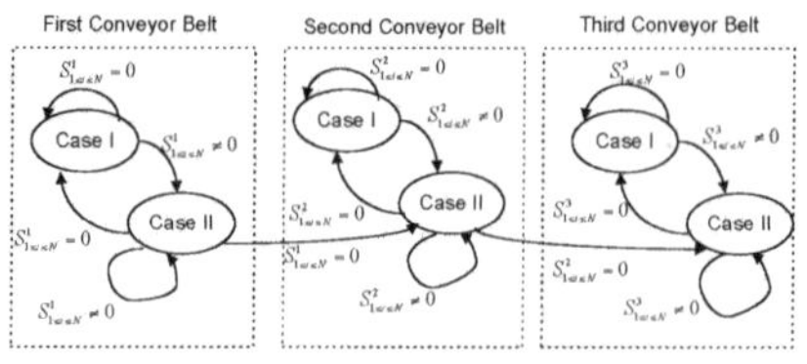

Figure 4.3. Fault-free scenario in terms of global structure

The switching effect is neglected in this fault-free scenario. For the case of a fault scenario, a new case appears for global control, as shown in Figure 4.4 where a new state appears that is related to the action pursued when a fault is presented. The necessary event for reaching such a state is $S^1_{1\le i\le N} \ne 0$, and the fault's last event is composed of local information given by each local sensor with a relation to the health condition measures.

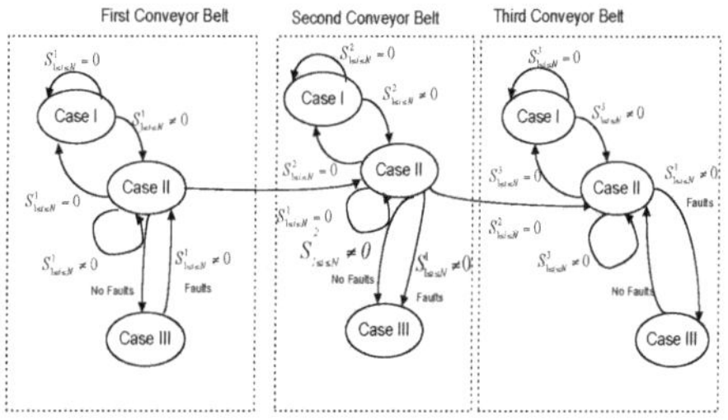

Figure 4.4. Local fault scenario for the global structure

5. RESULTS

Following results present weight matrix conformation for different type of scenarios (Fig. 5.1). In this graphic main areas are conformed around first patterns with values of 0.772.

Further on SOM response during online performance is shown in Fig. 5.2 where our approach tends to spend less than 10% of total time spent by planning scheduler. The quality valid response plans selected from SOM network is around 97% from total of proposed tests as shown in Fig. 5.3.

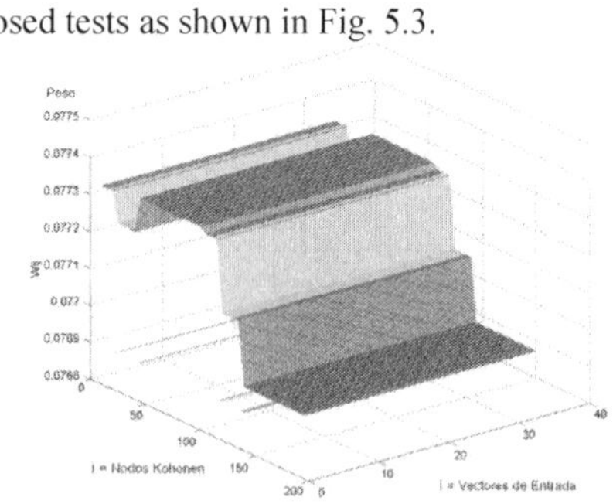

Fig. 5.1. Weight confirmation from SOM Network

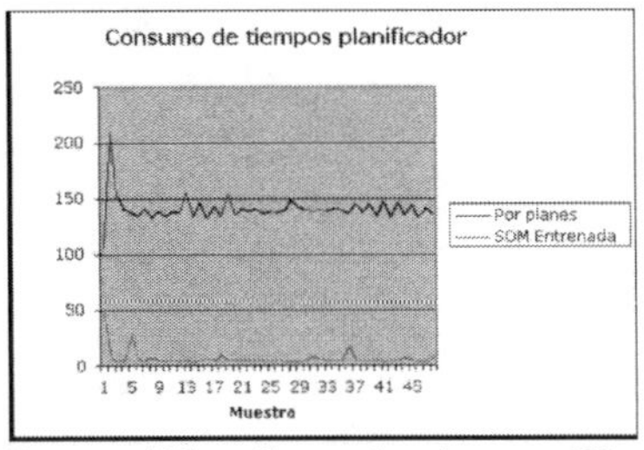

Fig. 5.2. Consumed Time Comparison between PS and SOM

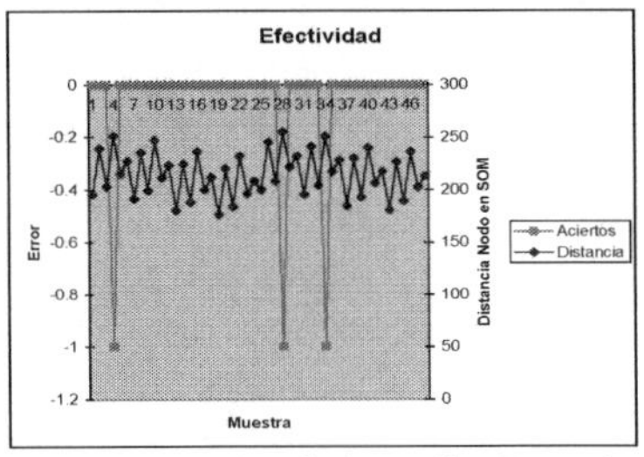

Fig. 5.3 Valid Plan selection during online Execution

6. CONCLUSIONS

Present approach follows the approximation of quasi dynamic algorithm by using neural network giving a high certainty of case study response where the proposal is divided in two stages. First stage (offline stage) allows SOM training using planning scheduler approach and case study response from measurement of steady state error. Second stage (online procedure) uses the already trained neural network to classify plan proposals according to case study response and plan analysis.

During online stage SOM network performs results according to design expectations since non valid plans an time spent by reconfiguration is less than 0.1% in both cases.

Future work is related to overcome time restrictions with respect to static time windows when online reconfiguration is possible. Furthermore, online learning need to be suitable even in unknown scenarios where classification becomes a challenge.

Acknowledgments

The authors would like to thank PAPIIT-UNAM (Num. 106100 and 105303) Mexico.

REFERENCES

Almeida L., Pedreiras P., and Fonseca J. A.; "The FTT-CAN Protocol: Why and How"; *IEEE Transactions on Industrial Electronics*, **Vol. 49**, No. 6, pp. 1189-1201, 2002.

Almeida, L., Pasadas, R., and Fonseca, J.A.; "Using a Planning Scheduler to Improve the Flexibility of Real-Time Fieldbus Networks"; *Control Engineering Practice*, **Vol.** 7, pp. 101-108, 1999.

Benítez-Pérez, H., and García-Nocetti, F.; "Reconfigurable Distributed Control"; *Springer-Verlag*, 2005.

Blanke, M., Kinnaert, M., Lunze, J., and Staroswiecki, M.; "Diagnosis and Fault Tolerant Control"; *Springer*, 2003.

Cheng A.; "Real-Time Systems: Scheduling, Analysis and Verification"; *Wiley- Interscience*, 2002.

Gudmundsson, D., and Goldberg, K.; "Tuning Robotic Part Feeder Parameters to Maximize Throughput"; *Assembly Automation Publisher: MCB University Press*, **Vol. 19**, No. 3, pp. 216-221, 1999.

Hassoum, H.; "Fundamentals of Artificial Neural Networks"; *Massachusetts Institute of Technology*, 1995.

Izadi-Zamanabadi R. and Blanke M.; "A Ship Propulsion System as a Benchmark for Fault-Tolerant Control"; *Control Engineering Practice*, **Vol.** 7, pp. 227-239, 1999.

Jiang J., and Zhao Q.; "Reconfigurable Control Based on Imprecise Fault Identification"; *Proocedings of the American Control Conference, IEEE*, pp. 114-118, San Diego, June, 1999.

Kohonen, T.; "Self-Organization and Associative Memory"; *Springer-Verlag*, Berlin, Germany, 1989.

Nelles, O.; "Non-Linear Systems Identification"; *Springer-Verlag*, 2001.

Nilsson, J.; "Real-Time Control Systems with Delays"; *PhD. Thesise, Department of Automatic Control, Lund Institute of Technology, Sweden*, 1998.

Thompson, H.;"Wireless and Internet Communications Technologies for monitoring and Control"; *Control Engineering Practice*, **vol. 12**, pp. 781-791, 2004.

Wu N.; "Reliability of Reconfigurable Control Systems: A Fuzzy Set Theoretic Perspective"; *Proceedings of the 36th Conference on Decision & Control, IEEE, TP15 5:10*, pp. 3352-3356, San-Diego, USA, 1997.

ELSEVIER
IFAC
PUBLICATIONS

INTERCONNECTING CAN BUSSES VIA AN ETHERNET BACKBONE

Jean-Luc Scharbarg * **Marc Boyer** *
Christian Fraboul *

* *IRIT - ENSEEIHT*
2, rue Camichel - BP 7122 31071 - Toulouse - France
Jean-Luc.Scharbarg@enseeiht.fr

Abstract: Embedded systems have specific real-time requirements that led to the development of dedicated communication protocols. Such systems must face increasing communication needs and the needed integration of switched Ethernet architecture. But moving from existing dedicated fieldbusses architectures to a new Ethernet based architectures is not always easily feasible, due to industrial constraints.
In this paper, we compare different solutions for integrating existing data busses (such as CAN, which is an important standard in automotive context) on a global architecture that respects increasing bandwidth requirements and preserves the respect of real-time constraints. In a first step, we study classical CAN/CAN bridging strategies. In a second step, we propose CAN/Ethernet bridging strategies that respect the real time behaviour of CAN End System when communicating through an Ethernet network that is also shared by (non CAN) applications.

1. INTRODUCTION

Fieldbusses (Thomesse, 1999), e.g. CAN, World-FIP, Profibus have been developed in the context of real-time applications (distributed computer control systems) that have specific communication requirements such as:

- bounded end-to-end transmission delays in order to guarantee respects of deadlines,
- the bounded and small jitter for periodic traffic.

However, the amount of information that is nowadays exchanged in such systems has been increasing steadily and is now reaching the limits of traditional fieldbusses, especially in terms of bandwidth (Dietrich and Sauter, 2000).

Switching from dedicated field-busses to Ethernet is a classical trend in embedded systems due to the wide acceptance of the Ethernet standard and its evolution toward a more predictable switched architecture.

However, successful experience with introduction of a switched Ethernet in avionic systems (AFDX, (Grieu *et al.*, 2003; Grieu, 2004)) is mainly due to the preservation of the applications communication model (periodic schemes) and the respect of the expected real time properties (bounded delay).

The goal of the study presented in this paper is slightly different, as the objective is to build an heterogeneous architecture obtained by interconnecting existing CAN data busses on an Ethernet backbone.

We have build a twofold study: on the one hand, we have developed a prototype with some Linux-PCs with CAN or Ethernet cards inside, on the other hand, we realised a more theoretical study which is shortly presented in this paper.

Section 2 presents CAN and Ethernet technologies and their extensions to cope more efficiently with real-time constraints. In section 3, we study classical pure CAN architectures. In section 4, we consider CAN/Ethernet architectures and propose CAN/Ethernet bridging strategies that respect the real-time behaviour of CAN End Systems under certain conditions on non-CAN Ethernet traffic. Section 5 concludes the paper and presents some ideas for future works.

2. COMMUNICATION TECHNOLOGIES

We present the two communication technologies we intend to use, i.e. the Controller Area Network and Ethernet. We summarise the proposed solutions to make those technologies deterministic, especially in terms of respect of deadlines and boundaries for jitter.

2.1 CAN

The Controller Area Network (CAN, (ISO, 1993)) is a serial communication protocol suited for networking sensors, actuators and other nodes in real-time systems. The CAN specification defines several versions of the protocols for the physical and the data link layer. In this paper, we focus on CAN 2.0 A. Several application layer protocols have been proposed.

The CAN addressing system is based on message identifier: a frame does not have a destination nor a source address. Frames are broadcasted on the bus. Stations get the frames they are interested in by a filtering process of the identifiers.

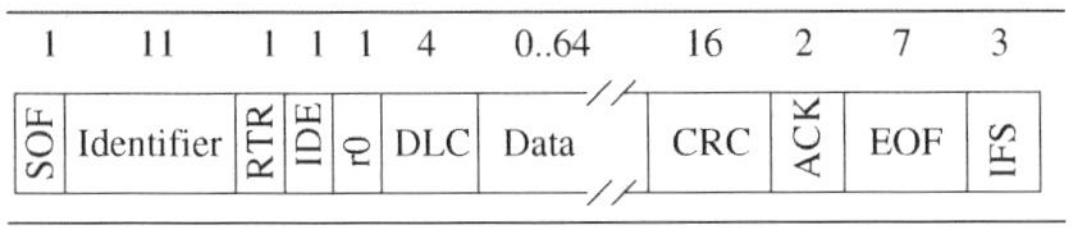

Fig. 1. CAN frame (sizes in bits)

The frame format is depicted in figure 1. The detail of each field will not be presented. The relevant fields for the remaining of the paper are the following:

- the identifier field, as mentioned earlier identifies the data contained in the frame,
- the DLC field gives the length (in bytes) of the data field,
- the data field is the payload of the frame.

Bit-stuffing is used to avoid the transmission of long sequences of bits with identical value.

The medium access method (MAC) is CSMA/CR: the starting of frame transmissions on the bus are synchronous. When two or more stations start a transmission simultaneously, the one with the smaller frame identifier wins and the others stop their transmission. This mechanism guaranties strict priority order on identifiers. It implies limitations of the bandwidth and the maximal length of the bus (e.g. 1 Mbs for 40 meters).

2.2 Enhancement of the CAN protocol

Some drawbacks of the CAN native MAC have been identified. First, it is event-triggered: when a station has a frame to transmit, it tries to. It will succeed as soon as no frame with a higher priority is being transmitted. This mechanism can induce large jitter on periodic frames. Second, identifiers are associated with frames statically. This imposes a scheduling algorithm using static priorities, e.g. rate monotonic (Liu and J.W., 1973) when periodic traffic is considered. It is well known that higher utilisation of the medium is obtained with a scheduling algorithm using dynamic priorities, e.g. Earliest Deadline First (Liu and J.W., 1973).

Solutions have been proposed to solve those drawbacks. Most of them add a protocol over CAN native MAC.

Time triggered CAN (TT-CAN) (ISO, 2000; Führer *et al.*, 2000) imposes a static scheduling on CAN. This scheduling is memorised in a table which is known by all the stations. This scheduling comprises in particular exclusive and arbitration windows. Each exclusive window is dedicated to exactly one frame identifier while an arbitration window is shared. The stations are resynchronised with a trigger message broadcasted periodically by the master station.

Flexible TTCAN (FTT-CAN, (Almeida *et al.*, 2002)) aims at introducing flexibility in the static scheduling of TTCAN. The table is memorised by the sole master. The trigger message contains the numbers of the messages that can be transmitted until the next trigger message. Moreover, time is reserved for non synchronous messages. Problems may occur if the master fails. Solutions to this problem have been proposed in (Fonseca *et al.*, 2002).

Implementations of EDF scheduling relying on the native MAC have been proposed (Zuberi and Shin, 1995; Di Natale, 2000). They use a part of the identifier to encode the (dynamic) priority of the message, implying a limited number of different messages. In (Nolte *et al.*, 2003), a server-based method is proposed.

This paper will only consider the native CAN MAC. However, it could be of great interest to evaluate the solutions proposed in this section in our context, as will be stated later.

2.3 *Ethernet*

The Ethernet link layer (Eth, 2002) is designed for computer local networks where high bandwidth and low cost hardware is more important than guaranteed deadlines and/or jitter.

The Ethernet addressing system is based on MAC addresses: each Ethernet entity has a unique MAC address. In each frame, the destination (unicast, broadcast or multicast) and source addresses are inserted. Frames are broadcasted on the physical layer. Entities get the frames there are interested in by a filtering process.

The Ethernet MAC is CSMA/CD: the time is divided into slots. Emission always begins at slot start. When two entities (or more) start to emit at the same time ($\pm$ the signal propagation delay), a collision is detected by the entities, they immediately stop to emit and goes into a *retransmission* state (in order to be sure that all entities have detected the collision, at least 64 bytes are sent). In retransmission state, a sender draws a random integer value n uniformly in $[0, 2^c - 1]$, where c is the number of collisions already observed for this frame (bounded to 10). It waits n slots before to re-emit the frame. The number of retransmissions is bounded to 15.

The Ethernet payload can vary from 46 to 1500 bytes.

The Ethernet traffic was 10Mbs, the most common is now 100Mbs and there also are 1Gbs and 10Gbs solution.

The Ethernet frame format for 10Mbs and 100Mbs is described in figure 2. In an Ethernet frame, there are (at least) 26 bytes of control and 0 up to 1500 bytes of data. Padding guarantees a minimum payload of 46 bytes.

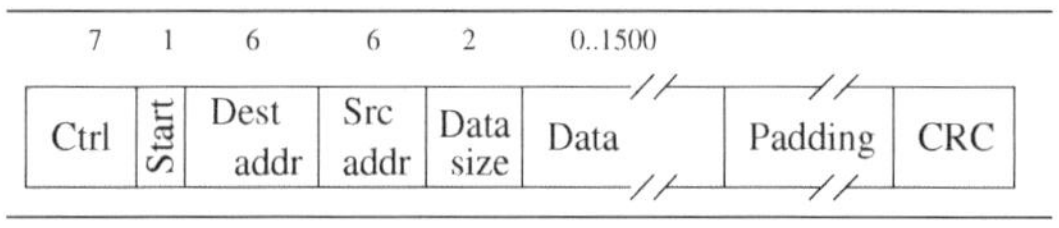

Fig. 2. Ethernet frame (sizes in bytes)

2.4 *Real-time Ethernet*

As stated earlier, the original goal of Ethernet was to maximise the bandwidth utilisation and to minimise the mean response time. Consequently, it is not well suited for real-time applications where the main challenge is to guarantee the respect of deadlines (to bound the response time) and to limit jitter. However, there has been many attempts to make Ethernet real-time (Pedreiras *et al.*, 2002).

A first class of approaches consist in modifying the medium access control to achieve a bounded access time to the bus e.g. (Le Lann and Rivierre, 1993). The worst-case transmission delay is frequently orders of magnitude greater than the average transmission time, leading to overscaling of the system. Furthermore, it often implies a modification of the firmware which forbids the use of standard Ethernet cards.

A second class of approaches consists in adding a control layer over Ethernet, in order to bound or even eliminate collisions. Some of those techniques are based on a master/slave architecture, which generate an important overhead, due to master messages. Some others use token-passing, which often induces large jitter and/or overhead. TDMA requires a costly precise clock synchronisation (Venkatramani and Chiueh, 1994). In the virtual time protocol (Malcolm and Zhao, 1995; Molle and Kleinrock, 1985), every frame waits for a specific amount of time before being transmitted and, in case a collision occurs, a probabilistic approach is used. Parameters of this technique are hard to optimise and worst-case transmission delays are often much greater than average ones.

A time-triggered approach has been recently proposed. It is an adaptation of FTT-CAN to an Ethernet link and is called FTT-Ethernet (Pedreiras *et al.*, 2002). A master transmits periodically a trigger message indicating the frames that should be transmitted until the next trigger message and the instant of transmission. It is a master/slave architecture where the master overhead is reduced.

Switched Ethernet is a way to bypass the medium access strategy of Ethernet: each station is directly connected to an Ethernet switch with a full duplex link. Then, the medium is always free. Consequently guaranteed performances are strongly connected to policies of the switch. Several approaches have been developed. One generic approach is the Network Calculus (Cruz, 1991*a*; Cruz, 1991*b*), that have been successfully applied in the AFDX network system for Airbus embedded networks (Grieu *et al.*, 2003; Grieu, 2004). Another is the Response Time Analysis (Liu and J.W., 1973). Both approaches have been compared in (Koubâa and Song, 2003).

As already mentioned, the use of switched Ethernet is out of the scope of this paper.

3. PURE CAN ARCHITECTURES

The real-time network architectures considered in this section are composed of CAN busses. They are shown as a reference for the remaining of the paper. Performance of those architectures will be illustrated with an application comprising the message set listed in table 1. It includes 148 periodic messages. The relative deadline of each

Type	Nb of mes.	Per. (ms)	Data (bits)	Lg. (bits)	trans. time (ms)	bw Mbs
M_1	13	4	64	135	0.135	0.43
M_2	13	4	48	115	0.115	0.37
M_3	13	4	32	95	0.095	0.31
M_4	13	4	16	75	0.075	0.24
M_5	12	10	64	135	0.135	0.16
M_6	12	10	48	115	0.115	0.14
M_7	12	10	32	95	0.095	0.11
M_8	12	10	16	75	0.075	0.09
M_9	12	15	64	135	0.135	0.11
M_{10}	12	15	48	115	0.115	0.09
M_{11}	12	15	32	95	0.095	0.08
M_{12}	12	15	16	75	0.075	0.06

Table 1. message sets of the application

message is equal to its period. The values for length and transmission time correspond to a 1 Mbs CAN bus. Let's have a look at the first line. It means there are 13 periodic CAN messages of period 4 ms. Those messages will be called M_1 messages in the following. Each occurrence of an M_1 message contains 8 bytes of data. The length of an occurrence of the message is 135 bits. It is computed using the following formula:

$$length = 47 + 8 \times DLC + \left\lfloor \frac{34 + 8 \times DLC}{4} \right\rfloor \quad (1)$$

47 is the number of control bits of a CAN frame, including the interframe space. $8 \times DLC$ is the number of data bits of the frame. The remainder of the formula is the maximum number of stuff bits inserted in the frame. So, it is a worst case length. The transmission time is 135 μs. The bandwidth needed by the M_1 CAN messages is 0.43 Mbs.

3.1 One shared CAN bus

This architecture considers a single CAN bus interconnecting all the stations. It is the simplest solution. It is impracticable in the following cases:

- the cumulative bandwidth needed by all the traffics exceeds the available bandwidth of the bus (bandwidth condition),
- the geographical dispatching of the stations is incompatible with the maximal length of the CAN bus, e.g. 40 meters at 1 Mbs (distance condition).

Considering a strictly periodic traffic and allocation of frame identifiers following a rate monotonic policy (the smallest the period of a message, the highest its priority), the bandwidth condition can be evaluated statically. The distance condition is independent of the traffic.

Concerning the application depicted in table 1, the cumulative bandwidth needed by all the traffics is 2.19 Mbs. It clearly violates the bandwidth condition. Thus, the application cannot be implemented using one shared CAN bus.

3.2 Several CAN busses interconnected by bridge stations

This architecture aims at satisfying the bandwidth condition. It is a classical solution frequently used in embedded systems. An example with 4 CAN busses is depicted on figure 3. Each

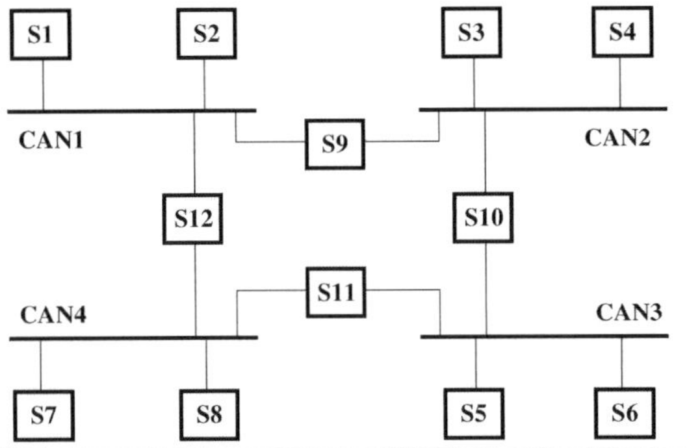

Fig. 3. CAN busses interconnected by bridge stations

CAN bus includes two local stations ($S1$ and $S2$ for CAN bus 1) and two bridge stations ($S9$ and $S12$ for CAN bus 1). In this example, each CAN bus is not directly connected to all the other ones. For instance, bus 1 is not directly connected to bus 3.

System operation is very simple. Different kinds of frames have to be considered:

(1) frames local to a CAN bus s: they only have to be transmitted over this bus,
(2) frames from a local station Sa of a CAN bus s to a local station Sb of a CAN bus d, s and d being directly connected by a bridge Sk: they have to be transmitted by Sa on bus s, received by Sk, transmitted by Sk on bus d and received by Sb,
(3) frames from a local station Sa of a CAN bus s to a local station Sb of a CAN bus d, s and d being not directly connected by a bridge: they have to be transmitted by Sa on bus s, received by a bridge connected to bus s, transmitted via busses and bridges to a bridge connected to bus d, and then to station Sb.

We will suppose that the network architecture is build so that the last kind of frame never occurs. Frames of the first type are called local frames, while frames of the second type are called distant frames.

As an example, we map the example of table 1 on the network architecture of figure 3, where

Type	Nb of mes.	Kind	Priority	Per. (ms)	Worst-(case) (ms)
M1	8	Local C1	21-28	4	2.262
M1	5	C1 → C2	1-5	4	2.114
M2	8	Local C2	29-36	4	2.302
M2	5	C2 → C3	6-10	4	2.569
M3	8	Local C3	37-44	4	1.942
M3	5	C3 → C4	11-15	4	2.229
M4	8	Local C4	45-52	4	1.542
M4	5	C4 → C1	16-20	4	2.624
M5	8	Local C1	69-76	10	6.312
M5	4	C1 → C2	53-56	10	6.144
M6	8	Local C2	77-84	10	6.392
M6	4	C2 → C3	57-60	10	6.184
M7	8	Local C3	85-92	10	3.522
M7	4	C3 → C4	61-64	10	5.184
M8	8	Local C4	93-100	10	2.822
M8	4	C4 → C1	65-68	10	5.824
M9	8	Local C1	117-124	15	14.410
M9	4	C1 → C2	101-104	15	14.284
M10	8	Local C2	125-132	15	14.570
M10	4	C2 → C3	105-108	15	11.874
M11	8	Local C3	133-140	15	6.932
M11	4	C3 → C4	109-112	15	9.874
M12	8	Local C4	141-148	15	5.552
M12	4	C4 → C1	113-116	15	11.154

Table 2. Pure CAN architecture

the number of local stations per CAN bus is not known (it is two on the figure). The messages are distributed as descibed in table 2 (3 first columns). Priorities are assigned to messages as shown in column 4 of table 2. The higher priority corresponds to the value 1. A rate monotonic policy is applied. For messages with the same period, the local ones have a lower priority. For distant messages with an identical period, priority allocation is made arbitrarly. The same applies for local messages with identical period.

In order to validate the system, it is necessary to guaranty that every frame of every message respects its deadline. One way to do that is to calculate a worst-case end-to-end transmission delay. Such a calculation is presented in (Scharbarg *et al.*, 2005). Results are given by the last column of table 2. We observe that, for every message, the worst-case delay is smaller than the period. As deadlines equal periods, we can conclude that every frame of every message will meet its deadline.

The architecture presented in this paragraph is a good solution to satisfy the bandwidth solution. However, it is of little efficiency concerning the distance conditon. The architecture proposed in the next section aims at being an answer to this distance condition.

4. SEVERAL CAN BUSSES INTERCONNECTED BY ENCAPSULATING BRIDGE STATIONS ON ETHERNET

The solution developed in this section consists in using Ethernet as a backbone between CAN busses. This Ethernet backbone is also shared with non-CAN applications. We first present the network architecture and the general operation of the system. We then evaluate several CAN/Ethernet bridging strategies.

4.1 System description

An example of such an architecture is depicted on figure 4. It includes four CAN busses and

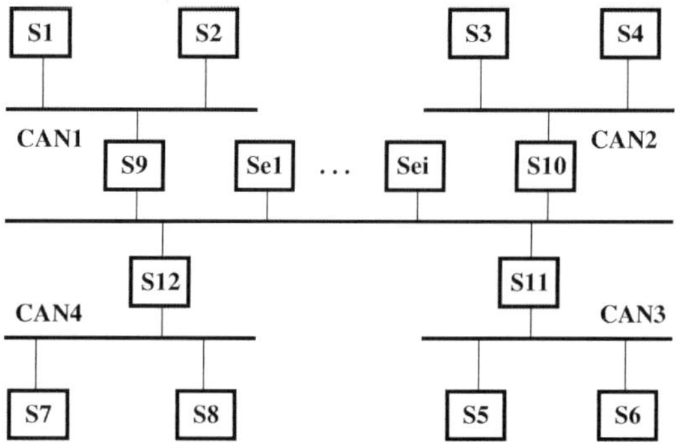

Fig. 4. CAN busses interconnected by encapsulating bridge stations on Ethernet

an Ethernet link (no switched Ethernet will be considered in this section). Each CAN bus shares a bridge station with Ethernet (S9, S10, S11 and S12 on figure 4). Two kinds of frames have to be considered:

(1) frames local to a CAN bus s : they only have to be transmitted over this bus,
(2) frames from a local station Sa of a CAN bus s to a local station Sb of a CAN bus d : they have to be transmitted by Sa on bus s, received by the bridge associated with s, transmitted over Ethernet, received by the bridge associated with d, transmitted by this bridge on bus d and received by Sb.

Moreover, the Ethernet link has to support non CAN traffic between pure Ethernet stations ($Se1$, ... , Sei of figure 4).

We consider applications such as that depicted in tables 1 and 2.

The question we have to answer is : what bridging strategy between CAN and Ethernet ? Answering this question, we have to keep in mind that characterisitics of CAN and Ethernet are very different :

- the available bandwidth :1 Mbs or less for CAN, 10 Mbs, 100 Mbs, 1Gbs for Ethernet,
- the addressing system : identifiers associated to data for CAN, MAC addresses of stations for Ethernet,
- the data encapsulted in a frame : between 0 and 8 bytes for CAN, between 46 and 1500 bytes for Ethernet,
- the collision resolution : deterministic and non destructive for CAN, non deterministic and destructive for Ethernet.

The very different addressing systems make an encapsulation bridge the most suitable solution : CAN frames are encapsulated in Ethernet frames. More precisely, Identifier, DLC and Data fields of CAN frames are put in the Data field of Ethernet frames (the other fields of CAN frames can be easily reconstructed). This means that a CAN frame occupies at most 10 bytes of the Data field of an Ethernet frame. Consequently, if one CAN frame is encapsulated in one Ethernet frame, there is at least 36 bytes of padding. This is clearly an important waste of bandwidth.

The worst-case transmission delay of each CAN frame depends on the type of frame. For a distant frame $F_{s,d,m,i}$ from a local station on bus s to a local station on bus d, the worst-case end-to-end transmission delay $T_{s,d,m,i}$ is :

$$T_{s,d,m,i} = T^{s}_{s,d,m,i} + 2 \times T_B + T^{eth}_{s,d,m,i} + T^{d}_{s,d,m,i} \quad (2)$$

$T^{s}_{s,d,m,i}$ and $T^{d}_{s,d,m,i}$ are the transmission delays for the frame $F_{s,d,m,i}$ on bus s and d. Worst-case values are calculated in the same manner as for the previous architecture. T_B is the overhead for one bridge (there are two bridges on the way). $T^{eth}_{s,d,m,i}$ is the transmission delay on the Ethernet link for the CAN frame $F_{s,d,m,i}$ encapsulated in an Ethernet frame.

For a local frame $F_{s,0,m,i}$ of bus s, we have

$$T_{s,0,m,i} = T^{s}_{s,0,m,i} \quad (3)$$

4.2 The "one for one" strategy

The more straightforward encapsulation strategy is to put each global CAN frame in a separate Ethernet frame and to transmit it as soon as possible. The expected benefit is a minimal delay. This strategy has been evaluated by a simulation model (queueing network implemented in QNAP2). We consider an Ethernet link at 100 Mbs and $T_B = 0.05\ ms$ (considering a modern microprocessor). The non-CAN Ethernet traffic is equally distributed between frames of 500, 1000 and 1500 bytes. There are two traffic sources for each frame length, one generating 40 % of the corresponding traffic and the other one the remaining 60 %. The curve $n = 1$ of figure 5 shows the results for the example application of table 1 and 2.

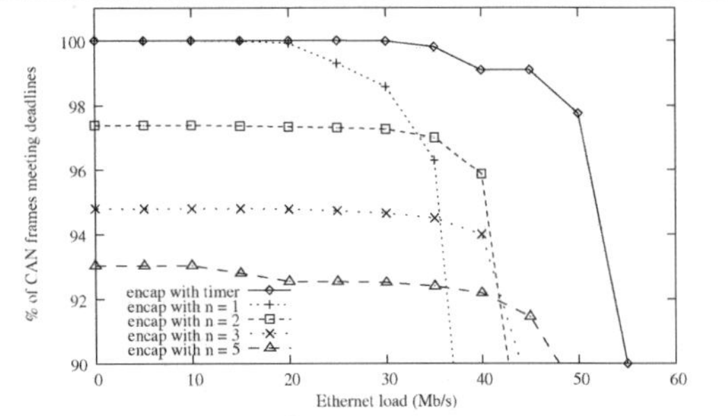

Fig. 5. Encapsulation strategies and Ethernet load

We notice that there are missed deadlines for CAN frames as soon as non-CAN Ethernet load is greater than or equal to 20 Mbs. When this load is 35 Mbs, 4 % of CAN frames miss their deadline. Above 35 Mbs, the percentage of temporal faults on CAN frames increases dramatically.

4.3 The "n for one" strategy

Missed deadlines are due to collisions on the Ethernet link. As explained above, a CAN frame occupies at most 10 bytes of the data field of an Ethernet frame. When there is only one CAN frame per Ethernet frame, padding is mandatory and it wastes Ethernet bandwidth. Conversely, if, for instance, we put 5 CAN frames of maximum length in one Ethernet frame, it represents 50 bytes of Ethernet Data and no padding is necessary. More formally, such a strategy consists in encapsulating n CAN frames in an Ethernet frame (frame bunching). Simulations have been made to evaluate this strategy. Configuration is identical to the one used for the simulation of the "one for one" strategy. Figure 5 shows results for $n = 2$, 3 and 5. It appears that:

(1) greater values of n are better when non-CAN Ethernet load increases (e.g. $n = 5$ is the only value depicted in figure 5 for which there are less than 10 % of CAN frames that miss their deadlines when non-CAN Ethernet load is 45 Mbs).
(2) for low non-CAN Ethernet loads, the percentage of CAN frames missing their deadline increases when n increases,

The first point is explained by the reduction of the load induced by CAN frames on Ethernet when n increases.

The second point is due to the maximal duration a CAN frame has to wait before being encapsulated and transmitted on the Ethernet link, which increases as n increases. This can lead to missing deadlines. Let's suppose for instance an architecture with two CAN busses, no local CAN messages and three global CAN messages M_1, M_2 and M_3 with period 10 ms that are produced on the first bus and consumed on the second. We consider that the three CAN messages have the same phase. If $n = 2$, figure 6 shows that the first frames of M_1 and M_2 are encapsulated in an Ethernet frame, while the first frame of M_3 has to wait the second frame of M_1 and, so, misses its deadline.

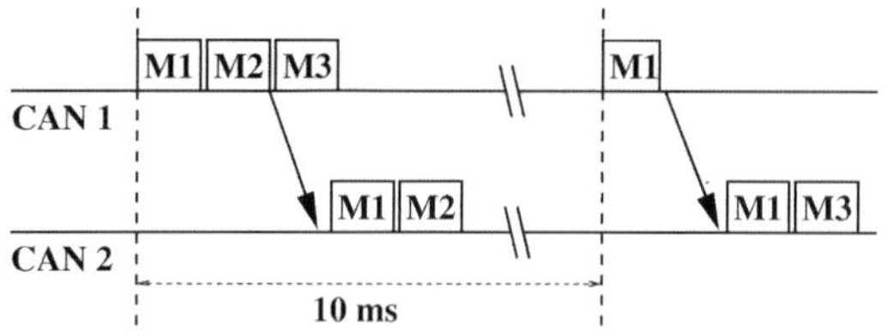

Fig. 6. Example of missed deadline for **n = 2**

4.4 *The "timed n for one" strategy*

In order to improve the results, we have to guarantee that no global CAN frame will wait more than a given amount of time before being encapsulated and sent over Ethernet. We propose a solution that associates a timer $WD_{s,d,m,i}$ with each distant CAN frame $F_{s,d,m,i}$. A bridge transmits an Ethernet frame encapsulating all pending CAN frames as soon as it has n pending CAN frames or a pending CAN frame $F_{s,d,m,i}$ has been initiated since a duration of $WD_{s,d,m,i}$. As an example, suppose that $n = 2$ and $WD_{s,d,m,i} = 0.5\ ms$ for all distant CAN frames $F_{s,d,m,i}$. A bridge transmits an Ethernet frame as soon as it has two pending CAN frames or one pending CAN frame initiating for more than 0.5 ms.

Curve *encap with timer* of figure 5 shows simulation results for the following set of $WD_{s,d,m,i}$ values. We consider $WD_{s,d,m,i} = T^s_{s,d,m,i} + 0.1 \times L_{s,d,m,i}$, for each distant CAN frame $F_{s,d,m,i}$ from a local station on bus s to a local station on bus d. $L_{s,d,m,i} = P(m) - (T^s_{s,d,m,i} + 2 \times T_B + T^d_{s,d,m,i})$. $L_{s,d,m,i}$ is the duration available for the transmission of $F_{i,j,k}$ over Ethernet, considering bridges overhead and worst-case transmission delays on CAN busses. We fix $n = 100$.

Results of simulation show that no CAN frames miss their deadline as long as the non-CAN load on Ethernet is not greater than 30 Mbs. When this load is 50 Mbs, 2 % of CAN frames miss their deadlines. This strategy clearly gives better results than the other ones presented on figure 5.

Figure 7 shows a histogram of the percentages of CAN frames that are encapsulated in an Ethernet frame for a non-CAN Ethernet load of 20 Mbs. This number is widely distributed between 1 and 13. It is of course highly application dependant.

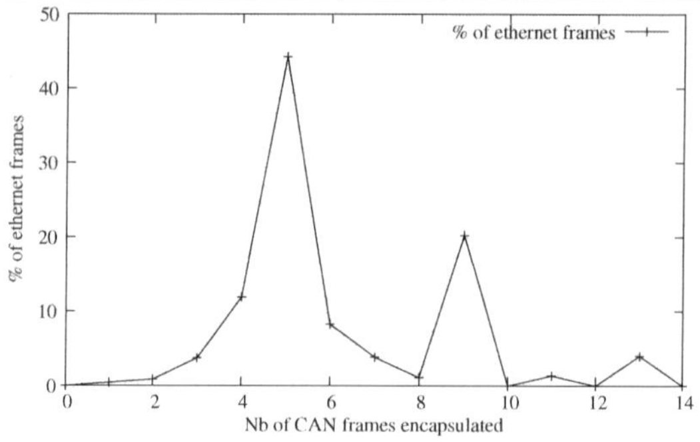

Fig. 7. Number of CAN frames in an Ethernet frame

This strategy gives a good solution for soft real-time applications and/or guarantees on the maximum non-CAN load on Ethernet. However, it is not well-suited for hard real-time applications, because there are still collisions on the Ethernet link and no guarantee on worst-case transmission delays can be given.

Furthermore, finding optimal $WD_{s,d,m,i}$ values is not an easy problem and those values are application dependant.

5. CONCLUSION AND FUTURE WORKS

In this paper, we mainly focused on two types of communication technologies:

- the first one is Controller Area Network (CAN), which is a good example of deterministic real-time communication system,
- the second one is Ethernet, which is the most popular non real-time communication system.

The aim of the paper was to study the use of Ethernet in conjunction with CAN for communications in a real-time system.

Pure CAN architectures have first been studied. They are limited in terms of available bandwidth (CAN maximum bandwidth is 1 Mbs) and area coverage (maximum length of a CAN bus at 1 Mbs is 40 meters). The use of several CAN busses directly interconnected by bridges partially solve the bandwidth limitation, but is quite inefficient against the area coverage limitation.

The use of an Ethernet link to interconnect the various CAN busses is a good alternative if a solution can be found to bound the transmission delay on Ethernet. We propose different CAN-Ethernet bridging strategies and compares their relative qualities. We show that a good strategy consists in bounding the time a CAN frame has to wait before being encapsulated in an Ethernet frame (the "timed n for one" strategy).

However, it is not possible to guarantee worst-case transmission delays with pure CSMA/CD Ethernet, since we have no control on collisions.

An important characteristic of real-time communication is the limitation of the jitter of periodic traffics. In this context, it would probably be judicious to study the use of the time-triggered paradigm on CAN (via TT-CAN or FTT-CAN) in conjunction with the time-triggered Ethernet solution.

Nowadays, full duplex switched Ethernet can be used for real-time applications. There are no more collisions on the medium and guaranteed transmission delays are strongly connected to potential congestion problems that may occur in output queues of the switches.

We intend to apply the time-triggered paradigm on switched Ethernet considering both the architecture of one switch and the global architecture of the network. The open question is: given their service disciplines, are switches able to support a time-triggered communication schema?

REFERENCES

Almeida, Luis, Paulo Pedreiras and José Alberto G. Fonseca (2002). The FTT-CAN protocol : why and how. *IEEE transactions on industrial electronics.*

Cruz, R.L. (1991*a*). A calculus for network delay, part I. *IEEE Transactions on Information Theory* **37**(1), 114–131.

Cruz, R.L. (1991*b*). A calculus for network delay, part II. *IEEE Transactions on Information Theory* **37**(1), 132–141.

Di Natale, M (2000). Scheduling the can bus with earliest deadline techniques. In: *Proceedings of the IEEE Real-Time Systems Symposium.*

Dietrich, D. and T. Sauter (2000). Evolution potentials for fieldbus systems. In: *IEEE Workshop on Factory Communication systems.* Porto.

Eth (2002). CSMA/CD access method. IEEE Standard 802.3. IEEE.

Fonseca, José A., J. Ferreira, M. Calha, Paulo Pedreiras and Luis Almeida (2002). Issues on task dispatching and master replication in ftt-can. In: *IEEE Africon.*

Führer, Thomas, Bernd Müller, Werner Dieterle, Florian Hartwich, Robert Hugel and Michael Walther (2000). Time triggered communication on can. In: *International CAN Conference.*

Grieu, Jérôme (2004). Analyse et évaluation de techniques de commutation Ethernet pour l'interconnexion des systmes avioniques. PhD thesis. Insitut National Polytechnique de Toulouse (INPT). INPT – Toulouse – France.

Grieu, Jérôme, Fabrice Frances and Christian Fraboul (2003). Preuve de déterminisme d'un réseau embarqué avionique. In: *Actes du 10ème Colloque Francophone sur l'Ingenierie des Protocoles.* Paris.

ISO (1993). *ISO International Standard 11898 - Road vehicles - Interchange of digital information - Controller Area Network for high-speed communication.*

ISO (2000). *ISO International Standard 11898-4 - Road vehicles - Controller Area Network - Part 4 : Time-Triggered Communication.*

Koubâa, Anis and Yé Qiong Song (2003). Evaluation et amélioration des bornes du temps de réponse pour des applications temps réel avec ordonnancement à priorité fixe et non-préemptif. In: *Actes du 4éme Colloque Francophone sur la Modélisation des Systèmes Réactifs.*

Le Lann, G. and N Rivierre (1993). Real-time communications over broadcast networks : the CSMA-DCR and the DOD-CSMA-CD protocols. Report RR1863. INRIA.

Liu, C.L. and Layland J.W. (1973). Scheduling algorithms for multiprogramming in hard real-time environment. *Journal of ACM* **20**(1), 46–61.

Malcolm, N. and W Zhao (1995). Hard real-time communications in multiple-access networks. *Real Time systems* **9**, 75–107.

Molle, M. and L Kleinrock (1985). Virtual time CSMA : why two clocks are better than one. *IEEE Transactions on Communications* **33**(9), 919–933.

Nolte, Thomas, Mikael Sjödin and Hans Hansson (2003). Server-based scheduling of the can bus. In: *IEEE International Conference on Emerging Technologies and Factory Automation.*

Pedreiras, Paulo, Luis Almeida and Gai Paolo (2002). The ftt-ethernet protocol : merging flexibility, timeliness and efficiency. In: *Euromicro conference on real-time systems.*

Scharbarg, Jean-Luc, Marc Boyer and Christian Fraboul (2005). Can-ethernet architectures for real-time applications. In: *International Conference on Emerging Technologies and Factory Automation.* IEEE. Catania, Italy.

Thomesse, Jean-Pierre (1999). Fieldbusses and interoperability. *Control Engineering Practice* **7**, 81–94.

Venkatramani, C and T Chiueh (1994). Supporting real-time traffic on ethernet. In: *IEEE Real-Time Systems Symposium.* San Juan.

Zuberi, Khawar M. and Kang G Shin (1995). Scheduling messages on controller area network for real-time cim applications. In: *Proceedings of Real-Time Technology and Applications symposium.*

Copyright © Fieldbus Systems and Their Applications
Puebla, Mexico, 2005

ELSEVIER
IFAC
PUBLICATIONS

A NOVEL SIMULATOR FOR CLOCK SYNCHRONIZED DISTRIBUTED SYSTEMS

Georg Gaderer, Patrick Loschmidt, and Thilo Sauter

*Austrian Academy of Sciences, Viktor Kaplan Strasse 2,
Wiener Neustadt, Austria*

Abstract: Although clock synchronization in packet-oriented networks is beneficial for distributed real-time systems only the behaviour of the used algorithms in static cases and idealised environments are well documented. Due to the great complexity in larger networks and the interrelation of hardware, software, and synchronization algorithms, transient states (e. g., start-up, loss of a node or new network topology) cannot be analysed using analytical methods. The present paper proposes a network simulation environment which allows to gain new knowledge about the synchonization behaviour in non-static real-world cases. Further, the simulator will give insights into the importance of the influence on the accuracy of the different network layers. Consequently the critical elements in time synchronized distributed systems should be identifiable. *Copyright ©2005 IFAC*

Keywords: Clock synchronization , Simulators, Networks, Communication protocols, Fault tolerance

1. INTRODUCTION

As shown in (Kopetz, 1997), clock synchronization is a crucial topic in distributed systems, since it is known that the analysis, design, and stable operation of such systems is considerably simplified if the participating nodes share a common notion of time. Essential benefits are the possibility of tasks like synchronous data acquisition or simultaneous triggering of events. A very common approach used in real-time systems is to establish a network-wide common time base and schedule tasks as well as exchange of data with respect to it. Examples for this can be found in (Verissimo and Rodrigues, 2001; Pedreiras and Almeida, 2005; Dana, 1997). As more and more powerful yet affordable μ-controllers are available for industrial applications, distributed control systems are increasingly attracting interest. For such tasks mere time synchronization alone is not sufficient. Thus a well-defined upper bound for the clock deviation at any given point of time is compulsory to implement traditional control algorithms. Clock synchronization is therefore a topic that gradually left the pure real-time research area and spread into automation and instrumentation as well, increasingly gaining economical significance (Kopertz *et al.*, 2005). Although basic clock synchronization is very well investigated (e.g. by (Fetzer and Cristian, 1997; Schmid, 1997)) , some important aspects like practical implementation have not been covered yet. One is the start-up behaviour of clock synchronization algorithms under any given boundary condition and with arbitrary start-up sequences comprising local clocks with accuracies varying over time. Other stuctures not investigated until now are large-scale, heterogeneous networks with asymmetric delays (Gaderer *et al.*, 2005*a*). In the same problem class are such fault tolerance aspects of clock synchronization algorithms in those networks. State-of-the-art clock synchronization strategies as well as network sizes have reached a complexity where a purely analytical investigation is no longer useful or possible, because it tends to get limited to steady-state analysis and thus cannot address the above mentioned problems with sufficient accuracy. Hence simulation (in close combination with accurate and fast modelling) of those environments remains as the only feasible solution (Weiss *et al.*, 1999) offering new possibilities and ways for a detailed and reliable analysis.

Nowadays integration of clock synchronization capabilities is already available or under development in a variety of products ranging from processors, dedicated network equipment up to protocol stacks for exchanging time information. Basic research has delivered important results, but when it comes to applications with a substantial number of nodes and products where tradeoffs with respect to clock synchronization have to be taken into account, reliable answers for the system behaviour system are still missing (e. g., investigations on cheap but unstable oscillators versus expensive but more stable ones, influence of cascading and hierarchy on the achievable clock synchronization accuracy, fault tolerance of the existing protocols and possible improvements). Furthermore, advanced features in clock synchronization, such as redundancy concepts, are disregarded or tackled by ad-hoc approaches which lack in-depth understanding of the response (and stability) of the system to transient events. Table 1 shows a collection of applications, which use TDMA based on a common notion of time or at least synchronized clocks for real-time distributed systems. For most of these approaches, the new IEEE 1588 standard forms the basis for clock synchronization. More advanced paradigms like SynUTC are not yet used in industry. Furthermore, practical applications for IEEE 1588 are presently limited to Ethernet based networks, even though the standard defines identifiers for a great variety of communication protocols. Still, for none of these protocols a tangible application of the standard has been investigated up to now.

The reaminder of this paper is structured as follows: Section 2 will adress the special requirements for a clock synchronization simulator and disuss existing network simulation technologies. This section will be followed by a description of the proposed architecture and use cases of the simulation environment. Finally, the paper will be rounded up with a conclusion and a short summary of further research activities.

2. PROBLEM DEFINITION AND STATE OF THE ART

The goal of the proposed approach is a simulator capable of modelling different clock synchronization protocols in various network topologies. The following overview of currently available simulation tools the design process lead to the proposed solution outlined in Section 3. The chosen approach takes advantage of existing technology in order to minimise development efforts while introducing new layer-overlapping simulation capabilities.

2.1 Clock synchronization simulation issues

It is a challenging property of the problem of clock synchronization over a packet-oriented communication network, that hardware aspects (e. g., implementation of the hardware clock) or even physical effects (e. g., long and short-term variations like temperature drifts of the oscillator driving the hardware clock) closely interrelate with the efficiency of the actual communication protocol for exchanging time synchronization packets (e. g., IEEE 1588 or SynUTC) and the synchronization algorithm running in software (e. g., the tentative more fault-tolerant democratic approach and master-slave). Tasks to be solved are therefore threefold:

(1) Accurate and fast modelling of both hardware and communication media
(2) Modelling of the protocol stack and the synchronization algorithm
(3) A generic, scalable, and flexible simulation environment able to handle complex time variant network structures

The accuracy of models for hardware and communication media heavily depends on parameters like delay jitter and delay skew. Furthermore, parameter stability evaluations have to be undertaken for the physical medium. Especially for the widely used Ethernet, basic input data can be acquired using an Ethernet network analyser. For other technologies like power grids or RS-485 links those parameters have to be measured manually using special test beds.

Existing protocol stacks and synchronization algorithms have to be adapted, to be used within the simulation environment to allow, e. g., multiple instantiation or tracing of internal states. Otherwise the protocol and the synchronization algorithm have to be abstracted and modelled to allow their simulation.

Setting up simulations with a fixed set of models and a well-defined network structure (e. g., solely for Ethernet networks) would not require extra effort for ensuring that the same simulation can be run with, e. g., the same protocol but a different network technology. But as clock synchronization is very likely to be used in a variety of applications, this flexibility is of outmost importance.

Product Name	Company	Market Segment
Ethernet-IP (Organization, 2005)	Rockwell Automation	general automation
Ethernet Powerlink (EPSG Organization, 2005)	B&R	machine automation
Profinet V3 (Popp *et al.*, 2005)	Siemens	motion control
SERCOS (SERCOS IGS, 2005)	SERCOS IGS	motion control
Ethercat (Jansen and Buttner, 2004)	Beckhoff	machine automation
SynqNet (Matheson, 2004)	MEI	semiconductor machine automation
JetWeb (Jetter GmbH, 2005)	JETTER	general automation
FlexRay (Millinger and Nossal, 2005)	Consortium	automotive
TTP (Kopertz *et al.*, 2005)	TTTech	airborne, automotive

Table 1. Collection of applications using any sort of clock synchronization

Furthermore, it is vital for the practical relevance of the results to investigate the scalability, which therefore has to be taken into account in the development of the simulation environment.

2.2 Discussion of existing technologies

For modelling and simulation of a networked, distributed system several simulation framework tools exist. The most promising representatives are discussed in the following.

C++Sim is a free C++ library, which supplies the environment for executing C++ models in parallel. The project SimUTC (Weiss *et al.*, 1999) used this environment for simulation of interval based clock synchronization algorithms. C++Sim offers a generic, open, and freely available simulation tool set, but has the disadvantage of a spartan support for state-of-the-art networks and limited visualisation support.

The NS2 (Network Simulator) tool set is another candidate for the simulation of clock synchronization networks. The models in NS2 are usually very coarse-grained: e. g. transmission lines in NS2 are typically represented only by the properties of a bandwidth, a delay, and an identifier. The improvement of the delay model, which is in fact perfectly suitable for simulations in IP-based networks, would need a complete refinement of the lower network models. Additional drawbacks in the so far mentioned simulation tools are tedious debugging as well as limited visualisation.

Other tools like PARSEC, SMURPH, Ptolemy, NetSim++ and CLASS offer practically the same functionality as the simulators mentioned above, but unfortunately do also have similar restrictions and drawbacks.

In the set of commercial tools, OPNET® is widely held to be the state of the art in network simulation. This event-based simulator uses finite state machines (FSM) with call-back functions as a programming model (in opposite to C++Sim, where entities have to be implemented in a multi-threaded fashion). The simulator comes with support for various predefined models, which are easy to reuse. Disadvantages of this simulation tool are the relatively expensive license fee and the unavailability of the source code. Modifications and extensions for special cases are not possible. More importantly, this also inhibits a fast and flexible kernel interfacing to other simulation tools.

The OMNeT++ simulator is available as source code under the *Academic Public License* offering both programming models: FSM-based as well as multi-threaded. The simulator is widely used (Varga, 1999; Wang and Keshav, 1999) and very well documented. OMNeT++ models are built upon hierarchically nested modules, which may be linked in an unlimited fashion. This allows the reflection of the logical structure of a system in the model structure. The so-called modules communicate by means of passing messages which may contain any complex data structure. Modules may always send messages either directly to their destination or along a self-defined, dynamic path. Since the whole simulation environment is available in source code interfacing to other specialised simulators with standard C/C++ APIs like Modelsim® and SystemC can be done in a very efficient way.

3. PROPOSED SIMULATION ENVIRONMENT

Figure 1 shows the simulation environment and the implementation of the different layers which are mapped to modules. Every network node consists of three modules:

- The *physical model*, which describes the behaviour of the communication channel. This model shall consider sending and receiving delays, respectively. Generally these delays also vary over time (Horauer, 2004) and thus a jitter and its time variant distribution has to be implemented in this layer.
- The *hardware layer* models which represent the receiving hardware with special emphasis on the timing properties. The purpose of this layer is on the one hand to simulate specialised clock synchronization entities like on-the-fly timestamping modules and priority queues as well as general hardware properties like influences of elastic buffers

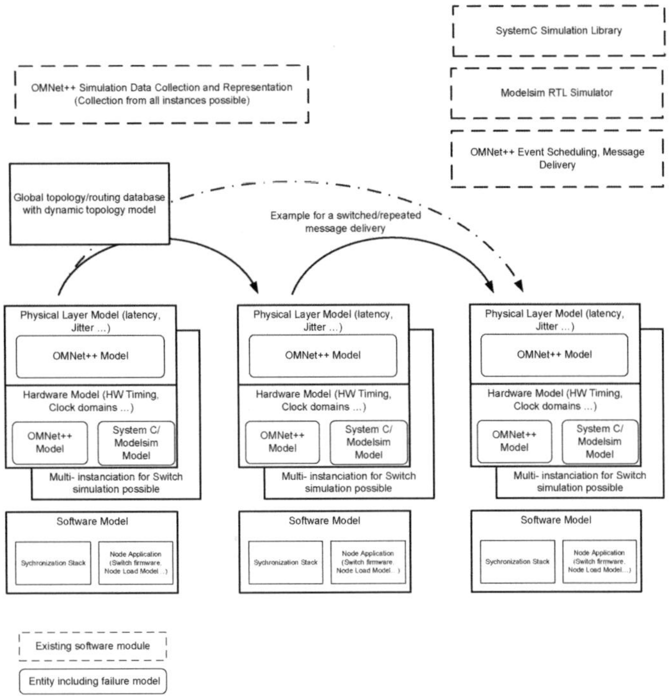

Fig. 1. Simulation Components and Communication Example

and synchronization effects between clock domains (e. g., between the reconstructed receiver clock and the node clock). For a fine-grained simulation of hardware on a register-transfer level the state-of-the-art hardware simulator Modelsim® is beneficial. This elegantly allows using and verifying the application specific VHDL/Verilog code of existing designs. The advantage of this simulation architecture is that models are taken from their actual implementation and not from abstracted models, and thus the simulated hardware will behave exactly like implementation in silicon. For the sake of reducing simulation times, modules are also be implemented in SystemC/SystemVerilog whenever possible without sacrificing accuracy. This approach offers the possibility to describe hardware modules in C with library extensions for hardware specific properties.

- The *software model* which finally consists of the clock synchronization stack as well as a model for the application at the network nodes. The latter is a rather simple model which simulates load. This, on the one hand limits the computing time for the synchronization stack and on the other hand generates overhead load within the network. Since OMNet++ uses C++ as description language, synchronization protocols like IEEE 1588 PTP or the SynUTC stack can be used by simply providing the necessary interfaces. In the particular case of the IEEE 1588 communication stack, all protocol enhancements towards fault tolerance or transparent master groups can be optimised by modifying the stack at this point.

In order to deal with entities multiplexing messages over the different segments of the network (e. g., switches or bridges) and sharing the code of the simple network nodes, the physical and the hardware layer shall be implemented in a way that it may be multi-instantiated. Thus only the software stack has to take care of the distribution of the messages within the multiple hardware interfaces. Furthermore, this modular architecture offers the possibility to evaluate multiple physical communication channels within one environment. For example, a synchronized powerline network with an Ethernet based backbone can be simulated as well. Figure 2a shows such a topology. The transparent masters (bridges) are needed by the communication channel. Other aspects to be investigated are meshed networks, where multiple transmission paths are needed due to redundancy reasons. The modular simulator concept, with the support of multiple physical layer instances allows the simulation of networks shown in Figure 2b.

The network topology is controlled by a central module which distributes the packets between all nodes. This entity is necessary to model dynamic topology changes. These topology changes need to be simulated both for Ethernet type networks with redundant but unreliable communication channels and for powerline networks spanning across transformer stations. Networks like this have to cope with a non-stable topology,

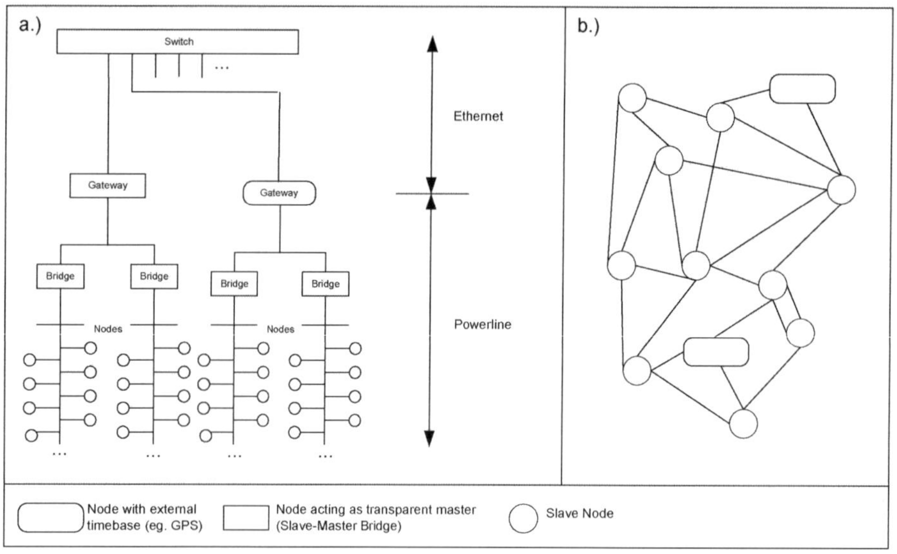

Fig. 2. (a) A typical powerline network with an Ethernet backbone (b) meshed network with multiple external reference-timebases.

since energy suppliers may switch whole subnets from one transformer station to another at any time. As standard OMNet++ simulation setups use merely an editor for graphical topology editing (NED), which can only be set up statically before simulation, a new concept for the global topology/routing database is needed.

Data inspection and simulation evaluation can be done either online by the OMNet++ GUI or – via ASCII files – with the A/GPL tools PLOVE and GNUPLOT as well as direct output of debug-files. All this analysing tools can be taken unchanged from the OMNet++ environment. Figure 3 shows the software structure of the simulation environment. The software tools needed are,

- a C++ Compiler/Linker
- The SystemC/SystemVerilog library, and
- The Modelsim® simulation environment.

Integrating all parts consequently ends up in a model for a particular implementation. By replacing single modules, different synchronization strategies may be compared with each other. For example performance, in terms of efficiency, synchronization speed, and fault tolerance of different synchronization stacks can be compared.

4. APPLICATIONS FOR THE SIMULATION ENVIRONMENT

The primary task of the proposed environment is the simulation of dynamic behaviour of clock synchronization in distributed applications and to provide means for its investigation. As clock synchronization is a typical problem of distributed and embedded systems, it is a requirement to be able to simulate hardware (e. g., the hardware clock in an application specific integrated circuit (ASIC)) and software (e. g., the precision time protocol (PTP) stack) simultaneously. It is therefore crucial to offer interfaces to other simulation tools and languages. This can be achieved by using the event based simulator OMNet++(Varga, 1999) for the core of the simulation environment. Interfaces to simulators of hardware description languages (HDL) (e. g., Modelsim®) or languages offering hardware-software co-simulation features like SystemC or SystemVerilog are thus easy to implement.

A potential drawback for clock synchronization in switched Ethernet together with effects of a proposed solution shall be evaluated as well: switch delay latency and transparent clocks (Jasperneite *et al.*, 2004). Any kind of standard network equipment potentially degrades the accuracy of network based time synchronization, because it may introduce unpredictable transmission delay latencies (Horauer, 2004). It is possible to eliminate this uncertainty if the time every packet resides on a switch is known to the receiving node. This can be accomplished by augmenting a network device with dedicated time stamping hardware, which adds a time stamp to every synchronization packet as soon as it enters and leaves the switch (Horauer *et al.*, 2000; Gaderer *et al.*, 2005*b*).

According to the overall use case of the simulator, two general groups of application areas can be defined. Firstly, the simulator is sufficiently comprehensive and flexible to answer questions relating to the dynamic behaviour of clock synchronization strategies in complex networks, such as

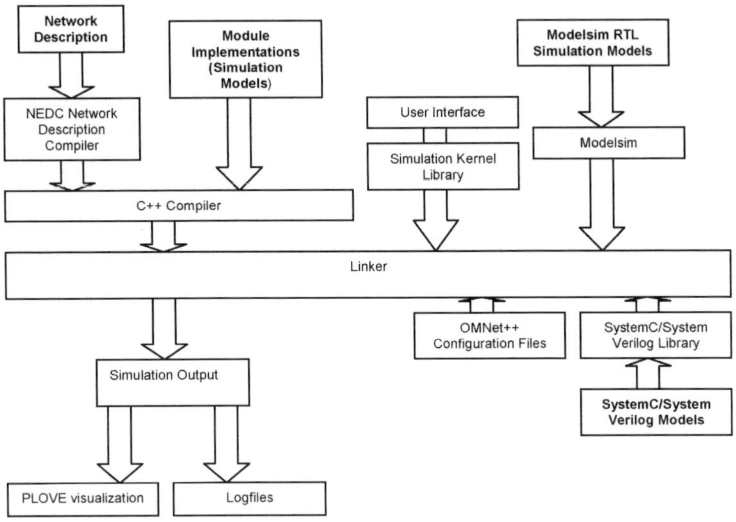

Fig. 3. Software Structure

- Improvement of fault tolerance and robustness in IEEE 1588-based synchronization schemes through the introduction of a master group
- Improvement of the synchronization accuracy by hardware-based on-the-fly timestamping as opposed to a handshake procedure
- Dynamic behaviour of a network-wide time base in the event of changing network topologies
- Possibilities and achievable accuracy of synchronization in heterogeneous networks
- Scalability and performance limitations depending on the number of nodes and network load
- Performance and accuracy limitations due to jitter introduced by cascaded switches

On the other hand, the tight coupling of the simulation environment to simulators used for hardware design as well as the possibility to use actual software modules will facilitate the transfer of the results to practical implementations. In fact, it are to a large extent the real-world implementations of hardware and protocols or algorithms which are embedded in the simulation, rather than abstract models which always bear the possibility of modelling errors. In particular, this concerns

- Flexible and extensible simulation framework allowing the inclusion of physical models for the network links, hardware models for network components, and behavioural models for application processes
- Time synchronization protocol stacks (like PTP defined in IEEE 1588) with extensions to improve robustness and accuracy
- Simulation models for Ethernet and powerline communication links

5. CONCLUSIONS AND FURTHER RESEARCH

Currently available simulation environments are not capable of addressing the specific problems that arise in the course of simulating network time synchronization. The presented approach specifically addresses an integrated simulation model, which allows to monitor effects arising from the interaction of the involved network layers. Furthermore, the usage of a combination of existing simulation technologies allows to run simulations directly with the source code which is also used in the final application. This reduces the risk of mismatches due to modifications of the time synchonization application in order to be run in the simulation environment.

The developed network simulator will be tested in detail on several real-world cases, putting special emphasis on the evaluation of master groups as an extension to IEEE 1588. This clock synchronization standard distributes time information following a master-slave principle – a well-known method, which is simple to implement. However, it lacks fault tolerance: If a master within a network fails, all nodes relying on its time information will deviate gradually. IEEE 1588 handles this drawback as follows: Within a network every node may possibly become a master. All nodes continuously observe the quality of all other local clock sources and choose a master via a best- master-selection algorithm. This improves fault tolerance to a certain extent, yet it does not tackle transient sync losses, which occur during the time it takes for the network to select a new master.

By introducing a so-called master group this problem can be solved. Within a normal IEEE 1588 network a group of nodes (most likely those with very accurate local clocks) are linked to each other. They use SynUTC clock synchronization technologies to maintain a fault tolerant common notion of time without affecting other network nodes. They select a single node which will act as an IEEE 1588 master to the residual network. If this node fails, any other node out of the master group may take over instantaneously and unnoticed by all IEEE 1588 slaves. The final simulation environment will be used to analyse the drawbacks and advantages of this approach compared to other synchronization mechanisms. Especially the transient behaviour of different architectures and the achievable accuracy in case of errors or during their recovery phase will be among the benefits derived from the simulations.

REFERENCES

Dana, Peter H. (1997). Global Positioning System (GPS) time dissemination for real-time applications. *Real-Time Systems* **12**(1), 9–40.

EPSG Organization (2005). The Ethernet-Powerlink Homepage. http://www.ethernet-powerlink.org.

Fetzer, Christof and Flaviu Cristian (1997). Integrating external and internal clock synchronization. *J. Real-Time Systems* **12**(2), 123–172.

Gaderer, Georg, Thilo Sauter and Gerd Bumiller (2005*a*). Clock synchronization in powerline networks. In: *Proceedings of the 2005 IEEE International Symposium on Power Line Communications and its Applications*. pp. 71–75.

Gaderer, Georg, Thilo Sauter and Roland Höller (2005*b*). Strategies for clock synchronization in powerline networks. In: *Proceedings of the 2004 International Workshop on Real-Time Networks*. p. 77.

Horauer, Martin (2004). Clock Synchronization in Distributed Systems. PhD thesis. Vienna University of Technology.

Horauer, Martin, Nikolaus Kerö and Ulrich Schmid (2000). A network interface for highly accurate clock synchronization. In: *Proceedings AUSTROCHIP'00*. Graz, Austria. (to appear).

Jansen, D. and H. Buttner (2004). Real-time Ethernet the EtherCAT solution. In: *Computing & Control Engineering Journal*. Vol. 15. pp. 16–21.

Jasperneite, J., K. Shehab and K. Weber (2004). Enhancements to the time synchronization standard ieee-1588 for a system of cascaded bridges. In: *Proceedings of the 2004 IEEE International Workshop on Factory Communication Systems*. pp. 239–244.

Jetter GmbH (2005). The Jetter Homepage. http://www.jetter.de.

Kopertz, Hermann, Günther Bauer and Wilfried Steiner (2005). *Dependable Time-Triggered Communication*. The Industrial Communication Technology Handbook. Taylor & Francis.

Kopetz, Hermann (1997). *Design principles for Distributed Embedded Applications*. Kluwer Academic Publishers.

Matheson, M. (2004). Synqnet: high performance motion control based on Ethernet. In: *Computing & Control Engineering Journal*. Vol. 15. pp. 32–38.

Millinger, Dietmar and Roman Nossal (2005). *FlexRay Communication Technology*. Chap. 30, pp. 30–1. The Industrial Communication Technology Handbook. Taylor & Francis.

Organization, ODVA (2005). The Ethernet-IP Homepage. http://www.ethernet-ip.org.

Pedreiras, Paolo and Luis Almeida (2005). *Approaches to Enforce Real-Time Behavior in Ethernet*. Chap. 20, pp. 20.1 – 20.28. The Industrial Communication Technology Handbook. Taylor & Francis.

Popp, Manfred, Joachim Feld and Ralph Bögen (2005). *Principles and Features of PROFInet*. Chap. 11, pp. 11–1. The Industrial Communication Technology Handbook. Taylor & Francis.

Schmid, Ulrich, Ed. (1997). *Special Issue on The Challenge of Global Time in Large-Scale Distributed Real-Time Systems*. J. Real-Time Systems 12(1–3).

SERCOS IGS (2005). The SERCOS Homepage. http://www.sercos.de.

Varga, A. (1999). Using the omnet++ discrete event simulation system in education. *IEEE Transaction on Educcation* **42**, 11.

Verissimo, Paolo and Luis Rodrigues (2001). *Distributed Systems for System Architects*. Kluwer Academic Publishers.

Wang, J. and S. Keshav (1999). Efficient and accurate ethernet simulation. In: *Proceedings of the Conference on Local Computer Networks*. pp. 182–191.

Weiss, Bettina, Günther Gridling, Ulrich Schmid and Klaus Schossmaier (1999). The SimUTC fault-tolerant distributed systems simulation toolkit. In: *Proceedings 7th International Symposium on Modeling, Analysis and Simulation of Computer and Telecommunication Systems (MASCOTS'99)*. College Park, MD, USA. pp. 68–75.

ELSEVIER
IFAC
PUBLICATIONS

Single Microprocessor Implementation for Safety-Related Networks Using CANopen

Thilo Schumann, CAN in Automation (CiA)
Cyrilla Jane Menon, CAN in Automation (CiA)
www.can-cia.org

Abstract

For most of history, safety-related systems used only mechanical backups. These systems are specialized. In only a few application fields or pilot projects the mechanical systems are replaced by electronic solutions. The time has come when more engineers want to design using electronic systems instead of mechanical. For this to be possible a solution must exist. The communication network must support this.

In its current usage, CAN, does not have all the requirements necessary for use in safety-related applications. Even after adding CANopen as the higher-layer protocol specific needs were not met. After exploration of these deficiencies, the CANopen Framework for Safety-related applications was created. This framework allows the use of safety-related and non-relevant communication on the same network. The basic idea is to transmit each safety-related message twice using different message identifier. The data is bit-wise inverted as well with a time-out attributes assigned. CANopen can now be used for safety-related applications. If CANopen is the solution, then a one microprocessor can be the single solution for the application within a device.

Introduction

The increasing usage of bus systems in the automation industry is generated by cost. The first savings came from the reduce wiring. Next it was realized that the time to install and troubleshoot the installation more than balanced the extra cost of having the network interface on the device. But the bus systems used in automation were mainly used for control purposes and not for safety-related devices. Such safety-related devices, e.g. emergency push buttons, are still additionally hardwired in a conventional and expensive way. They utilize their own system. Their primary requirement is more than just control - no person should be endangered while working with an industrial machine.

It is possible to network the safety-related devices with existing networks, but that would not necessarily reduce cost. Safety-related bus systems have been developed as proprietary solutions, typically with additional wiring in parallel to the control bus system. For the end user this means another system has to be installed and maintained. The clearest way to reduce cost is to look at the existing systems and see if they can support safety-related applications.

A consortium was formed within CAN in Automation (CiA), the international manufacturers and users group, to find a solution to make CANopen safe and obtain certification from the German authorities - BIA (Berufsgenossenschaftliches Institut für Arbeitssicherheit, Institute for Occupational Safety) and TÜV (Technischer Überwachungsverein, Association for Technical Inspection).

Theory of Safe Operation

Some embedded systems controlling applications, like safety mats, emergency stops, and two-handed controls, absolutely need a safe state to exist as a reaction to an emergency command such as an alarm or an error. These are not at the safety-critical level, but at a safety-related level. This means they must have a safe, secure state that can be easily obtained, but does not need excessive redundancy.

For these, functionality must include safeguard measures that are regularly checked and can not allow a single defect during safety-related communication to override the safety circuitry. If such errors do occur, they need to be detected quickly before a second error occurs.

All the systems, especially the safety-related circuitry, must have high reliability in order to extend the time-span between the safety-tests and minimize the down time of the whole system (e.g. if one redundant components fails, the system has to be shut off). The need for safety decreases the availability of a system.

In Europe, requirements have been outlined to standardize levels of safety-relevance (Table 1). The BIA has adopted the Safety Integrity Levels (SIL) described in IEC 61508, while other organizations accept EN 954-1. A system is defined as a safe system if it fulfills the requirements of SIL 2 or better SIL 3. This requires a probability of non-detected faulty messages to be lower than 10^{-14} for safe communication.

Probability of non-detected faulty messages	10^{-12}	10^{-13}	10^{-14}	10^{-15}
SIL (IEC 61508)	0	1	2	3
CAT (EN 954-1)	B	2	3	4

Table 1: References between safety-classes and the probability of non-detected faulty messages

These standards also outline possible failures:

- Repeated message - an old message will be repeated by a damaged device at the wrong time. During this occurrence, a receiver can be critical disturbed. For example a safety door could remain closed, when it should open.
- Lost message - important information will be deleted by a damaged device. For example a request for safety-stop.
- Inserted message - a new message could be inserted by a damaged device. For example a request for continue from safety-stop.
- Renumbered message - the sequence of the messages could be changed by a damaged device. For example before the request of a safety-stop there is a need to reduce the velocity, but the messages are changed in the sequence and the machine runs on.
- Falsified message - a message could be changed by a damaged device or a disturbed communication medium.
- Delayed message
  - [1] The transmission path is overloaded during the normal operation.
  - [2] A damaged device causes an overload by transmitting wrong messages and a safety-related message can not be sent.
- Manipulated message - There is a node that manipulates messages to its needs. (This can be excluded because CANopen is mainly used in a closed system configuration.)

There are several common methods to recognize these failures and create a safe system:

(1) Running number in safety-related messages
(2) Relative, absolute or double time-marks
(3) Time-out
(4) Confirmation of message
(5) Identifying producer and consumer

(6) Application CRC
(7) Redundancy with crosschecking
(8) Different data checking for safety-related and non safety-related messages

When put in a cross chart, the BIA selected certain combinations as recommendations for providing a safety-related system (Table 2).

Failure	Methods: Numbering	Time stamp	Time expectation	Reply	Identification (for sender and receiver)	Data security	Redundancy with cross-examination
Repeated message	√	√					√
Lost message	√			√			√
Inserted message	√			√	√		√
Renumbered message	√	√					√
Falsified message				√		√	Only serial bus systems
Delayed message		√	√				

Table 2: Common failure detection methods

The hardware must also be analyzed in safety-related systems. Shown are four possible design models. The Model A shows a system where the microprocessor is redundant, but only one network interface exist (Figure 1). Model B is a completely redundant system with duel microprocessors and complete network interfaces. Model C employs the redundant microprocessors, but only partial redundant network interfaces (Figure 2), while Model D only has the redundant microprocessors with a single network interface.

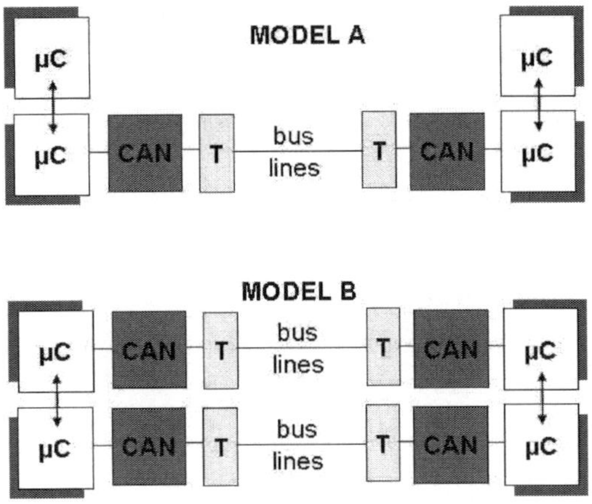

Figure 1. BIA Model A and B

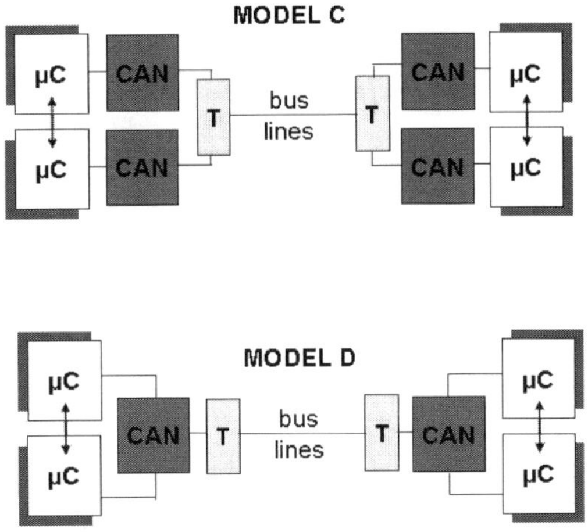

Figure 2. BIA Model C and D

CAN and CANopen

Controller Area Network (CAN) was invented by Bosch in the early 1980's for use on vehicles as a control system which makes it robust enough for many automation environments, as well. Since the CAN protocol (performed by off-the-shelf controllers) only defines the data link layer in the ISO Open Systems Interconnect (OSI) model, higher layer protocol (HLP) specifications outline the rest – physical layer, application layer, etc. CANopen is one such HLP. The focus of CANopen's design was as a standardized embedded network with highly flexible configuration capabilities that unburdens the control system developer from dealing with CAN-specific details. It provides enough flexibility for tailoring to their specific needs, but remains a standard application layer with available off-the-shelf configurable products.

CANopen version 4 (CiA DS 301) is standardized as EN 50325-4. The CANopen specifications cover the application layer and communication profile (CiA DS 301), a framework for programmable devices (CiA 302), and recommendations for cables and connectors (CiA 303-1). These documents set up the framework, or object dictionary, for dealing with CAN-specific details such as bit-timing and implementation-specific functions. They provide standardized communication objects for real-time data (Process Data Objects, PDO), configuration data (Service Data Objects, SDO), and special functions (Time Stamp, Sync message, and Emergency message) as well as network management data (Boot-up message, NMT message, and Error Control). The CANopen specifications, frameworks, and profiles are freely available from the CiA (www.can-cia.org).

CAN Failure Mechanisms

All CAN controllers support a protocol which includes a mechanism to detect errors and globalize them, in the case of local failures. Safety in CAN communication is not to ensure that there are absolutely no errors and faults, but to detect all possible errors and react in a predictable (safe) way.

In a safe CAN system there are producers of safe information (e.g. safety switches, light barriers, emergency stop buttons) and consumers of such information (e.g. relay, valve or drive controlling a possibly dangerous movement, safety PLC). As the "consumers" control the possible dangerous situation they are responsible for entering the safe state after any safety-related interference. They also have to check the data integrity of the safety-related communication. The "consumers" are the safety masters in a possible safe CANopen system.

As the producers (safety inputs) are the origin of the safe communication objects, their numbers are limited. The number of safety masters is not limited in theory, as CAN allows many consumers to listen to the same safe communication objects, e.g. many actuator devices can use the same information.

CAN's Five Error Detecting Mechanisms

CAN has its own error detection mechanism. Because of the CAN protocol, system elasticity and content-based identification as provided by the CAN data link layer complicates error handling. For example, the classical method for error handling is to send back a response message from the receiver to the transmitter upon correct reception of a message. This requires the address of the receiver to be known. CAN does not support local addresses which means the identifier labels a message, not the device sending it. In lieu of this information in every message, CAN error detection required local assignment of errors to stations.

In CAN, a combination of positive and negative acknowledge is used. If a CAN node detects an error, it discards the received bits of the current message and transmits an error flag, which is six consecutive bits of the same polarity. All CAN nodes error check all messages. To detect failures there are five mechanisms implemented: bit error, stuff error, CRC error, acknowledgement error, and form error.

Globalization of Failures

While nearly all global errors are detectable, there are also local errors, which may occur because:

- nodes with different sampling points interpret the same bit value differently, or
- in single nodes there are different thresholds due to tolerances and drifts, or
- some nodes interpret different bit values due of line desperations.

In order to provide network-wide data consistency the CAN protocol uses the bit-stuff method to signalize errors globally. An Error Flag contains six bits of the same value. For any node that has not yet detected an error, this will be interpreted as a Stuff Error. An overlapping second Error Flag may prolong the Error Frame. Each node sends 'recessive' bits after the Error Flag transmission. If the node monitors a 'recessive' bit, it continues transmission of an additional seven 'recessive' bits. This is the eight-bit long Error Delimiter.

CANopen Safety

The mechanisms to detect failures in CAN messages are not enough for safety-related communication, because they are handled locally and focused for one message. There are also some failures which must be handled that are application driven. Even though CANopen supports more error detection and has emergency messaging, it too is not enough.

For CANopen, adding the functionality of safety-related applications was defined under CiA DS 304 - CANopen Framework for Safety-related Communication. It allows for safety-related communication use within CANopen networks, while allowing the co-existence of normal data on the same network (figure 3). To ensure compatibility, the usage of identifiers and pre-defined objects has been coordinated with the CANopen standard and existing device profiles.

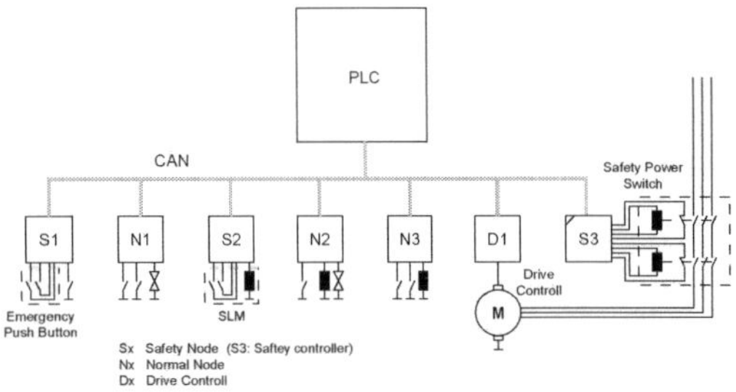

Figure 3: Example of a CANopen network with safe nodes

Transport of safety-related data

To satisfy safety-related requirements it was decided to use a redundancy with cross-examination. Additionally, special identifiers and a time expectation are used (Table 3).

	Methods						
Failure	Numbering	Time stamp	Time expectation	Reply	Identification	Data security	Redundancy with cross-examination
Repeated message							√
Lost message							√
Inserted message					√		√
Renumbered message					(√)		√
Falsified message							√
Delayed message			√				

Table 3: Failure detection methods in CANopen

The safety-related data is transmitted in an SRDO (safety-related data object). An SRDO consists of two CAN messages where the identifiers differ in at least two bit positions (Figure 4). The content of the second CAN SRDO message is bit-wise inverted to the content of the first message.

This concept, called serial redundancy, allows the use of the entire 8-byte data field. This means the PDO (process data object) mapping defined in standard CANopen device and application profiles may be adapted to the SRDO mapping with no specific safety profiles required. The SRDO is transmitted periodically with a SCT (safeguard cycle time).

In a CANopen network the data interface to the application program within a certain node is only via the CANopen object dictionary. The application itself has to transfer the data correctly, in time and in sequence to the CANopen kernel.

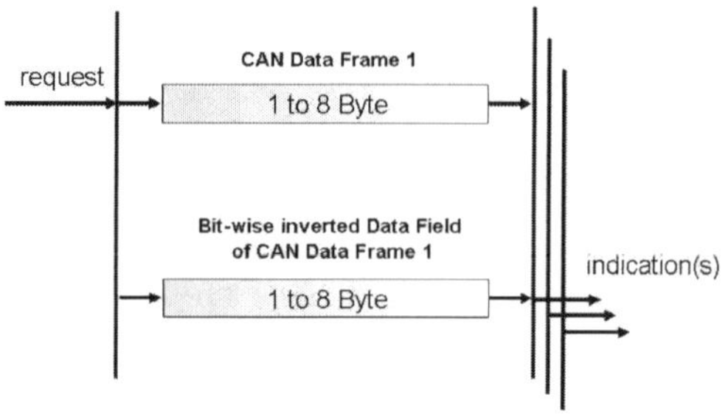

Figure 4. SRDO messages

SRDOs are typically transmitted periodically. If required, SRDOs may also be transmitted event-driven, e.g. to ensure fast reaction after a change on the input. SRDOs are only allowed in the network state "operational".

There are two uses for SRDOs - data transmission and data reception. It is distinguished by the information direction. Devices where the information direction is set to transmit (tx) are SRDO producer and devices where the information direction is set to receive (rx) are SRDO consumer.

SRDOs are described by the SRDO communication parameter (26h) and the SRDO mapping parameter (Figure 5). The SRDO communication parameter describes the communication capabilities of the SRDO. The SRDO mapping parameter contains information about the content of the SRDOs (device variables).

Index	Object	Name	Type	Acc.[1]	M/O
1300h	VAR	GFC parameter	UNSIGNED8	rw	O
SRDO Communication Parameter					
1301h	RECORD	1st SRDO parameter	SRDO Parameter (26h)	rw	M
1302h	RECORD	2nd SRDO parameter	SRDO Parameter (26h)	rw	M/O*
...	...	...	...	...	...
1340h	RECORD	64th SRDO parameter	SRDO Parameter (26h)	rw	M/O*
1341h	reserved				
...	...				
1380h	reserved				
SRDO Mapping Parameter					
1381h	ARRAY	1st SRDO mapping	UNSIGNED32	rw	M
1382h	ARRAY	2nd SRDO mapping	UNSIGNED32	rw	M/O*
...	...	...	...	...	...
13C0h	ARRAY	64th SRDO mapping	UNSIGNED32	rw	M/O*
13C1h	reserved				
...	...				
13FDh	reserved				
13FEh	VAR	Configuration valid	UNSIGNED 8	rw	M
13FFh	ARRAY	Safety Configuration Checksum	UNSIGNED16	ro	M

Figure 5. Object map

SRDOs are transmitted periodically in order to ensure the safety-relevant communication is functioning. This periodic time is referred to as the safeguard cycle time (SCT).

A second check is done with a second timer. The maximum time between the two CAN SRDO messages is referred to as SRVT (safety-related validation time). If any of the configured timers expires on the SRDO consuming devices, it causes a failsafe situation. In case of SRDO receiving, the application has to collect and check SRDO data so frequently, that the time expectation can be fulfilled (figure 6).

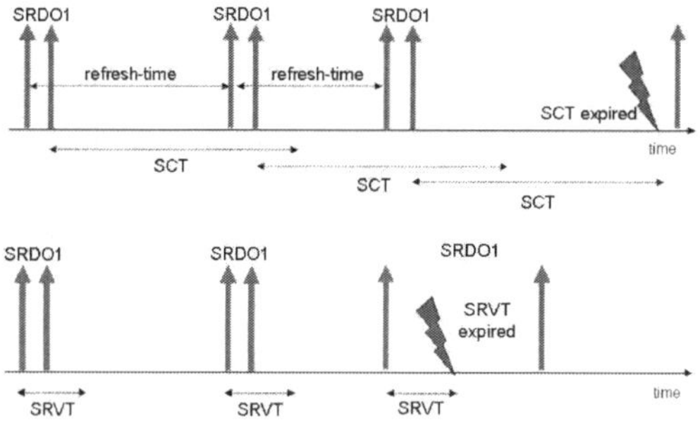

Figure 6. SRDO Validation Time

The CANopen Safety protocol also defines the GFC (Global Failsafe Command). The GFC may be used to speed up the system reaction time. Since it is transmitted event-triggered, it is not considered safe. But when coupled with the transmission of a corresponding SRDO the safety issue is resolved.

Due to CANopen application layer compatibility, the safety-related communication is limited to transmitting 64 SRDOs. The number of SRDO consumers is not limited. It is assumed that a device with the need of safety-related communication may use all other CANopen functionality including PDO (process data object), SDO (service data object), NMT (network management), and Emergency services. SDO access to safety-related application and configuration objects is allowed only while in NMT pre-operational state.

As the safety controllers are responsible for the data integrity and actuality, every safety-related output device has to survey all corresponding sources of safety data.

CSC01 – CANOPEN SAFETY CHIP

To realize safety-related systems within a CANopen system, the consortium opted to achieve SIL 3. They also opted to create a silicon / software combination solution that employed the following physical layer model (Figure 7):

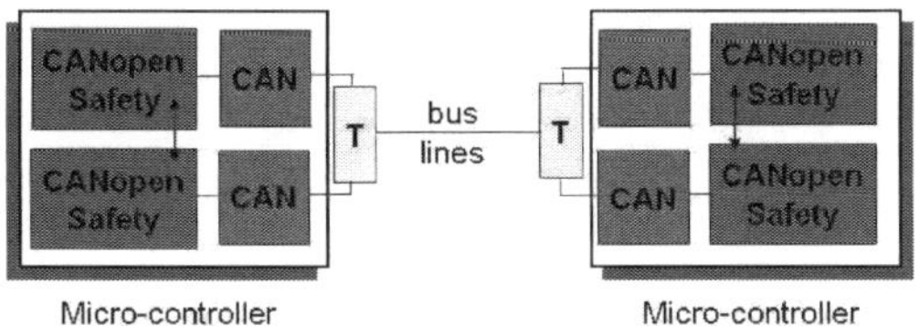

Figure 7. CANopen CSC model

The result was CSC01, the CANopen Safety Protocol chip. The chip is based on the M306NAFGTFP step D 16-bit microcontroller by Renesas and features two independent on-chip CAN modules.

The CSC01 hardware provides two transmit SRDOs and two receive SRDOs. In addition, the chip supports the GFC function. It implements the CANopen NMT slave functionality including default SDO server, heartbeat producer, two transmit PDOs and two receive PDOs. The on-chip CAN modules support all CANopen bit-rates up to 1 Mbit/s.

The CSC01 is clocked at 16 MHz. It provides 10 KiB of SRAM and 256 KiB of Flash memory. The processor comes in a QFP-100 housing, and is specified for a temperature range of -40 °C to +85 °C. The on-chip periphery includes A/D and D/A converters, timers, synchronous and asynchronous serial links, as well as digital I/O ports.

The CANopen Safety firmware consists of two parts:

- CSC main function - comprised of the certified CANopen Safety protocol stack and the certified diagnostic functions.
- safety application program - to be developed by the CSC user.

The CSC main function controls the internal program flow, all used peripherals, system stacks, registers, and op-codes. The implemented diagnostic functions include the 16-bit CRC (cyclic redundancy check) of the memory. When an error is detected, the sending of SRDOs is stopped, and the external watch-dog is triggered in order to switch the outputs to the safe state.

The CSC01 provides 2-ms computing time for the safety-application program within the 20-ms safety cycle time. The safety-application software is periodically called eight times (2.5 ms with a jitter of ±0.6 ms). The CSC main function requires 2 KiB RAM and 32 KiB Flash memory. In addition, the system stack needs 512 byte of RAM.

In keeping with the original intent of the consortium, the application has been certified by the TÜV (Technischer Überwachungsverein - German Association for Technical Inspection). It is designed to be used in simple safety-related senor and actuator applications. It may be used to connect emergency switches, light grids, and position transducer directly to the CANopen network. One unique feature is the ability to configure a direct communication between safety-related sensors and actuators with no safety-related controller or monitor needed.

CONCLUSION

Networks are an important part of embedded systems. Safety-related applications are equally important. Specific networks applications for use in safety-related embedded applications have become available.

The CANopen solution is based on proven technology and as an off-the-shelf solution. This is important because it now saves developers' time and money in creating a solution for their application. The solutions allow the user to include their critical applications with non-critical devices for a one-network solution. Hardware / software solutions are also available, as shown with the CSC01 chip. These are easy to integrate into product solutions.

REFERENCES

1. R. Bosch GmbH, "Controller Area Network Specification: Version 2," 1991.
2. CAN in Automation, "CANopen Specification."
3. ISO / 7498 – 1984, "OSI Basic Reference Model."
4. Pfeiffer, O., Ayre, A., and Keydel, C., *Embedded Networking with CAN and CANopen*, RTC Books, San Clemente, 2003.
5. Jungandreas, F., "The CANopen Safety Chip," *CAN Newsletter*, 4/2003.
6. CAN in Automation, "CiA DSP 304-v1.0.1: Framework for safety-relevant communication," 2004.

ELSEVIER
IFAC
PUBLICATIONS

Virtual Automation Networks – Start Conditions for a European Integrated Project

Peter Neumann

Institut für Automation und Kommunikation Magdeburg
Steinfeldstraße 3, 39179 Barleben, Germany
peter.neumann@ifak-md.de

Abstract: In 2005 the European Integrated Project "Virtual Automation Networks" has been established. This project deals with heterogeneous communication networks and their use for geographically distributed Automation applications. The heterogeneous character of the used network with different network transitions and used business models results in specific requirements necessary for the automation domain to guarantee a specific behaviour of the end-to-end connection between the distributed applications: scalable real-time, scalable safety and security mechanisms. Additionally, the influence of the wireless communication technology on the heterogeneous communication networks has to be investigated. The complexity of the heterogeneous networks requires the investigation of the given approaches, the recent research and standardisation activities, and products under development.

Keywords: Ethernet, Communication Systems, Decentralised Control Systems, Industry Automation, Real-time, Performance, Safety, Security.

1 INTRODUCTION

Today's industrial manufacturing is confronted with a high and further growing degree of fast changing, customised production. As a result flexible management structures handling the activities e.g. commissioning, diagnosis, maintenance and asset management have to be developed. This requires a flexible and scalable company-internal data exchange and exchange with any other kind of involved remote companies. Within the process industry, the improvement of flexibility within the lifecycle of a process plant (about 25 years) is going on. Especially the engineering and asset management activities require the introduction of modern Web-based technologies to enable local and remote access to process data as well as parameters describing the features of installed devices. Following these requirements an interconnection of all parts of the information system architecture (horizontal and vertical, local and remote) of an enterprise has to be realised. The necessary research aims to provide the needed technologies to build up widely distributed, flexible, virtual automation networks (Bratoukhine, A. et al., 2003).

At the office level of a company many standardised technologies are used to handle the data and its exchange. At the factory floor special, different, mostly incompatible technologies are used. By the overall aim of an easy data exchange between office and factory floor and to use the high potential of well-developed office (IT) technologies (including Internet and Web-based technologies) these office technologies conquer the factory floor. But these technologies and concepts do not reach all domain-specific industrial requirements and standards in areas as security, wireless, safety, and real-time. Thus it is necessary to adopt, modify and extend common office/IT solutions according to industrial standards.

The domain-specific requirements are:

- Guaranty for real-time behaviour. There are different classes with different requirements,
- Guaranty for functional safety. This means protection against hazards caused by incorrect functioning including communication via heterogeneous networks,
- Guaranty for security. This means a common security concept for distributed automation using heterogeneous networks.

This means that heterogeneous networks consisting of local and wide area and wired and wireless communication systems will play an increasing role. However, there is not only a need for real-time, safe and secure communication. The desired context awareness leads to the usage of location-based communication services and context-sensitive applications.

A *Virtual Automation Network* (VAN) is a heterogeneous network consisting of wired and wireless local Area Network, the Internet, and wired or/and wireless telecommunication systems. This means that geographically distributed application programmes, co-operating to fulfil a control application, are connected via this VAN accessed by remote connection endpoints. Worldwide distribution of Internet offers the Automation domain a good infrastructure, but introduces many additional problems, which need to be solved.

To fill the existing gap, 15 partners coming from six European countries applied for a European Integrated Project "Virtual Automation Network" (VAN project, 2005) and started the project in September 2005. The project is co-ordinated by SIEMENS Automation & Drives. This paper gives a survey of the project targets and research areas and describes the start conditions.

2 THE PROJECT TARGETS

A typical enterprise can be roughly cut into two areas: an administrative office and a manufacturing or process area.

The VAN project will provide innovative solutions, extensions and standards dedicated to industrial environments, to fill the existing gap between office technologies and industrial automation technology, focused on a new dimension of uniform networking of production and manufacturing processes. The objectives will be achieved by a merging of existing and emerging IT and automation standards covering embedded architectures, fieldbus technologies, Ethernet, Internet technologies, real-time technologies, wireless LAN technologies, security and safety concepts.

VAN aims at an approach to integrate heterogeneous (multi-stakeholder) network concepts to an applicable Virtual Automation Network that could be widely used throughout the industry, with particular consideration to networked embedded systems. An end-to-end connection through a heterogeneous network has to guarantee location awareness, required scalable real-time behaviour, security and privacy, a very high degree of intrusion protection, and safety. Since there are much stronger requirements within the automation domain, focused within this paper[1], the Virtual Private Networks VPN known from the office domain do not suit the demands mentioned above. A target scenario of an "Industrial System Environment" could be:

a) Local Industrial Domain consisting of

- Industrial backbone connecting various industrial segments
- The industrial segments meet the different industrial requirements, e.g. "normal" field data transfer, intrinsic safety, hard (isochronous) real-time. These segments can be realised also as wireless segments.

b) Remote Industrial domain consisting of the mentioned parts of the local domain.

c) Local/remote Office Domain consisting of the well-known LAN and WLAN technologies.

d) Heterogeneous Wide Area Network consisting of different provider-oriented or provider-less telecommunication networks/Internet with many network transitions.

e) Direct Single Device Integration (telecontrol) via public telecommunication networks.

Existing IT solutions will be utilised and extended in a compatible way to accomplish the network concept while respecting the unique and harsh industrial environment. Moreover the VAN project aims to utilise the latest advances in wireless communications for the purposes of the industry.

Concluding the research and development results of VAN will be:

- Adoptions of office IT technologies extended by the required new functionalities in automation systems,
- Scalable real-time, safety and security technologies and capabilities over all levels of a (virtual) network,
- Integration concepts and guidelines for private and public Ethernet and Internet based networks
- Concepts and development of corresponding engineering tools for development and installation of embedded industrial communication devices in industrial plants
- Proposals for European and International standardisation of the VAN results

To reach the project targets the following technical Work Packages has been defined (Figure 1).

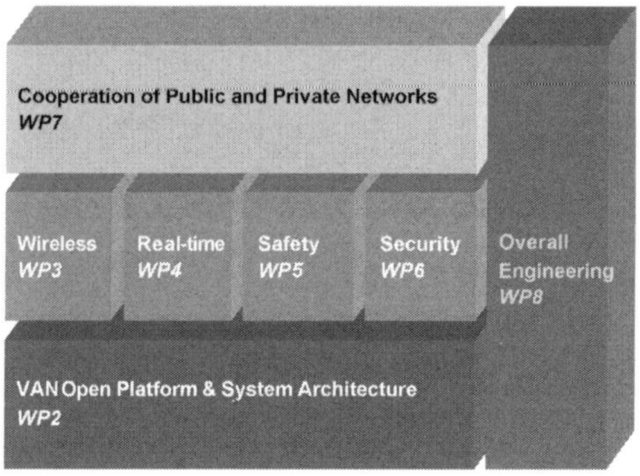

Figure 1. Work package structure

The following sections discuss the start conditions of the technology-oriented work packages. The results of the project will be published later.

3 REAL-TIME COMMUNICATIONS

3.1 Local Domain

Nowadays, there is a large community inventing the usage of Ethernet based communication systems to be used in the industrial automation domain, e.g. in the real-time and safety-critical world. However, opposite to that, Fieldbus systems are the most important communication systems used in commercial control installations. In the future, both the Fieldbus systems and Ethernet-based real-time communication systems will co-existent over a mid-term period. Thus, concepts for migration of standardised Fieldbus systems into Ethernet-based real-time communication systems become important. For the last 7 years, many scientific results has been published (e.g. Alves, M., et al., 2000; Baek-Young et al., 2000; Dong Sung Kim et al., 2005; Georges, J.-P. et al., 2002; Jasperneite,F., 2002; Lo Bello, L., O. Mirabella, 2000, 2001; Lüder, A. et al., 2004; Palensky, P. T.Sauter, 2000; Pereira, P. et al., 2004). The following considerations are directed to Ethernet-based real-time communication systems.

There are three real-time classes guaranteeing response time:

[1] The requirements of other application domains (e.g. Military, Banking Systems, Airplanes) are not under consideration within this paper

- Class 1: soft real-time (scheduling on top of UDP/TCP): scalable cycle time; used in factory floor and process automation
- Class 2: hard real-time (scheduling on top of MAC): cycle time 1...10ms. Used for control
- Class 3: isochronous real-time (with time/clock synchronisation and routing with time schedule): cycle time 250µs...1ms; jitter less than 1 µs. Used for motion control.

Additionally, there is a class "non real-time" not considered here.

Regarding real-time class 1,
There are many investigations regarding temporal behaviour related to Ethernet-TCP/IP based local networks (e.g. Härtig, H. J.Löser, 2004; Hoang,H., M.Jonsson, 2003; Lo Bello, L. et al., 2005; Kanghee Kim et al., 2002); Pedreiras,P. et al., 2005). They include mainly the response aspect of data packet transmission, which is very important within the industrial automation domain. The synchronous video or audio stream transmission is of secondary interest. In contrast to the given infrastructure in the industrial automation application, the *data packet transmission* has to have the highest priority.
The systems, which are using Ethernet-TCP/IP, offer response time in the millisecond range. The data transmission is based on the best effort principle. To use these systems within the automation domain, mechanisms are needed to monitor time limits, to use substitution values, to optimise the transmission (using records of many values within one MAC-PDU) as well as time- and event-triggered data transmission. The following three systems are part of the international Fieldbus standard IEC 61158 (IEC, 2003a; IEC 2003b):

- **Ethernet/ IP** (Rockwell, ControlNet International, Open DeviceNet Association) uses a Control and Information Protocol CIP (Ethernet/IP, 2001),
- **High Speed Ethernet HSE** (Fieldbus Foundation) (HSE, 2001),
- **PROFINET** using the application model « Component-Based Architecture" **(CBA)** (PROFIBUS user organisation, Siemens) (PROFIBUS , 2002). An open source code and various exemplary implementations/portations for different operating systems are available on the PNO Website.

All the mentioned approaches are able to support the office domain protocols, e. g., SMTP, SNMP, HTTP, some of them BOOTP, DHCP, for Web access and/ or for Engineering data exchange.
Additional systems using Ethernet TCP (UDP)/IP has been introduced in the Public Available Specification "Real-Time Ethernet" in 2004 (IEC, 2004)):

- **Modbus RTPS,** based on Interface for Distributed Automation **IDA** (MODBUS-IDA Group, Schneider) (IDA, 2002),
- **P-Net on IP** (Proces Data) (PAS).

(Pöschmann, 2003) compares the first four approaches. (Pedreiras, P. et al., 2005b) and (Schwager, J., 2004) give an overview of Real-time Ethernet approaches.

Regarding class 2,
Many research activities deal with a middleware on top of the MAC layer of Ethernet, scheduling the hard real-time and soft real-time/ non real-time traffic. (Baek-Young Choi, Sejun Song, et al., 2000; Lo Bello, L. and O.Mirabella, 2000 and 2001; Kanghee, K. et al.; 2002 and 2005; Georges, et al., 2002; Jasperneite, 2002; Pedreiras, P. and L.Almeida, 2003, 2004 and 2005a; Hoang, H. et al., 2003; Haertig, H. et al., 2004) deal with the usage of Ethernet in the automation domain. In academic and industrial research, different scheduling strategies and smoothing concepts has been investigated (Alves et al, 2000; Bonaccorsi et al., 2003; Carpenzano et al., 2002; Pereira, N. et al., 2004; Lo Bello, L. et al., 2005; Kanghe Kim et al., 2002, 2005).
Industrial examples as part of the Public Available Specification (IEC, 2004) are:

- Time-critical Control Network **Tcnet**(Toshiba),
- **Vnet** (Yokogawa),
- **PROFINET** with the application model **IO** (Input/Output) (PROFIBUS International, Siemens).

(Neumann, P., Pöschmann, A., 2005) describes the fundamentals of PROFINET IO.

Regarding class 3,
The standardisation of a synchronisation mechanism has been done and used for star and tree topology (IEC, 2003). (Höller, R. et al., 2004) investigate its application, (Jasperneite, J. et al., 2004) proposes a mechanism suitable for line topology.
Within the PAS (IEC, 2004) there are the following systems:

- **Ethernet/IP with time synchronisation** (ODVA, Rockwell Automation),
- **Powerlink** (Ethernet PowerLink Standardisation Group EPSG, Bernecker & Rainer), developed for Motion Control (Pfeiffer, A, 2004),
- **EtherCAT** (EtherCAT Technology Group (ETG), Beckhoff) developed as a fast backplane communication system (Janssen, D., 2004),
- **PROFINET/IRT** (PROFIBUS International, Siemens) developed for Motion Control, but suitable for any industrial applications,
- **SERCOS III** (IG SERCOS Interface e.V.), developed for Motion Control.

Ethernet/IP with Time Synchronisation uses on the basis of Ethernet/IP technology the CIP Synch protocol to enable the isochronous data transfer. Since the CIP Synch protocol is fully compatible to standard Ethernet, additional devices without CIP Synch features can be used in the same Ethernet system. The CIP Synch protocol uses the Precision Clock Synchronisation Protocol (IEC, 2002b) to synchronise the node clocks using an additional hardware function. CIP Synch can deliver time-synchronisation accuracy of less than 500 nanoseconds between devices, which meets the requirements of the most demanding real-time applications. The jitter between Master and Slave clocks can be less than 200 nanoseconds.
Powerlink uses a proprietary real-time protocol on top of the shared Ethernet. The scheduling mechanism is a time-division scheme. Using 100 Mbps

Ethernet Powerlink allows real cycle times of 400 microseconds or less in applications. The network jitter has been proven to be below 1 micro second. The drives (less than 50 with cycle times in the range of 2 ms) can communicate synchronously.

EtherCAT distinguishes two modes: direct mode and open mode. Using the direct mode, a Master Device uses a standard Ethernet port between the Ethernet Master and an EtherCAT segment. EtherCAT uses a ring topology within the segment. The medium access control adopts the Master/Slave principle, where the Master node (typically the control system) sends the Ethernet frame to the Slave nodes (Ethernet device). One single Ethernet device is the head node of an EtherCAT segment consisting of a large number of EtherCAT Slaves. Using the open mode, one or several EtherCAT segments can be connected via switches with one or more Master devices and Ethernet-based "Basic Slave" devices. Each segment can be addressed using a "Segment Address Slave" device (the head station of the segment.).

The technical background of EtherCAT is 100Base-TX and –FX Ethernet used within an EtherCAT segment and between Master devices and Slave devices. Within EtherCAT segments can also be used Low Voltage Differential Signals (LVDS) (IEEE 803-3ae-2002).

The Application Layer follows the CANopen model.

PROFINET/IRT (IRT means Isochronous Real-Time) uses a middleware on top of Ethernet MAC layer. The layer 7 functionality is directly linked to that middleware. The middleware itself contains the scheduling and smoothing functions. A special Ethertype is used to identify real-time PDUs (only one PDU type for real-time communication). That enables an easy hardware support for the real-time PDUs. The technical background is a 100 Mbps full duplex Ethernet (switched Ethernet). PROFINET IRT adds an isochronous real-time channel to the RT channels of class 2 option channels. This IRT channel enables a high-performance transfer of cyclic data in an isochronous mode (Jasperneite et al., 2004). The time synchronisation and node scheduling mechanism is located within and on top of the Ethernet MAC Layer. The offered bandwidth is separated in bandwidth for cyclic hard real-time and soft/non real-time traffic. . The cycle time should be in the range of 250 µsec (35 nodes) to 1 msec (150 nodes) when simultaneously TCP/IP traffic of about 6 Mbps is transmitted. The jitter will be less than 1 µsec.

A *SERCOS* network consists of Masters and Slaves. Slaves contain integrated repeaters, which have a constant delay time Trep (input àoutput). The nodes are connected via point-to-point transmission lines. Each node (participant) has two communication ports. The ports are interchangeable. The topology can be either a ring structure or a line structure. The ring structure consists of a primary and a secondary channel. All slaves work in forwarding mode. Through this ring, redundancy against cable break is achieved. It is also possible to open the ring and insert/remove slaves during operation (hot plug). The line structure consists of either a primary or secondary channel. The last physical slave performs the loop-back function. All other slaves work in forwarding mode. No redundancy against cable break is achieved. It is also possible to insert and remove slaves during operation (hot plug). This is restricted to the last physical slave. The cycle time can be set to 31,25ms, 62,5ms, 125ms, 250ms and integer multiples of 250ms. The jitter is limited to 1ms (high performance class) or 50ms (low performance class).

3.2 Wide Area Domain

Allowing remote mechanisms (remote supervisory, remote operation, remote service) using Wide Area Networks, the stock of existing communication technology becomes broader:

- All appearances of the Internet (mostly with best effort quality of services)
- Public digital wired telecommunication systems (ISDN, DSL etc.)
- Public digital wireless telecommunication systems (GPRS-based, UMTS-based)
- Private wireless telecommunication systems, e. g. trunk radio systems.

Using these technologies within the automation domain there are many private protocols over leased lines, tunnelling mechanisms etc. There exists a long-term experience regarding telematic systems in the domain of utility automation. Recent activities are directed to the usage of telecommunication networks and the Powerline technology for the energy management (European project REMPLI, see Sauter, T. et al, 2005; Treytl, A. and T. Sauter, 2005). The behaviour of the end-to-end connection via these telecommunication systems depends on the recently offered quality of service and cannot be guaranteed in many cases. It strongly limits the use of these systems within the automation domain.

Within the automation domain, there are other classes of real-time behaviour of the WAN than for other domains (office, e-commerce etc). The most important requirement is to guarantee application-specific deadlines, whatever the required value of the deadline in the various applications is. Using Wide Area Networks, the actual offered mechanisms support mainly the best effort QOS and privilege the video and audio stream transmission. There is a need for offering QOS levels, which are required for a suitable response behaviour of data packet transmission. The IPv6 approach offers real-time mechanisms, user priorities and a broader range of addresses. This is very important for the overboarding need of address space using embedded Web servers within the automation devices. In the last few years, a stable IPv6 network supported by different providers was established.

The usage of wide area networks within the automation domain has to be investigated just as the uninterrupted commercial availability. Since the Internet or other telecommunication systems are general-purpose communication systems, the infrastructure and business model preconditions for the selection of requested QOS within a spectrum of available communication services of various providers have to be developed. This means, in analogy to the "switched" Ethernet in LANs, a WAN switching mechanism has to be developed for this selection, i.e. choosing dynamically the network type and/ or network provider, which guarantees the required QOS.

4 WIRELESS INDUSTRIAL COMMUNICATIONS

The recent development in manufacturing requires more flexible production systems and therefore more mobile and movable parts of the production system. Adequate solutions for supporting automation systems are based on wireless communication. Thus, the interest in radio-based communication in industrial automation is growing. Previous activities were directed to the extension of Fieldbus systems. A wireless Fieldbus system is a wireless communication network suitable for use at the device level of an automation system. For that purpose, Wireless Local Area Networks WLAN and Wireless Personal Area Networks WPAN can be employed. Inline with the development of radio technologies different vendors of Fieldbus systems (e.g. CAN, Interbus, PROFIBUS) investigated the replacement of wired transmission lines by radio front ends (see e.g. Rauchhaupt L, 1999; Rauchhaupt L and Hähniche J, 1999; FUNBUS, 2000; Pohlmann T, 2005), followed by the European project "RadioFieldbus" (RFieldbus) (Rauchhaupt, 2002; Rauchhaupt, 2003; Ferreira et al, 2002).

Interesting approaches/standards in the context of Industrial Wireless communications may be grouped as follows:

- Proprietary protocols for radio technologies, e.g. Wireless Interface for Sensors and Actuators2 WISA (Scheible, 2005),
- Lower layer standards (IEEE 802.11 and 802.15) as a basis of Wireless Local Area Networks, Pico Networks and Sensor/Actuator Networks ,
- Higher layer standards (specific Application Layer on top of IEEE 802.15.4, e.g. Wireless Fidelity, Bluetooth, ZigBee),
- Complete standards of mobile communications (GSM, GPRS, UMTS) and wireless telephones (DECT), not discussed here,
- Ultra Wideband technology UWB.

Most of these Wireless Radio Networks can be used in non real-time applications, some of them in soft real-time applications (but industrial environments and ISM band limit their applications). (Hyung, S.K. et al., 2004; Facchinetti, T. et al., 2004; Lo Bello, L. et al., 2005; Willig, A., 2005; Soo Young Shin et al, 2005) deal with different aspects of wireless communication in the automation domain.

The *WLAN technology* is more and more used in the higher architecture levels of the automation hierarchy, but also in the shop floor.

Bluetooth has been successfully introduced in industrial short-range applications, operating in the same local area as WLAN with good results (e.g. Weczerek, J., 2005; Lührs, C., 2005; Sikora, 2005; Esch, 2005).

ZigBee should be introduced to connect the automation devices at the field level, especially in the process automation, because it will operate on a lower baud rate. But the specification of the higher layer protocols has not been finished. Thus, many vendors experiment with the available hardware to test their properties. There are positive but also negative results (Hansen, Grohmann, 2005). There are significant technical limitations in version 1.0. Basic features (e.g. support of portable and mobile nodes, handling PAD ID conflicts, multicast messages, support of bulk data, device hardware fault recovery, node power failure recovery), required for industrial & commercial applications, will not be available before version 1.1 (fall 2006). The certification program is not fully defined. The ZigBee alliance seems to offer strong marketing activities but little product development. The recent activities are directed to the application domain Building Automation. Nevertheless, ZigBee has the potential to become a standard for wireless Monitoring & Control in industrial & commercial environments.

Ultra Wideband Systems are becoming more and more important for sensors and indoor location-based services.

The development in the field of digital radio communication (mobile communication, wireless LAN, Bluetooth, short-range devices) offers interesting features. Recently, it can be noticed that manufacturers of radio modules discover the market of industrial communication. Here not the number of pieces per year is important, instead it is more the stability, which makes the market segment attractive. Common practice is to use radio modems or radio gateways to industrial communication systems or to integrate single radio modules in selected automation devices. Examples are described in (Schildknecht, 2005); Meier et al., 1999; RT, 2005). This approach is acceptable for a number of applications, where e.g. mobility is obligatory. For a broad use of radio communication in order to increase the efficiency of industrial systems and their automation this approach is not suitable.

However, the specific requirements of industrial automation do not belong to the design criteria of these technologies. Thus, the manufacturers of automation components and systems as well as system integrators are responsible for the integration of available radio implementations into automation components and systems. Figure 2 illustrates the different perspectives of the integration of radio based communication components into industrial automation systems focussing on a single distributed component.

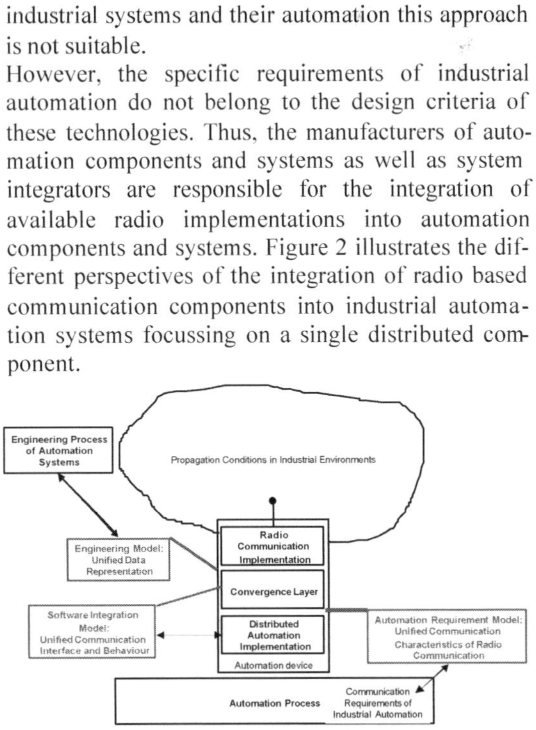

Figure 2: Aspects of integrating radio based implementation into automation applications

It is obvious that the problem in an entire distributed automation system with gateways between wired and wireless communication systems is much more complex.

The largest effort requires the connection between the automation software and the software and firmware of the radio modules. This was one of the experiences made in the European RFieldbus project in the 5th IST program (Rauchhaupt, 2003). One of the RFieldbus results is the integration of one certain radio technology into a Fieldbus system (PROFIBUS). The specification was made in accordance to IEC standardisation rules. However, in the RFieldbus project it was not possible to take into account several radio technologies. This is necessary since depending on the requirements concerning robustness, time behaviour, coverage etc. different radio technologies are suitable. Today for each technology the radio implementation is to be individually integrated into the automation software system. Additionally, regarding the short innovation cycles and the related frequent replacement of radio communication, the potential of savings is obvious, which could be achieved by replacing the individual integration by using a general approach with unified convergence layer models (see figure 2). Moreover, the process of integration has to be recognised as a general approach including requirements, design and maintenance engineering. This task was also not part of the RFieldbus project.

Nowadays, there are many activities in wireless technology for Automation both in Europe and in Overseas. In Germany, the related work started in 1999 and lead to the establishment of a GMA Technical Committee "Wireless Communication in Automation". In 2003, this committee published the first guide (VDI, 2003).

In Europe, preparing the 7th Frame Programme the European Commission is working nowadays with experts from Europe coming from industry and research to find the main topics in Wireless Communication. The use of wireless communication in the automation domain will become a crucial point.

In 2002, the Wireless Industrial Networking Alliance WINA started its business to focus the US activities in Industrial Wireless Communication. In 2004, ISA established a committee SP 100 to work out a standard for sensor networks.

The Hart Foundation decided in April 2005 to start the work for wireless in Hart technology.

All the mentioned activities show the strong interest and high potential of wireless in industrial environment.

Radio modules are originally developed for telecommunication, office and consumer markets. Single implementations are produced in huge amounts, thus solitary features are more important than standardisation. Standardisation is only concentrated on the communication layers, necessary for interoperability. Furthermore, the focus has to be directed to engineering methods and tools for wireless industrial communication. Other necessary investigations are as follows:

- Safety over wireless communications. It means: functional safety and intrinsic safety,
- Security using wireless communications (problem of the open door)
- Own frequency band for the automation domain.

5 SAFETY AND SECURITY

5.1 Functional Safety

Safety means protection against hazards (movement, heat, radiation, electrical shock, etc.) (IEC 61508). "Functional Safety" means protection against hazards caused by incorrect function. Safety includes the communication via heterogeneous network. (EN 50159; Diedrich et al., 2003) deal with different aspects of safety in communication and computer systems. Caused by the distribution of data via the communication networks, the safety of these networks becomes more and more important regarding the functionality of an automation system.

There is a need to meet defined Safety Integrity Levels (SIL), see (IEC 61508), e.g. Residual Error Probability <= 10^{-7} errors/h for SIL 3. The communication part requires a residual error probability of <= 10^{-9} errors/h for SIL 3 (1% of 10^{-7}; the other 99% are required for sensors, PLCs, and actuators etc.).

Measures to avoid the influence of the failures (falsified addresses, loss or damage of data, delay) could be: numbering transmitted data, observation of transmission time (time expectation), authentication using passwords (identification), optimised CRC (redundancy).

One principle to use communication systems for safety applications is to consider the communication channel as a so-called "Black Channel". That means all safety-related functions will be realised outside (on top of) the communication layers. The safety measures will be realised in a separated safety layer that is situated between communication protocol and application. With this principle, an existing communication system can be used as it is – and with existing components like ASICs, cables, connectors, repeaters, links etc. as far as they meet the electrical requirements. Such solutions exist for several fieldbus protocols. For example, the PROFIBUS user organisation has specified safety profiles for PROFIBUS and PROFINET, the Interbus Club specified a safety layer for Interbus, and CAN in Automation for CAN. This background has to be mapped to Internet-based Virtual Automation Networks using heterogeneous networks. A safety layer for PROFINET is under specification. Investigations are necessary on how TCP/IP-oriented local as well as wide area networks can be extended for the transfer of safety-related data in automation systems. It is necessary to prove that the used VAN safety codes fulfil the SIL requirements - especially for different types of communication media. The results should be introduced in the IEC standardisation activities on "Communications for functional safety" (IEC SC65C WG 12).

5.2 Security

Today's automation islands are relatively secure against attacks. The usage of Internet-based technologies compared with today's automation islands requires a common security concept for distributed automation using Virtual Automation Networks. (Naedele *et al*., 2001; Fritz *et al*., 2002; Adamczyk, 2004) deal with several aspects of security in communication systems. In the past, various investigations regarding security in Fieldbus systems have been investigated (Palensky, P. Sauter, T., 2000; Dietrich, D. et al.; 2003; Schwaiger, D. et al., 2003; Palensky, P., 2003; Dietrich, D. et al., 2004; Treytl, A., 2004 and 2005; Fischer, P., P. Palensky, 2004). Typical types of assaults are: eavesdropping, hacker attacks, data corruption, unauthorised access, espionage, sabotage, loss, theft, and operator errors. Important is that the threats are coming from inside and outside an enterprise.
Main security targets are:

- Confidentiality of information (to protect against access by unauthorised third parties),
- Integrity of information (detection of unauthorised modification or manipulation of message contents with a specified level of confidence),
- Timeliness or sequentiability of message delivery (detection of any unauthorised message re-timing, re-sequencing or replay of prior messages),
- Authentication and authorisation of communication peers (Verification of identity of a communication peer and that peer´s entitlement),
- Availability of information.

The security of control systems distinguishes three aspects:

- Data communication security: protection of data transfer between components of control systems, operator stations, and with external remote devices. That leads to the three zones model (Naedele, 2003),
- Physical security: protection against local manipulation, destruction and unauthorised physical access (e.g. memory sticks),
- Operational security: protection by administrative measures as authentication, availability, desired operation conditions etc.

Basic requirements (from the user´s point of view), especially in the automation domain, are:

- Ability to use hierarchical concepts of control systems architecture. It means that there are segments which can be protected separately,
- Control system underlies different changes during the operation phase. The security concept has to support reconstructions,
- Availability should not be influenced by security mechanisms,
- Security mechanisms have to enable security updates,
- "open door" effect of wireless communications, which cannot be influenced by administrative measures, has to be compensated by technical measures,
- New business models (new scenarios) for external commissioning and maintenance/asset management by external service organisations lead to new security risks (or conditions). There are links of external (heterogeneous) communication networks to internal enterprise networks with sensible data exchange,
- Documentation of security mechanisms (configurations, parameters) to support back-up strategies.

The question is how TCP/IP-oriented networks have to be secured under specific automation conditions such as low resources and easy to use configuration in production environments. For that purpose security risk analysis for production networks is necessary (figure 3).

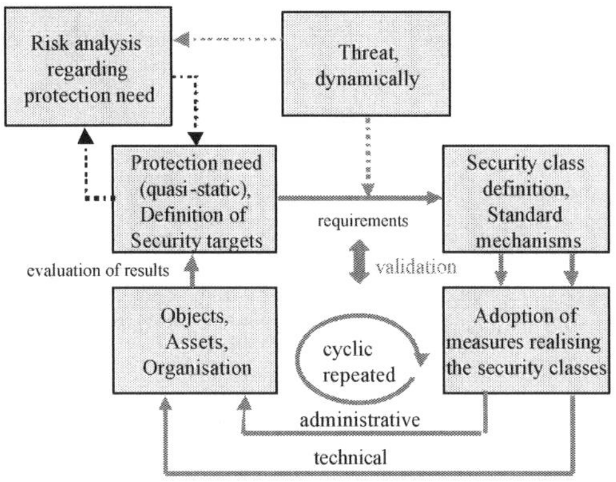

Figure 3. Security risks, classes and measures (Adamczyk, 2004)

The risk analysis regarding above mentioned protection needs has to be carried out cyclicly repeated on the base of use cases, describing the different application scenarios. During the actual cycle the protection needs and security targets can be accepted as static. But the threats are really dynamically. Thus, the recent security measures have to be evaluated cyclicly repeated. That is important, because the automation system underlies several phases of the life cycle for many years.
Many arguments are given why defence-in-depth with multiple, staged, complementary security mechanisms is the suitable approach. There are different abstraction levels and security zones within a Virtual Automation Network. Different types of mechanisms against attacks by remote access are used concurrently around and inside each security zone to defend it. The outer zones contain less valuable targets. The safety-critical automation system is in the innermost zone. Investigations should lead to the designation of VAN security zones (e.g. Remote Access Zone, Local Plant Operator Zone, Automation Device Zone), and to existent security mechanisms used in all or specific zones (e.g. authorisation, intrusion detection, response, mechanism protection, VAN specific security mechanisms). (Naedele, 2003) describes the approaches in principle. The security concept of the PROFIBUS user organisation (PNO, 2005) offers the mechanisms to protect the innermost automation device zone, called "cell". All the automation devices inside the cell could be protected by a

special security network component (security module).

In addition to defence mechanisms, also detection mechanisms could be helpful, which allow the automation system operators to detect attacks, and reactive mechanisms and processes to actively defend against them. Each zone also has to spend time to detect and fend off the attacker.

The situation regarding standardisation of security in automation is as follows:

- There is a number of basic standards for *IT security* to be also used in the industrial environment,
- The US has the outrider role for the standardisation of security in automation. The committee ISA SP99 "Manufacturing and Control Systems Security" focuses on improvement of confidentiality, integrity, and availability of components or systems used for manufacturing or control and provides criteria for procuring and implementing secure control systems. The first two technical reports TR1 and TR2 (ISA, 2004a; ISA, 2004b) has been published in 2004,
- The actual activities in IEC TC 57 "Power Systems Control and associated communications" and IEC SC65C WG 13 "Cyber Security" regarding the use of security mechanisms in the automation have a very early status. Additionally, the ISO/IEC JTC/SC27 "IT Security Techniques" is of interest for the security development in the automation domain,
- Stabilising the IEC standards needs time until 2006/2007. The IEC TC 65 decided to shift the IEC SC65C WG13 "Cyber Security" (regarding communication systems) to the TC 65 and extended the focus to "System Security". That procedure will consume so much time, that there will be a noticeable delay.

6 CONCLUSIONS

Future scenarios of distributed automation require more mechanisms for the geographical distribution of automation functions. It requires a special communication network – a Virtual Automation Network. This means that geographically distributed application programmes, co-operating to fulfil a control application, are connected via this VAN accessed by remote connection endpoints. Worldwide distribution of Internet offers the Automation domain a good infrastructure but introduces many additional problems, which need to be solved. The end-to-end communication includes all communication approaches within the local and remote area, all approaches to fulfil the real-time requirements of the three real-time classes. Additionally, the safety approaches have to be mapped to the Wide Area Networks, and the security requirements are fundamental extensions of those of the legacy Fieldbus systems. Most of the requirements are known, but there are many problems to be solved.

REFERENCES

Adamczyk, H. (2004): Security in Automation (in German). Study for the Research Society Computer Automation (FGCA), ifak, Sept. 2004.

Alves, M.; Tovar, E.; Vasques, F. (2000): Ethernet Goes Real-Time : a Survey on Research and Technological Developments. Techn. Rep. HURRAY-TR-0001, Polytechnic Institute of Porto, Jan. 2000.

Baek-Young Choi, Sejun Song, Birch, N. and Jim Huang (2000). Probabilistic approach to switched Ethernet for real-time control applications, Seventh International Conference on Real-time Computing Systems and Applications, 2000, Proceedings pp. 384-388.

Bonaccorsi, A.; Lo Bello, L.; Mirabella, O.; Neumann, P.; Pöschmann, A. (2003): A Distributed Approach to Achieve Predictable Ethernet Access Control in Industrial Environments. 5th IFAC International Conference on Fieldbus Systems and their Applications (FET 2003), Aveiro, Proceedings, pp. 173 - 176.

Bratoukhine, A., T. Sauter, J. Peschke, A. Lüder, A. Klostermeyer (2003): Distributed Automation: PABADIS vs. HMS. IEEE International Conference on Industrial Informatics (INDIN), 2003, Banff, Canada, pp. 294 - 300.

Carpenzano, A.; Caponetto, R.; Lo Bello, L.; Mirabella, O. (2002): Fuzzy Traffic Smoothing: an Approach for Real-Time Communication over Ethernet Networks. 4th IEEE International Workshop on Factory Communication Systems. WFCS'02, Västeras.

Diedrich, Ch.; Wollschläger, M.; Hintze, E.(2003): UML specification and developement of safety-relevant industrial communication systems. 4th MATHMOD - 4th IMACS Symposium on Mathematical Modelling, Proceedings.February 5-7, 2003, Vienna University of Technology, Austria.

Dietrich, D., Palensky, P., Treytl, A. (2003): Communication in Automation with the Emphasis on Security. EUSAS Workshop on Integrated Systems, Rome 2003, pp.89-104

Dietrich, D., A. Treytl, T. Sauter (2004): IT-Sicherheit für Feldbus-Netzwerke. In: VDE Kongress 2004, Band 2, Fachtagungsbericht DGBMT*DMM*GMA, 2004, Berlin, pp. 227 - 232.

EN 50159: Railway Applications: Requirements for Safety-Related Communication in Closed Transmission Systems

Esch, W. (2005): Traps developing Bluetooth products (in German). GMA conference Radio-based Communication in Industrial Automation, Düsseldorf, May 2005, pp.43-47.

Ethernet/IP (2001): Ethernet/IP specification, Release 1.0. June 5, 2001, ©2000, 2001 by ControlNet International and open DeviceNet Vendor association

Facchinetti, T., L. Almeida, G. C. Buttazzo, C. Marchini (2004): Real-Time Resource Reservation Protocol for Wireless Mobile Ad Hoc Networks. 25th IEEE Real-Time Systems Symposium (RTSS), Lisbon, Portugal,2004. pp.382-391.

Fischer, P., P. Palensky (2004): Safety and Security in Communication Standards for Building Automation and Control Systems. 13th International Conference on Automatic Fire Detection, 2004, Duisburg, Germany, pp. 112 – 120.

Fritz, R.; Halang, W. (2002): Sichere Abwehr von Viren. 2002, ISBN 3-89577-266-6

FUNBUS (2000). Das BMBF Verbundprojekt „Drahtlose Feldbusse im Produktionsumfeld". Förderkennzeichen: 02PV4060, Report (in German), Okt. 2000, INTERBUS CLUB, Order No. TNR5121324.

Georges, J.-P., E.Rondeau and T. Divoux (2002). Evaluation of switched Ethernet in an industrial context by using the network calculus. IEEE 4th Workshop on Factory Communication Systems WFCS'02, Västeras, 2002, pp. 19-26.

Hansen, S., Grohmann, B. (2005): Experience with Mesh Network Stacks in Industrial&Commercial environment. GMA conference Radio-based Communication in Industrial Automation, Düsseldorf, May 2005.

Haertig, H. and Loeser, J. (2004): Low-latency Hard Real-Time Communication over Switched Ethernet. 16th Euromicro Conference on Real-Time Systems (ECRTS 04), Catania, Italy, June 2004

Hoang, H. and Jonsson, M. (2003): Switched Real-Time Ethernet in Industrial Applications - Deadline Partitioning. 9th Asian

Pacific Conference on Communication, Penang, Malaysia, 2003
Höller, R., T. Sauter, N. Kerö (2003): SynUTC and IEEE 1588 Clock Synchronization for Industrial Ethernet. 2003 International Workshop on Real-Time LANs in the Internet Age, 2003, Porto, Portugal, S. 67 – 70.
HSE (2001): High Speed Ethernet Specification documents FF-801, 803, 586, 588, 589, 593, 941. Fieldbus Foundation, Austin 2001
Hyung Seok Kim, Wook Hyun Kwon (2004): Spatial and Temporal Multi-Aggregation for State-Based Sensor Data in Wireless Sensor Networks. Telecommunication Systems 26(2-4). pp. 161-179.
IDA (2002): Interface for Distributed Automation, Architecture Descriptio n and Specification, Revision 1.1, Nov. 2002, IDA Group e. V., Blomberg
IEC (2000). IEC 61508: Functional Safety of Electrical/Electronic/Programmable El. Safety-Related Systems.
IEC (2002). IEC 61588: Precision clock synchronization protocol for networked measurement and control systems.
IEC (2003a). IEC 61158 series, Edition 3: Digital data communication for measurement and control - Fieldbus for use in industrial control systems.
IEC (2003b). IEC 61784-1. Digital data communications for measurement and control – Part 1: Profile sets for continuous and discrete manufacturing relative to fieldbus use in industrial control systems.
IEC (2004a). IEC 65C/359/NP. Real-Time Ethernet: Public Available Specification
ISA (2004a): "Security Technologies for Manufacturing and Control Systems", Technical Report ISA-TR99.00.01-2004, ISA SP99, 2004.
ISA (2004b): Integrating Electronic Security into the manufacturing and Control Systems Environment. Technical Report ISA-TR99.00.02-2004, ISA SP99, 2004.
Janssen, D. (2004): EtherCAT als Antriebsbus. Synchronität ist nicht genug. IEE - Das Automatisierungsmagazin 49(11)2004. pp.18-22.
Jasperneite, J. (2002): Leistungsbewertung eines lokalen Netzwerkes mit Class-of-Service Unterstützung für die prozessnahe Echtzeitkommunikation. (Performance evaluation of a local network with Class-of-Service support for process-coupled real-time communication). Shaker Verlag, Aachen 2002. ISBN 3832208321.
Jasperneite, J., Shehab, K., Weber, K. (2004). "Enhancements to the Time Synchronization Standard IEEE-1588 for a System of Cascaded Bridges", WFCS2004, 5th IEEE International Workshop on Factory Communication Systems, Vienna, Austria, Sep. 2004.
Kanghee Kim, L. Lo Bello, Sang Lyul Min, O. Mirabella (2002): On Relaxing Task Isolation in Overrun Handling to Provide Probabilistic Guarantees to Soft Real-Time Tasks with Varying Execution Times. ECRTS 2002. pp. 193-202
Kanghee Kim, J. L. Diaz, L. Lo Bello, J. M. Lopez, Chang-Gun Lee, and Sang Lyul Min. (2005): An Exact Stochastic Analysis of Priority-Driven Periodic Real-Time Systems and Its Approximations, IEEE Transactions on Computers, 2005.
Lo Bello, L., O. Mirabella (2000): To What Degree will Ethernet Penetrate the World of Real-Time Distributed Process Control? In: Developments in Engineering Computational Technology - Chap. 2 - Ethernet Technology, Civil-Comp Press, Edinburgh, United Kingdom. pp. 17-23.
Lo Bello, L., O. Mirabella (2001): Design Issues for Ethernet in Automation. ETFA 2001, 8th International Conference on Emerging Technologies and Factory Automation 2001, Antibes Juan-les-Pins, France.
Lo Bello, L., M. Collotta, O. Mirabella, T. Nolte (2005): Approaches to Support Real-Time Traffic over Bluetooth Networks. 4th International Workshop on Real-Time Networks (RTN'05) in conjunction with the 17th Euromicro International Conference on Real-Time Systems (ECRTS 2005).
Lo Bello, L., G. A. Kaczynski, and O. Mirabella (2005): Improving the Real-Time Behavior of Ethernet Networks Using Traffic Smoothing. IEEE Transactions on Industrial Informatics, Volume 1, Number 3, August 2005. pp. 151-161.
Lührs, C. (2005). HMI with Bluetooth- from application to integration (in German).). Workshop on Radio-based Communications in Industrial Automation. Düsseldorf, Mai 2005.
Meier, U. et al. (1999): Drahtloser Interbus über DECT (Wireless Interbus via DECT). Elektronik, 1999, Vol. 19. pp.85-89.
Naedele, M. (2003): IT Security for Automation Systems, Motivations and Mechanisms. Automatisierungstechnische Praxis 45 (2003), H. 5, S 84-91
Naedele, M.; D. Dzung; M. Stanimirov (2001): Network Security for Substation Automation Systems. In Computer Safety, Reliability and Security. Proceedings Safecomp 2001, LNCS 2187. Springer, Berlin 2001, pp. 25-34.
Neumann, P. Pöschmann, A. (2005): Ethernet-based Real-Time Communications with PROFINET IO. WSEAS Transactions on Communications. Issue 5, Vol.4, May 2005, pp. 235-245.
Palensky, P, Sauter, T. (2000): Security Considerations for FAN-Internet Connections. IEEE Int. Workshop on Factory Communication Systems. Porto 2000, pp. 27-35.
Palensky, P. (2003): Smart Card Security for Field Area Networks. IEEE-Siberian Conference on Control and Communications SIBCON, 2003, Tomsk, pp. 135 – 138
Pedreiras, P., L. Almeida (2003): The Flexible Time-Triggered (FTT) Paradigm: An Approach to QoS Management in Distributed Real-Time Systems. 17th International Parallel and Distributed Processing Symposium (IPDPS), Nice, France 2003. pp.123.
Pedreiras, P., L. Almeida (2004): Message Routing in Multi-Segment FTT Networks: The Isochronous Approach. 18th International Parallel and Distributed Processing Symposium (IPDPS), Santa Fe, New Mexico 2004. pp.
Pedreiras, P., P. Gai, L. Almeida and G. C. Buttazzo (2005a): FTT-Ethernet: A Flexible Real-Time Communication Protocol That Supports Dynamic QoS Management on Ethernet-Based Systems. IEEE Transactions on Industrial Informatics, Volume 1, Number 3, August 2005 pp. 162-172.
Pedreiras, P., L. Almeida, J. A. Fonseca (2005): The Quest for Real-Time Behavior in Ethernet. The Industrial Information Technology Handbook 2005. pp. 1-14.
Pereira, N., Tovar, E., Pinho, L.M. (2004): INDEPTH: Timeliness Assessmant of Ethernet/IP based Systems. MASCOTS 04. 12thIEEE Int. Symposium on Modelling, Analysis abd Simulation of Computer and Telecommunication Systems 2004, pp.192-201
Pfeiffer, A. (2004): Real-Time Industrial Ethernet is Real. PowerLink Lecture on the Industry Day on Industrial Ethernet of the University Wien, Sept. 21, 2004.
Pöschmann, A.; Werner, Th. (2003): Studie "Ethernet in der Automation", ZVEI, Frankfurt, 2003
PROFIBUS (2003): PROFInet Architecture Description and Specification, Version V 2.0, Januar 2003, PNO Karlsruhe
PNO (2004a): PROFInet IO Application Layer Service Definition, Version. 1.0, Karlsruhe: PROFIBUS Nutzerorganisation e. V., 2004.
PNO (2004b): PROFInet IO Application Layer Protocol Specification, Version. 1.0, Karlsruhe: PROFIBUS Nutzerorganisation e. V., 2004.
PNO (2005): PROFINET Security Guideline. Draft TC3-04-0004, PROFIBUS user organisation, 2004.
Pohlmann T (2005). Radio technology for CAN networks (in German). Workshop on Radio-based Communications in Industrial Automation. Düsseldorf, Mai 2005.
Rauchhaupt, L. (1999). Wireless CAN Extensions. 6th International CAN Conference (iCC' 99), Turin, 1999: 02-07 to 02-14.
Rauchhaupt, L., Hähniche, J. (1999): Opportunities and Problems of Wireless Fieldbus Extensions. Fieldbus Technology; Systems Integration, Networking, and Engineering. Fieldbus Conference FeT'99 in Magdeburg, 1999: 48-54.
Rauchhaupt, L. (2003): RFieldbus a Survey. 5th IFAC International Conference on Fieldbus Systems and their Applications (FET 2003), Aveiro, Portugal, 2003, Proceedings, pp. 105-114.
RT (2005): rt-solutions.de GmbH - Homepage. URL: www.rt-solutions.de (Access: 17.03.2005).
Sauter, T., C. Schwaiger (2002): Achievement of secure Internet access to fieldbus systems. In: Microprocessors and Microsystems, Bd. 26, 2002, pp. 331 – 339.
Sauter, T., G. Pratl, A. Treytl, G. Bumiller (2005): Secure and Reliable Wide-Area Power-Line Communication for Soft-Real-Time Applications within REMPLI. 2005 International Symposium on Power Line Communications and Its Applications, 2005, Vancouver, pp. 57 – 60.
Scheible, G., Endresen, J., Frey, J.E. (2005). Reliability and interference resistance as basis of the selection of radio technologies for factory automation (in German). Workshop

on Radio-based Communications in Industrial Automation. Düsseldorf, Mai 2005.
Schildknecht (2005): Industrieelektronik Schildknecht - Homepage. URL: www.schildknecht.info (Access: 17.03.2005).
Schwager, J. (2004): Comparing real-time Ethernet –Are fieldbuses going to disappear? Lecture on the International Industrial Ethernet Symposium, Berlin, 2004, June 14.
Schwaiger, C., Treytl, A. (2003): Smart Card Based Security for Fieldbus Systems. IEEE Int Conference on Emerging Technologies and Factory Automation, Lisbon 2003, pp. 398-406.
Sikora, A. (2005): IEEE802.15.4 and ZigBe for Sensor and Actuator Networks. Embedded World Conference, 2005, WEKA-Verlag Poing, procedings, pp. 189-197.
Soo Young Shin, Sunghyun Choi, Hong Seong Park, Wook Hyun Kwon (2005): Lecture Notes in Computer Science: Packet Error Rate Analysis of IEEE 802.15.4 Under IEEE 802.11b Interference. Wired/Wireless Internet Communications Conference 2005, Xanthi, Greece). pp.279-288
Treytl, A., T. Sauter, C. Schwaiger (2004): Security Measures for Industrial Fieldbus System - State of the Art and Solutions for IP-based Approaches. 2004 IEEE International Workshop on Factory Communication System, 2004, Wien, pp. 201 - 209.
Treytl, A., T. Sauter (2005): Security Concept for a Wide-Area Low-Bandwidth Power-Line Communication System. 2005 International Symposium on Power Line Communications and Its Applications, 2005, Vancouver, S. 66 – 70
Treytl, A., C. Schwaiger (2005): Security Topics and Solutions for Automation Networks. In: The Industrial Communication Technology Handbook., CRC Press, 2005
VAN project (2005): Virtual Automation Network. Description of Work. EC No. 016969, May 2005.
VDI (2003): Wireless Communication in Automation (in German). VDI/VDE Guide 2185, Düsseldorf, Nov. 2003
Weczerek, J. (2005). Experience with Bluetooth in industrial applications (in German). Workshop on Radio-based Communications in Industrial Automation. Düsseldorf, Mai 2005
Willig, A. (2005): Redundancy Concepts to Increase Transmission Reliability in Wireless Industrial LANs. IEEE Transactions on Industrial Informatics, Volume 1, Number 3, August 2005 pp. 173-182.
Xinggang Fan, Zhi Wang and Youxian Sun (2002). How to guarantee factory communication with switched Ethernet: survey of its emerging technology, , IECON'02 IEEE 2002 28th Annual Conference of Industrial Electronics Society, Nov. 2002. Proceedings, pp. 2525-2530.

ELSEVIER
IFAC
PUBLICATIONS

STATISTICAL MODEL-BASED SENSOR DIAGNOSTICS FOR AUTOMATION SYSTEMS

Brian Sallans * Dietmar Bruckner ** Gerhard Russ *

* *{brian.sallans,gerhard.russ}@arcs.ac.at*
Information Technologies,
ARC Seibersdorf research GmbH, Vienna, Austria
** *bruckner@ict.tuwien.ac.at*
Institute of Computer Technology,
Vienna University of Technology, Vienna, Austria

Abstract: A method for the automatic detection of abnormal sensor values in automation systems is described. The method is based on statistical generative models of sensor behavior. A model of normal sensor behavior is automatically constructed. Model parameters are optimized using an on-line maximum-likelihood algorithm. Incoming sensor values are then compared to the model, and an alarm is generated when the sensor value has a low probability under the model. Model parameters are continuously adapted on-line. The system automatically adapts to changing conditions and sensor drift, as well as detecting isolated abnormal sensor values.

Keywords: automation, statistical inference, sensor failures, diagnostic tests

1. INTRODUCTION

Automation systems have seen widespread deployment in modern buildings, and include systems for environmental control, energy management, safety, security, access control, and remote monitoring. As the cost of automation systems falls, and the technology converges towards standardized protocols, we can expect automation to move from the office into the home. It will also encompass not just building management technology, but also entertainment, kitchen appliances and communications devices.

Todays sensor and control systems are primarily based upon the processing of sensor information using predefined rules. The user or operator defines, for example, the range of valid temperatures for a room by a rule – when the temperature value in that room is out of range (e.g. caused by a defect), the system reacts (for example, with an error message). More complicated diagnostics require an experienced operator who can observe and interpret real-time sensor values. However, as systems become larger, are deployed in a wider variety of environments, and are targeted at technically less-sophisticated users, both possibilities (rule-based systems and expert users) become problematic. The control system would require comprehensive prior knowledge of possible operating conditions, ranges of values and error conditions. This knowledge may not be readily available, and will be difficult for an unsophisticated user to input. It is impractical for experienced operators to directly observe large systems, and naive users can not interpret sensor values.

The goal of this work is to automatically recognize error conditions specific to a given sensor, actuator or system without the need of pre-programmed error conditions, user-entered parameters, or experienced operators. The system observes sensor and actuator data over time, constructs a model of "normality", and issues error alerts when sensor or actuator values vary from normal. The result is a system that can recognize sensor errors or abnormal sensor or actuator readings, with minimal manual configuration of the system. Further, if sensor readings vary or drift over time, the system can automatically adapt itself to the new "normal" conditions, adjusting its error criteria accordingly.

2. BACKGROUND

The diagnostic system is based on statistical "generative" models (SGMs). Statistical "generative" models attempt to reproduce the statistical distribution of observed data. As an example, one of the simplest and most widely used statistical generative models is the Gaussian distribution. The mean and standard deviation are parameters which can be fit to data. The model can then be used to determine how likely a new data value is (its probability under the model), or to "generate" new data (i.e. draw samples from a Gaussian distribution with the same mean and standard deviation). [1]

The Gaussian distribution is not appropriate for many modeling tasks, because much data does not actually come from a Gaussian-distributed source. Because of this, recent work in generative models has focused on non-Gaussian models (see for example (Hinton *et al.*, 1995; Hinton *et al.*, 1998; Lee *et al.*, 1999)). In addition to the Gaussian distribution, the diagnostic system uses a number of different SGMs to capture the distribution of sensor data, including histograms, mixture models (Bishop, 1995), hidden Markov models (Rabiner and Juang, 1986) and Bayesian networks (Pearl, 1988).

Sensor and control data poses several challenges. The data can be anything from very low level temperature readings to higher level "fault" detectors or occupancy data from building entry systems. This very different data must be fused into a single system. The volume of data requires fast algorithms (Jaakkola, 1997; Frey, 1997; Jordan *et al.*, 1999), and algorithms that can work with on-line data as it arrives (Neal and Hinton, 1998). The data values are time-dependent, so a model must (explicitly or implicitly) take time into account (Kalman, 1960; Rabiner and Juang, 1986; Ghahramani and Jordan, 1997; Ghahramani and Hinton, 1998; Sallans, 2000).

Previous work in applying statistical methods for fault detection includes the use of methods from statistical quality control (SQC) (see Fasolo and Seborg (1994) and Schein and House (2003), for example). These methods compare sensor values to prepared statistical "charts". If the sensor values vary from their expected values over a prolonged period of time, the charts can detect this variation.

[1] For probability density models, such as the Gaussian, which are functions of real-valued variables, the probability of the input value is not directly computed from the model. Rather the probability of either exceeding the given value, or of generating a value within a small neighborhood of the given value is computed. See section 3.3 for more details.

These methods differ from the SGM method in a number of important respects. First, and most importantly, because the charts are designed to detect small variations over a long time period, SQC methods require collection of data and detection of faults over a relatively long time window (for example, several faults detected over the course of 1000 sensor readings) in order to detect faults reliably. The SGM approach is designed to detect individual sensor errors, by comparing each new sensor value to a statistical model. Because of this, abnormal sensor values are detected immediately by the SGM-based system.

Second, because SQC methods are designed for long-term use, they can detect but not adapt to system changes or sensor drift. The SGM method was specifically designed to automatically accomodate system behavior changes and long-term sensor drift.

There are several other approaches to fault detection. In classical model-based detection, detailed domain knowledge is used to build a model of the system. Deviations between model predictions and system behavior are flagged as faults (see Isermann (1984) for a survey). In pattern matching detection, faults are induced in a system, and the resulting sensor values are recorded. A classifier, such as a neural network, is trained using this data set of normal and abonormal behavior to detect failures (see House *et al.* (1999) for example). These methods require either a working system for experimentation, or an in-depth knowledge of the system in question, both of which are lacking for large building automation systems.

Despite their success in other domains, SGMs have not been applied to error detection for building automation sensor and control data. There are two reasons for this. First, it has only recently become possible (and economical) to collect a wide range of cross-vendor sensor data at a central location. Second, most algorithms to optimize the parameters of SGMs are quite compute-intensive. Many algorithms have been of only theoretical interest, or are restricted to small toy problems. Only recently have powerful approximation algorithms and powerful computers become available that can handle large quantities of data in real-time. The combination of fast, powerful optimization algorithms, fast computers, and available multi-sensor data makes the time ripe for probabilistic modeling of sensor and control data.

3. THE DIAGNOSTIC SYSTEM

The goal of the diagnostic system is to automatically detect sensor errors in a running automation system. It does this by learning about the behavior of the automation system by observing data flowing through the system. The diagnostic system builds a model of the sensor data in the underlying automation system, based on the data flow. From the optimized model, the diagnostic system can identify abnormal sensor and actuator values. The diagnostic system can either analyze historical data, or directly access live data (see Figure 1).

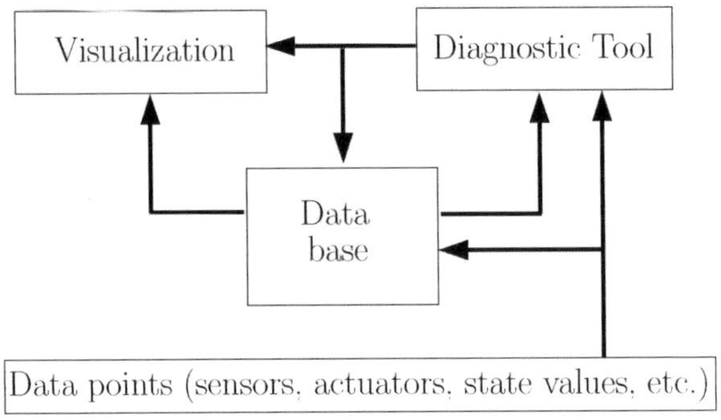

Fig. 1. The Automation System. Sensor and actuator values are stored in a database, and simultaneously analyzed on-line by the error detection system. Connections between the diagnostics system and other system components are either direct, or via the Database.

We use a set of statistical generative models to represent knowledge about the automation system. A statistical generative model takes as input a sensor value, status indicator, time of day, etc., and returns a probability between zero and one. For example, the most widely used example of a SGM is the Gaussian distribution. For each input value, a Gaussian computes a weighted distance from the Gaussian mean, and returns a decreasing probability as the value moves away from the mean.

Using SGMs has several advantages. First, because the model encodes the probability of a sensor value occurring, it provides a quantitative measure of "normality", which can be monitored to detect abnormal events. Second, the model can be queried as to what the "normal" state of the system would be, given an arbitrary subset of sensor readings. In other words, the model can "fill in" or predict sensor values, which can help to identify the source of abnormal system behavior. Third, the model can be continuously updated to adapt to sensor drift.

3.1 *The Statistical Models*

The system implements a number of SGMs (see Table 1).

Table 1. Statistical Generative Models

Model	Variable Type	Parameters
Gaussian	Real	μ,σ^2
Histogram	Discrete, Real	Bin counts
Mixture of Gaussians	Real	μ_i,σ_i^2,π_i
Hidden Markov model	Real	T_{ij},μ_i,σ_i^2
Hidden Markov model	Discrete	T_{ij},Bin counts

The more complex models add additional capabilities, or relax assumptions in comparison to the Gaussian model:

Histogram: This is a very general model that is appropriate for discrete sensor values, as well as real-valued sensors with an arbitrary number of modes. One drawback is that a histogram requires a rather large quantity of data before it becomes usable or accurate.

Mixture of Gaussians: This model relaxes the Gaussian assumption that the distribution has only one mode. A mixture of Gaussians is composed of a number of Gaussian models, and each data value is attributed to the Gaussian modes with a weighting given by a "posterior probability". See for example (Bishop, 1995).

Hidden Markov Model: This model is the equivalent of the Mixture of Gaussians model or the histogram model, but with the addition that the current sensor value can be dependent on previous values.

3.2 *Error Detection*

Given an SGM, implementation of this functionality is straight-forward. The system assigns to each newly-observed data value a probability. When this probability is high, the system returns that the new data value is a "normal" value. When the probability falls below a specific threshold, the system rates the value as "abnormal". The SGM system generates alarm events when it observes abnormal sensor values. This leaves open the question of how to assign the threshold for normality. In practice, the user sets the threshold using a graphical interface. Initially, before the system has learned normal system behavior, many alarms are generated, and the user may decide to set the threshold to a value near zero. As the system acquires a better model of the sensor system, the threshold can be raised. In any case, the threshold parameter tells us how improbable an event should be to raise an alarm. The system can also use a log-probability scale, so that the threshold can easily be set to only register extremely unlikely events.

3.3 *Statistical Generative Models*

The diagnostic system uses SGMs of automation data points. For any given data value x, model M assigns a probability to x: $P_M(x) \rightarrow [0, 1]$.

Note that, for discrete distributions such as a histogram, the value assigned to x by the model $P_M(x)$ is a well-defined probability, since the set of possible assignments to x is finite. For a probability density, such as a Gaussian or mixture of Gaussians, the probability value assigned to x by the model is the probability *density* at that value. In order to convert this density to a probability, the probability of generating a value within a neighborhood $\pm\delta$ around x is computed as $\int_{-\delta}^{\delta} P_M(x+\phi)d\phi$, and approximated as $2\delta P_M(x)$ for small δ. Alternatively the probability under the model of equaling or exceeding the observed value can be computed: $P_M(x' \geq x) = \int_{\phi}^{\phi+\infty} P_M(x+\phi)d\phi$.

The data x can be a sensor reading such as a motion sensor, door contact sensor, temperature sensor, and so on. Given a new data value, the system can assign a probability to this value. When the probability is above a given threshold, the system concludes that this data value is "normal".

Given a sequence of sensor readings $\boldsymbol{x} = \{x_1, ..., x_T\}$ from times 1 to T, the system must create a model of "normal" sensor readings. The system uses maximum-likelihood parameter estimation. We describe this method below.

Assume that model M has parameters θ. For example, the standard Gaussian distribution has two parameters, a mean μ and a variance σ^2. In this case, $\theta = \{\mu, \sigma^2\}$. The log-likelihood of the model parameters given the data $\boldsymbol{x}$ is given by:

$$\mathcal{L}(\theta) = \log P_M(\boldsymbol{x}|\theta) \tag{1}$$

where the notation $P(\boldsymbol{x}|\theta)$ denotes a conditional probability: The probability assigned to $\boldsymbol{x}$ depends on the current values of the parameters θ. The log-likelihood tells us how "good" the parameters of the model are. When it is high, the data values that we usually see have a high probability under the model. That is, the model captures what are "normal" data values.

The maximum-likelihood parameters are defined as the parameter values which maximize the log-likelihood over the observed data:

$$\theta_{ML} = \underset{\theta}{\operatorname{argmax}} \{\log P_M(\boldsymbol{x}|\theta)\}$$

Again, take the Gaussian distribution as an example. The log-likelihood of the 1D Gaussian is given by:

$$\mathcal{L}_{\text{Gaussian}}(\mu, \sigma^2) = -\sum_i \left[\frac{1}{2}\log(2\pi\sigma^2) + \frac{1}{2\sigma^2}(x_i - \mu)^2\right]$$

where $x_i \in \boldsymbol{x}$ are independent, identically-distributed data values. Taking the derivative of the log-likelihood with respect to the parameters yields:

$$\frac{\partial \mathcal{L}_{\text{Gaussian}}(\mu, \sigma^2)}{\partial \mu} = \sum_i \frac{(x_i - \mu)}{\sigma^2} \tag{2}$$

$$\frac{\partial \mathcal{L}_{\text{Gaussian}}(\mu, \sigma^2)}{\partial \sigma^2} = \sum_i \left(-\frac{1}{\sigma} + \frac{(x_i - \mu)^2}{\sigma^3}\right) \tag{3}$$

We can now solve for the maximum-likelihood parameters. Setting the derivatives in Eq.(2) and Eq.(3) to zero and solving the system of two equations yields:

$$\mu_{ML} = \frac{1}{T}\sum_i x_i \tag{4}$$

$$\sigma^2_{ML} = \frac{1}{T}\sum_i (x_i - \mu_{ML})^2$$

which are the familiar textbook definitions of mean and variance of a Gaussian.

3.4 *On-line Parameter Updates*

In order for the system to continually adapt the model parameters, we need a parameter update algorithm that can incrementally change the parameters based on newly observed sensor values. Such methods are called "on-line" parameter update rules. On-line updates have the advantage that there is no time during which the system is in an "optimization" phase, and unavailable for diagnostics.

One simple on-line method for computing locally-optimal maximum-likelihood parameters is "stochastic gradient ascent". As each new data value x_i is observed, the parameters are adjusted so as to increase the log-likelihood:

$$\theta^{\text{new}}_{ML} = \theta^{\text{old}}_{ML} + \epsilon \frac{\partial \log P_M(x_i|\theta)}{\partial \theta}$$

where ϵ is a "learning rate". Given appropriate conditions on ϵ and the log-likelihood function, this algorithm will find a local maximum of the log-likelihood. In practice, the success and speed of the algorithm depends on the form of the log-likelihood and the order and frequency with which parameters are updated. Good results can be achieved for many generative models, although convergence can be slow (see for example (Dempster *et al.*, 1977; Russell *et al.*, 1995; Olshausen and Field, 1996). Our diagnostic system makes extensive use of this parameter update rule.

For the example of a Gaussian model, the on-line parameter update rules are given by:

$$\mu^{\text{new}}_{ML} = \mu^{\text{old}}_{ML} + \epsilon_\mu \frac{\partial \mathcal{L}_{\text{Gaussian}}(\mu, \sigma^2)}{\partial \mu}$$

$$\sigma^{2\,\text{new}}_{ML} = \sigma^{2\,\text{old}}_{ML} + \epsilon_{\sigma^2} \frac{\partial \mathcal{L}_{\text{Gaussian}}(\mu, \sigma^2)}{\partial \sigma^2}$$

In some cases, as with the Gaussian, where parameters can be explicitly solved for, another on-line parameter update method can be used. The alternative to a gradient update is the on-line moving average, or "decaying average". Consider for example the expression for the mean of a Gaussian (equation 4):

$$\mu_{ML} = \frac{1}{T}\sum_i x_i$$

If the system receives new data values on-line, we can not use the expression as given, because we do not want to store the entire history of values $\{x_i\}, i = 1, ..., T$. However, we can use a sliding window such that:

$$\mu_{SW}^{\text{new}} = \frac{1}{N} \sum_{i=t_1}^{t_2} x_i$$

where we only store the last $N = (t_2 - t_1)$ values. As a final alternative, we can recompute the mean at each step without storing previous values:

$$\mu_{MA}^{\text{new}} = \frac{1}{N} x_i + \frac{(N-1)}{N} \mu_{MA}^{\text{old}}$$

which is a "decaying average" update, and approximates a sliding window of length N. We call this update rule an "on-line moving average" update rule. This rule can be used for some models but is not as generally applicable as the gradient update rule.

Figure 2 shows an example of parameter optimization for the Gaussian model. The initial mean and variance were set to $\{-5, 9\}$ (the "X"). The contour lines show lines of equal log-likelihood for the Gaussian model. The "O" shows the true parameters for the model $\{5, 4\}$. The curved (upper) line shows the path in parameter space followed by the gradient-ascent rule, from the original guess $\{-5, 9\}$ to the true parameters $\{5, 4\}$. The parameters are updated as each new data value (sampled from a Gaussian with the true parameters $\{5, 4\}$) arrives. Because the gradient is computed using a single value, the "up-hill" direction is noisy, resulting in the slightly "wandering" line we see. Also, when the guess reaches the true parameter values, it wanders around in this area, because for each new data value, the "optimal" parameters are different.

Similarly, the straight (lower) line shows the path in parameter space followed by the on-line moving average rule. As we can see, the two rules follow completely different paths in parameter space, but arrive at the same answer. Which method works best is an empirical question, and depends on the model used, the learning rate, how much data is available, and so on. We tend to use moving-average rules where possible, because of their fast initial convergence, and gradient-based rules otherwise.

4. RESULTS

The diagnostics system has been implemented in an office environment with 20 rooms, an entrance area, and two long corridors. In this environment, motion, door state, luminosity, temperature and humidity sensors have been installed. All sensors are wireless, and report sensor values to a central computer with an attached receiver station. The motion and door state sensors are standard consumer "X10"-brand sensors. The temperature, luminosity and humidity sensors are custom-built wireless sensors (Mahlknecht, 2004).

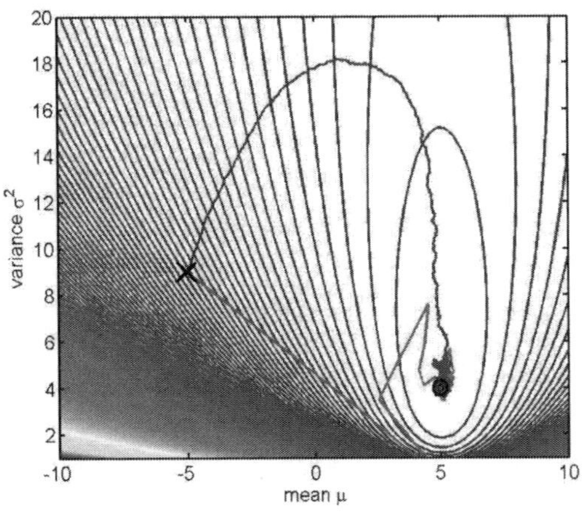

Fig. 2. Example of gradient-based optimization and on-line averaging for the Gaussian model.

The sensors are battery-powered, and the custom sensors have built-in voltmeters which report the current battery state. All data is stored in an SQL database. At the time of writing, data has been collected over the course of more than three weeks. The diagnosis system can initialize its models by analyzing stored historical data, or simply optimize on-line as it reports error conditions.

4.1 *Parameter Optimization*

If the system begins reporting errors without first analyzing historical data, it will initially report many errors. This will continue until the system has acquired enough knowledge of the automation network and sensors to at least have rudimentary models of normal sensor values. As analysis continues, the number of alarms generated falls (see Figure 3).

The model log-likelihood, given by Eq.(1), is a measure of model quality. As the parameter values are optimized on-line, the log-likelihood of the sensor models increases. Figure 4 shows the average log-likelihood during parameter optimization, on average, for the sensor models in the test system. The log-likelihood increases with time, indicating that the models improve over time. The log-likelihood does not consistently increase, however, due to the on-line fitting of parameters simultaneously with reporting of abnormal sensor values. If sensors receive abnormal values, the log-likelihood decreases, until the values return to their normal range or the sensor model adapts to the new range of values.

4.2 *Optimized Sensor Models*

After analyzing a quantity of data the system has models of typical sensor values. In the current implementation, the system uses one model per sensor, for every

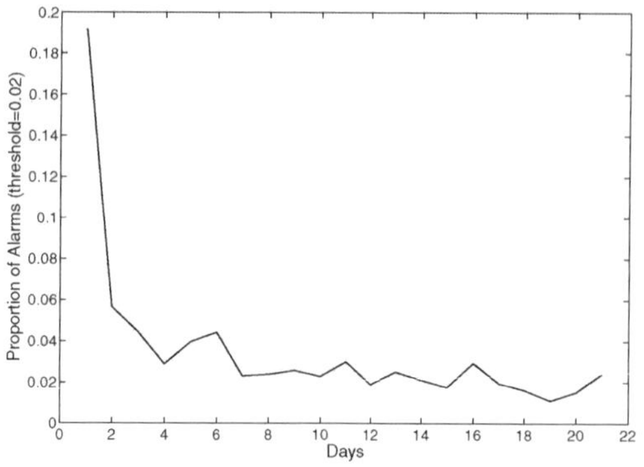

Fig. 3. Proportion of alarms generated (at a threshold of $\log(P_M(x)) = -4$) as a function of training time. The proportion of alarms is the ratio of the number of alarms generated to the number of sensor values received by the diagnostic system. With time, the alarm rate drops to approximately match the chosen threshold, as the statistical models used by the system match themselves to the actual behavior of the system.

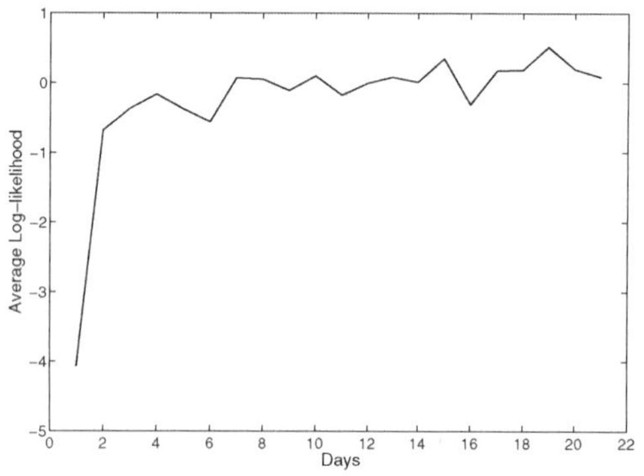

Fig. 4. Average learning curve for sensor models. On average the log-likelihood for the model improves over time. There is a rapid improvement in the first days, followed by a slower improvement over weeks.

half-hour interval. That is, there is a distinct model for each half-hour interval.

The system also has models for "data period" (the typical number of seconds between data packets from a sensor). For sensors such as motion and door state, it is informative to know how often the sensor state changes, in order to detect sensor failures or abnormal conditions. Models are fit to the typical time interval between arrival of sensor data packets. Figure 5 shows models for three sensors: Motion frequency, temperature and luminosity. The first two are examples of "mixture of Gaussians" models, while the third is a simple Gaussian. Note that the temperatures, for example, would not be well-characterized by a simple Gaussian model.

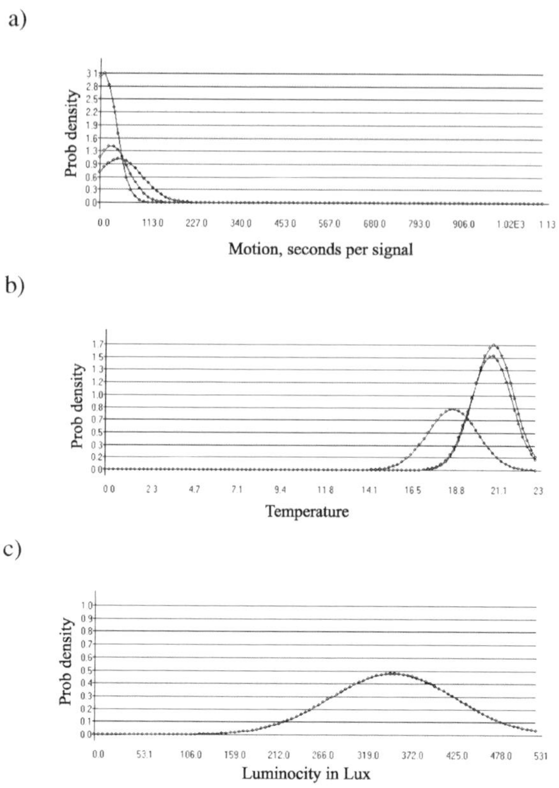

Fig. 5. Examples of models optimized by the system. In some cases, the Gaussian model is sufficient, and in other cases the model has (a) a non-Gaussian shape or (b) multiple modes.
a) A motion sensor from 5:00pm-5:30pm;
b) A temperature sensor from 8:00am-8:30am;
c) A luminosity sensor from 6:00pm-6:30pm.

4.3 *Error Detection and Drift Adaptation*

We can see the trade-off between adaption to sensor drift and detection of short-term behavior changes by examining an individual model more closely. The wireless sensors are powered by small batteries, which leaves open the possibility that an unusual power drain can deplete the batteries and cause a sensor failure. The sensors are equipped with voltage sensors, indicating the voltage currently available from the battery. One of the sensors in the test system did in fact have an unusual power drain during the test period. Figure 6 and 7 show the diagnostic system behavior for two different speeds of parameter updates. In Figure 6, the model parameters are optimized during the initial days, and then held fixed. In this case, the drift is detected relatively quickly.

In Figure 7, the model parameters are continuously optimized on-line. In this case, the drift is detected by the model, but does not result in an alarm until the drift rate increases beyond the adaptation rate of the model.

In the case of battery voltage level, the initial behavior may be more desirable. For other sensors (luminosity,

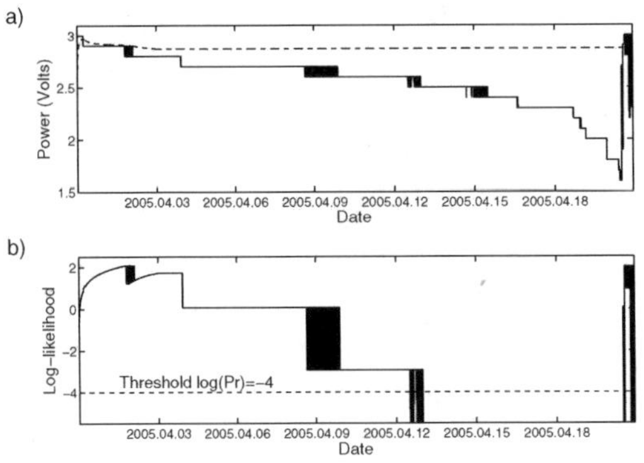

Fig. 6. a) Voltage as a function of time for a wireless sensor. The sensor was faulty, resulting in battery drain and a decrease in system voltage over time. The available voltage increased at the end, when the batteries were changed. The solid line is the actual voltage, and the dashed line is the expected voltage under the voltage model.
b) Log-likelihood curve for the voltage model. Model parameters are quickly optimized for the observed voltage level. Because the model parameters are not adapted after the initial period, the model detects the slow voltage drift relatively quickly.

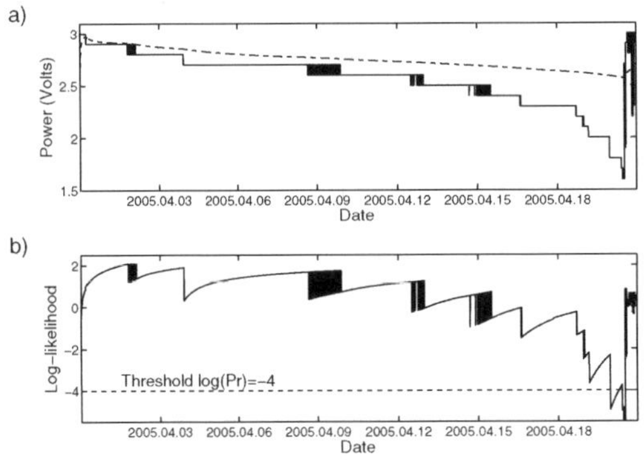

Fig. 7. a) Voltage as a function of time for a wireless sensor. The sensor was faulty, resulting in battery drain and a decrease in system voltage over time. The available voltage increased at the end, when the batteries were changed. The solid line is the actual voltage, and the dashed line is the expected voltage under the voltage model.
b) Log-likelihood curve for the voltage model. Model parameters are quickly optimized for the observed voltage level. As the batteries are drained, there is a slow drift downward. The model continues to adapt, resulting in no alarm. As the rate of drift accelerates, the change is too rapid for the model to adapt, and an alarm is triggered.

temperature) it may be preferable for the system to adapt to drifting sensor values without generating an alarm. Behavior such as shown in Figure 7 may be the most preferable, where some drift can be accommodated, but rapid changes will be detected.

5. CONCLUSION

This paper describes a method for the automatic detection of abnormal sensor values, based on statistical generative models of sensor behavior. The system can automatically build a model of normal sensor behavior. It does this by optimizing model parameters using an on-line maximum-likelihood algorithm. Incoming sensor values are then compared to the model, and an alarm is generated when the sensor value has a low probability under the model. The model parameters are continuously adapted on-line. The result is a system that automatically adapts to changing conditions and sensor drift. As well as detecting isolated abnormal sensor values, the system can either detect or adapt to slower sensor drift, depending on the rate of model parameter adaptation.

With the current system, some knowledge of the expected rate of sensor drift is required, in order to correctly set the parameter adaptation rate. An alternative would be to use two error-detection mechanisms. For isolated abnormal values, the current system would be used. For the detection of long-term drift, changes in model parameter values themselves would be monitored over a longer time-scale. The result would be a hierarchical model, with sensor values at the lower level, and model parameters at the higher level.

The current system only detects abnormal values for single sensors. The system is currently being extended to include multivariate models which capture correlations between sensor values. In this way, abnormal groups of sensor values can be automatically detected.

6. REFERENCES

Bishop, C. M. (1995). *Neural Networks for Pattern Recognition*. Oxford University Press Inc.. New York NY.

Dempster, A. P., N. M. Laird and D. B. Rubin (1977). Maximum likelihood from incomplete data via the EM algorithm. *J. Royal Statistical Society Series B* **39**, 1–38.

Fasolo, Paul and Dale E. Seborg (1994). An SQC approach to monitoring and fault detection in HVAC control systems. In: *Proceedings of the American Control Conference*. Baltimore, Maryland.

Frey, Brendan (1997). Continuous sigmoidal belief networks trained using slice sampling. In: *Advances in Neural Information Processing Systems* (Michael C. Mozer, Michael I. Jordan and Thomas Petsche, Eds.). Vol. 9. The MIT Press, Cambridge. pp. 452–458.

Ghahramani, Z. and G. E. Hinton (1998). Variational learning for switching state-space models. *Neural Computation* **4**(12), 963–996.

Ghahramani, Z. and M. I. Jordan (1997). Factorial hidden Markov models. *Machine Learning* **29**, 245–273.

Hinton, G. E., B. Sallans and Z. Ghahramani (1998). A hierarchical community of experts. In: *Learning in Graphical Models* (M. I. Jordan, Ed.). pp. 479–494. Kluwer Academic Publishers.

Hinton, G. E., P. Dayan, B. J. Frey and R. M. Neal (1995). The wake-sleep algorithm for unsupervised neural networks. *Science* **268**, 1158–1161.

House, J.M., W. Y. Lee and D. R. Shin (1999). Classification techniques for fault detection and diagnosis of an air-handling unit. *ASHRAE Transactions* **105**(1), 1987–1997.

Isermann, R. (1984). Process fault detection based on modeling and estimation methods – a survey. *Automatica* **20**(4), 387–404.

Jaakkola, T. S. (1997). *Variational Methods for Inference and Estimation in Graphical Models*. Department of Brain and Cognitive Sciences, MIT. Cambridge, MA. Ph.D. thesis.

Jordan, M. I., Z. Ghahramani, T. S. Jaakkola and L. K. Saul (1999). An introduction to variational methods for graphical models. *Machine Learning* **37**, 183:233.

Kalman, R. E. (1960). A new approach to linear filtering and prediction problems. *Trans. ASME, Series D, Journal of Basis Engineering* **82**, 35–45.

Lee, T-W, M. Girolami and T. Sejnowski (1999). Independent component analysis using an extended infomax algorithm for mixed sub-gaussian and super-gaussian sources.

Mahlknecht, S. (2004). *Energy-Self-Sufficient Wireless Sensor Networks for the Home and Building Environment*. Institute of Computer Technology, Technical University of Vienna. Vienna, Austria. Dissertation thesis.

Neal, R. M. and G. E. Hinton (1998). A view of the EM algorithm that justifies incremental, sparse, and other variants. In: *Learning in Graphical Models* (M. I. Jordan, Ed.). pp. 355–368. Kluwer Academic Publishers.

Olshausen, B. A. and D. J. Field (1996). Emergence of simple-cell receptive field properties by learning a sparse code for natural images. *Nature* **381**, 607–609.

Pearl, J. (1988). *Probabilistic Reasoning in Intelligent Systems: Networks of Plausible Inference*. Morgan Kaufmann. San Mateo, CA.

Rabiner, Lawrence R. and Biing-Hwang Juang (1986). An introduction to hidden Markov models. *IEEE ASSAP Magazine* **3**, 4–16.

Russell, Stuart, John Binder, Daphne Koller and Keiji Kanazawa (1995). Local learning in probabilistic networks with hidden variables. In: *Proc. Fourteenth International Joint Conference on Artificial Intelligence*. Montreal, Canada.

Sallans, B. (2000). Learning factored representations for partially observable Markov decision processes. In: *Advances in Neural Information Processing Systems* (S. A. Solla, T. K. Leen and K-R Müller, Eds.). Vol. 12. The MIT Press, Cambridge. pp. 1050–1056.

Schein, J. and J.M. House (2003). Application of control charts for detecting faults in variable-air-volume boxes. In: *ASHRAE Transactions*. Vol. 109. pp. 671–682.

ELSEVIER
IFAC
PUBLICATIONS

INTEGRATION OF BUILDING AUTOMATION NETWORK DESIGN AND 3D CONSTRUCTION TOOLS BY IFC STANDARD

Alexander Karavan[1], Mario Neugebauer[1,2], Klaus Kabitzsch[1]

[1]*Institute of Applied Computer Science, Dresden University of Technology, 01062 Dresden, Germany*
{s2700047,mn7,kk10}@inf.tu-dresden.de

[2]*SAP Research Group Dresden, Dürerstrasse 24, 01307 Dresden, Germany*

Abstract: Currently, different design tools and databases are used for building automation networks and construction projects. Effort (design time, tools) can be reduced if these two design spaces can be merged. In this paper an approach for the integration of building automation network design in IFC is introduced. Based on the resulting model, existing tools are to enhance in order to accomplish a holistic design approach for construction projects. *Copyright © 2005 IFAC*

Keywords: integration, information integration, models, automation, fieldbus, design systems.

1. INTRODUCTION

Today, building automation networks are designed using standardized tools and data bases. The model structure of a building automation network with all required components (e.g., sensors, controllers, actuators) is created with graphically supported tools (e.g., LonMaker, see Figure 1). All necessary connections between the components on different layers (application and physical layer) are carried out within these tools.

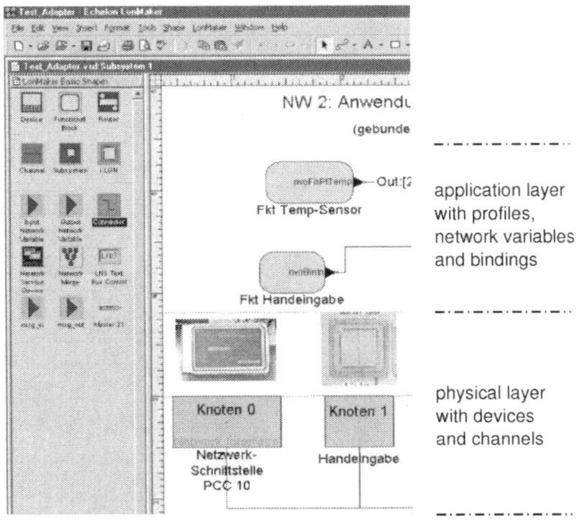

Fig. 1. LonMaker for graphically supported design of building automation networks.

Even deployment (configuration of real components according to model structure) and maintenance (network monitoring, deployment of additional functionalities and devices, etc.) are facilitated by these tools.
There are certain similarities between the approaches used for designing building automation networks and construction projects. The LNS Network Operating System (Echelon 1999) is the quasi standard data base for design and deployment of building automation networks used by different parties involved. Therein, the complete information about the structure of the network on different layers is contained. The aim of Industry Foundation Classes (IFC) of International Alliance for Interoperability (IAI) (IAI 2005) is about the same in a different application area: a common standard to exchange information among various parties involved in a construction projects (e.g. spatial structure, walls, building services elements).
So far, the two design approaches for building automation networks and construction projects are carried out in two different tool worlds. According to the data on which the structures rely on there is no connection, except a drawing delivered from the architect to the building automation network designer. There are no basic approaches for interoperability guidelines that could enable seamless integration of building automation network structures in construction projects.

In this article we show an approach for future design methodologies and tools to integrate building automation network design and therewith to enable a comprehensive building model including the automation facilities (Schach et al. 2002). Especially, the similarities in current tool domains (building automation networks and construction projects) are elaborated in order to enable merging of different design approaches. Hence, the existing model structures in building automation networks have to be reused and integrated in IAIs IFC.

In Section 2 and 3 LonWorks technology and IAIs IFC are introduced. Concepts for the integration the two design methodologies are discussed in Section 4 and explained in more detail with an example in Section 5.

2. USING THE LONWORKS TECHNOLOGY IN A BUILDING AUTOMATION

Building automation is a fastly growing market. It proposes now open and interoperable solutions (Dietrich at al. 2001). One of the leading networking standards in this shift is the LonMark standard, based on LonWorks control network technology.

A LonWorks network consists of intelligent devices that communicate with each other using a common protocol over one or more communication channels (Fig. 2). Each device includes one or more processors that provide its intelligence and implement the protocol. Each device also includes a component called a transceiver to provide its electrical interface to the communication channel. Transceivers are available for a variety of communication media including single twisted-pair cable, power line, radio frequency (RF), infrared, fiber optics, and coax cable (Echelon 1999).

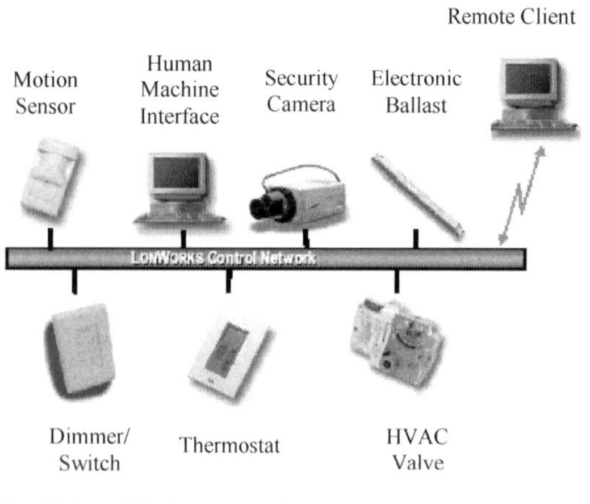

Fig. 2. LonWorks network.

The LonWorks protocol, also known as the LonTalk protocol and the *ANSI/EIA 709.1 Control Networking Standard*, is the heart of the LonWorks system. The LonWorks protocol is a layered, packet-based, peer-to-peer communication protocol. The seven layers along with the corresponding services are provided by the LonWorks protocol.

The LonWorks protocol implements the concept of network variables. Network variables simplify the tasks of designing LonWorks application programs for interoperability with multivendor products and facilitate the design of information-based, rather than command-based, control systems.

Via a process that takes place during network design and installation called binding, the device firmware is configured to know logical addresses of the other devices or groups of devices in the network. The binding process creates logical connections between an output network variable in one device and an input network variable in another device or group of devices. Connections may be thought of as "virtual wires" (Fig. 11).

Every network variable has a type that defines the units, scaling and structure of the data contained within the network variable. A set of standard network variable types (SNVTs) is defined for commonly used types.

LonWorks Network Services (LNS) architecture is important for the LonWorks protocol. LNS provides an object-oriented programming model and brings the power of client-server architecture and component-based software design into control networks (Fig. 3). LNS enables to install, maintain, monitor and control LonWorks networks (Echelon 2004).

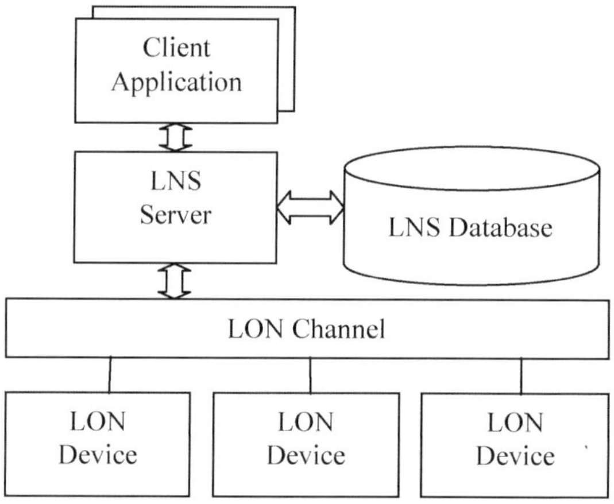

Fig. 3. LNS architecture.

The LNS Object Server defines a set of objects, properties, methods and events that represent the physical attributes of networks and their configurations. LNS Object Server is a COM server that provides an interface, independent of any programming language, to the LNS Server and the LNS database.

The objects are grouped together in a hierarchical fashion (Fig. 4), such that the *ObjectServer* object is at the top of the hierarchy. The *ObjectServer* object contains a collection of *Network* objects, each of which represents a network defined in the database. Each *Network* object contains a *System* object representing the network's system, and each *System* object contains a set of *Subsystem* objects that represent logical or physical partitions of that particular network. Each *Subsystem* object contains a collection of *AppDevice* objects which represent the application devices defined in that subsystem.

For example, the user must set up a network to control a building with three floors. The user can define three subsystems within the network, so that he can logically group the devices on each floor separately. He can also set up additional subsystems representing different rooms on each floor of the building.

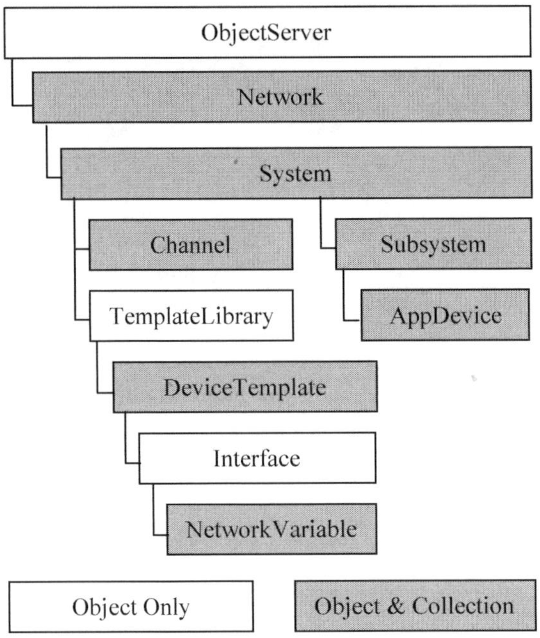

Fig. 4. LNS Object Server hierarchy.

As the subsystems are logical divisions of a network, and devices can belong to multiple logical divisions, the user can create multiple subsystems that cross-reference a network, and add individual devices to each of them. In this manner, devices can appear in multiple subsystems. For example, the user can create one subsystem to represent the physical layout of network, and another one to represent the functional layout (e.g. HVAC, lighting, etc.).

3. DATA EXCHANGE STANDARD IN THE BUILDING INDUSTRY

IFC is the set of internationally standardized object definitions (ISO 2005) for use in the building industry developed by the IAI (IAI 2005).

IAI is an alliance of organizations dedicated to bring about a coordinated change for the improvement of productivity and efficiency in the construction and facilities management industry.

IFC is an universal language to improve the communication throughout the design, construction, operation and maintenance life cycle. The IFC concept is based on the idea of objects brought together in an integrated model. These objects are defined to support the whole lifecycle of facility development from inception through design, then facility management and finally demolition (Fig. 5).

Objects in IFC in addition to the full geometric description in 3D have also location and relationships, as well as all the properties of each object.

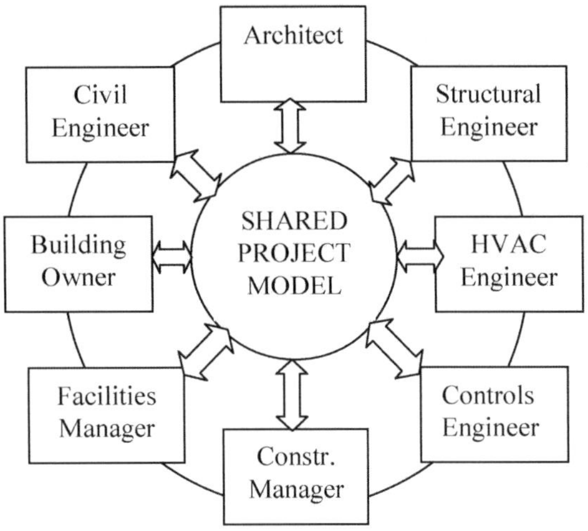

Fig. 5. Shared project model for building.

IFC contains not only physical properties, but also spatial concepts such as floors and rooms (Liebich 2004).

The following three different concepts are subsumed under the *IfcSpatialStructureElement* (Fig. 6):

- Building
- Building storey
- Space

These different entities contain each other such as they provide a clear hierarchical structure for the building project. The three subtypes *IfcBuilding*, *IfcBuildingStorey*, *IfcSpace* are used to represent the levels of the spatial structure. The spatial structure is created by using the fundamental decomposition relationship. The subtype *IfcRelAggregates* is used to link the instances and establishes a hierarchical structure.

IfcBuilding is used to provide additional information about a building itself. The building includes the references to the building storeys belonging to that building or it can have building sections.

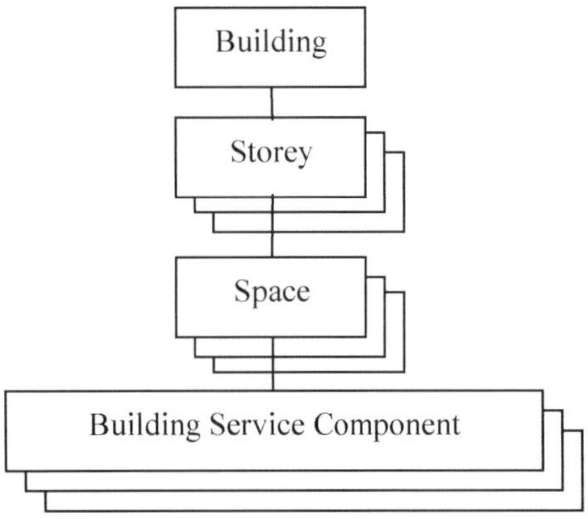

Fig. 6. Spatial structure of a building.

IfcBuildingStorey is used to provide information about a building storey itself, a vertical structure normally used with building and construction. The building storey includes the references to the spaces belonging to it.

IfcSpace is used to provide information about the space as a functional area or volume with a spatial structure. Spaces normally contain all building services or interior design elements (such as distribution elements, electrical elements, furniture). The space boundaries define the relationship between the space and its bounding elements.

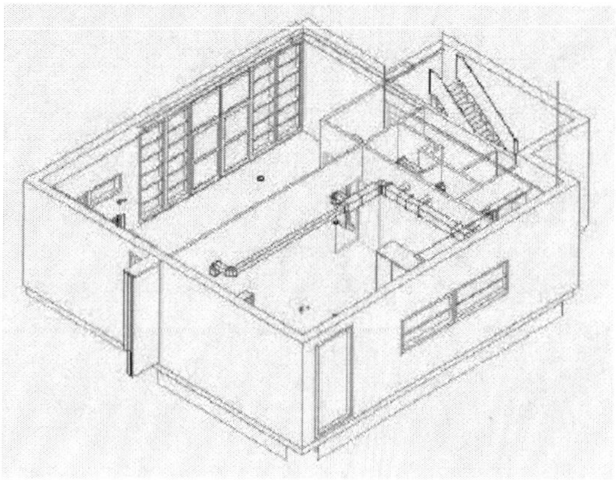

Fig.7. Spatial structure and building service components of a 3D construction model developed by the CAD tool.

The IFC defines basic object concepts required for interoperability within the heating, ventilating and air conditioning (HVAC) and lighting (Fig. 7).

The methodology to specialize building service components in IFC follows the function of a component within a distribution system, and not the discipline which handles the components. Therefore the first level decomposition follows the structure:

- element within a distribution flow system
    - energy conversation devices
    - flow controlling devices
    - flow moving devices
    - flow segments
    - flow terminals
- control element.

IfcEnergyConversionDevice defines the occurrence of a device used to perform energy conversion or heat transfer and typically participates in a flow distribution system (e.g., boiler, chiller, coil, electric motor, space heater, etc).

IfcFlowController defines the occurrence of elements of a distribution system that are used to regulate flow through a distribution system (e.g., damper, valve, switch, relay, etc.).

IfcFlowMovingDevice defines the occurrence of an apparatus used to distribute, circulate or perform conveyance of fluids, including liquids and gases, and typically participates in a flow distribution system (e.g., pump, fan).

IfcFlowSegment defines the occurrence of a segment of a flow distribution system that is typically straight, contiguous and has two ports (e.g., a section of pipe, duct or cable).

IfcFlowTerminal defines the occurrence of a permanently attached element that acts as a terminus or beginning of a distribution system (e.g., air outlet, electric appliance or heater, light fixture, etc.).

IfcDistributionControlElement defines occurrence elements of a building automation control system that are used to impart control over elements of a distribution system (Fig. 8). These are typically used to control distribution system elements to maintain temperature, humidity, pressure, flow, power, lighting levels, etc., through the modulation, staging or sequencing of mechanical or electrical devices. The three general functional categories of ControlElements as defined for this release of the IFC model are as follows:

- Impart control over flow control elements (*IfcFlowController*) in a distribution system such as dampers, valves, relays, etc., typically through the use of actuation *IfcActuator*.
- Sensing elements *IfcSensor* that measure changes in the controlled variable (e.g., temperature, humidity, pressure, flow, etc.).
- Controllers *IfcController* typically classified according to the control action they seek to perform and generally responsible for making decisions about the elements under control.

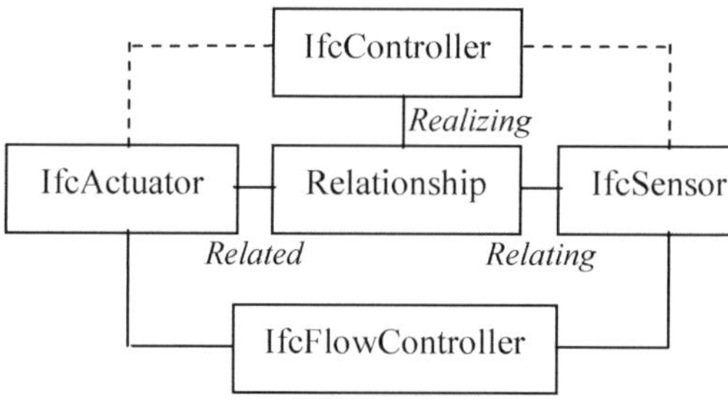

Fig. 8. Control loop in the IFC model.

4. INTEGRATION PLATFORM CONCEPT

As it has been mentioned in Section 2, the LNS model represents the structure of the LonWorks automation network, and IFC model - the structure of the building. These models are created by means of different tools saved in various databases and have different interfaces to data.

By creation of a new project of the building automation network, the user repeats operations which have already been done by the user of the CAD system and results are kept in IFC model (definition of subsystems: floors, rooms and systems of lighting, ventilation, heating). For the process acceleration of the building automation network design, it is expedient to integrate LNS model in IFC model, as more general and universal one. The holistic model also can be used for more detailed modelling of various processes in the building.

For the planning architecture of the building and its life-support systems various CAD systems are used. These systems support import and export of the IFC data model.

For the integration of two models we propose the following approach:

1. data import from Architect CAD to Building Automation Network Tool;
2. data import from HVAC and Electro CAD to Building Automation Network Tool;
3. data export from Building Automation Network Tool.

4.1 Data import from Architect CAD to Building Automation Network Tool.

Architect CAD is used for development of a building architecture. From the available IFC model it is possible to receive the spatial structure of a building (floors and rooms) for use in the LNS model (Fig. 9). The additional information about characteristics of rooms (e.g. volume, area, recommended temperature and light exposure) and internal elements (windows and door) can be useful for developers of the building automation networks.

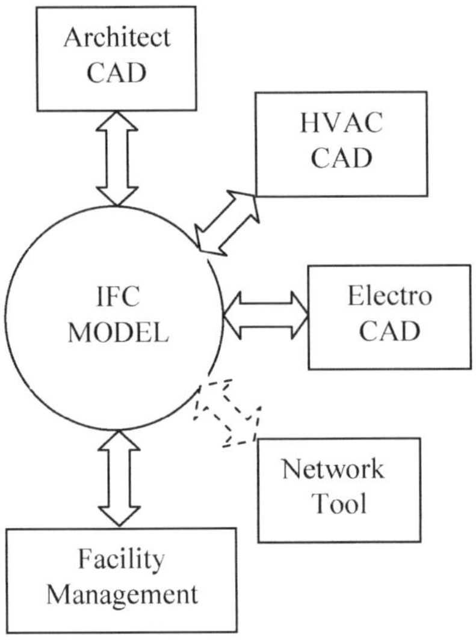

Fig. 9. Data exchange through the IFC model.

4.2 Data import from HVAC and Electro CAD to Building Automation Network Tool.

HVAC and Electric CAD are used for development of building services. This information is fundamental for development of building automation networks and specifies what processes need to be automated. From the IFC model it is possible to receive objects of automation (e.g., boiler, coil, electric motor, space heater, damper, valve, switch, etc) and their characteristics (e.g., capacity, range) (Fig. 9). The IFC model can have the information about control loops (sensor, actor and controller) and characteristics (type of the controller, time constant of the sensor). On the basis of this information the building automation network can be designed automatically.

4.3 Data export from Building Automation Network Tool.

After the completion of the building automation network designing, the information about the network is stored in the LNS database. In order to make this information accessible for other participants of the building life-cycle, it is necessary to transform (import) it into the IFC model (Fig. 9). The information about the building automation network can be used in Electric-CAD for placement of network nodes and tracing of cables. Within the maintenance time of the automation network at the level of the building facility management, the information about current condition of processes is necessary. For this, the measured values from the network should be transferred to the IFC server.

4.4 Practical realization of the integration concept.

A converter which can transform the IFC model into the LNS model and back is necessary for realization of the integration process. Therefore the converter consists of two parts (Fig. 10):

- IFC to LNS
- LNS to IFC.

In practice the process of the integration looks as follows. The converter "IFC to LNS" receives the necessary information from the IFC file by means of an IFC Toolbox.

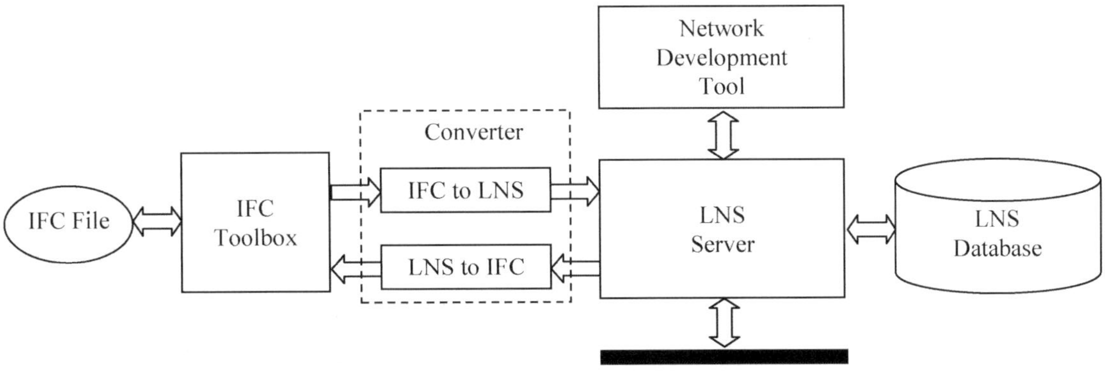

Fig. 10. Conversation between IFC and LNS models.

The IFC Toolbox it is the program component which realizes operations of reading, writing and search IFC objects in the IFC file that have ASCII format.
Then the converter transforms the IFC model into the LNS model and through the LNS Server saves it in the LNS database.
Users of Network Development Tool applying this basic information from the LNS database design the full model of a building automation network. The result is kept in the LNS database and loaded into the real network. All these operations are carried out using the LNS Server. After the network has been designed, the converter "LNS to IFC" receives the necessary information from the LNS database, transforms it into the IFC model and supplements the IFC file with the new information about a automation network through the IFC Toolbox.

5. INTERGATION IFC AND LNS MODELS

In the previous section the general integration scenarios also have been described.
The following example provides more detailed explanation of the integration process.

5.1 Integration the IFC model into the LNS model.

We shall assume that there is the IFC model of a building created by means of various CAD systems (Architect CAD, HVAC CAD, Electro CAD) which contains the information about the spatial building structure, objects and automation elements. This information is fundamental for the designing of a building automation network and should be integrated into the LNS model.
At the first stage, the information about the spatial building structure is extracted from the IFC model. On the basis of classes *IfcBuilding*, *IfcBuildingStorey* and *IfcSpace* the hierarchical structure is constructed on the basis of class *Subsystem* in the LNS model.
At the second stage, the information about automation objects and elements is extracted from the IFC model. For example, the controlled object *IfcFlowController* and its control loop elements *IfcSensor*, *IfcActuator* and *IfcController* (fig. 8) are found.

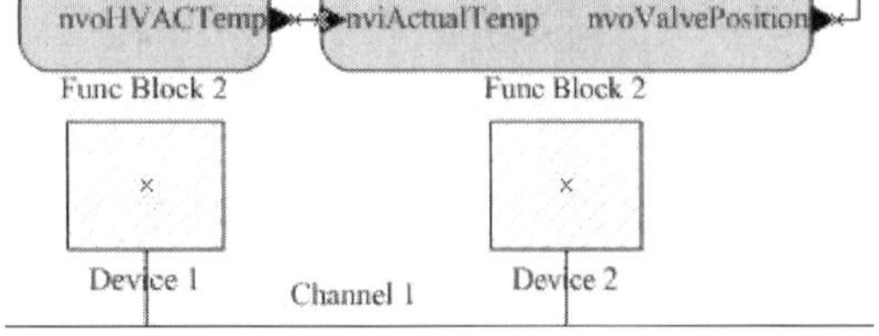

Fig. 11. Objects communications in the automation network.

On the basis of this information, standard functional blocks (Sensor, Actuator, Controller) can be chosen in the Network Development Tool and the software for them can be automatically generated using design patterns.
The data for the adjustment of functional blocks are parameters of control elements in the IFC model. Then connections between functional blocks are built and devices (*AppDevice*) are chosen which will contain functional blocks. The devices are connected to channels (*Channel*). After the integration of all control loops and addition of the new ones, we have the full model of a building automation network (fig. 11) which is stored in the LNS database.

5.2 Integration the LNS model into the IFC model.

At the third stage, when designing of a building automation network is completed, the information about it can be integrated into the IFC model. The example of the integration the LNS model into the IFC model follows below.

For the realization of these scripts, it is necessary to select IFC subtypes and relationships which would represent an essence of objects and their communications in LNS model.

Communications between objects of the automation network model are presented by two levels: physical and logical (Fig. 11). The IFC provides the mechanism of connectivity management between objects (Liebich 2004).

Physical layer of connections in the network. In LonWorks networks a physical connection exists between network nodes and data channels (e.g. cable, powerline, radio channel), and between network nodes and elements of control systems (sensors, actuators).
In the IFC model there are classes by means of which it is possible to realize the physical level of the connection in a building automation network.
Class *IfcDistributionFlowElement* (with subclasses *IfcFlowTerminal* and *IfcElectricApplianceType*) corresponds to the physical network node, class *IfcDistributionFlowElement* (with subclasses *IfcFlowSegment* and *IfcCableSegmentType*) - to the data channel, class *IfcDistributionControlElement* (with subclasses *IfcSensorType*, *IfcControllerType* and *IfcActuatorType*) – to the network variable and the element of control systems.
In the IFC it is possible to operate physical connectivity through ports which are presented by classes *IfcDistributionPort*, *IfcConnectsPort* and *IfcRelConnectsPortToElement*. Here the port has no physical representation and is a virtual object for other object communications (Fig. 12).
All physical objects of the model can be connected with *IfcDistributionPort* by means of the relationship *IfcConnectsPort*. *IfcDistributionPort*'s can be connected among themselves through *IfcRelConnectsPortToElement*. Presence of the attribute *IfcConnectsPort.RealizingElement* means that connection is realized by means of some other element.

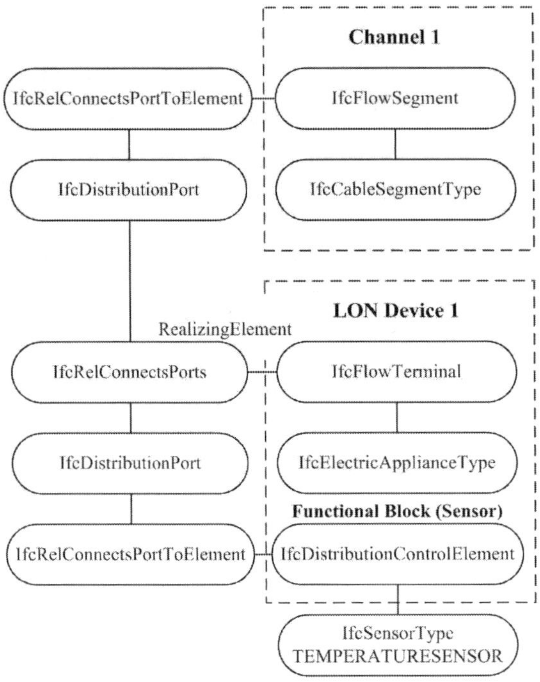

Fig. 12. Definition of physical connections with ports.

Thus elements of control systems are physically connected to the data channel through LonWorks network nodes.

Logical layer of connections in the network. In LonWorks networks a logical connection is present at the level of communications between network variables (*IfcDistributionControlElement*). In our IFC model of a building automation network connections between network variables are aggregated to connections between functional blocks.

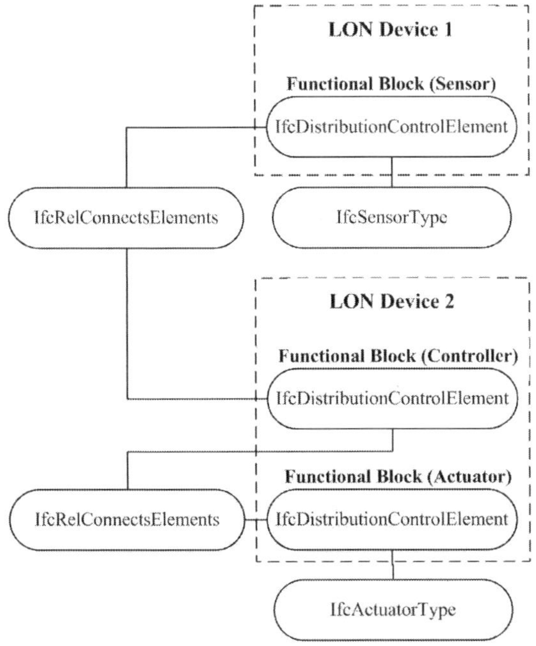

Fig. 13. Definition of logical connections.

At the logical connection it is possible to do without use of ports that can be reached by introduction of the relationship *IfcRelConnectsElements* which will connect directly *IfcDistributionControlElement* objects (Fig. 13).

Measurement of network variables. Network variables are used for the measurement of process conditions in a building and for the control of these processes. For this purpose, the object *IfcMesureWithUnit* can be attached to a network variable (object *IfcDistributionControlElement*) with the help of relationship *IfcRelAssignsToProcess*. Object *IfcMesureWithUnit* has connectivity with object *IfcValue* which contains the numerical value of a network variable (Fig. 14).

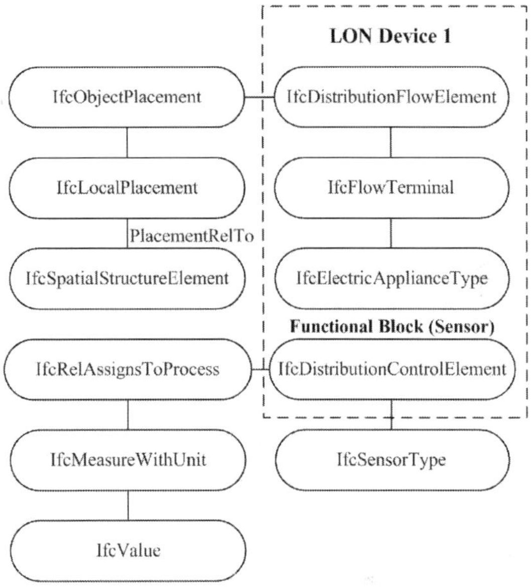

Fig. 14. Definition of network variable value and device placement.

LonWorks objects placement. All physical objects of an automation network (e.g. nodes, cable, sensor) have their certain placement in the building. They are defined by *IfcLocalPlacement* (a subtype of *IfcObjectPlacement*) which can determine an absolute placement, relative placement, or grid reference, with each defining the local coordinate system referenced by all geometric representations. The attribute *PlacementRelTo* of *IfcLocalPlacement* points to the *IfcSpatialStructureElement* (Storey, Space) (Fig. 14).

6. CONCLUSION

This paper has presented an approach for the integration of building automation network structures in construction projects. The integration concept is to merge the LNS model into the IFC model.

The question about a full list of data which are necessary for the integrating still remains open. For this purpose we try to cooperate with industrial organizations both in the field of automation networks and in the field of CAD designing of building services. In the given paper only the most necessary components of the integration, in authors' opinion, are shown.

It is necessary to define a reasonable border of the integration of the LNS model into the IFC model, to select the information which will be useful for other participants of the building life-cycle. In case of occurrence of some limitations of the IFC model, it is possible to expand it as it is the open standard.

The practical realization of the offered integration concept has already begun; data from the IFC model are integrated into the LNS model.

Development of the more detailed mechanism of the models converting using the scientific bases is planned to be made in the future.

REFERENCES

Boed, V. (1999). *Networking and Integration of Facilities Automation Systems.* Yale University, Connecticut, ISBN: 084930699X

Bozany, A. (2003). Integration of Building Automation Systems and Facility Information Systems. In: *Hungarian Electronic Journal of Sciences, HU ISSN 1418-7108.* http://heja.szif.hu/INF/INF-030530-A/inf030530a.pdf

Dietrich, D., D. Loy and H. Schweinzer. (2001). *Open Control Networks.* Kluwer Academic Publishers , Boston/Dordrecht/London.

Dietrich, D., W. Kastner and T. Sauter. (2000). *EIB Gebäudebussystem.* Hüthig, Heidelberg.

Echelon Corporation. (1999). *Introduction to the LonWorks System.* Echelon Corporation, Palo Alto.

Echelon Corporation. (2004). *LNS Programmer's Guide.* Echelon Corporation, Palo Alto.

Emmrich, F., E. Penetti and T. Lost. (2001). *Industrial Foundation Classes (IFC) for networks in buildings.* Final report. IIEF GmbH. (In German).

Hong, S. and W. Song, (2003). Study on the performance analysis of building automation network. In: *Industrial Electronics*, No. 6, 2003, pp. 359-364

IAI. (2005). International Alliance for Interoperability (IAI). http://www.iai-international.org/

ISO/PRE PAS 16739. (2005). IFC 2.x Platform specification.

Liebich, T. (2004). IFC 2x Edition 2 Model Implementation Guide. International Alliance for Interoperability.

Neuberg, F, E. Rank and C. Ekkerlein. (2002). Internet based simulation of the resource requirement of buildings. In: *European Conference of Product and Process Modeling – eWork and eBusiness in AEC; Proc. ECPPM,* Portoroz, 2002

Nisbet, N. and T. Liebich. (2005). ifcXML Implementation Guide. International Alliance for Interoperability.

O' Sullivan, D., M. Keane and D. Kelliher. (2002). Specification of a STEP Compliant Integrated Environment to Support Performance Based Assessment and Control of Building Energy Systems. In: *International Conference on Systems Simulation of Buildings SSB2002.* Liege.

Palensky, P. (2004). Requirements for the Next Generation of Building Networks. In: *Proceedings of the International Conference on Cybernetics and Information Technologies, Systems and Applications (ISAS CITSA 2004).* Orlando, Florida, pp. 225 - 230.

Salsbury, T. and R. Diamond. (2000). Performance Validation and Energy Analysis of HVAC Systems Using Simulation. In: *Energy and Buildings.* LBNL-43638, pp 5-17.

Schach, R., K. Kabitzsch and G. Knabe. (2002). Bridges between building automation and facility management. ISBN 3-86005-354-X: TU Dresden. (In German).

Copyright © Fieldbus Systems and Their Applications
Puebla, Mexico, 2005

ELSEVIER
IFAC
PUBLICATIONS

AN OPEN APPROACH TO EIB/KNX SOFTWARE DEVELOPMENT

Wolfgang Kastner, Georg Neugschwandtner, Martin Kögler

TU Wien, Institute of Automation, Automation Systems Group,
Treitlstraße 1, A-1040 Vienna, Austria,
{k,gn,mkoegler}@auto.tuwien.ac.at

Abstract: EIB/KNX is a field bus used in home and building automation. When building application programs for EIB/KNX nodes, one was hitherto faced with low-level constructs. To improve this situation, a RAD (Rapid Application Development) like approach was adopted. This model encapsulates the system software entities in a way which is inspired by the object-oriented paradigm. It also makes use of functional blocks to describe the application behaviour. To allow the roles of software developer and project engineer to be separated, the tool chain is designed for interfacing with an integration tool. The article discusses the work flow when building an EIB/KNX system and the resulting requirements on the tool chain. The GCC-based solution which was developed is presented. Specific challenges in porting the GNU tool chain to the standard microcontroller for EIB/KNX nodes are sketched. The implementation also includes an open PC-based EIB/KNX network access and management server. *Copyright © 2005 IFAC*

Keywords: Embedded Systems, Software Engineering, Fieldbus, Configuration, Programming Approaches

1. INTRODUCTION

The structure of EIB/KNX systems requires that node applications can be customized without low-level programming skills. A common tool software is available for this purpose. Still, the first-time creation of these applications remains a highly technically involved process. The goal of the present project is to apply an RAD approach to this step while respecting these particular work flow requirements.

1.1 EIB/KNX

The European Installation Bus *EIB* (Kastner *et al.*, 2005) is a home and building automation bus system. It is optimized for low-speed control applications like lighting and shading. A European standard since 1998, EIB now forms a subset of the KNX specification (Konnex Association, 2004).

EIB/KNX is specified over various physical media, including powerline, with twisted pair (TP) being the mainstay. Tunnelling EIB/KNX frames over IP networks is known under the term EIBnet/IP. EIBnet/IP provides a point-to-point mode, e. g. for remote configuration, as well as a multicast mode for coupling multiple EIB/KNX subnets via IP. These modes are – in a slightly confusing fashion – labelled Tunnelling and Routing, respectively.

TP and powerline EIB/KNX devices are frequently split into two parts. Generic modules providing bus connectivity (so-called Bus Coupling Units, BCUs) are combined with application specific hardware modules. The standard 10-pin connector between BCUs and these application modules is called the Physical External Interface (PEI). It can be operated in a variety of configurations covering parallel and serial digital I/O and analog input.

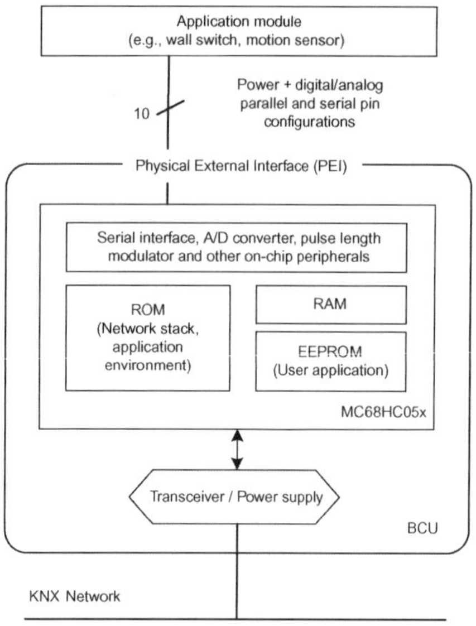

Fig. 1. EIB/KNX Bus Coupling Unit

BCUs are based on Freescale M68HC05 family microcontrollers, which are complemented with a transceiver and link power supply (cf. Fig. 1). They contain an implementation of the entire EIB/KNX network stack and application environment in their ROM. The application software customizing the BCU for a particular application module is stored in the EEPROM.

Currently, two major BCU variants referred to as BCU 1 and BCU 2 exist for the TP medium, plus another one for powerline. Their key differences lie in the system software implementation and the amount of available memory. While the BCU 1 is a standard M68HC05B6, the BCU 2 is a special variant with EIB/KNX-specific on-chip-peripherals.

For more complex user applications which use a separate microcontroller, BCUs offer high-level access to the network stack via the PEI, using it as an asynchronous serial interface. In such cases, however, the TP-UART IC often is a more attractive alternative. It only handles the physical and most of the data link layer, instead of the entire EIB/KNX protocol stack as BCUs do.

For Microsoft Windows based systems, a common high-level API ("Falcon") for accessing functionality throughout the network stack by way of various interfaces is available. For the Linux operating system, which is of ever-increasing importance for embedded platforms, kernel level drivers for BCU access as well as TP-UART based serial interfaces have been implemented (Kastner and Troger, 2003).

1.2 EIB/KNX software development and deployment

Obviously, writing all BCU application programs from scratch is typically not a feasible approach towards building an EIB/KNX system. Therefore, manufacturers provide ready-made applications matching their hardware. The behaviour of these applications can be customized by the project engineer by modifying manufacturer defined parameters. The final configuration is then downloaded to the BCU. The work flow of creating an EIB/KNX system is thus divided into three steps:

(1) *Development:* A *software developer* writes a BCU application program for a particular hardware configuration. He documents its behaviour and defines the parameters available to influence it. The application is brought into a format suitable for distribution to EIB/KNX project engineers. This format also includes the necessary meta information to allow a software tool to display the application parameters. Moreover, it provides this tool with the necessary knowledge on how to apply these changes to the program code.
(2) *Project planning:* A *project engineer* selects appropriate EIB/KNX devices to fulfil the requirements of a particular project. Using a (typically PC-based) integration tool, he makes the necessary adjustments to the application parameters of the chosen devices. He also sets up their communication relationships. While the software developer defines the behaviour (or set of possible behaviours) of one single node, the project engineer thus defines the behaviour of the entire system. This step is entirely off-line, i.e., no target devices are required yet.
(3) *Installation and download:* The BCUs are combined with the appropriate application modules (if not already delivered in a common housing by the manufacturer) and installed to their final location. This step is often carried out by a *site technician*. Before or after installation, the configuration is downloaded to the BCUs. This can be done via the network. Targets are identified via their *Individual address* (a configurable identifier which is unique within the installation) or via a special button on the BCU if they have not yet been assigned such an address.

The second and third step shall together be referred as *system integration*, the software tool which assists the work of the project engineer (and possibly site technician) as *integration tool*. For EIB/KNX systems, only one single integration tool software is necessary. This tool, called *ETS* (EIBA s.c.r.l., n.d.), handles every certified EIB/KNX device, no matter from which manufacturer. This solution significantly lowers the effort involved in setting up a multi-vendor system.

1.3 Project goals

In an effort to ensure that a project built with ETS will always work as desired, ETS will not accept any device application which has not passed compliance certification. This creates an issue for projects which want to explore new approaches to EIB/KNX node software development and deployment.

The goal of the project presented was to remedy this situation by creating an open-source framework for developing BCU applications. The framework should support the use of high-level languages, in particular C. It should make obtaining the desired behaviour as easy as possible, allowing to focus on the desired functionality rather than how to achieve it. Applications which are easier to understand are also less error prone. Still, the programming model should not be overly restricted in its expressiveness regarding access to the features of the BCU system software. The framework should support both BCU 1 (TP version) and BCU 2.

The separation of development and integration as outlined above was to be supported. For this purpose, a suitable, open interface between the development framework and an integration tool had to be defined. The interface should be generic enough to accommodate hardware platforms other than BCUs. The format should also describe the behaviour of the application, which we believe to be necessary for any advanced approach to configuration. Still, the RAD approach in the context of the BCU SDK was to focus on the development step. Although a suitable interface was desired to support automating the system configuration (project planning) process, the actual automation of this step itself was out of scope.

2. BCU SDK OVERVIEW

Fig. 2 illustrates the approach the BCU SDK (software development kit) developed during the present project takes towards implementing the work flow discussed in Section 1.2. The software developer uses the *preparation build script* to prepare the source code of the new application for distribution to project engineers. The build script controls the execution of the necessary tools. First, the input files are run through a custom preprocessor which implements the RAD programming model. An example program illustrating this model is presented in Section 3. The code is checked for errors and transformed into a format suitable for distribution, which will in the following be referred to as *program text*. Also, a description of the customizable aspects of the application (and its behaviour) is generated (*application meta data*).

The application meta data are collected in some form of *product data base*. From this collection, the project engineer retrieves the descriptions of the devices he has decided to use in a new project. The exact design of this database, which will also hold the matching program texts, and its interfaces are not relevant for the purposes of the BCU SDK. The project engineer may base his choice on external documentation (as it is the case with ETS now) or the integration tool may assist him using the information provided in the application meta data. Either way, he defines the customization options and binding data (communication relationships) for every device. The integration tool generates the appropriate interchange data format for these *configuration meta data*.

It does not, however, touch the contents of the program text. The integration tool is merely responsible for passing it to the finalization build script together with the associated configuration meta data. The finalization build script will then take all necessary actions to make the necessary changes to the program text. How this is done precisely is hidden from the integration tool, which therefore has no need to interpret the program text.

To match up program texts and their associated application meta data, the meta data description contains a reference attribute which uniquely identifies the associated program text. Since it is not limited in length, the current SDK implementation exploits the fact that the most unique identifier is the program text itself. It thus simply stores the (appropriately encoded) program text as the attribute value.

The *finalization build script* generates a final *binary image*, which is ready for download to the BCU. This image can be stored in a *project data base* together with the associated configuration meta data. Finally, the *loader* downloads this image to the device with the matching individual address. Again, the integration tool controls the loader (and the address assignment procedure), but is not required to have any knowledge of the format of the binary image. For these online steps, a *bus daemon* was implemented to provide the necessary EIB/KNX network access. It is presented in detail in Section 4.

The design of the project data base is again not relevant to the BCU SDK. It would equally be possible to have it store the program text and generate the final image on the fly before download. This would, for example, allow to change parameters later without the need to keep the product data base. However, since creating the final image is an offline step, it was included in the middle column of Fig. 2.

Although the data formats for application and configuration meta data have been fully defined, no matching integration tool is available yet. Therefore, the BCU SDK contains a stub implementation. It creates a configuration skeleton which has to be completed manually.

Actually, a developer will always need a special integration tool, since he should not have to manually reapply configuration information for every debug iteration of the program text. This is addressed in the current version by allowing to specify the program text to the finalization build script separately from the configuration meta data.

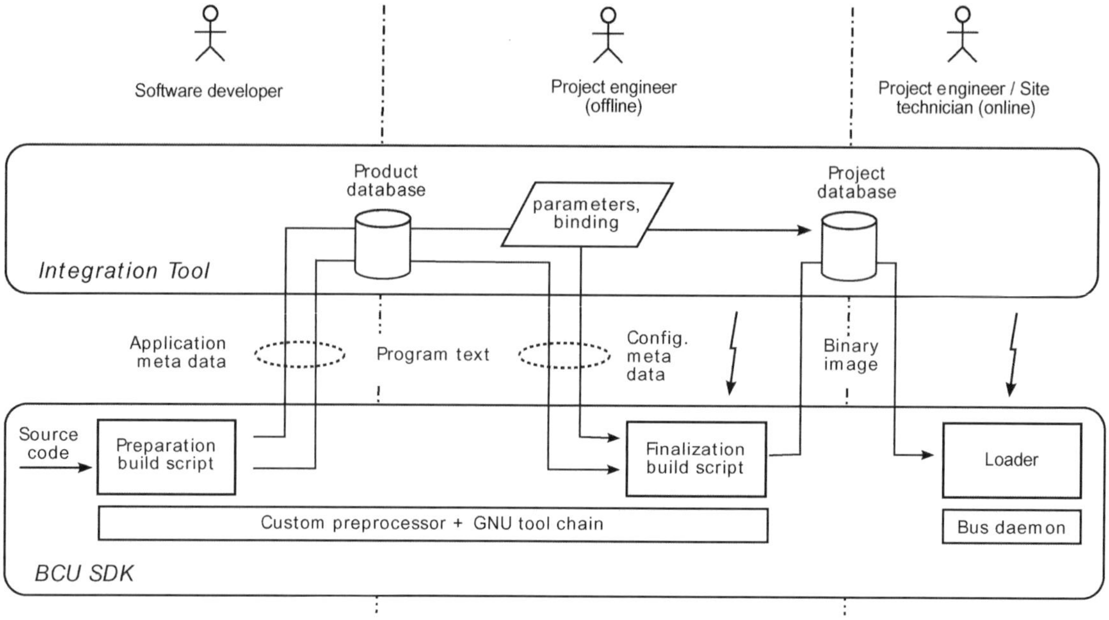

Fig. 2. Work flow and data flow using the BCU SDK

In the BCU SDK, the program text is not a binary image yet. Instead, it contains the preprocessed source code and mapping information between its identifiers and the ones used in the application meta data. The final compiler run is made only when all configuration settings are known to reduce image size.

Still, the preparation build script already builds a binary image for test purposes. This is done to check the source code for syntax errors. It also gives an upper limit on the code size, as no optimizations which could be made in response to certain parameter settings are applied.

This is in contrast to ETS, which modifies parameters and binding information in the executable binary. This approach supports only relatively minor modifications in response to a parameter change (e.g., changing a single byte). Nevertheless, such an approach would be possible using the BCU SDK as well. Experimental patching programs for the test image generated by the preparation build script were implemented.

Code generation relies on the GNU tool chain, which was ported to the Freescale/Motorola M68HC05 family microcontroller architecture. Key features of this port are presented in Section 5.

2.1 Meta data interchange formats

The application and configuration meta data formats are XML based. XML Schema definitions are provided. Fig. 3 shows an example application description (only selected nodes are shown). At the top level, it contains the program text identifier (which, as discussed, currently stores the program text itself). A global description block describes aspects such as the type of BCU the application is designed for. This is necessary since BCU 1 and BCU 2 support different system entities. Parameters can be numeric values (both integers and floating point values are supported), strings, or an enumeration of discrete choices. They carry descriptive text (including the engineering unit for numeric values) as well as tags constraining the range of valid choices.

The communication endpoints of EIB/KNX nodes relevant for process data exchange are referred to as *group objects*. A group object can be regarded as an application related variable exposing a particular aspect of the functionality of an EIB/KNX node. Group objects can be data sources, which provide information to other devices, or data sinks, which carry out certain actions according to the information received. XML nodes describing a group object include information about the supported data flow direction.

The project engineer forms network-wide shared variables by assigning group objects of multiple nodes to a common logical group. The values of all group objects belonging to such a group will be held consistent with the help of the BCU system software. As an example, to have a wall switch control a relay, the project engineer assigns the the group object representing the switch position to the same group as the one the relay node exposes as its control input. These groups are the *data points* (cf. (Kastner *et al.*, 2005)) of an EIB/KNX installation.

Obviously, group objects forming a group need to use a common encoding to ensure that the shared value will be interpreted in a consistent way. The KNX specification refers to these encodings as *data point types* (DPT). For example, DPT 1.001 stands for a Boolean value with the state labels "On" and

```xml
<?xml version="1.0" encoding="UTF-8"?>
<!DOCTYPE DeviceDesc SYSTEM [...]>
<DeviceDesc xmlns:xsi="http://[...]>
  <ProgramID>213C6172[...]</ProgramID>
  <Description>
    <MaskVersion>0012</MaskVersion>
    <Title>Simple Timer Switch</Title>
  </Description>
  <FunctionalBlock id="id7">
    <ProfileID>417</ProfileID>
    <Title>Light Switching
       Actuator Basic</Title>
    <Interface id="id5">
      <DPType>1.001</DPType>
      <Abbreviation>SOO</Abbreviation>
      <Reference idref="id2"/>
    </Interface>
    [...]
    <Interface id="id4">
      <DPType>7.005</DPType>
      <Abbreviation>TOD</Abbreviation>
      <Reference idref="id1"/>
    </Interface>
  </FunctionalBlock>
  <GroupObject id="id2">
    <Title>Switch Input</Title>
    <Receiving>true</Receiving>
    [...]
  </GroupObject>
  [...]
  <Parameter>
    <IntParameter id="id1">
      <Title>Time-on duration</Title>
      <Unit>seconds</Unit>
      <Default>180</Default>
      [...]
    </IntParameter>
  </Parameter>
</DeviceDesc>
```

Fig. 3. Application meta data format example

"Off". Yet, the expressiveness of DPTs with regard to describing processing rules (i.e., how the change of one data point affects others) is limited. For this reason, the functionality of the application is described by declaring one or more functional blocks. The KNX specification already defines several functional blocks, but they are not yet being applied to EIB/KNX. Every XML node describing a functional block contains an external reference to the precise definition of its behaviour. In the example, the Interface Object type of a functional block from the KNX specification was chosen as reference. However, this reference could as well be a URL pointing to a custom description.

Display rules can be defined for all entities depending on selected parameter values to allow revealing the complexity of a particular application to the project engineer step by step. The configuration meta data format looks similar. For every configurable entity from the application information, it contains the value selected by the project engineer. The relationship is established via the *id* attributes. This allows the easy translation of the descriptive texts contained in the application information. An example configuration description is shown in Fig. 4.

```xml
<?xml version="1.0"?>
<DeviceConfig xmlns:ns1="http://[...]>
  <ProgramID>213C6172[...]</ProgramID>
  <IndividualAddr>1.3.3</IndividualAddr>
  <GroupObject id="id3">
    <Priority>low</Priority>
    <SendAddress>1/1/2</SendAddress>
    <ReadAddress>
      <GroupAddr>1/1/2</GroupAddr>
    </ReadAddress>
  </GroupObject>
  <GroupObject id="id2">
    <Priority>low</Priority>
    <ReceiveAddress>
      <GroupAddr>1/1/1</GroupAddr>
    </ReceiveAddress>
  </GroupObject>
  <Parameter>
    <IntParameter id="id1">
      <Value>120</Value>
    </IntParameter>
  </Parameter>
</DeviceConfig>
```

Fig. 4. Configuration meta data format example

In EIB/KNX, group membership is managed using what is described in (Thomesse, 1999) as the *producer-consumer* mechanism. Thus, every node is provided with the information which logical groups each of its group objects belongs to. Groups are identified using *group addresses*. In the configuration information, this assignment is made separately for the different service types of transmitting and receiving a shared value update as well as requesting and receiving its current value. Group objects can also belong to multiple groups. In addition, a transmission priority is assigned to every data source group object.

The meta data interchange formats are designed to be open for EIB/KNX node architectures other than BCU 1 and BCU 2 as they are aligned with the wire protocol. They also support the EIB/KNX client-server communication scheme for node management (referred to as *interface objects* and *properties*) and TP1 polling groups.

3. PROGRAMMING MODEL

For interacting with the BCU system software, the BCU SDK offers an increased level of abstraction. Instead of accessing configuration data and library functions at specific memory addresses, the developer uses a simple specification language to define which BCU system entities he would like to use and how he would like to refer to them in his C code. The specification file is plain text. Its syntax is inspired by modern RAD environments, where objects are instantiated from a range of available classes and customized by changing properties and assigning event handlers.

An example is shown in Fig. 5. Here, a standard functional block from the KNX specification is implemented. The functional block describes a light switching actuator. Basically, it just closes and opens a relay *timer.config*

```
Device {
  BCU bcu12;
  PEIType 4;
  Title "Simple timer switch";
  include { "timer.c" };

  FunctionalBlock {
    Title "Light Switching
             Actuator Basic";
    ProfileID 417;

    Interface {
      Reference { IN };
      Abbreviation SOO;
        // Switch On Off
      DPType 1.001;  // DPT_Switch
    };

    Interface {
      Reference { OUT };
      Abbreviation IOO;
        // Info On Off
      DPType DPT_Switch; // 1.001
    };

    Interface {
      Reference { DELAY };
      Abbreviation TOD;
        // Timed On Duration
      DPType TimePeriodSec; // 7.005
    };
  };

  GroupObject {
    Title "Switch input";
    Name IN;  Type UINT1;
    StateBased false;
    on_update input_changed;
  };

  GroupObject {
    Title "Status output";
    Name OUT;  Type UINT1;
    Sending true;
    StateBased true;
  };

  Timer {
    Name TIMER1;
    Type EnhUserTimer;
    Resolution RES_4266ms;
    on_expire switch_off;
  };

  IntParameter {
    Title "Time-on duration";
    Name DELAY;
    Unit "seconds";
    Default 180;  Precision 4;
    MinValue 4;  MaxValue 544;
  };
};
```

timer.c

```
#define PEI3_ON 0x11
#define PEI3_OFF 0x10

void switch_on() {
  _U_ioAST(PEI3_ON);
  OUT = 1;
  OUT_transmit();
}

void switch_off() {
  _U_ioAST(PEI3_OFF);
  OUT = 0;
  OUT_transmit();
}

void input_changed() {
  if (IN == 1)
  {
    TIMER1_set
      (DELAY*10/43);
    switch_on();
  }
  else
  {
    TIMER1_cancel();
    switch_off();
  }
}
```

Fig. 5. Source code example

by switching a PEI output on and off in response to the change of state of a group object. Two optional functions are added. First, the actuator shall autonomously switch off after a certain duration (e.g., for use as stairway or hallway lighting). Second, a status output is added which reflects the current state of the relay.

The desired behaviour is as follows. The relay shall be closed when an "On" message is received on the `Switch input` group object. The length of this message is one bit. This length is assigned the type name UINT1 (Siemens AG, 1996; Siemens AG, 2004). It shall open again when an "Off" message is received on the same group object or after a specified timeout period. Whenever the state of the relay changes, its current state shall be transmitted via a second group object (`Status output`, again of type UINT1). The status output shall be readable from the network. Finally, the time-on duration shall be customizable via the integration tool.

The definition of the group objects in the RAD specification (*timer.conf*) is straightforward. The `Title` is what will appear in the respective meta data description XML tag. The group objects are assigned the internal names of `OUT` and `IN`. `OUT` has the `Sending` attribute set. This makes the function `OUT_transmit` available in the C code, which is used to transmit new values via the `SendAddress` associated with this group object. `OUT` is also `State Based`, which tells the BCU to answer read requests. Since `IN` contains no useful data for reading, its `StateBased` attribute is set to false. When a new value is received for `IN`, the `input_changed()` function will be called.

The time-on duration is specified as an integer parameter together with its useful range and unit. To autonomously switch off after a certain time interval, a timer is used. Specifying a `Timer` block sets up the necessary framework. It is of type `EnhUserTimer` (which provides additional functionality over the standard BCU user timers). One timer tick corresponds to 4.266 s. This resolution is reflected in the precision attribute of the `DELAY` parameter. When the timer expires, the handler `switch_off()` will be called.

The corresponding C code is located in *timer.c*. The group object values are available as global variables which will always contain the last value received for the associated group address. The UINT1 type is mapped to an unsigned 8 bit integer. Note the automatically generated functions for transmitting the current value of `OUT`, starting the timer with a specified timeout period, and canceling a runnning timer. The timer interval is scaled to its resolution (it must not exceed 127). The relay output is controlled using the BCU API function for accessing the PEI `_U_ioAST()`. C wrappers like this one are provided for all BCU API calls.

Back in *timer.config*, the target BCU is defined using its system software version (mask version) 1.2, which corresponds to a BCU 1. The desired PEI configuration type for this application is 4 (parallel I/O with 2 inputs and 3 outputs). The program contains a single functional block. The reference code (as discussed in the previous section) is taken out of the KNX specification, as are the interfaces of the functional block (i.e., their `Abbreviation` code and *DPT*). References are made to the appropriate system entities. The preparation build script will automatically transform this information into the application meta data format as shown in Fig. 3.

Other relevant BCU system functionality such as initialization and power-failure handlers and interface objects and their properties can be used in the same way. As far as the RAD specification is concerned, all differences between the BCU 1 and BCU 2 low-level APIs are hidden by the SDK.

In the C code, signed and unsigned integer types are available in any byte width from 1 to 8 to optimize RAM usage. As a special feature, transparent EEPROM access is conveniently available by declaring a variable with the appropriate storage class and attribute, e.g., `int x EEPROM_SECTION EEPROM_ATTRIB;`. Support for interface object parameters located in EEPROM is provided as well.

4. NETWORK ACCESS AND MANAGEMENT

For access to the EIB/KNX TP1 network, the BCU SDK includes a Unix daemon (called *eibd*, Fig. 6). It allows multiple clients to connect simultaneously via IP or Unix domain sockets. Clients can invoke various services of the EIB/KNX protocol stack. This includes sending and receiving unicast, multicast and broadcast telegrams. Also, eibd handles the protocol state machine for the client endpoint of a reliable connection. Based upon this, eibd can also autonomously execute various device and network management procedures, such as assigning individual addresses. Also, a bus monitor can be opened, which optionally can decode EIB frames. A "Raw" interface is available for clients that wish to implement the server endpoint for unicast communication, including reliable connections. This allows the host eibd resides on to be managed remotely via EIB/KNX management procedures.

The method of access to the EIB/KNX network is entirely hidden by the backends. The BCU 1 supports a protocol with RTS/CTS handshake ("PEI 16") which is best handled via a kernel driver. The FT1.2 based protocol of the BCU 2 ("PEI 10") can be implemented using the plain serial driver. Moreover, the respective backends have to adapt the message formats (EMI). For lab use, the TP-UART serial protocol is supported as well. EIBnet/IP Tunnelling and Routing can also be used for network access. Implementation of USB interfacing is currently being added.

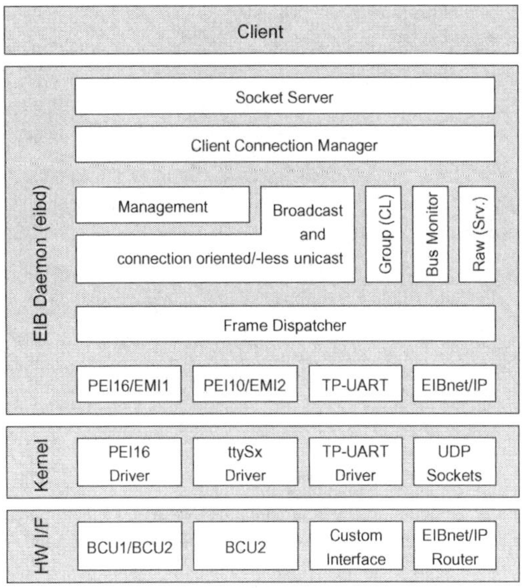

Fig. 6. Network access/management daemon

Higher-level tasks within eibd register with the frame dispatcher and state which frames (based upon addressing mode and destination address) they are prepared to process. To be able to serve multiple clients simultaneously, backends should deliver as many incoming frames as possible and leave filtering to the frame dispatcher.

In principle, this allows one client to maintain a point-to-point connection – where only frames from one single source are relevant – and another to operate in bus monitor mode. Yet, since bus monitor operation entails switching the hardware into a read-only mode, a special "best-effort" monitor mode was introduced which builds upon the fact that telegram filtering is not performed in the access hardware, but in the frame dispatcher. It forwards all frames the backend will provide in normal operation mode. For the BCU SDK, it will also act as a loader for writing applications into the BCU.

5. IMPLEMENTATION

To avoid duplicating work already done in various other places, it was decided to port an existing C compiler to the M68HC05 architecture. GCC (GNU Compiler Collection) was selected for its proven front end and optimizer. GCC is in wide-spread use as it is the standard compiler on most free operating systems. Its core parts are maintained by a large community. The GNU tool chain (Free Software Foundation, n.d.*b*) is supplemented by the GNU binutils (assembler and linker).

Several creative measures had to be taken to make GCC cope with the resource limitations of BCUs ($\ll$ 1 kb EEPROM, $\ll$ 100 bytes RAM). This constituted a key point of the porting activities. Full details are available in (Kögler, 2005). SDCC (SDCC

Project, n.d.) was considered as an alternative candidate. It supports different architectures and implements some optimizations. Compared to GCC, SDCC is much simpler and has less features. If SDCC had been the first choice, the structure would have been much simpler. Yet, some parts, like the automatic removal of unused static variables, would have had to be implemented from scratch.

The port can be used for any member of the M68HC05 microcontroller family. For automated regression tests, a CPU core simulator for the M68HC05 architecture was developed. It is based on its counterpart in the M68HC11 port of the GNU tool chain (Free Software Foundation, n.d.*a*). It is also accessible via a limited interactive GNU debugger frontend.

Since GCC is designed to work with a considerably different hardware architecture, certain missing features need to be emulated. This includes providing a larger number of general purpose registers (using RAM), multiplication and division operations, a data stack, and pointers which can cover the entire address space. The GCC floating point simulator for single and double precision values is available. However, many of its functions need too much memory, which causes them to fail on a BCU. Some functions even need a larger stack than the M68HC05 GCC port supports. The tight memory constraints also affect the expressiveness of the standard regression test suite. Approximately five per cent of the test cases yield unexpected outcomes, but we believe these cases to be unsuitable for the target. Due to the huge overall number of test cases, a detailed assessment was not possible yet.

The port had to make special provisions to allow GCC to automatically distribute variables over the non contiguous RAM segments of a BCU 2. Additionally, transparent access to the EEPROM was added. Special effort was directed towards obtaining small sized output. Still, the compiler output will in most cases be larger than well optimized, hand written assembler code. For the exceptionally resource-constrained BCU 1 environment, this considerably limits the amount of functionality which can be realized.

The interplay of an assembler-level operating system API and an optimizing compiler can bring about interesting effects which prove to be a challenge when implementing the compiler. For example, the BCU API provides optimized functions for multiplication, division, shift and bit operations. Yet, their use may in certain cases prevent M68HC05-GCC from applying optimizations and thus actually result in larger code.

6. CONCLUSION

A set of software development tools for EIB/KNX nodes was presented. It adopts a RAD like programming model, which we believe to be a novel approach for field devices. The separation of development and deployment is supported via an integration tool interface. Its implementation involved porting GCC to the MC68HC05 microcontroller family and the development of an API for EIB/KNX access from PC-based programs. All relevant parts of the SDK are fully implemented, albeit partially still in the testing phase. The entire SDK is placed under the GPL (GNU General Public License) and can be downloaded together with further documentation from `http://www.auto.tuwien.ac.at/projects/hba/`.

Numerous improvements are still possible. Specifically, target specific optimizations within GCC provide ample possibilities for further work. Also, the handling of the build process should be improved. This includes adding a graphical interface (e.g., using Eclipse or KDevelop) for the developer. For system integration, we intend to enhance the open BAsys tool (BASys Project, n.d.) to enable interaction with the BCU SDK.

REFERENCES

BASys Project (n.d.). BASys 2003. `http://www.basys2003.org`.

EIBA s.c.r.l. (n.d.). The ETS 3 Software Family. `http://www.eiba.com/ets`.

Free EibIDE Project (n.d.). Free EibIDE for Linux. `http://sourceforge.net/projects/freeeibide/`.

Free Software Foundation (n.d.*a*). GNU development chain for 68HC11 & 68HC12. `http://www.gnu.org/software/m68hc11/`.

Free Software Foundation (n.d.*b*). GNU tool chain documentation. Distributed with the respective program sources, also available from `http://ftp.gnu.org/gnu/`.

Kastner, Wolfgang and Christian Troger (2003). Interfacing with the EIB/KNX: A RTLinux device driver for the TPUART. In: *5th IFAC Intl. Conf. on Fieldbus Syst. and Appl. (FeT'2003)*.

Kastner, Wolfgang, Georg Neugschwandtner, Stefan Soucek and H. Michael Newman (2005). Communication systems for building automation and control. *Proceedings of the IEEE* **93**(6), 1178–1203.

Kögler, Martin (2005). Free development environment for Bus Coupling Units of the European Installation Bus. Master's thesis. TU Wien.

Konnex Association (2004). *KNX Specifications, Ver. 1.1*.

SDCC Project (n.d.). SDCC – Small device C compiler. `http://sdcc.sourceforge.net/`.

Siemens AG (1996). *BCU 1 Online Help*.

Siemens AG (2004). *BCU 2 Online Help, Version 1.1*.

Thomesse, Jean Pierre (1999). Fieldbuses and interoperability. *Control Engineering Practice* **7**(1), 81–94.

ELSEVIER
IFAC
PUBLICATIONS

PASSIVE MONITORING OF CONTROL LOOPS IN BUILDING AUTOMATION

Volodymyr Vasyutynskyy, Joern Ploennigs, Klaus Kabitzsch

Faculty of Computer Science
Dresden University of Technology,
D-01062 Dresden, Germany
Fax: ++49 351 463 38460
E-mail: {vv3, jp14, kk10}@inf.tu-dresden.de

Abstract: The monitoring of control loop performance in building automation is important for decreasing the maintenance costs. Passive monitoring with network analyzers is inexpensive and does not influence the network load. However, control signals that are not transmitted over the network are unobservable. This paper purposes an approach to the reconstruction of missing control signals and the computation of control loop performance parameters. The analysis algorithms are adapted to the non-uniform sampling that is widely spread in building automation. Further, criteria of loop performance and network traffic for passive monitoring are introduced.

Keywords: control loops, performance monitoring, network analyzers, signal reconstruction

1. INTRODUCTION

Building automation systems consist of up to several thousand heterogeneous, intelligent nodes connected by networks. They control not only heating, ventilation and air conditioning (HVAC) but also lighting, fire protection and access control. Precise models of the plants are usually unknown during the design of a building and cannot be used for fine-tuning of the control parameters. Instead of that, the developers use default values they know from experience as universal. This may lead to poor control loop performance, with the following disadvantages (Salsbury and Diamond, 2001):

- User discomfort: deviations from setpoint, noticeable oscillations, long reaction times.
- Higher energy consumption due to poor setpoint tracing.
- Increased wear of control elements because of more control actions, damaged control elements in consequence of wrong control actions.

All this results in increasing the maintenance costs. The effect is rather small for a single control loop, but the large number of controls in a building could lead to significant economies by their optimization. The first step in control loop optimization is the monitoring and analysis of *control loop performance* (CLP) (Dexter and Pakanen, 2001), which aims to detect the badly tuned loops. In this step the plant models are usually not available. CLP-monitoring is followed by the plant model identification for the detected loops, based on the monitoring information. Finally, the controller can be retuned or replaced with another type.

CLP-monitoring can be realized either with distributed or with centralized monitoring. The *distributed* approach requires the implementation of monitoring functions directly in the different automation nodes, which allows effective data preprocessing and fast, autonomous reaction. However, the distributed approach is limited in building automation to the simple diagnostic functions provided by the device manufacturers, since devices are usually not open for reprogramming. This limits the possibilities of a system wide optimization.

The *centralized* approach relies on one powerful monitor to observe a whole network segment. The monitor can either collect information by polling messages actively or listen to the network traffic passively. *Active monitoring* increases the network load and, therefore, may provoke instabilities (Soucek and Sauter, 2004), but it is able to request information not intended by the developer to be transmitted over the network. *Passive monitoring* does not load the network additionally and will not interfere with the network load, and, consequently, with message transmission time and process stability. Therefore, it is predestined for loop observation.

Two effects have to be considered concerning the passive monitoring. At first, some signals cannot be observed directly, because only messages transmitted over the network can be acquired. Secondly, the transmitted messages are influenced by the network itself, i.e. they are delayed or can even get lost. As a consequence, control signals are unobservable or distorted. Moreover, non-uniform sampling is widely spread in building automation, either intended by the developer to reduce network load or caused by message delay jitter. However, usual CLP-methods presume the uniform sampling, see (Qin, 1998) and need to be adapted.

These issues are considered in the proposed CLP-analysis on the basis of passive monitoring. Common control loop performance and network traffic criteria are introduced in Section 2 and opposed to the special conditions of monitoring in building automation. The reconstruction of unobservable signals is demonstrated in Section 3. In Section 4 important CLP criteria are adapted to the introduced special conditions. Further, the quality of control in networked loops is influenced by the network traffic (Lian, 2001). To estimate the coherences between these two issues, it is necessary to know the traffic parameters. The possibilities of passive monitoring in this task are investigated in Section 5.

2. MONITORING OF CONTROL LOOPS IN BUILDING AUTOMATION

2.1. Observed parameters

The main objectives of the monitoring are to identify the unacceptably tuned control loops. Afterwards the measurements are gathered needed for identification of the control loop models, which is necessary for later tuning. The first objective is achieved by control loop performance analysis. Salsbury (1999) proposed a quite simple method for building automation that exploits the integral absolute error (IAE) between the plant response and the setpoint. Also the CLP-parameters used in industrial applications can be adapted to building automation systems. In general, the *control loop performance parameters* are divided in deterministic and stochastic ones (Qin, 1998):

- *Deterministic* parameters depict the control loop behavior by changes of the setpoint. Those are the times characterizing the step response (rise time t_r, settling time t_s, time delay t_d) and stability parameters (overshoot, excessive oscillations in loops).
- *Stochastic* parameters characterize the variance of the control loops in steady state (Harris, *et al*, 1999) and represent the stability of setpoint tracking. For networked control systems, they also include the message transmission rates and number of control actions.

Some of these parameters are depicted in Figure 1. They are assessed to identify the badly tuned control loops (e.g. if they exceed required limit) and the plant models (e.g. on the basis of the rise time t_r and settling time t_s as demonstrated in (Swanda and Seborg, 1999; O'Dwyer, 1999)).

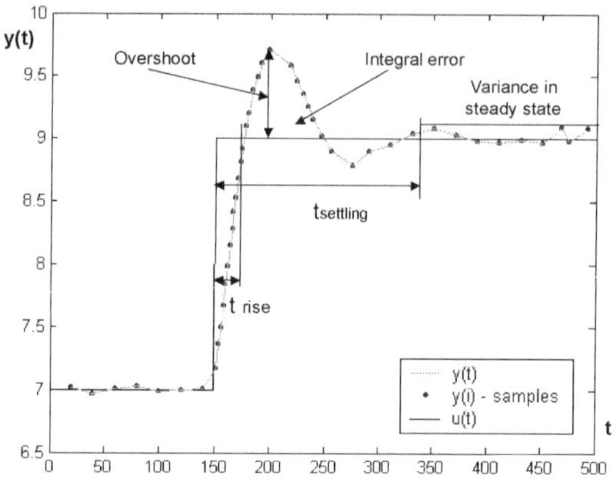

Fig. 1. Some control loop performance parameters.

2.2. Networked control loop simulation

For purpose of testing of proposed methods the simulator of networked control loops was developed. It implements the p-persistent CSMA/CD LonTalk protocol (LonTalk, 1994). The plants are presented in the simulator as a linear time-invariant system in state space with

$$x(k+1) = \mathbf{A}\,x(k) + \mathbf{B}\,u(k),$$
$$y(k) = \mathbf{C}\,x(k) + \mathbf{D}\,u(k) + \xi(k). \qquad (1)$$

A, **B**, **C**, **D** are constant matrixes using parameters from the lab setup of a room automation system described in (Dementjev and Kabitzsch, 2004) and $\xi(k)$ is a white noise. Such a simple and static plant behavior is not typical for building automation. Nevertheless, many processes change slowly and the complex plant models can be linearized in limited time intervals. Further, the control loops are controlled by PID controllers, which are widespread in building automation. It is assumed that the PID controller parameters are known (e.g. from the design databases (Ploennigs, *et al*, 2004)). The parameters of the plants and controllers have been varied during simulations experiments to get the different cases of loop behavior, e.g. slow, quick, oscillating or noised loops.

2.3. Passive monitoring of networked loops

The passive monitoring is carried out by network analyzers, also called protocol analyzers or sniffers. They are connected to the fieldbus as shown in Figure 2 and observe online all messages passing by as an event stream. The obtained events can be interpreted in a time series of signal values according to rules formulated with help of monitoring data model purposed in (Vasyutynskyy and Kabitzsch, 2004). This enables to combine simple system events, time intervals and hierarchical sets of events (complex events) in one structure. The rules can be automatically created from the design databases (Ploennigs, *et al*, 2005). The observation can be provided either constantly with stationary analyzers, e.g. in combination with other maintenance services (Dementjev and Kabitzsch, 2004)); or at regular intervals, e.g. during maintenance checks.

The different steps of loop optimization imply multiple levels of complexity of the monitoring functions. In the simplest level of CLP-monitoring it should be distinguished between well-performing and badly tuned control loops. The control loops are only observed for performance problems, without precise knowing of plant models. Only for detected non-optimal controls the sophisticated and resource-consuming analysis functions try to identify the plant models for tuning. The different monitoring levels allow observing and tuning of all control loops connected to one channel simultaneously. This paper focuses only the first level to facilitate a comprehensive description.

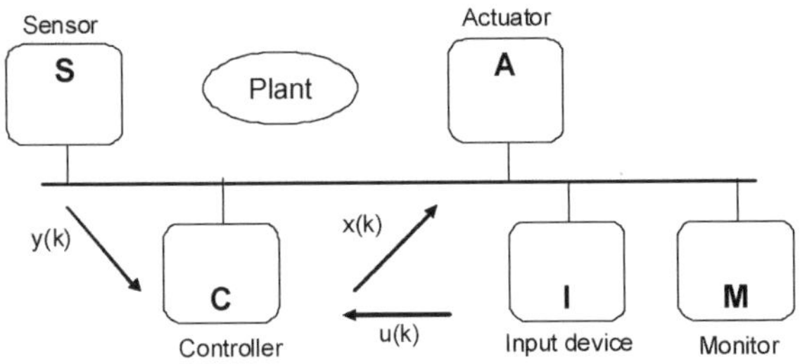

Fig. 2. Nodes and variables of a networked control loop.

The control variables needed for CLP-monitoring can be observed without problems by network analyzers as long as their functionality is completely separated as depicted in Figure 2. But multiple functions are usually combined in one device and their control signals are unobservable. Further, the control signals may be sampled non-uniformly or distorted by communication delays. Both obstacles of passive monitoring of control loops are explained in the next two sections.

2.4. Unobservable signals

In general, a control loop consists of four basic elements: a sensor (S), a controller (C), an actuator (A) and an input unit (I), where user can enter the desired setpoint. The transmitted variables are the plant input x, the plant output y and the setpoint (SP) u. Depending on the implementation several functions may be integrated in a single device, e.g. when the input unit is combined with the controller. In this case the setpoint is not transmitted over the network and is unobservable for the network analyzer. Table 1 lists the most common combinations of devices along with the observable signals.

Table 1 Node configurations

No.	Node configuration	Observable	u known?
1	S, C, A, I	u, x, y	known
2	S, C-A, I	u, y	known
3	S, C-I, A	x, y	recoverable
4	S, C-A-I	y	recoverable
5	S-C-I, A	x	partly recoverable
6	S-C-I-A	-	unknown

The last column indicates whether the setpoint signal is observable or can be reconstructed with the later introduced algorithm. The notation "C-A" means that the controller and actuator are implemented in one node.

The combination of functions in one device reduces hardware costs and accelerates the communication, as the communication delays between both nodes drop out. However, it reduces the observability of control loops. In the cases 1 and 2 the setpoint signal is still known and allows applying the CLP algorithms immediately. But, for the cases 3 and 4 the setpoint signal needs to be reconstructed before the analysis algorithms can be applied. The Section 3 will develop such a reconstruction algorithm.

2.5. Node communication and signal sampling

Although the most processes in building automation are rather slow, high network utilization may cause serious message delays, jitters or losses, due to a large number of communicating devices. This leads to such effects like performance degradation or instability of the control loops (Lian, 2001; Koller, *et al*, 2003). These problems aggravate with the increasing popularity of wireless sensor networks and power line transmission, because of their smaller bandwidths and higher error rates.

This influences also the CLP analysis, since the signals are in general non-uniformly sampled, that may be caused either by communication delay jitter or non-periodic sampling. In the first case, the time stamp of message record at the monitor is different from the sending time of the device, which is again different from the event time in the process. The times are shifted and jittered by all delay times during processing and transmission. As a consequence, the arrival time of messages is stochastically jittered. Even though a constant time shift can be omitted in the signal reconstruction, the jitter cannot and needs to be considered in the algorithm.

Non-periodic sampling is intended to prevent the network problems by reducing the number of the transmitted messages to the minimum necessary for required control quality. The non-periodic *send-on-delta*, also called deadband, sampling (Otanez, *et al*, 2002) is broadly used in building automation. Using it, a message is transmitted only if the measured signal has changed more than a significant value Δ since the last transmitted value, i.e.

$$\left|y_{actual} - y_{sent}\right| > \Delta \quad (2)$$

Though other types of non-uniform sampling have been proposed (Yook, *et al*, 2000; Rehbinder and Sanfridson, 2004), the send-on-delta sampling is the most widespread. Actually, it is accepted in LONWorks standard (LON, 1994). Figure 3 compares the uniform and send-on-delta sampling for a step change of the setpoint.

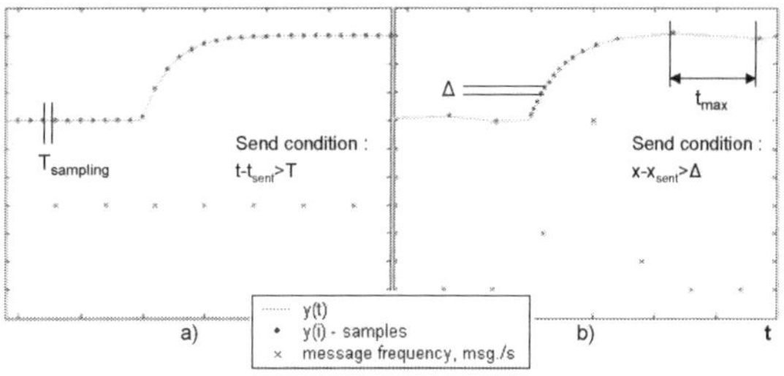

Fig. 3. Step response of the control loops with uniform (a) and send-on-delta (b) sampling with the same IAE.

The send-on-delta sampling results in a message silence in periods without process changes. To avoid uncertainty in this case, the parameter max-send-time t_{max} defines the maximum time between two messages. The allowed minimum time between subsequent messages can be limited by the min-send-time t_{min}.

Send-on-delta sampling is able to reduce the network load, but it also may cause instabilities in case of an error-prone connection because of the lack of the correcting messages. Further, incorrectly adjusted parameters Δ, t_{max} or t_{min} can cause aliasing and quantization oscillations near the setpoint, which not only impact the control loop performance but also increase the network load.

To conclude, control signals in a networked loop are distorted by non-uniform sampling, stochastic message delay jitter and sensor noise. This requires an appropriate adjustment of the CLP-algorithms.

3. DETECTION OF SETPOINT CHANGES

3.1. Detection of setpoint change

The knowledge of the setpoint signal is essential for the CLP analysis. A missing setpoint signal can be reconstructed from the process output in steady state for the cases 3 and 4 in Table 1 with the following assumptions. (A1) The setpoint needs to be changed in a step. (A2) The changes have to be infrequent with an event period significant larger than the reaction time of the control cycle. (A3) The jitter of the monitoring is negligible in comparison to the settling time. Then, the process variable mean value μ in steady state corresponds to the setpoint value u. This is quite simple, but it is necessary to identify the steady state after a step, as well as the time moment of the step change. This is possible with the algorithm introduced in subsection 3.3, but first the sampled signal needs to be reconstructed.

3.2. Signal interpolation

The reconstruction of the original signal from the observed non-uniform and jittered samples is performed by interpolation. This is reasonable, as complex reconstruction algorithms require plant models, which are unknown in this step. The Shannon reconstruction presumes periodical samples. Therefore, four simple interpolation algorithms were compared:

- Hold-on of zero order (ZOH), where the value of signal is hold until the next sample comes;
- Linear interpolation (first-order hold-on), where two samples are connected with a direct line;
- Spline interpolation with different spline types, particularly B-splines and NURBS, and
- Polynom interpolation of different grades.

The methods of interpolation were compared on a set of simulated signals obtained from control loops with varied parameters like plant model parameters, PID-settings and noise variance. The methods were compared according to two parameters. The first one is the absolute interpolation error, defined as sum of the absolute differences between real and interpolated signal. The second parameter is the accuracy of the process time estimation, defined as difference between the real process times and times defined with the help of interpolation. The results of the comparison are listed in Table 2.

Table 2 Comparison of interpolation methods

Method	Absolute interpolation error	Accuracy of process times	Computa-tional efforts
stand-on	0.17-0.34	0.2-0.3	low
linear	0.14-0.29	0.1-0.2	low
spline	0.27-1.5	0.1-0.2	medium
polynomial	0.29-1.45	0.1-0.2	high

Hold-on interpolation is the simplest one but results in a large interpolation error and low accuracy of the estimated process times. The spline and polynomial interpolations are too sensitive to the grade of interpolation function, signal continuity and noise variance and require significant computational efforts. The linear interpolation is the most applicable one as it combines simplicity with the small interpolation error for a large class of signals. The approximative formulae for the iterative calculation of signal mean μ_i and variance σ_i with linear interpolation are:

$$\mu_i = \frac{(t_i - t_{i-1})(x_i + x_{i-1})/2 + (t_{i-1} - t_1)\mu_{i-1}}{t_i - t_1} \tag{3}$$

$$\sigma_i^2 = \frac{|(t_i - t_{i-1})(x_i + x_{i-1} - 2\mu_i)/2| + (t_{i-1} - t_1)\sigma_{i-1}^2}{t_i - t_1} \tag{4}$$

3.3. Detection of the time parameters

Basseville and Nikiforov (1993) compared several statistical jump detection algorithms in their precision to detect the change time t_c, which is the time moment when u changes. The two-sided CUSUM algorithm was chosen out of these methods, because it combines robust change detection with modest computational efforts.

Assume, the signal mean value μ_0 before the jump is known from the previous signal observations and the minimal detectable change υ is defined. Then the jump time is defined as

$$t_c = \min\{i : (g_i^+ \geq \bar{h}) \cup (g_i^- \geq \bar{h})\}, \tag{5}$$

which is the first moment when any decision function g_i^+, g_i^- exceeds the limit h. The upper and lower decision functions are given respectively for increasing and decreasing process value as:

$$g_i^+ = \left(g_{i-1}^+ + y_i - \mu_0 - \frac{\upsilon}{2} \right)^+ ,$$
$$g_i^- = \left(g_{i-1}^- - y_i + \mu_0 - \frac{\upsilon}{2} \right)^- . \quad (6)$$

The detectable change can be selected as $\upsilon=\Delta$ for send-on-delta sampling and $\upsilon=\sigma$ (variance) for uniform sampling. The detection limit is chosen as $h=0.2\text{-}0.3\cdot\sigma$.

To define the new signal mean and the transition time, the new process steady state must be detected. The transition process is not abrupt and may take a long or a short time, according to controller settings and plant time constants. A known steady state detection criterion (Cao and Rhinehart, 1995) uses the filtered variances (mean-square deviations) of the original samples and the filtered ones. As it could be stated on the real process and simulated data, this approach is very sensitive to loop time constants, noise variance, controller settings and the parameter Δ in send-on-delta. The iterative calculation of criterion parameters as proposed by (Bhat, *et al*, 2003) requires too large computational efforts.

The proposed criterion uses a growing window of the size k which starts at the change detection time t_c. The mean values μ_1, μ_2 and mean squared deviations σ_1, σ_2 are calculated in the first and second halves of the window. The intuition of the criterion is that the mean value and deviation are changing intensively within the transition period. The difference between these values in two periods is calculated as

$$d_k = \left| \frac{\mu_2 - \mu_1}{\mu_2 - \mu_0} \cdot \frac{\sigma_2}{\sigma_0} \right| \quad (7)$$

If $d_k < 1$ then the transition end time t_s is detected, which is adopted as the settling time. The time period between t_c and t_s corresponds to change state, the time period from t_s till the next t_c corresponds to steady state. The signals and the criterion are shown in Figure 4.

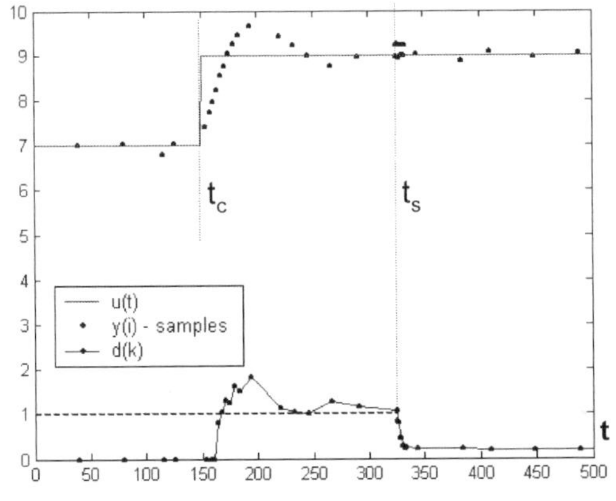

Fig. 4. Example of sensor values, change detection criterion d and detected change and settling times.

After detection of the transition end time the algorithm is reset. The newly calculated values of μ_2 and σ_2 are used as μ_0 and σ_0 correspondingly to detect the next change with the same algorithm.

Large disturbances can be misinterpreted as step changes with the purposed algorithm, which need to be removed. This is done by comparison of the reconstructed setpoints before and after the change. If the setpoint has not changed significantly, e.g. the detected step amplitude $\Delta_{SP}=|\mu_2\text{-}\mu_0|$ is $\Delta_{SP}<\sigma$, a disturbance is detected. Also, if several step changes occur in succession, then these steps will be recognized as disturbance corresponding to the assumption A2.

The application of the algorithm to the simulated data have shown its robustness to noise as well as various plant and controller parameters. It brings better results in a broad class of signals and slightly worse results in case of fine tuned parameters in the method of Cao and Rhinehart (1995).

However, the approach has its limitations. If only the controller output x is observable (case 5 in Table 1), the approach is limited to detect the change time t_c and settling time t_s correctly. The computed setpoint signal u will contain a constant error, as the amplitude of the setpoint jump is not recoverable if the plant model is unknown. A constant error is also obtained, if the control algorithm implies a constant control error, as it is a case by P-controllers applied to P-plants.

The reconstruction of the setpoint is a coarse estimation and the results may be inaccurate. To receive more precise plant models active monitoring can be used in uncertain situations to generate test setpoint signals unrecognizable for the user.

4. CONTROL LOOP PERFORMANCE PARAMETERS

The calculation of CLP-parameter adapted to the non-uniform sampling will be demonstrated in this section on some examples. The most general loop performance indicator is the integral absolute error (IAE), which is defined as the integral of the absolute difference between setpoint and actual process variable:

$$\text{IAE} = \int_{t_0}^{t_1} |\,(y(t) - u(t))\,|\, dt \quad (8)$$

In case of the non-uniform sampling this integral is calculated approximate via linear interpolation as:

$$IAE_i = \frac{(t_i - t_{i-1}) \cdot |e_i + e_{i-1}| / 2 + (t_{i-1} - t_1) \cdot IAE_{i-1}}{t_i - t_1} \quad (9)$$

where $x_i=y_i\text{-}u_a$, and u_a is the last value of setpoint, or, if unknown, $u_a=\mu_0$. It should be mentioned that the IAE needs to be distinguished between steady state and settling time. In the first case, the IAE is determined by stochastic causes like noise and, in the second case, by deterministic control loop characteristics.

The deterministic parameters of a loop are the rise time, settling time, peak time and the overshoot (see Figure 1). These times are used to evaluate the loop performance and perform the controller tuning. Since the signals are not continuous, the time values are also estimated with the help of linear interpolation. For example, the rise time is defined as the time when the signal reaches $x(t_{\mathrm{rise}}) = 0.9 \cdot \Delta_{SP}$ at first time. The algorithm starts with t_c and proceeds until the samples (x_0, t_0) and (x_1, t_1) fulfill the condition $x_0 < 0.9\ \Delta_{SP} < x_1$. The rise time is then interpolated by

$$t_{rise} = t_o + \frac{0.9\Delta_{SP} - x_0}{x_2 - x_1}(t_1 - t_0) \qquad (10)$$

Further useful process times like $t_{0.1}$, $t_{0.283}$ or $t_{0.632}$ (O'Dwyer, 1999) are defined in analogous way. These times can be used for the identification of plant models, if the changes are well defined. To receive reliable time values the amplitude of change must be high enough, e.g. $\Delta_{SP} > 2\text{-}3 \cdot \sigma$. The process times are compared with the ergonomic requirements. For example, the user does not want the room temperature to change slower than $v_{\mathrm{change}} = 1°\mathrm{C}/t_{1°\mathrm{C}}$, then the rise time needs to be $t_{\mathrm{rise}} < v_{\mathrm{change}} \cdot \Delta_{SP}$.

The overshoot characterizes the aggressiveness of the loop tuning. The overshooting index is defined as

$$\eta_{overshoot} = \frac{\max_{c<i<s} (x_i - \mu_2)^2}{\sigma_2^2}, \qquad (11)$$

where c and s are samples corresponding to change and settling time. The criterion compares the overshoot amplitude with the noise variance to ensure that the overshoot was not caused by noise. The optimal overshooting index is about 1.1, while larger values indicate aggressive tuning. Alternatively, the overshoot can be estimated by comparing with the step amplitude Δ_{SP}.

The variance in steady state is calculated accordingly to (4). In contrary to industry controls (Harris, *et al*, 1999) the quality of control in building automation does not influence the product quality like in chemical plants, where minimum variance approach may be used. However, the loop behavior in steady state influences the message rates in the network and the number of control actions. In case of the send-on-delta sampling a large variance $\sigma >> 2..3 \cdot \Delta$ indicates that the send-on-delta parameter Δ should be increased to decrease the amount of sent messages. The impact of variance on message transmission rates is discussed in Section 5.

The oscillation index is used to differ between noise and loop oscillations (Forsman and Stattin, 1999). The approach allows to identify the oscillations on the basis of the above defined integrated absolute error. The advantage of this approach is that it allows detection of oscillations with diffused limits, since oscillations in real data are not strongly periodical. In our experiments a number of successive oscillations switching detector of N=3 proved good results. If the oscillations are detected after the setpoint change, they are commonly caused by aggressive tuned controllers. If they are detected in steady state, they may be caused by process instability or false selection of the delta parameter Δ. The frequency and amplitude of oscillations can be derived from this criterion.

5. NETWORK TRAFFIC PARAMETERS

The observation of network traffic allows to evaluate the influence of network delays on the control loops and, on the other hand, the influence of the control loop sampling rates on the network (Soucek and Sauter, 2004). However, the message delays cannot be acquired directly since the synchronized sending times stamps are commonly not available in the messages. Though the delay can be estimated by performance analysis (Ploennigs, *et al*, 2005), those forecasts should be validated by observations. One possibility is to use special test messages to determine the network characteristics. However, as active requests interfere with the network load they should be used cautiously giving preference to the passive monitoring.

The *overall* network load is depicted by statistics of the time intervals between subsequent messages, the so-called message inter-arrival time t_{ia}. The sending behavior of *single* nodes is monitored in the same way, while the control loops of the most active nodes are observed in more detailed way. The following network parameters can be observed for single nodes:

- For uniform sampling the message delay jitter σ_{delay} can be determined from the variance of the inter-arrival time σ_{ia}. Since a process with sample time T can be adopted as two processes with sample time 2T, the variance of these processes is $\sigma_{ia}^2 = \sigma_{delay}^2 + \sigma_{delay}^2 = 2\sigma_{delay}^2$. So the real variance of the message delays can be approximated by $\sigma_{delay} = \sigma_{ia} / \sqrt{2}$. The approximated maximal message delay is then $t_{ia\max} = \max(t_{ia}(i); i = 1:N) - \sigma_{delay}$. The number of messages with $t_{ia} > (1+\sigma_{delay}) \cdot T$ indicates the message loss rate. The constant communication delay can be estimated with network performance analysis or test messages (Ploennigs, *et al*, 2005).
- For send-on-delta sampling the arrival rates indicate the effectiveness of setting the parameter Δ. An incorrectly chosen Δ may cause the quantization oscillations around the setpoint which can be detected by observing the inter-arrival times t_{ia}, compared with t_{max} or historical values. In general, the oscillation vanishes with an increased Δ. The histograms of the inter-arrival times showed in our experiments usually two peaks: one for the steady state and one during the settling time, as it is recognizable in Figure 5.
- A further delay in the control cycle is inserted by the controller itself. It needs time to calculate the new plant input after the arrival of a sensor message. This think time can be computed from the time stamps at the monitor of the sensor and controller message by $t_{think} = t_{sensor} - t_{controller}$ and includes variations caused by the network arbitration algorithm.

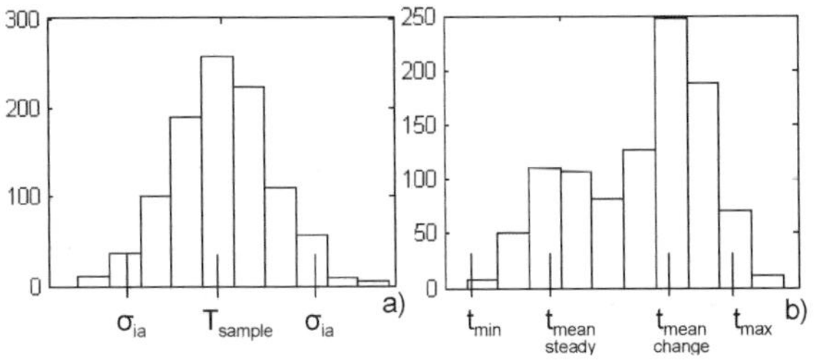

Fig. 5. Histograms for transmission rates for uniform (a) and send-on-delta (b) sampling.

The network traffic statistics depends strongly on the accuracy of the network analyzer, which varies from 1 ms down to 50μs for different devices.

6. CONCLUSIONS AND FUTURE WORK

This paper considered passive monitoring of building automation systems with the help of network analyzers. The problems of observability of control signals, non-uniform sampling and network delays were discussed. An approach to reconstruct the missing setpoint signals was proposed. Some criteria of evaluation of the performance and the network traffic of the networked control loops were adapted to passive monitoring, particularly for non-uniform sampling. The approach was approved by simulating the characteristic examples of building automation systems. Although the proposed simple criteria are limited in the accuracy, they are applicable to detect the performance problems without knowing of plant models.

REFERENCES

Basseville, M. and I. Nikiforov (1993). *Detection of Abrupt Changes*. Prentice Hall, Englewood Clis, New Jersey.

Bhat, S., Chatterjee, T. and D.N. Saraf (2003). Online Data Processing and Product Properties Prediction for Crude Distillation Units. *AIChE 2003 Spring National meeting New Orleans*, Louisiana.

Cao, S. and R. Rhinehart (1995). An efficient method of on-line identification of steady state. *Journal on Process Control*, 5(6), 363-374.

Dementjev, A. and K. Kabitzsch (2004). A Consulting Module in Room Automation. *Proc. IFAC Symposium on Telematics Applications In Automation and Robotics*. Espoo, Finland, pp. 37 – 42.

Dexter, A. and J. Pakanen, eds. (2001). Demonstrating Automated Fault Detection and Diagnosis Methods in Real Buildings. *VTT Building Technology, Symposium 217*, Espoo, pp. 11-123.

Forsman, K. and A. Stattin (1999). A new criterion for detecting oscillations in control loops. *Proc. ECC'99, European Control Conference*, Karlsruhe, Germany.

Harris, T., C.T. Seppala and L.D. Desborough (1999). A review of performance monitoring and assessment techniques for univariate and multivariate control systems. *Journal on Process Control*, 9(1), pp. 1-17.

Koller, G., T. Sauter and T. Rauscher (2003). Effects of Network Delay Quantization in Distributed Control Systems. *Proc. FeT2003, 5th IFAC International Conference on Fieldbus Systems and their Applications,* Aveiro, Portugal, pp. 297 – 304.

LonTalk Protocol Specification (1994). 3.0 ed. Palo Alto. http://www.echelon.com.

O'Dwyer, A. (1999). PI and PID controller tuning rules for time delay processes: a summary. *Proc. Irish Signals and Systems Conference 1999*, Galway, Ireland, pp. 331-349.

Otanez, P., J. Moyne, and D. Tilbury (2002). Using deadbands to reduce communication in networked control systems. *Proc. of American Control Conference 2002*, Vol. 4, pp. 3015 - 3020.

Ploennigs, J., M. Neugebauer and K. Kabitzsch (2004). A Traffic Model for Networked Devices in the Building Automation. *Proc. 5th IEEE International Workshop on Factory Communication Systems*, Vienna, pp. 137-145.

Ploennigs, J., P. Buchholz, M. Neugebauer and K. Kabitzsch (2005). Automated Modeling and Analysis of CSMA Type Access-Schemes for Building Automation Networks. *Transactions on Industrial Informatics.* To appear.

Qin, S.J. (1998). Control performance monitoring - A review and assessment. *Comput. and Chem. Eng.*, Vol. 23, pp. 178-186.

Rehbinder, H. and M. Sanfridson (2004). Scheduling of a limited communication channel for optimal control. *Automatica*, Vol. 40, pp. 491-500.

Salsbury T.I (1999). A practical algorithm for diagnosing control loop problems. *Energy and Buildings*, Vol. 29, No. 3, pp. 217-227.

Salsbury, T.I. and R.C. Diamond (2001). Fault detection in HVAC systems using model-based feedforward control. *Energy and Buildings*. Vol. 33, No. 4, pp. 403-415.

Soucek, S. and T. Sauter (2004). Quality of Service Concerns in IP-Based Control Systems. *IEEE Trans. on Industrial Electronics*, V. 51, No. 6, pp. 1249-1258.

Swanda, A.P. and D.E. Seborg (1999). Controller Performance Assessment Based on Setpoint Response Data. *Proc. American Control Conference*, San Diego, pp. 3863-3867.

Yook, J.K., D.M. Tilbury, H.S. Wong and N.R. Soparkar (2000). Trading computations for band-width: State estimators for reduced communication in distributed control systems. *Proc. Japan-USA Symposium on Flexible Automation.*

Vasyutynskyy, V. and K. Kabitzsch (2004). Architecture and data model for monitoring of distributed automation systems. *Proc. IFAC Symposium on Telematics Applications In Automation and Robotics*. Espoo, Finland, pp. 19 – 24.

Author´s Index

Printed in Dunstable, United Kingdom

91898333R10159